D1289232

HUMAN STEM CELL
TECHNOLOGY AND BIOLOGY

Companion DVD

This book includes a companion DVD with:

- Protocols from the text in a searchable format
- Video clips showing procedures discussed in the text
- All video clips are referenced in the text where you see this symbol:

Updates and additional resources will be posted on a companion website at:

www.wiley.com/go/stein/human

Human Stem Cell Technology and Biology

A RESEARCH GUIDE AND LABORATORY MANUAL

Edited by

Gary S. Stein

Maria Borowski

Mai X. Luong

Meng-Jiao Shi

Kelly P. Smith

Priscilla Vazquez

WILEY-BLACKWELL

A John Wiley & Sons, Inc., Publication

Copyright © 2011 by Wiley-Blackwell. All rights reserved

Published by John Wiley & Sons, Inc., Hoboken, New Jersey

Published simultaneously in Canada

No part of this publication may be reproduced, stored in a retrieval system, or transmitted in any form or by any means, electronic, mechanical, photocopying, recording, scanning, or otherwise, except as permitted under Section 107 or 108 of the 1976 United States Copyright Act, without either the prior written permission of the Publisher, or authorization through payment of the appropriate per-copy fee to the Copyright Clearance Center, Inc., 222 Rosewood Drive, Danvers, MA 01923, (978) 750-8400, fax (978) 750-4470, or on the web at www.copyright.com. Requests to the Publisher for permission should be addressed to the Permissions Department, John Wiley & Sons, Inc., 111 River Street, Hoboken, NJ 07030, (201) 748-6011, fax (201) 748-6008, or online at http://www.wiley.com/go/ permission.

Limit of Liability/Disclaimer of Warranty: While the publisher and author have used their best efforts in preparing this book, they make no representations or warranties with respect to the accuracy or completeness of the contents of this book and specifically disclaim any implied warranties of merchantability or fitness for a particular purpose. No warranty may be created or extended by sales representatives or written sales materials. The advice and strategies contained herein may not be suitable for your situation. You should consult with a professional where appropriate. Neither the publisher nor author shall be liable for any loss of profit or any other commercial damages, including but not limited to special, incidental, consequential, or other damages.

For general information on our other products and services or for technical support, please contact our Customer Care Department within the United States at (800) 762-2974, outside the United States at (317) 572-3993 or fax (317) 572-4002.

Wiley also publishes its books in a variety of electronic formats. Some content that appears in print may not be available in electronic formats. For more information about Wiley products, visit our web site at www.wiley.com.

Library of Congress Cataloging-in-Publication Data:

ISBN 978-0-470-59545-9

Printed in the United States of America

oBook: 978-0-470-88990-9
ePub: 978-0-470-88989-3

10 9 8 7 6 5 4 3 2 1

Contents

SECTION III

LABORATORY GUIDE FOR HUMAN STEM CELL CULTURE: CHARACTERIZATION OF PLURIPOTENT STEM CELLS

SECTION IV

PERSPECTIVES IN HUMAN STEM CELL TECHNOLOGIES

Companion DVD

This book includes a companion DVD with:

- Protocols from the text in a searchable format
- Video clips showing procedures discussed in the text
- All video clips are referenced in the text where you see this symbol:

Updates and additional resources will be posted on a companion website at:

www.wiley.com/go/stein/human

Foreword

In recent years interest in stem cells has become intense, not only within the scientific and medical communities but also amongst politicians, religious groups and ethicists. Stem cells offer tremendous potential to alleviate human disease, yet opinions diverge about how that potential can best be realized. In *Human Stem Cell Technology and Biology: A Research Guide and Laboratory Manual*, Gary Stein and his colleagues provide a timely introduction to stem cells and step-by-step protocols for working with them in the laboratory. The book is based on the highly successful practical courses provided at the University of Massachusetts Center for Stem Cell Biology & Regenerative Medicine.

This book achieves two important goals: to provide reliable laboratory protocols for culturing pluripotent stem cells and to present current perspectives and applications of stem cells. It begins with a historical account of stem cell research, explaining the different types of stem cell that have been characterized. The second section describes techniques for culturing, maintaining and characterizing pluripotent stem cells, including the pros and cons of different methods. The final section provides an in-depth perspective on the current status of the stem cell research field. Special features include downloadable protocols and narrated videos that present the principle techniques for human pluripotent cell culture.

Human Stem Cell Technology and Biology: A Research Guide and Laboratory Manual will greatly benefit both established investigators and newcomers in the stem cell field. It helps to demystify and explain the different types of pluripotent cell and will be an important tool for stem cell research for years to come.

FIONA M. WATT

Cambridge, Massachusetts
January 2010

Preface

Human Stem Cell Technology and Biology: A Research Guide and Laboratory Manual was developed to serve two distinct but important functions: to provide laboratory techniques and protocols that have been tested and proved to be effective for culture of pluripotent stem cells, and to present current perspectives and applications of stem cells by the leaders in the field. The first section of the text begins with an overview of the research contributions that established the foundation for current initiatives in stem cell research and goes on to describe the best methods for researching and obtaining pluripotent cell lines: both blastocyst derived human embryonic stem cells and reprogrammed pluripotent stem cells.

The second and third sections provide skills and techniques necessary for the culture, maintenance and characterization of pluripotent stem cells. These sections have been developed as a multimedia stem cell course in an effort to present the most comprehensive instruction possible, including the following features:

- **Videos demonstrating step-by-step laboratory protocols.** While reading the correct procedure is important to understanding technique, there is no substitute for watching the procedure done by an expert. To ensure this objective is met, laboratory procedures for the culture of human embryonic and induced pluripotent stem cells are presented in narrated video format via an accompanying CD/DVD and on the web at www.wiley.com/go/stein/human Look for the video camera icon to locate text associated with video format.

- **Online updates.** Electronic updates will be made available on a scheduled basis. Availability of the rapidly developing strategies and research protocols that support stem cell research will extend the "life" of the book.

- **Printable laboratory protocols.** The CD/DVD also contains PDF files of lab protocols to be printed and taken into the lab. In this way, the files can become the basis for a lab notebook to provide a seamless transition from reading the text to performing the experiments.

We recognize and appreciate that the stem cell field is rapidly evolving. Concepts and strategies are emerging that will advance understanding of genetic and epigenetic control for development, tissue renewal, and regenerative medicine. The challenges and opportunities are formidable and we are committed to providing online updates on a regular basis to maximize the effectiveness of this book.

GARY S. STEIN
MARIA BOROWSKI
MAI X. LUONG
MENG-JIAO SHI
KELLY P. SMITH
PRISCILLA VAZQUEZ

Acknowledgments

The editors of *Human Stem Cell Technology and Biology: A Research Guide and Laboratory Manual* gratefully acknowledge the dedication and efforts of the following, without whom this manual would not have been possible:

Klaus Becker

Mariluci Bladon

Peter Bodine

Amy Briggs

Joseph Buckley

Jennifer Colby

Michael Collins

Patricia Crowley-Larsen

James Evans

Tera Filion

Dana Fredricks

Rachel Gerstein

Antonio Giordano

Jonathan Gordon

Mohammad Hassan

Kathleen Hoover

Lan Ji

Marissa Johnson

Stephen Jones

Maribeth Leary

Xue-Jun Li

Barry Komm

Alex Lichtler

Teneille Ludwig

Matthew Mandeville

Ricardo Medina

Kerri Miller

Patricia Miron

Thomas Owen

Arthur Pardee

Sheldon Penman

Meng Qiao

Jie Song

Jeffrey Spees

Kenneth Soprano

Arun Srivastava

Jeffrey Stoff

Viktor Tepulyuk

Fiona Watt

Ren-He Xu

Sayyed K. Zaidi

The editors would also like to thank the researchers of the UMass community who contributed the use of their stem cell images for the cover of the textbook:

John Butler

Jeanne Lawrence

Stephen Lyle

Janet Stein

Editors and Contributors

Editors

Maria Borowski
Department of Cell Biology
Center for Stem Cell Biology & Regenerative Medicine
University of Massachusetts Medical School
Worcester, Massachusetts

Mai X. Luong
Department of Cell Biology
Center for Stem Cell Biology & Regenerative Medicine
University of Massachusetts Medical School
Worcester, Massachusetts

Meng-Jiao Shi
Center for Stem Cell Biology & Regenerative Medicine
University of Massachusetts Medical School
Worcester, Massachusetts

Kelly P. Smith
Department of Cell Biology
Center for Stem Cell Biology & Regenerative Medicine
University of Massachusetts Medical School
Worcester, Massachusetts

Gary S. Stein
Department of Cell Biology
University of Massachusetts Medical School
UMass Memorial Cancer Center
Worcester, Massachusetts

Priscilla Vazquez
Center for Stem Cell Biology & Regenerative Medicine
University of Massachusetts Medical School
Worcester, Massachusetts

Contributors

Alicia Allaire
Center for Stem Cell Biology & Regenerative Medicine
University of Massachusetts Medical School
Worcester, Massachusetts

Daniel G. Anderson
Harvard-MIT Division of Health and Science Technology
Massachusetts Institute of Technology
Cambridge, Massachusetts

Mark Burcin
Cardiovascular and Metabolism Disease Area
Novartis Institutes for Biomedical Research
Cambridge, Massachusetts

Jen-Fu Chiu
Department of Biochemistry
The Open Laboratory for Tumor Molecular Biology
Shantou University Medical College
Shantou, Guangdong, People's Republic of China

Ji-Hoon Cho
Institute for Systems Biology
Seattle, Washington

Li-Fang Chu
Center for Cell and Gene Therapy
Baylor College of Medicine
Houston, Texas

Jeremy Micah Crook
Stem Cell Medicine
O'Brien Institute
Department of Surgery, St. Vincent's Hospital
University of Melbourne
Melbourne, Victoria, Australia

James R. Davie
Department of Biochemistry and Medical Genetics
Manitoba Institute of Cell Biology
University of Manitoba
Winnipeg, Manitoba, Canada

Geneviève P. Delcuve
Manitoba Institute of Cell Biology
University of Manitoba
Winnipeg, Manitoba, Canada

Tanja Dominko
Biology and Biotechnology Department
Bioengineering Institute
Worcester Polytechnic Institute
Worcester, Massachusetts

Bojan Drobic
Department of Biochemistry and Medical Genetics
Manitoba Institute of Cell Biology
University of Manitoba
Winnipeg, Manitoba, Canada

Terence R. Flotte
Gene Therapy Center and Department of Pediatrics
University of Massachusetts Medical School

Christopher C. Ford
Department of Biomedical Engineering
Boston University Pulmonary Center
Center for Regenerative Medicine (CReM)
Boston University School of Medicine
Boston, Massachusetts

Prachi N. Ghule
Department of Cell Biology
University of Massachusetts Medical School
Worcester, Massachusetts

Gustavo Glusman
Institute for Systems Biology
Seattle, Washington

Anne Higgins
Clinical Cytogenetics Laboratory
University of Massachusetts Medical School
Worcester, Massachusetts

Tan A. Ince
Assistant Professor of Pathology
Director, Tumor Stem Cell Division
Interdisciplinary Stem Cell Institute
University of Miami Miller School of Medicine
Miami, Florida

Irina Klimanskaya
Advanced Cell Technology, Inc.
Worcester, Massachusetts

Darrell N. Kotton
Boston University Pulmonary Center
Center for Regenerative Medicine (CReM)
Department of Medicine
Boston University School of Medicine
Boston, Massachusetts

Thomas P. Kraehenbuehl
Department of Chemical Engineering
Massachusetts Institute of Technology
Cambridge, Massachusetts

Burak Kutlu
Institute for Systems Biology
Seattle, Washington

Arnaud Lacoste
Developmental and Molecular Pathways
Novartis Institutes for Biomedical Research
Cambridge, Massachusetts

Robert S. Langer
Department of Chemical Engineering
H. Koch Institute for Integrative Cancer Research
Harvard—MIT Division of Health Science and
    Technology
Cambridge, Massachusetts

Robert Lanza
Stem Cell & Regenerative Medicine International and
    Advanced Cell Technology, Inc.
Worcester, Massachusetts

David Lapointe
Information Services
University of Massachusetts Medical School
Worcester, Massachusetts

Andy T. Y. Lau
Department of Anatomy
Li Ka Shing Faculty of Medicine
The University of Hong Kong
Pokfulam, Hong Kong Special Administrative Region
    People's Republic of China

M. William Lensch
Department of Pediatrics
Harvard Medical School
Division of Hematology/Oncology
Children's Hospital Boston
Boston, Massachusetts

Shi-Jiang Lu
Stem Cell & Regenerative Medicine International
Worcester, Massachusetts

Abigail K. R. Lytton-Jean
David H. Koch Institute for Integrative Cancer Research
Massachusetts Institute of Technology
Cambridge, Massachusetts

Bruz Marzolf
Institute for Systems Biology
Seattle, Washington

Edmund Mickunas
Advanced Cell Technology, Inc.
Worcester, Massachusetts

Christian Mueller
Gene Therapy Center and Department of Pediatrics
University of Massachusetts Medical School
Worcester, Massachusetts

Beatriz Pèrez-Cadahía
Department of Biochemistry and Medical Genetics
Manitoba Institute of Cell Biology
University of Manitoba
Winnipeg, Manitoba, Canada

Shirwin M. Pockwinse
Department of Cell Biology
University of Massachusetts Medical School
Worcester, Massachusetts

Mojgan Rastegar
Department of Biochemistry and Medical Genetics
Department of Immunology
Regenerative Medicine Program
University of Manitoba
Winnipeg, Manitoba, Canada

Kimberly Stencel
Center for Stem Cell Biology & Regenerative Medicine
University of Massachusetts Medical School
Worcester, Massachusetts

Qiang Tian
Institute for Systems Biology
Seattle, Washington

Kai Wang
Institute for Systems Biology
Seattle, Washington

Yan-Ming Xu
Department of Biochemistry and Molecular Biology
The Fourth Military Medical University
Xi'an, People's Republic of China

Janet Zoldan
Department of Chemical Engineering
Massachusetts Institute of Technology
Cambridge, Massachusetts

Thomas P. Zwaka
Center for Cell and Gene Therapy
Baylor College of Medicine
Houston, Texas

Introduction

Human Stem Cell Technology and Biology, edited by Stein, Borowski, Luong, Shi, Smith, and Vazquez
Copyright © 2011 Wiley-Blackwell.

INTRODUCTION TO PLURIPOTENT STEM CELLS: BIOLOGY AND APPLICATIONS

1

Maria Borowski and Gary S. Stein

Current advances in human stem cell research utilize the latest tools of cell biology, molecular biology, chemistry, biomedical imaging, genomics, proteomics, and bioinformatics. Pluripotent stem cells, such as human embryonic stem cells (hESCs) and reprogrammed cells (induced pluripotent stem (iPS) cells) are defined as cells with the capacity to proliferate indefinitely as well as differentiate into all specialized cells, tissues, and organs of the body. These cells have enormous potential to offer therapies for diseases that have not proven treatable by conventional strategies.

HISTORICAL PERSPECTIVE

The exploration and utilization of cellular differentiation can be traced back more than half a century (Fig. 1.1). Visionary experiments from the 1950s provided a compelling foundation for two principal parameters of biological regulation. These studies established the concepts of pluripotency (as defined above) and epigenetic control, which is the transmission of regulatory information during cell division that is not encoded by DNA. At that time, developmental biologists were testing whether the process of cellular specialization or differentiation involved permanent changes in the DNA. Irreversible changes at the DNA level in a differentiated cell would prevent that cell's DNA from directing embryonic development. In a series of elegant experiments, Robert Briggs and Thomas King established a method for nuclear transfer, which John Gurdon subsequently used to transfer nuclei from frog intestinal cells to enucleated frog. These egg cells successfully divided, eventually developing into a tadpole.[1,2] These experiments demonstrated that changes to the DNA during differentiation were reversible and laid the groundwork for the discovery of pluripotent cells.

In mammals, work with embryonal carcinoma cells (malignant germ cell tumors) in the 1950s and 1960s established the presence of populations of mammalian cells that are capable of unlimited self-renewal and differentiation into many cell lineages.[3] These experiments set the stage for the identification, characterization, and isolation of stem cells.

Human Stem Cell Technology and Biology, edited by Stein, Borowski, Luong, Shi, Smith, and Vazquez
Copyright © 2011 Wiley-Blackwell.

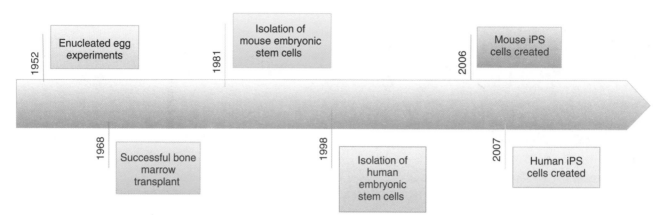

FIGURE 1.1. *Abbreviated timeline of stem cell research.*

Adult Stem Cells

Adult stem cells, or ASCs, were the first type of stem cells to be explored for use in clinical therapy. ASCs are present in many tissues and organs, including bone marrow, skin, muscle, and fat, in which they produce new, healthy cells to replace those that have been damaged (Fig. 1.2).

Hematopoietic stem cells (HSCs) are specialized adult stem cells. The existence of a common HSC was a topic of debate for several decades until definitive evidence was provided by the work of James Till, Ernest McCulloch, and others in the 1960s. Till and McCulloch demonstrated that a single bone marrow cell could give rise to different types of blood cells.[4,5] Pioneering research on bone marrow transplantation by E. Donnall Thomas established that injecting bone marrow cells into the bloodstream could repopulate the bone marrow and produce more blood cells.[6] On the basis of this pivotal demonstration, Dr. Robert A. Good at the University of Minnesota performed the first successful bone marrow transplant in

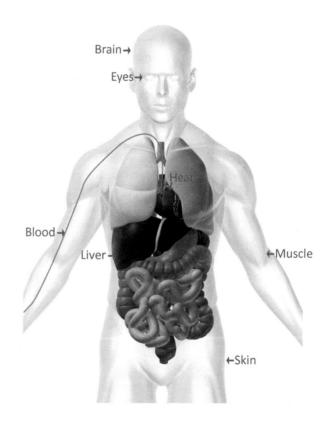

FIGURE 1.2. *Known sites of adult stem cells (ASCs).*

1968.[7] Bone marrow transplantation remains a therapy that effectively treats many cancer patients every year.

Unfortunately, ASCs do not have limitless potential, but are currently thought to be restricted to becoming only one or a few specific types of cells. While more types of ASCs are known to exist than previously thought, they have not been found in every organ and tissue in the human body. Additionally, they are difficult to maintain and do not multiply indefinitely in culture.

Embryonic Stem Cells

While ASCs are classified as "multipotent," meaning they have the potential to become many, but not all cell types, embryonic stem cells (ESCs) are categorized as "pluripotent," meaning they can give rise to all the cells in the body. Four to six days after fertilization of an egg, a human embryo is a blastocyst, a hollow ball of cells, that contains an inner cluster of cells designated the inner cell mass (ICM). When the cells of the ICM are removed from the blastocyst and cultured, they retain pluripotency and can be directed through controlled interventions to become specialized cells (Fig. 1.3a).

Murine embryonic stem cells were first isolated from mice in 1981.[8,9] It was not until 1998 that James Thompson at the University of Wisconsin obtained embryonic

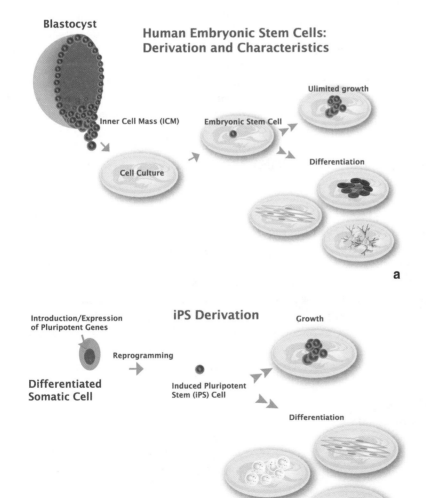

FIGURE 1.3. *Derivation of embryonic stem cells and iPS cells.*

stem cells from humans.[10] After extensive testing, Thomson and colleagues verified that they had isolated human stem cells characterized by two important properties; they could multiply indefinitely and they could develop into specialized tissue types. However, the most compelling proof of stem cell pluripotency is their ability to produce an entire organism, which, due to ethical concerns, has only been successfully demonstrated with mouse ES and iPS cells.[11,12]

Most blastocysts used for human stem cell research are donated by couples who have an excess of embryos after completion of an in vitro fertilization (IVF) treatment. The ethical concerns associated with the derivation of human ES cells have been expressed by some groups and has influenced the availability of federal funding for embryonic stem cell investigation. Nevertheless, ES cells have the potential to provide insights into human development as well as offer a wealth of potential therapies.

iPS Cells

The use of ES cells in research has created emotionally charged bioethical issues that have resulted in loss of US federal funding, therefore inhibiting scientific progress. The development of human iPS cells in 2007 helped to circumvent some of these obstacles. Unlike ES cells, iPS cells will have the potential to overcome compatibility issues when transplanted.

The iPS cells are somatic cells that have been reprogrammed to resemble embryonic stem cells (Fig. 1.3b). Mouse iPS cells were first reported in 2006,[13] and human iPS cells followed soon thereafter in 2007.[14] Both mouse and human iPS cells appear nearly identical to embryonic stem cells in terms of defining criteria: mouse iPS cells express stem cell markers and form tumors containing cells from all three germ layers, which are precursors of all cell types. In addition, mouse iPS cells are able to give rise to an entire mouse when injected into a mouse. Human iPS cells also express stem cell markers and are capable of generating cells characteristic of all three germ layers. Although these cells meet the defining criteria for pluripotent stem cells, more research will need to be done to determine whether iPS cells and embryonic stem cells differ in significant ways.[15]

While there is great hope in the potential of clinical therapies utilizing iPS cells, a number of questions must first be answered. Pluripotency is often reestablished in differentiated cells by expression of several nuclear proteins that are preferentially expressed in ES cells. The use of viruses to introduce genes encoding the pluripotency factors may cause mutations by viral insertion into the genome. Induced expression of the pluripotency factors can result in aberrant control of proliferation. However, recent advances in inducing pluripotency directly with proteins under controlled conditions may eliminate these risks. In the future, in situ programming may provide a window of opportunity for repair and replacement of cells and tissues in order to correct structural or metabolic defects that are associated with cancer, kidney, and neurological diseases (Table 1.1).

TABLE 1.1 SUMMARY OF TYPES OF STEM CELLS

Cells	Ethical Concerns	Immune Rejection Concerns	Cancer Concerns	Cell/Tissue Types	Availability
hESCs	Yes	Yes	Yes	All	Many lines
Adult stem cells	No	No	No	Limited	Many (limited types)
iPS cells	No	No	Yes	All	Many

OUTLOOK OF THE FIELD

There are many expectations and hopes for potential applications of embryonic stem cells and iPS cells. These therapeutic applications include tissue and organ regeneration, the potential for drug screening and toxicity assessment, and interfaces with gene therapy and tissue engineering.

To translate these expectations to reality, there are a number of challenges that the field will need to surmount. While a defining characteristic of both ES and iPS cells is their pluripotency, this same characteristic may result in the formation of tumors. Thus for therapeutic applications, a blueprint of instructions for differentiation that results in pure cultures of differentiated cells will be necessary. In vitro genetic and epigenetic modifications that direct the ES cells to the required cell type may provide solutions. Another challenge to using pluripotent cells in transplantation is the certainty that they will function correctly once transplanted. This is particularly relevant if the transplant is to damaged or diseased organs.

There is still much to be learned about the properties of pluripotent stem cells. The stem cell field is positioned to build on the enormous progress that has been made over the past five decades. It is recognized that a more in depth and systematic investigation is essential to establish an understanding of the maximal potential that stem cells hold for our future. Given advances that are emerging in stem cell biology and technologies, there is confidence that stem cells will provide opportunities to maximize function and enhance quality of life.

REFERENCES

1. Briggs R, King TJ. Transplantation of living nuclei from blastula cells into enucleated frogs' eggs. *Proc Natl Acad Sci USA*. 1952;38:455–463.

2. Gurdon JB, Elsdale TR, Fischberg M. Sexually mature individuals of *Xenopus laevis* from the transplantation of single somatic nuclei. *Nature*. 1958;182:64–65.

3. Kleinsmith LJ, Pierce GB Jr. Multipotentiality of single embryonal carcinoma cells. *Cancer Res*. 1964;24:1544–1551.

4. Till JE, McCulloch E. A direct measurement of the radiation sensitivity of normal mouse bone marrow cells. *Radiat Res*. 1961;14:213–222.

5. Till JE, McCulloch EA, Siminovitch L. A stochastic model of stem cell proliferation, based on the growth of spleen colony-forming cells. *Proc Natl Acad Sci USA*. 1964;51:29–36.

6. Thomas ED, Lochte HL Jr, Lu WC, et al. Intravenous infusion of bone marrow in patients receiving radiation and chemotherapy. *N Engl J Med*. 1957;257:491–496.

7. Gatti RA, Meuwissen HJ, Allen HD, et al. Immunological reconstitution of sex-linked lymphopenic immunological deficiency. *Lancet*. 1968;2:1366–1369.

8. Martin GR. Isolation of a pluripotent cell line from early mouse embryos cultured in medium conditioned by teratocarcinoma stem cells. *Proc Natl Acad Sci USA*. 1981;78:7634–7638.

9. Evans MJ, Kaufman MH. Establishment in culture of pluripotential cells from mouse embryos. *Nature*. 1981;292:154–156.

10. Thomson JA, Itskovitz-Eldor J, Shapiro SS, et al. Embryonic stem cell lines derived from human blastocysts. *Science*. 1998;282:1145–1147.

11. Nagy A, Rossant J, Nagy R, et al. Derivation of completely cell culture-derived mice from early-passage embryonic stem cells. *Proc Natl Acad Sci USA*. 1993;90:8424–8428.

12. Okita K, Ichisaka T, Yamanaka S. Generation of germline-competent induced pluripotent stem cells. *Nature*. 2007;448:313–317.

13. Takahashi K, Yamanaka S. Induction of pluripotent stem cells from mouse embryonic and adult fibroblast cultures by defined factors. *Cell*. 2006;126:663–676.

14. Takahashi K, Tanabe K, Ohnuki M, et al. Induction of pluripotent stem cells from adult human fibroblasts by defined factors. *Cell*. 2007;131:861–872.

15. Yamanaka S. Strategies and new developments in the generation of patient-specific pluripotent stem cells. *Cell Stem Cell*. 2007;1:39–49.

RESEARCHING AND OBTAINING ESTABLISHED STEM CELL LINES

2

Mai X. Luong, Kelly P. Smith, and Gary S. Stein

Since the original derivation of human embryonic stem cells (hESCs) in 1998,[1] and with the development of new technologies such as cellular reprogramming,[2,3] the stem cell field has expanded rapidly. Currently, it is estimated that over 1000 different hESC lines have been derived worldwide. In addition, reprogramming has made it possible for most labs to develop pluripotent stem cells tailored to their research, such as disease–specific induced pluripotent stem (iPS) cells. This rapid expansion of the stem cell field has necessitated the development of valuable resources to aid the researcher. Registries, which serve as repositories of stem cell information, as well as banks, which are physical repositories for stem cells, have become necessary tools for stem cell research.[4–6]

RESEARCHING STEM CELL INFORMATION

As research into the properties and therapeutic potential of human pluripotent stem cells accelerates, it becomes more difficult for the researcher to remain current with the available information. Thus there is a need to organize and integrate current knowledge of pluripotent stem cells in a manner that is comprehensive and readily accessible. Registries are databases of information intended to assist researchers by providing extensive and up-to-date information on human pluripotent stem cells. To provide this information, registries must be aware of and constantly adapt to the volume of research findings and the complexity of issues in a rapidly expanding field.

Technical Questions

There are several key characteristics that an hESC line should possess. These properties include karyotype stability, retention of an undifferentiated state through repeated cell division cycles and prolonged culture, competency for lineage commitment, and the ability for reproducible terminal differentiation into a variety of cell types. However, there are frequent reports of heterogeneity between cell lines and within the same cell lines in different laboratories. Differences between cell lines can be attributed to causes that include embryo quality and stage, and variations in the method and reagents (such as feeder cells) used in the derivation. Genetic variation, medical history of donors, and even the maternal diet can influence the phenotype of cell lines.[7] Characteristics of the same cell line can vary significantly depending on the culture conditions used.[8] All of these variations can influence the ability of the

Human Stem Cell Technology and Biology, edited by Stein, Borowski, Luong, Shi, Smith, and Vazquez
Copyright © 2011 Wiley-Blackwell.

ES cells to differentiate. Furthermore, approaches and reagents used to induce differentiation have varying degrees of success. Thus a challenge of hESC research is to obtain comprehensive knowledge of the cell lines and the consequences of disparate technical tools utilized for isolation, maintenance, propagation, and differentiation.

Intellectual Property Issues

The intellectual property environment in stem cell research, especially for hESCs, is extremely complex. The Wisconsin Alumni Research Foundation (WARF) and James Thomson of the University of Wisconsin were awarded a very broad US patent (#6,200,806) on the isolation of human embryonic stem cells on March 13, 2001. The patent claims are sufficiently broad–based that any use of hESCs for any purpose may fall under the WARF patent.

Currently, WiCell Research Institute Inc. (a subsidiary of WARF) requires a licensing agreement, or Memorandum of Understanding (MOU), acknowledging WARF's patent rights, for the distribution of any human ES cell lines in the United States, regardless of their source or NIH approval status. In addition, any university receiving human ES cells for research is expected to sign a MOU with WiCell. As WARF patents are not recognized outside the United States, US investigators may be at a disadvantage in pursuing commercial applications of hESCs. In addition to the WARF patents, the increased patent protection for stem cells and related technologies in the United States has raised concerns about the emergence of a patent thicket in which overlapping claims block therapeutic applications of hESCs and the pathways to market—both by causing uncertainty about Freedom To Operate (FTO) and by imposing multiple transaction costs.[9]

hESC Research Guidelines: A Regulatory Maze

The regulatory environment surrounding hESC research is complex in the United States and globally. Various governments around the world and individual states in the United States have their own regulations, which range from permissive to an outright ban on hESC research. In an effort to simplify the complex patchwork of guidelines within the United States and around the world, several groups have produced, or are in the process of developing, guidelines for hESC research that would provide standards for the derivation, procurement, banking, distribution, and applications of human ES cells and create universally accepted documents such as informed consent forms and material transfer agreements (Table 2.1). In addition, groups such as the Interstate Alliance on Stem Cell Research (IASCR, www.iascr.org) are working to facilitate collaborative hESC research across state lines within the Unites States, whereas others, such as the International Society for Stem Cell Research (ISSCR, www.isscr.org), have focused on facilitating collaboration across international borders. It has been suggested that the solution to the lack of cohesion across regulatory frameworks may reside in reciprocal policy agreements. For example, the California Institute for Regenerative Medicine (CIRM, www.cirm.ca.gov) regulations allow funding for hESC research that utilizes cell lines that were derived in the United Kingdom under the Human Fertilization and Embryology Authority license or in accordance with the Canadian Institutes of Health Research Guidelines.[10] CIRM also has a reciprocal agreement with Japan and would consider entering into an agreement with any country that has a body established for active certification of hESC lines. Despite these efforts, the various guidelines shown in Table 2.1 each have their own set of rules regarding provenance of hESC lines, including specific requirements about informed consent and donor reimbursement.

Choosing from among the hundreds of existing hESC lines requires much more than knowledge of their scientific qualities. When obtaining hESC lines, the researcher must consider the guidelines, intellectual property issues, and legislation

TABLE 2.1 PUBLISHED GUIDELINES FOR HESC RESEARCH

Organization	Title	Date of publication
National Institutes of Health (U.S.)	National Institutes of Health Guidelines on Human Stem Cell Research http://stemcells.nih.gov/policy/2009guidelines.htm	2009
National Academy of Sciences (U.S.)	Guidelines for Human Embryonic Stem Cell Research http://books.nap.edu/catalog.php?record_id=11278	2005
	Amendments to the National Academies' Guidelines http://books.nap.edu/catalog.php?record_id=11871	2007
	Amendments to the National Academies' Guidelines http://www.nap.edu/catalog.php?record_id=12260	2008
	Final Report of The National Academies' Human Embryonic Stem Cell Research Advisory Committee and 2010 Amendments to The National Academies' Guidelines for Human Embryonic Stem Cell Research http://www.nap.edu/catalog.php?record_id=12923	2010
California Institute for Regenerative Medicine	The CIRM Medical and Ethical Standards Regulations http://dev.cirm.ca.gov/?q=Guidance Documents	2008
Steering Committee for the Stem Cell Bank and for the Use of Stem Cell Lines (UK)	Code of Practice for the Use of Human Stem Cell Lines http://www.ukstemcellbank.org.uk Documents/Code%20of%20 Practice%20for%20the%20 Use%20of%20 Human%20Stem%20Cell%20 Lines.pdf	2006
International Society for Stem Cell Research	The ISSCR Guidelines for Human Embryonic Stem Cell Research http://www.isscr.org/guidelines/index.htm	2006
The Canadian Institutes of Health	Guidelines for Human Pluripotent Stem Cell Research http://www.cihr-irsc.gc.ca/e/34460.html	2006

that control hESC research in their country, locality, and institution as well as regulations adopted by specific funding agencies. Cell line provenance information must be interpreted in light of these considerations to determine if a cell line may be used for a particular research project.

CURRENT REGISTRIES

A comprehensive human stem cell registry would serve as a repository of all vital aspects of hESC lines including:

- Information on the provenance of hESC lines
- Intellectual property information
- Availability
- Published and unpublished characterization data
- Methodologies for each cell line

There are several obstacles to establishing and maintaining a comprehensive hESC databank or registry. General challenges include efficient initial data collection and subsequent maintenance of the databank to ensure that the information remains current. Difficulties in gathering information for every cell line derived to date include the availability of published data and unpublished provenance information (i.e., donor information that is not made available to the public due to their confidential nature). Although more than 1000 cell lines have been derived, characterization of only a fraction of these cell lines has been published in peer-reviewed journals. This is partly due to lack of a cost-effective approach for characterization, and the length of time required for characterization. In addition, as derivation of hESCs becomes more common, publication of these studies becomes more difficult.

Although a truly comprehensive hESC registry does not currently exist, several independent registries cover most of the relevant information mentioned above. Each registry listed provides useful information and can be considered a valuable resource for the stem cell researcher. Current stem cell registries include the International Stem Cell Registry, the International Stem Cell Forum, the European hESC Registry, the International Society for Stem Cell Research Provenance Registry, and the National Institute of Health hESC Registry.

International Stem Cell Registry (ISCR)

The ISCR was established in 2008 and is funded by the Massachusetts Life Sciences Center. The mission of ISCR is to provide a searchable, comprehensive database that includes published and validated unpublished information on all human ES cell lines as well as other pluripotent stem cell lines. For each cell line in the registry, curators gather data from multiple sources, such as publications, online resources (NIH SCU, ISCF, etc.), and unpublished data from investigators. In addition, the ISCR also provides information about the provenance of over half of the lines listed in the registry. This information can include a blank consent form and/or a letter that is intended to serve as documentation of provenance from the institution where each cell line was derived. This documentation gives assurance that the cell lines were derived under protocols that were reviewed by an Institutional Review Board, proper informed consent was obtained for the donation of embryos, and there were no financial inducements for the donations. The ISCR features a unique searchable hESC literature database that is indexed by cell line and displays search results as links to published studies via PubMed. Links to the data sources are also provided for most of the registered information. (Readers can access the website at www.umassmed.edu/iscr.)

International Stem Cell Forum (ISCF)

The ISCF was founded in 2003 with funding from 21 international medical research agencies around the world. Its objectives are to encourage international collaboration, fund research, and promote global good practice in the stem cell field. In 2006 the ISCF published the results of its first initiative, the International Stem Cell Characterization Initiative (ISCI), which characterized 59 hESC lines from 17 laboratories worldwide using specified protocols and a common pool of antibodies.[11] The Initiative was formed to systematically study hESCs in an effort to establish an international set of standards for characterization. Cells and embryoid bodies were cultured under specified conditions and surface antigen and gene expression patterns were established by FACS analysis, immunofluorescence (IF) microscopy, and Taqman low-density array (LDA) based assays. Additional data include microbiological status (mycoplasma contamination), imprinting data, and assessment of xenograft tumors. Information regarding origins (e.g., embryo status, derivation method), karyotype, and culture conditions was also gathered from the developer of each cell line. The ISCF study results are organized in two ways: as several files of analyzed data for each cell line or as aggregated data for all cell lines. Information for individual cell lines can be accessed using a keyword search feature or by browsing the list of cell lines or of participating laboratories. (Readers can access the website at http://www.stem-cell-forum.net/ISCF/.)

European hESC Registry (hESCreg)

The hESCreg was launched in January 2008 and is funded by the 6th Framework Programme for Research and Technological Development of the European Commission. The European hESC Registry is intended to provide information on hESC lines derived and used in the European Union. The Steering Committee of hESCreg, composed of national contacts from the European Union, Switzerland, and Israel, provides updates to the registry on scientific and legal developments in their countries as well as on the cell lines that are available. In addition to basic information, hESCreg currently provides results from expression assays such as FACS, rt-PCR, IF, ELISA, and arrays. Each listed cell line has a rating that is based on registered information such as hESC line availability, expression of ISCI core set of markers (Nanog, TDGF, Pou5F1, GABRB3, GDF3, and DNMT3B), culture conditions, provenance documents, and legislative information. In addition, hESCreg aims to increase the transparency of human stem cell research and to standardize hESC research by providing links to other repositories, cell banks, regulatory bodies, and, notably, specific research projects. The registry primarily depends on data submission from stem cell banks and research projects. Thus the amount of information and the rate of its accrual are not directly determined by the registry. (Readers can access the website at http://www.hescreg.eu/.)

ISSCR Registry of Human Embryonic Stem Cell Line Provenance

The ISSCR is in the process of developing a Registry of Human Embryonic Stem Cell Provenance. This registry is intended to provide information pertaining to the derivation of specific hESC lines by assembling documentation regarding regulatory and ethical standards used for these cell line derivations as well as the permitted uses of the lines. In order to work with hESCs, researchers must receive approvals from institutional and regional oversight bodies, as well as funding agencies. The ISSCR Registry provides access to the information needed by the researchers and oversight agencies in order to determine which lines are suitable for research under their guidelines. (The website at www.isscr.org provides more information on the development of this registry.)

NIH Human Embryonic Stem Cell Registry

From 2001 to 2009, the NIH hESC Registry reflected the US government's position on human ES cell research, listing and providing information for about 78 derivations of stem cell lines that according to President Bush's moratorium met the criteria for federal funding. In 2009, the moratorium was lifted and new guidelines for human ES cell research were adopted (Table 2.1). It is anticipated that these new guidelines will make many more cell lines eligible for federal funding. The current NIH registry provides lists of approved hESC lines and those that are pending approval for use with NIH funding. Unlike the other registries, which are databases of information about various aspects of hESC cell lines, the NIH registry will only provide the name of the cell line and information about how to obtain each cell line. (Readers can access the website at www.grants.nih.gov/stem_cells/registry/current.htm.)

OBTAINING STEM CELLS

While stem cell registries provide a wealth of information about pluripotent stem cells, the research community also needs the means to obtain cell lines that are of high quality and relevance. To fulfill this need, the researcher has several options, including cell line developers, institutional stem cell core facilities, and stem cell banks.

Cell Line Developers

Perhaps the simplest means of obtaining a particular cell line is to contact the cell line developer directly. This direct contact has the advantage of access to expertise and advice from the developer and the cell line may be available at little or no charge except shipping. However, especially for heavily used lines, developers are often unwilling or incapable of devoting valuable resources to expand, characterize, and distribute large quantities of cells to researchers.

Core Facilities

Many large research centers now have stem cell core facilities, which provide a limited number of cell lines to researchers within the institution or university system. These core facilities offer several advantages. First, they provide actively growing cultures, often in numbers to meet the specific requirements of the investigator. These lines have already cleared the intellectual property and regulatory hurdles for use in that institution. Second, these core facilities can provide onsite expertise for troubleshooting and technical training. However, many smaller institutions may not have the resources needed to develop a stem cell core. Additionally, core facilities often carry a limited number of cell lines, which may not be suitable for certain studies.

Stem Cell Banks

The large numbers of pluripotent stem cell lines that continue to be developed worldwide have necessitated the creation of stem cell banks, which serve as centralized repositories for many different stem cell lines. While core facilities can deliver a higher level of personalized service, only large banks have the capacity to gather, expand, characterize, and distribute large numbers of cell lines. Also, strict adherence to protocols for cell culture and extensive characterization assure the researcher of obtaining high-quality cells. Also, careful side-by-side characterization of many lines by each bank may aid in the development of standards for the field. Many different states and countries are either considering or actively developing stem cell banks. Current banking initiatives include large international stem cell banks in the United Kindom and the United States as well as a number of smaller national and institutional stem cell banks in various countries.

International Stem Cell Banks

These banks supply numerous cell lines, derived in many countries, to investigators throughout the world. These larger banking efforts serve to standardize the field by providing quality cell lines that have undergone a uniform series of characterization and quality control tests.

US National Stem Cell Bank

Mission Statement. "The National Stem Cell Bank (NSCB) is a repository for the pluripotent stem cells lines listed on the NIH Human Pluripotent Stem Cell Registry. These cells were derived prior to August 2001 using excess IVF embryos and were eligible for use in federally funded research under previous presidential policy. The eligibility of these lines will not be known until the NIH issues final stem cell guidelines in July 2009. The goal of the NSCB is to grow, characterize and distribute the cell lines listed on the registry, and to provide comprehensive technical support to stem cell researchers around the world." Note: As of February, 2010, the NIH no longer supports the NSCB. All NSCB cell lines are currently distributed through the Wisconsin International Stem Cell (WISC) Bank at WiCell. However, only a subset of the lines available from the WISC Bank is listed on the current NIH registry. (Readers can access the website at http://www. wicell.org/.)

UK Stem Cell Bank

Mission Statement. "UK Stem Cell Bank was established to provide a repository of hESC lines as part of the UK governance for the use of human embryos for research. Its role is to provide quality-controlled stocks of these cells that researchers worldwide can rely on to facilitate high quality and standardised research. It is also ready to prepare stocks of 'clinical grade' cell lines as seed stocks for the development of therapies. The UKSCB now holds non-hESC lines, leads in the development of best practice for the operation of banks of stem cell lines, and maintains an active engagement with the broader stem cell research community through its national and international collaborations. Through these activities the Bank is also developing expertise that is enabling it to play an important role in providing advice on the scientific and technical aspects of delivering cell therapies." (Readers can access the website at http://www.ukstemcellbank. org.uk/.)

Massachusetts Human Stem Cell Bank

Mission Statement. "The Massachusetts Human Stem Cell Bank provides the biomedical research community with expertly maintained human ES (hES) and reprogrammed (iPS) cell lines to facilitate studies into the properties and potential therapeutic applications of pluripotent stem cells. The Bank cultures, characterizes and distributes quality controlled hES and iPS cell lines derived in Massachusetts and beyond. The Bank is a 15,000 square foot facility that contains research and training space for visiting investigators. In addition, the Education and Training division provides technical training and programs to educate the community." (Readers can access the website at http://www.umassmed.edu/MHSCB/index.aspx.)

National and Institutional Banks

Many other countries (e.g., Spain, France, Korea, Australia, Japan, China, Canada, and Taiwan) have stem cell banks either in operation or planned. These banks primarily supply stem cell lines developed within a specific country to researchers of that country. Since countries have various regulatory constraints on stem cell research, the national banks can supply lines that meet the specific standards of each country.

In addition to these banks, many institutions have established banking efforts in order to provide cell lines that were created at these institutions to the research community. Examples of such banks are:

Harvard Stem Cell Institute	http://www.mcb.harvard.edu/melton/hues
Stem Cell Bank of Barcelona	http://www.cmrb.eu/banco-lineas-celulares/en_que_es.html
Singapore Stem Cell Bank	http://www.sscc.a-star.edu.sg/stemCellBank.php

Whatever source of cells is used, important issues to be considered when selecting cell lines include the quality of cells and the nature of agreements under which they are supplied. In addition, cell line passage number, culture history, genotypic and phenotypic characteristics, and ethical sourcing of embryonic cells should also be taken into account. These factors are addressed in more detail in the following chapters.

REFERENCES

1. Thomson JA, Itskovitz-Eldor J, Shapiro SS, et al. Embryonic stem cell lines derived from human blastocysts. *Science*. 1998;282:1145–1147.

2. Takahashi K, Tanabe K, Ohnuki M, et al. Induction of pluripotent stem cells from adult human fibroblasts by defined factors. *Cell*. 2007;131:861–872.

3. Yu J, Vodyanik MA, Smuga-Otto K, et al. Induced pluripotent stem cell lines derived from human somatic cells. *Science*. 2007;318:1917–1920.

4. Luong MX, Smith KP, Stein GS. Human embryonic stem cell registries: value, challenges and opportunities. *J Cell Biochem*. 2008;105:625–632.

5. O'Rourke PP, Abelman M, Heffernan KG. Centralized banks for human embryonic stem cells: a worthwhile challenge. *Cell Stem Cell*. 2008;2:307–312.

6. Isasi RM, Knoppers BM. Governing stem cell banks and registries: emerging issues. *Stem Cell Res*. 2009;3:96–105.

7. Martin DI, Ward R, Suter CM. Germline epimutation: a basis for epigenetic disease in humans. *Ann NY Acad Sci*. 2005;1054:68–77.

8. Allegrucci C, Young LE. Differences between human embryonic stem cell lines. *Hum Reprod Update*. 2007;13:103–120.

9. Saha K, Graff G, Winickoff D. Enabling Stem Cell Research and Development. eScholarship Repository, University of California; 2007.

10. Lomax G, McNab A. Harmonizing standards and coding for hESC research. *Cell Stem Cell*. 2008;2:201–202.

11. Adewumi O, Aflatoonian B, Ahrlund-Richter L, et al. Characterization of human embryonic stem cell lines by the International Stem Cell Initiative. *Nat Biotechnol*. 2007;25:803–816.

Laboratory Guide for Human Stem Cell Culture: Pluripotent Stem Cell Culture

SECTION II

Human Stem Cell Technology and Biology, edited by Stein, Borowski, Luong, Shi, Smith, and Vazquez
Copyright © 2011 Wiley-Blackwell.

BASICS OF CELL CULTURE 3

Alicia Allaire, Mai X. Luong, and Kelly P. Smith

ASEPTIC TECHNIQUE

Aseptic technique is a crucial component of the successful culture of human embryonic stem cells (hESCs). It is a set of specific practices and procedures intended to maintain sterility or asepsis, which is the state of being free from biological contaminants such as bacteria, viruses, fungi, and parasites. Practicing aseptic technique will protect the scientist from infectious agents in the culture, shield the hESC culture from contamination by the researcher, avoid cross-contamination with cells from another culture, and prevent the spread of pathogens between different hESC cultures. Aseptic technique is especially important in human embryonic stem cell culture because, unlike many types of cultured cells, many hESC cultures are maintained without any antibiotic or antimycotic reagents.

Preparation for Working with Stem Cell Culture

The following steps should be taken as precautionary measures to ensure the sterility of the work area:

- It is strongly recommended that all jewelry be removed from hands and arms.
- Hands and forearms up to the elbow are thoroughly scrubbed with an antiseptic soap and rinsed with warm water, then dried thoroughly as pathogens prefer moist environments.
- Personnel working with stem cell culture must wear gloves, which should be sprayed with 70% ethanol prior to cell culture activities. If the gloves touch anything nonsterile or have a tear, it is recommended that they be replaced.
- A dedicated lab coat should be worn at all times in the laboratory. It is suggested that lab coat sleeves be rolled up above the elbow in order to prevent accidental contact between the sleeves and sterile items within the biosafety cabinet. Alternatively, protective disposable sleeves can also be worn over the rolled up sleeve on the forearm, or gloves can be pulled over the lab coat sleeve.
- Consumables such as pipettes, tubes, and cell culture plates should be purchased presterilized.
- Glassware such as bottles or Pasteur pipettes should be clean and autoclaved.
- Wash glassware for cell culture with only water, as residual soap will decrease cell viability.

Human Stem Cell Technology and Biology, edited by Stein, Borowski, Luong, Shi, Smith, and Vazquez
Copyright © 2011 Wiley-Blackwell.

The Biosafety Cabinet

The following measures will help ensure the sterility of the biosafety cabinet and the cell culture.

- Sterilization of the biosafety cabinet by exposure to UV light for 15–20 minutes is recommended before beginning work.
- Prior to working in the biosafety cabinet, wipe the work surface with 70% ethanol.
- While working inside the biosafety cabinet keep hands well inside the work area and perform culture manipulations in the center of the biosafety cabinet. Working too close to the front of the biosafety cabinet will risk exposure to nonsterile conditions (Fig. 3.1a).
- The biosafety cabinet is engineered to maintain a sterile environment by creating a laminar flow of filtered air that enters the back of the biosafety cabinet and flows toward vents in the front; therefore only essential reagents and supplies should be kept inside the biosafety cabinets to reduce disruption of the laminar airflow. Do not place items over the airflow grids (Fig. 3.2b).
- Arrange materials toward the back of the biosafety cabinet so that accessing them does not require reaching over other items (Fig. 3.1a).
- Spray any materials brought into the biosafety cabinet with 70% ethanol and dry with a Kimwipe®.
- Treat the outer packaging of consumables as nonsterile and discard as soon as possible.
- Always make sure that the integrity of the packaging of sterile materials has not been compromised.

Aseptic Technique in Cell Culture

Keep in mind these points to limit exposure to contaminants:

- Culture vessels and media bottles are only to be opened inside the biosafety cabinet immediately before accessing them, and then capped immediately afterwards.

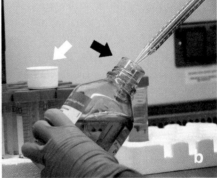

FIGURE 3.1. *Proper aseptic technique. This figure shows examples of proper aseptic technique. (a) The layout of items in the biosafety cabinet has essential items placed at the back. Work is done in the center of the biosafety cabinet. (b) The bottle cap (white arrow) has been placed at the back of the work area with the opening facing up. The bottle is tilted (black arrow) when pipetting and the pipette is not in contact with the lip of the bottle.*

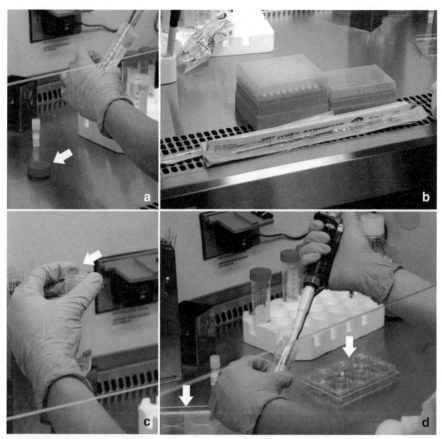

FIGURE 3.2. *Poor technique. This figure shows a few examples of poor aseptic technique in the biosafety cabinet. (a) Do not place caps or lids (arrow) with the opening facing down on the cabinet — the lip of the lid may become contaminated. (b) Do not place items over the airflow grate at the front of the cabinet. This can disrupt airflow needed to maintain a sterile environment within the biosafety cabinet. (c) Do not touch the lip of a container (arrow) — the lip of the opening could become contaminated; (d) do not work over open plates or lids (arrows) as contaminants may fall in.*

- Contact between other objects and the open end of the lid or cap should be avoided (Figs. 3.2a), and care should be taken so that nothing passes over this surface during culture manipulations (Fig. 3.2d). This can be accomplished by either working with the lid or cap in one hand, or alternatively, placing the lid or cap with the open-end facing up on top of another item (Fig. 3.1b) toward the back of the biosafety cabinet.

- The edge of any lids should be kept as sterile as possible, but remember that any medium in this lid is considered to be nonsterile. For this reason, do not shake or invert any bottles or tubes to avoid contact of sterile medium with the cap or lid.

- Medium inside the lid or cap should be aspirated to prevent the spread of contamination.

- It is good practice to aliquot medium that is needed from the stock bottle before beginning culture manipulations in order to prevent medium stock contamination.

- When transferring large volumes, it is best to pour liquids from one container to another, rather than pipetting.

- When pouring, do not allow the openings of the containers to touch.
- Any pipette or tip entering a bottle or tube of medium should be sterile.
- While pipetting from bottles or flasks, it is best to tilt the bottle at an angle to avoid positioning hands over the open bottle (Fig. 3.1b).
- Do not allow a pipette or pipette tip to come in contact with any nonsterile surface, such as the top edge of a tube, inside the lid of a medium bottle, or the workspace counter.
- Change pipettes and pipette tips frequently during culture manipulations.
- When aspirating medium from cells, be sure to change Pasteur pipettes frequently to reduce the risk of contamination.
- To prevent cross-contamination between different cell lines, work with only one cell line at a time in the biosafety cabinet.

Accidents Happen

- If any spills occur in the biosafety cabinet, immediately clean up the spill with 70% ethanol and a Kimwipe.
- If any cell culture reagent is accidentally aspirated into the cotton wool plug of a pipette, immediately discard the pipette and all of its contents.
- If any cell culture reagent has made contact with the filter of the pipettor, change the filter before continuing with cell culture activities.

When Works Are Complete

- Wipe the work surface with 70% ethanol.
- The glass sash should be closed and the UV lamp turned on for at least 15–20 minutes.
- Hands should be thoroughly washed.

This chapter has provided a basic set of procedures for preventing contamination of stem cell cultures. However, there are many other aspects to aseptic technique, including personal hygiene, and environmental factors such as air quality and laboratory cleanliness. Although aseptic technique can be very daunting initially, these procedures will become second nature with practice.

QUALITY CONTROL OF CELL CULTURES

It is essential that researchers ensure the sterility, authenticity, and stability of cell lines used in their work in order to publish and provide reproducible and informative experimental data. Upon receipt of a new cell line, it is highly recommended that cells are assayed for the criteria outlined below and monitored at regular intervals to confirm these characteristics.

Cell Line Sterility

Cells in continuous culture are generally vulnerable to microbial contaminants. Bacterial and fungal contamination cause cell death and eventual loss of entire cultures. Human embryonic stem cell (hESC) cultures are particularly susceptible since they are commonly cultured in enriched media without antibiotics. Significant contamination by bacteria or fungi is easily observed by turbidity of the culture media (Figs. 3.3 and 3.4). Detection of a minor or early contamination involves inoculation of bacterial broth medium with samples of the hESC culture medium. These samples are incubated overnight at both cell culture temperature (37°C) and at room temperature to reveal contaminants that may have different optimal growth

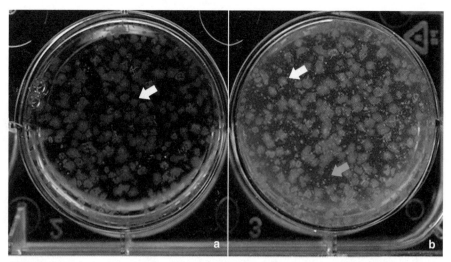

FIGURE 3.3. *Contaminated media and cells. (a) This figure shows hESC colonies (white arrows) grown in a healthy culture and (b) culture contaminated with bacteria and fungal aggregates (orange arrow). Microbial contamination causes the media to appear turbid. Cells were grown in a six-well plate.*

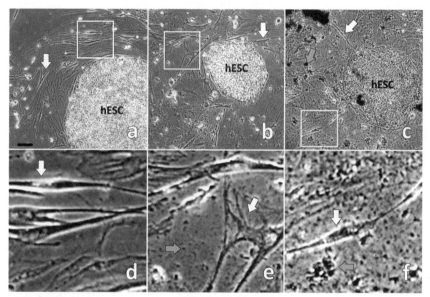

FIGURE 3.4. *Contaminated hESC culture. These micrographs show cultures of hESC colonies co-cultured with MEFs (white arrows): (b) and (e) contamination with bacteria (orange arrow), which appear as small dark dots in the spaces between cells; (c) and (f) heavy bacterial and fungal contamination with significant cell death and floating fungal aggregates (orange arrows). Scale bar represents 100 μm; bottom panels are enlargements of boxed areas in top panels.*

temperatures. Contaminated cultures should be discarded. Various methods to test for these contaminants are recommended by agencies such as the US Food and Drug Administration. In addition, a sterility testing service is offered by vendors for a reasonable fee.

In addition to bacteria and fungi, another common contaminant is mycoplasma. These microorganisms are generally smaller than bacteria and may affect cell growth, particularly at high levels of contamination. However, persistent infections of cells can result in genetic and phenotypic changes. Common sources of mycoplasma include contaminated materials of animal origin such as serum, trypsin, and primary feeder cell cultures. As mycoplasma infection of cell cultures does not always cause media turbidity or readily apparent cell damage, it often persists undetected for long periods of time. It is estimated that 10–15% of all cell cultures may have mycoplasma contamination. Thus it is important to test for the presence of mycoplasma on a regular basis and discard contaminated cultures. If a culture is deemed irreplaceable, elimination of mycoplasma contamination can be attempted with the use of certain antibiotics. However, the success rate for mycoplasma eradication is low and exposure to antibiotics may lead to phenotypic alterations or selective clonal outgrowth. Common methods used to detect mycoplasma include enzymatic assays, polymerase chain reaction (PCR), culture in selective media, and DNA staining of test cells to visualize mycoplasma that grow in close association with the cell membrane. Although the least sensitive, DNA staining with 4',6-diamidino-2-phenylindole (DAPI), a fluorescent stain that binds strongly to DNA, is a simple method that can be employed in most laboratories to detect mycoplasma. Commercially available kits for detection of mycoplasma include the rapid detection system from Lonza called MycoAlert®. Several other companies also offer mycoplasma detection services.

Another form of contamination is viruses, which alter a cell line's characteristics to varying degrees through multiple activities. These activities include utilization of host cellular resources for viral replication and integration into the host genome. Viral infection may affect experimental data and could result in misleading interpretations. In addition, cultures that are contaminated with bloodborne viruses capable of human infection pose a serious health risk. Common sources of viral contamination include animal products and preparations such as bovine serum, antibodies, and mouse embryonic fibroblasts.[1,2] Although testing for viral contamination is conducted by stem cell banks and repositories on a regular basis, this testing is not common practice for individual research laboratories. It is recommended that laboratories test cells for viral infection prior to their distribution and that recipient laboratories request documentation of this testing. Viral testing may include typical contaminants of blood products such as HIV and hepatitis, and cellular products of cell culture such as human herpes viruses.[3] Several companies offer a wide range of viral testing services; their fees depend on the breadth and types of tests performed.

It is highly recommended that cells received from any provider should be tested for microbial contamination before initial use and at regular intervals during routine culture in the laboratory. Since microbes are everywhere in the culture environment, stringent practice of aseptic technique is essential for preventing culture contamination.

Cell Line Authenticity

Cells in culture have a high risk for cross-contamination since most laboratories routinely culture multiple cell lines simultaneously. Studies have shown that up to 30% of cell lines donated to public repositories are contaminated by rapidly growing cell types,[4,5] such as HeLa cells, which can outgrow and replace the original cell lines. Thus cross-contamination can lead to false data and misleading conclusions, which could have catastrophic consequences for downstream clinical applications.

In addition to accidental cross-contamination, purposeful misrepresentation of the identity of cell lines can be an issue. The fraudulent replacement of the "federally approved" stem cell line (Miz-hES1) with another cell line when it was submitted to the NIH[6] highlights the need for providers such as stem cell banks to carefully identify all lines submitted for banking.

Clearly, authentication of cell lines obtained from outside sources prior to their use in experiments is essential and can be achieved by comparing the unique features of received cell lines against those of the original isolate. Several approaches can be employed to authenticate cell lines; the most common of which is genotyping, which takes advantage of the small genetic variations between individual cell lines. Current DNA typing employs PCR-based techniques to analyze similar hypervariable satellite DNA sequences and single nucleotide polymorphisms. Multiple companies offer genotyping services for a small fee.

Cell Line Stability

Cells grown in culture have a tendency to accumulate genetic and/or phenotypic changes. Studies have shown that long-term culture of hESCs results in genetic abnormalities including changes at the chromosomal level, which can be detected by karyotyping methods.[7] The karyotypes of human embryonic stem cell cultures should be frequently tested by a certified cytogenetics laboratory.

Multiple techniques are used to ensure that hESC lines retain their stem cell phenotype, and include expression of stem cell markers and the ability to form the three embryonic germ layers. It is highly recommended that several methods be used to monitor cell line stability (see Chapters 16 and 17).

All cultured cells, including hESCs, are at risk for contamination and prone to genetic and phenotypic changes. Consequently, it is vital that precautionary measures be taken to verify cell line identity and monitor their purity and stability. Although these tests require time and some expense, these measures will ensure that the data generated using cultured cells is reliable and informative.

CELL COUNTING USING A HEMACYTOMETER

NOTES

Many cell culture methods, such as cryopreservation, require an accurate determination of the number of cells in a culture. The simplest and most readily available method for cell counting is performed using a hemacytometer (or hemocytometer). Alternatively, there are automated counters that can provide faster, more accurate cell counting but are not as economical, portable, or easy to use.

A hemacytometer is a thick glass slide that has two chambers, each of which is divided into a grid of nine 1-mm^2 squares (Fig. 3.5). These squares are subdivided into smaller units of 0.0625, 0.04, and 0.0025 mm^2. A coverglass positioned 0.1 mm above the chamber creates the volumes listed in Table 3.1. After filling the hemacytometer with a known dilution of cells in suspension, the cells on the grid can be counted using a microscope and the number of cells in the culture can easily be calculated.

Although easy to use, there are several issues that can lead to errors in counting. These include inadequate mixing of the sample, incorrect dilution of the cell suspension (too few or too many cells), improper chamber filling, excessive cell clumping, and lack of consistent methods for counting cells on the grid lines. With care, these errors can be minimized and reasonably accurate counting is achieved. This protocol describes how hemacytometers can be used to determine the number of cells in freshly thawed mouse embryonic fibroblast cultures and plated human stem cell cultures.

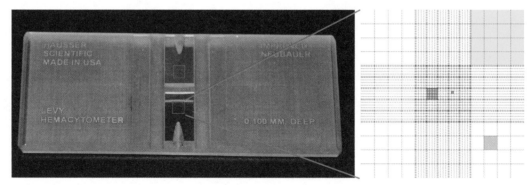

FIGURE 3.5. *The hemacytometer. A standard hemacytometer is pictured (left). The red squares shown on the hemacytometer denote the approximate positions of the counting grids. A diagram of a counting grid is shown on the right. Grid squares of different sizes are color coded. Yellow, 1 mm^2; green, 0.0625 mm^2; blue, 0.04 mm^2; red, 0.0025 mm^2 (see Table 3.1).*

TABLE 3.1 RELATIVE DIMENSIONS AND VOLUMES OF SQUARES IN THE HEMACYTOMETER GRID

Square Dimensions	Area	Number/mm^2	Volume
1×1 mm	1 mm^2	1	1×10^{-4} mL
0.25 × 0.25 mm	0.0625 mm^2	16	6.25×10^{-6} mL
0.20 × 0.20 mm	0.04 mm^2	25	4×10^{-6} mL
0.05 × 0.05 mm	0.0025 mm^2	400	2.5×10^{-7} mL

Overview

Prepare hemacytometer.

Prepare cell suspension.

Load hemacytometer.

Count cells.

Calculate cell number.

General Notes

- Read the entire chapter before initiating activities described in Procedure.
- Use dyes such as Trypan blue to distinguish between living and dead cells.

Abbreviations

EDTA: ethylenediaminetetraacetic acid

FBS: fetal bovine serum

HBSS: Hank's balanced salt solution

hESC: human embryonic stem cell

MEF: mouse embryonic fibroblast

μL: microliter

mL: milliliters

mm^2: square millimeter

PBS: phosphate buffered saline

Equipment

- Cell culture microscope with $4\times$ and $10\times$ objectives
- Handheld counter
- Hemacytometer with glass coverslip
- Incubator

Materials

- 15-mL and 50-mL Sterile conical tubes
- 70% Ethanol
- Absorbent paper towels (or Kimwipes)
- Disposable gloves
- Micropipettors P200 (20 to 200 μL), P1000 (200 to 1000 μL)
- Pasteur pipettes

Reagents

- Cell culture medium
- Fetal bovine serum (FBS)
- Hank's balanced salt solution (HBSS)
- MEF culture medium
- Phosphate buffered saline (PBS)
- Trypan blue (0.4% in PBS)
- Trypsin-EDTA (0.05% trypsin with EDTA)

Note: The next two sections describe the preparation of freshly thawed MEFs (Section I) and plated hESCs (Section II) for cell counting.

NOTES

I. PREPARATION FOR COUNTING A FROZEN VIAL OF INACTIVATED MEFs

NOTES

Preparation

1. Place the following items near the biosafety cabinet:
   - 70% Ethanol spray
   - Absorbent paper towels (or Kimwipes)
   - Disposable gloves

2. Ensure that the following items are in the biosafety cabinet:
   - 15-mL and 50-mL Tubes
   - Micropipettors P200 (20 to 200 µL), P1000 (200 to 1000 µL)
   - Pasteur pipettes
   - Trypan blue (0.4% in PBS)
   - MEF culture medium

3. Thaw a vial of inactivated MEFs:
   3.1. Label a sterile 50-mL tube (both on the cap and side) in the biosafety cabinet as "MEF."
   3.2. Transfer 9 mL MEF culture medium per cryovial to be thawed into the "MEF" tube.
   3.3. While wearing safety glasses and insulated gloves, remove the inactivated MEF vial(s) from the LN_2 freezer and verify the label on the cryovial(s).
   3.4. Immerse the vial(s) in 37°C water bath without submerging the cap in the water and swirl gently.
   3.5. After 30 seconds, check to see whether frozen medium has started to melt. Keep checking intermittently as this process may take up to 1 minute.
   3.6. When there is only a small piece of ice floating in the cryovial(s), remove the vials(s) from the water bath.
   3.7. Proceed to Section II, step 4 of Preparation: Concentration of Cells.

II. PREPARATION FOR COUNTING hESCs ALREADY IN CULTURE

Preparation

1. Place the following items near the biosafety cabinet:
   - Absorbent paper towels (or Kimwipes)
   - 70% Ethanol spray
   - Disposable gloves
2. Ensure that the following items are in the biosafety cabinet:
   - 15-mL and 50-mL Sterile conical tubes
   - Cell culture medium with FBS
   - HBSS
   - Micropipettors P200 (20 to 200 µL), P1000 (200 to 1000 µL)
   - Pasteur pipettes
   - PBS (calcium-and magnesium-free to help detachment)
   - Trypan blue (0.4% in PBS)
   - Trypsin-EDTA (0.05%)

   Note: *When counting adherent cells, trypsinization is required to produce cells in suspension.*

3. Prepare a single cell suspension:

   Note: *This procedure is intended for preparing a cell suspension for one well of a six-well plate. In order to accurately count hESCs using a hemacytometer they must be made into a single cell suspension. Once hESCs are in a single suspension or exposed to FBS or Trypan blue they can no longer be returned to regular culture and should be discarded.*
   Volumes should be adjusted for more wells or different plate sizes.

   3.1. Place six-well culture plate in the biosafety cabinet and aspirate the culture medium.
   3.2. Rinse and aspirate the well 2× with 3 mL of HBSS or PBS.
   3.3. Add 1 mL of 1× trypsin-EDTA to the well.
   3.4. Incubate at 37°C for 2–4 minutes. Using a microscope, confirm that cells have begun to detach from the plate.
   3.5. Within 5–6 minutes, add 2–3 mL of cell culture medium containing 10% FBS to each well.
   3.6. Completely detach cells from each well by drawing the medium from each well into a pipette and forcefully ejecting it back into the well. Systematically sweep the well with ejected medium to ensure detachment of all cells.
   3.7. Pipette up and down to break up the cell clumps.
   3.8. Pool the detached cells into one 15-mL or 50-mL tube.

   Note: *Typically, there are approximately (0.25–0.5) × 10^6 cells/mL in a single well of a six well plate. As this is too dilute for counting, a centrifugation step as described below is recommended.*

   3.9. Concentration of cells:
      3.9.1. Centrifuge the tube for 5–7 minutes at 200*g* (1000 rpm).
      3.9.2. Return the tube to the biosafety cabinet and aspirate supernatant.
      3.9.3. Resuspend the cell pellet in 1 mL of culture medium.

NOTES

4. Prepare hemacytometer:

   4.1. In the biosafety cabinet, clean hemacytometer and coverglass with 70% ethanol and dry with Kimwipe. The hemacytometer should be completely dry and free of dust and dirt.

   4.2. Place coverglass over the center of the hemacytometer.

Procedure

1. Follow preparations according to Aseptic Technique on page 19 especially when pipetting from cell culture stock bottles.

   Note: *Trypan blue is a dye that indicates the viability of a cell. However, it cannot distinguish apoptotic cells. Cells that stain blue are considered dead and should not be counted.*

2. Make a 0.08% solution of Trypan blue in PBS by diluting 0.2 mL of 0.4% Trypan blue solution in 0.8 mL of PBS.

3. Using a micropipettor, mix the cell suspension well and transfer 0.5 mL to a 15-mL tube then add 0.5 mL of diluted Trypan blue/PBS to the tube.

4. With a Pasteur pipette or micropipettor, fill both chambers of the hemacytometer without overflow (usually approximately 10 μL per chamber) by capillary action. Do not introduce bubbles.

   Note: *This must be done quickly as cells will settle in the pipette and the suspension will not be evenly distributed.*

5. Allow cells to settle in the hemacytometer for 2 minutes.

6. Place the hemacytometer on the microscope and count the cells in 1-mm^2 squares (Fig. 3.6, yellow) using the handheld counter. Count at least 100 cells. For most cells, counts should be done in all four of the yellow corner squares and the center square. For a dense suspension of very small cells, count the cells in the four 0.04-mm^2 orange corner squares plus the middle square in the central square (Fig. 3.6, orange).

   Note: *Use a specific counting pattern. The commonly used standard is to count cells that overlap the top or right sides of the square, but not those overlapping the bottom or left sides.*

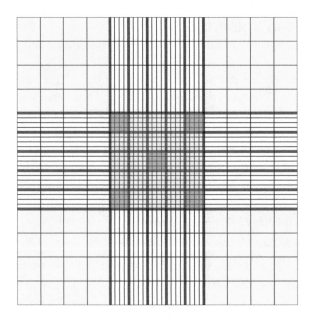

FIGURE 3.6. *The counting grid. Normally, cells in each of five 1-mm^2 squares (yellow) are counted. For very small cells, counting may be done in five of the 0.04-mm^2 squares (orange).*

NOTES

The cell suspension should be diluted so that each 1-mm² square has between 20 and 50 cells. If fewer than 20 cells are counted per 1-mm² square, repeat centrifugation (Section II, step 4 of Preparation: Concentration of Cells) and resuspend cells in a smaller volume of medium. For more than 50 cells, add more medium as necessary.

Cells should be evenly distributed, without overlaps or clumping. If more than approximately 10% of cells are in clumps, wash the hemacytometer and repeat the procedure.

7. Repeat counting on the other chamber.

8. Rinse coverglass and hemacytometer with water and 70% ethanol. Air-dry and store.

9. Calculate the number of cells per milliliter and the total number of cells in the original culture:

   9.1. Average the number of cells counted for each square.

   9.2. For the 1-mm² squares, cells/mL = average count per square $\times 10^4$.

   9.3. For the 0.04-mm² squares, cells/mL = average count per square $\times 10^4 \times 25$.

   9.4. Total cells = cells per mL × dilution factor (2, in this case) × total volume of cell preparation (in this case, 1 mL after centrifugation). For instance, if you counted an average of 30 cells per 1-mm² square, the calculation would be: cells/mL = $30 \times 10^4 = 3 \times 10^5$. Total # cells = 3×10^5 cells/ml $\times$ 2 ml $\times$ 1 (dilution factor).

10. Repeat count to check reproducibility.

Note: *To count hESCs cultured with feeders, you must subtract the feeder cells from the count. To do this, count a well of feeder cells alone with no hESCs. As each lot should be consistent, this would only need to be recounted with each new lot of feeder cells unless culture conditions are changed, which may affect MEF viability.*

REFERENCES

1. Nicklas W, Kraft V, Meyer B. Contamination of transplantable tumors, cell lines, and monoclonal antibodies with rodent viruses. *Lab Anim Sci*. 1993;43:296–300.

2. Stacey GN, Cobo F, Nieto A, et al. The development of "feeder" cells for the preparation of clinical grade hES cell lines: challenges and solutions. *J Biotechnol*. 2006;125:583–588.

3. Cobo F, Stacey GN, Hunt C, et al. Microbiological control in stem cell banks: approaches to standardisation. *Appl Microbiol Biotechnol*. 2005;68:456–466.

4. Nelson-Rees WA, Daniels DW, Flandermeyer RR. Cross-contamination of cells in culture. *Science*. 1981;212:446–452.

5. MacLeod RA, Dirks WG, Matsuo Y, et al. Widespread intraspecies cross-contamination of human tumor cell lines arising at source. *Int J Cancer*. 1999;83:555–563.

6. Cyranoski D, Check E. Koreans admit disguising stem-cell lines. *Nature*. 2006;441:790–791.

7. Hanson C, Caisander G. Human embryonic stem cells and chromosome stability. *APMIS*. 2005;113:751–755.

SUGGESTED READING

Cote, RJ. Aseptic Technique for Cell Culture: Current Protocols in Cell Biology 2003; Unit 1:1.3

Freshney, RI. *Culture of Animal Cells: A Manual of Basic Technique*. 5th ed. Hoboken, NJ: John Wiley & Sons; 2005:73–85.

Phelan MC, Lawler G. Cell Counting: Current Protocols in Cytometry 2003; Appendix 3A.

THE STEM CELL LABORATORY

4

Alicia Allaire, Mai X. Luong, and Kelly P. Smith

CONCEPTS IN STEM CELL CULTURE

Cell culture encompasses a broad range of techniques designed to maintain and expand cells under controlled conditions. Cells can be cultured from a variety of organisms including plants, insects, and mammals. Mammalian cell culture, especially culture of human cells, has been an enormous advantage to biomedical research, allowing basic research to be performed with fewer complexities of whole organism studies. Thus cell culture has become an essential part of most biological investigation.

Mammalian cell culture was first developed approximately a century ago.[1] Since that time, almost all cell types in the human body have been successfully cultured. Cells derived directly from an organism are referred to as primary cells. Most normal primary cells, excluding cancer cells, have limited life spans and stop growing after a finite number of cell divisions. Primary cells can be immortalized so that they gain the capacity for unlimited growth. Immortalization can occur spontaneously by random mutation, but cells can also be immortalized by expression of exogenous factors such as telomerase or SV40 T antigen. Unlike primary cells, cancer cell lines are immortal and are among the most widely cultured cells due to their unlimited, robust growth under a variety of conditions. Human embryonic stem cells (hESCs) are also distinct from primary cells in that they have unlimited life spans in culture until they differentiate.

In the whole organism, each type of cell grows under a unique set of conditions referred to as the microenvironment or niche. These conditions may be evident in the substrate in which the cells grow, surrounding cell types, and exposure to specific growth factors. Understanding the microenvironment is important to successful cell culture. For example, some cell types grow in suspension while others require attachment to a substrate for growth. For the latter, different substrates may be needed. Many cell types grow well on untreated cell culture vessels, while others require a coating of proteins found in the extracellular matrix, such as collagen (gelatin), laminin, or fibronectin, for proper growth. For example, hESCs have been found to grow best on a surface that has been coated with gelatin.

Human Stem Cell Technology and Biology, edited by Stein, Borowski, Luong, Shi, Smith, and Vazquez
Copyright © 2011 Wiley-Blackwell.

BASIC CELL CULTURE GROWTH CONDITIONS

Within the organism, cells grow under a strict set of physicochemical conditions. To mimic these conditions in the laboratory, cells are grown in a specialized cell culture incubator, which is equipped to maintain this environment as stringently as possible. The following environmental parameters are essential for all mammalian cell cultures.

Temperature

Temperatures used in the culturing of specific cell types are determined by the optimal body temperature of the source organisms. For human and most other mammalian cells, 37°C is the most commonly used temperature setting. While cells in culture can withstand fairly significant drops in temperature, raising the temperature by as little as 2°C can lead to cell death in a short period of time. Maintaining a constant temperature when the incubator is opened and closed often can be a challenge. Water-jacketed incubators, which surround the incubator chamber with water, are commonly used as the heated water helps to maintain temperature stability and uniformity in the incubator.

Humidity

Cell culture containers such as plates and flasks are normally not sealed so that there can be an exchange of gases between the culture and the atmosphere. However, at normal temperatures for mammalian cell culture, evaporation from the cell culture media can be a problem. Excessive evaporation can increase the concentration of salts and other components in the medium, which can be detrimental to cell viability. Most cell culture incubators use water reservoirs to maintain >95% humidity levels.

pH and Carbon Dioxide (CO_2)

Another environmental factor that must be carefully controlled in cell culture is media acidity or pH. Cells generally grow well at the physiological pH of 7.4; however, it is difficult to maintain this pH level as cells metabolize nutrients in the medium and produce waste products. In general, pH of the medium is maintained by addition of buffers such as sodium bicarbonate and HEPES (4-(2-hydroxyethyl)-1-piperazineethanesulfonic acid) as well as maintaining a constant level of CO_2 in the incubator.

Carbon dioxide (CO_2) concentration can dramatically affect the acidity of the medium. In solution, CO_2 interacts with water molecules to form carbonic acid (H_2CO_3). Carbon dioxide can cause fluctuations in pH, either by loss of CO_2 from the medium in open containers, which causes the pH to rise, or by overproduction of CO_2 by highly metabolic cells, causing the pH to drop. To guard against these pH changes, the amount of CO_2 in the atmosphere and sodium bicarbonate ($NaHCO_3$) in the medium need to be carefully controlled. These two factors must reach equilibrium in order to maintain a constant pH. For most cell cultures, the medium is formulated to maintain pH 7.4 in 5% CO_2. Modern cell culture incubators use sensors, which monitor CO_2 concentration and add CO_2 from an external tank when needed.

Beyond these three basic parameters, the culture of specific cell types requires unique media formulations and other conditions such as specific substrates or supporting cells for optimal growth.

Unique Characteristics of Pluripotent Stem Cell Culture

In addition to the standard culture conditions discussed previously, successful culture and maintenance of human embryonic stem cells (hESCs) in an undifferentiated state require the recreation of the biochemical and mechanical niche that regulates stem cell survival, self-renewal, pluripotency, and differentiation in vivo. Key components of the in vivo stem cell microenvironment include growth factors, cell-to-cell interactions, and cell-to-matrix adhesions. Standard culture of hESCs involves exposure to media enriched with growth factors found in fetal bovine serum (FBS). In recent years significant progress has been made toward defining the components of serum necessary to maintain self-renewal and pluripotency of hESCs.[2,3] Currently, basic fibroblast growth factor (bFGF) and serum replacer medium are in widespread use as an alternative to FBS to limit differentiation. It is generally accepted that bFGF inhibits hESC differentiation by bone morphogenetic proteins (BMPs), although other mechanisms may be involved.[4-6] In addition, the use of support cells such as an inactivated mouse embryonic fibroblast (MEF) feeder layer is critical to support hESC growth and prevent differentiation. These cells provide necessary intercellular interactions, extracellular scaffolding, and factors important to creating a niche for hESC growth and maintenance of the undifferentiated state. These supporting cells are so important that hESCs grown in feeder-free conditions will generate their own supporting cells to recreate the niche.[7] The requirement for cell-to-cell contacts is also met by culturing hESCs in aggregates or colonies.

Human embryonic stem cells possess enormous potential for regenerative medicine. However, any clinical use of these cells will require elimination of animal products that pose a risk of exposure to retroviruses and other pathogens from the culture environment. Many approaches have been published for culturing hESCs in an entirely animal-free environment, including the use of human fibroblasts and serum, or replacement of serum with defined growth factors.

There are several basic techniques needed for the culturing of mammalian cells, including thawing frozen stocks, plating cells in culture vessels, changing media, passaging, and cryopreservation. Passaging refers to the removal of cells from their current culture vessel and transferring them to one or more new culture vessels. Passaging is necessary to reduce the harmful effects of overcrowding and for expansion of the culture. Protocols provided in this book describe the standard culture of hESCs. Variations of these methods are widely employed across many laboratories and satisfy the needs of basic research. These procedures have been shown to be reliable and generate reproducible experimental data.

LABORATORY LAYOUT

There are many key steps to the successful culture of human embryonic stem cells (hESCs). Among these is the quality of the culture laboratory itself. The establishment of a well-equipped, ergonomically arranged, and properly maintained culture laboratory will improve workflow and cell viability, as well as significantly reduce the potential for contamination.

This section gives an overview of how to set up the laboratory to achieve a favorable environment for cell culture activities, including recommendations for the optimal laboratory layout and proper laboratory maintenance.

Abbreviations

CO_2: carbon dioxide

hESCs: human embryonic stem cells

HEPA: high-efficiency particulate air

lfpm: linear feet per minute

LN_2: liquid nitrogen

UV: ultraviolet

Laboratory Safety

All personnel working in the stem cell laboratory should follow all safety precautions required by their institution's safety office. Additionally, in the United States, research on human stem cells is governed by biosafety regulations established by the National Institutes of Health (NIH) and the Centers for Disease Control and Prevention (CDC). According to these regulations, stem cells require biosafety level 2 (BSL-2) containment, which is required when "work is done with any human-derived blood, body fluids, tissues, or primary human cell lines where the presence of an infectious agent may be unknown" (http://www.cdc.gov/od/ohs/biosfty/bmbl5/BMBL_5th_Edition.pdf). Furthermore, standards developed by the Occupational Safety and Health Administration (OSHA) for bloodborne pathogens (http://www.osha.gov/pls/oshaweb/owadisp.show_document?p_table=STANDARDS&p_id=10051) should be followed.

The following are several standard recommendations for cell culture laboratory safety:

- Access to the laboratory should be restricted. The door should remain closed to limit access by casual visitors.
- Eating and drinking are prohibited.
- Laboratory personnel should wear personal protective equipment that includes a lab coat, gloves, and eye protection.
- Cell culture work should be performed in a class II, type B2 biosafety cabinet.
- Pipetting should be done using a mechanical pipetting device.
- Waste should be decontaminated by disinfection or autoclaving before disposal.
- Care should be taken in handling any sharp materials such as needles or broken glass.

Ideally, stem cell lines used in the laboratory should have been tested for contamination by human viruses (such as HIV, HPV, or EBV) as discussed in Chapter 3, Quality Control of Cell Cultures on page 22.

Equipment

The following is a list of equipment necessary for setting up an embryonic stem cell culture laboratory.

- *Biosafety Cabinet.* Class II type B2 Biological Safety Cabinet with ultraviolet (UV) light option. Biological safety cabinets of this class have 100 linear feet per minute (lfpm) inward and 80 lfpm, high-efficiency particulate air (HEPA) filtered downward airflow. All inflow and downflow air is exhausted via air ducts after HEPA filtration to the outside of the cabinet without any recirculation within the cabinet (Fig. 4.1).
- *Cell Culture Incubators.* Accurate monitoring and maintenance of humidity, temperature, and carbon dioxide (CO_2) content is crucial. All incubators must be equipped to input CO_2 and monitor the CO_2 levels. Filters are to be used on the CO_2 gas line and ambient air input line prior to entry into the incubator to prevent contamination of cultures. Small incubators are preferred because they tend to have less temperature fluctuation. Most cell culture incubators are equipped with stainless steel racks and water pans. Although more expensive, copper components are worth considering as they can reduce contamination in the incubator. Some manufacturers produce incubators with copper racks, and copper lining the interior and housing around the HEPA filtration system.

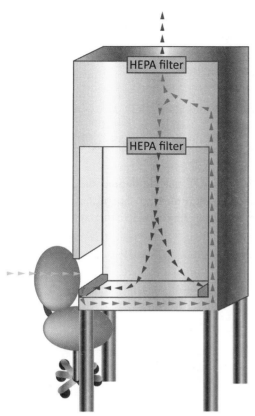

FIGURE 4.1. *Diagram of a biosafety cabinet. This diagram shows the flow of air in a standard class II, type A biosafety cabinet. Room air (yellow) flows into the grates at the front of the cabinet base. This nonsterile air (red), along with the air recirculated from the cabinet interior, flows through the back of the cabinet to the top, where it is either vented back into the room or into the cabinet interior as HEPA filtered air (blue). In the cabinet, the air is forced down to the cabinet floor, where it flows along the floor of the cabinet to intake grates at the front and back.*

- *Cell Culture Microscope.* An inverted microscope equipped with phase contrast ability and at least $4\times$ and $10\times$ objectives is essential. It is important to record colony morphology and growth rate, which can be facilitated by a small digital camera mounted on the microscope with an imaging software program.

- *Refrigerators ($4°C$) and Freezers ($-20°C$, $-80°C$).* For storage of medium and reagents.

- *Liquid Nitrogen (LN_2) Freezer with a Tank of LN_2.* For long-term storage of hESCs and feeder cells.

- *Tabletop Centrifuge with Swing-out Bucket Rotors.* Required for concentrating cells into a pellet with the objective of separating cells from their media. The centrifuge should work with 15 and 50-mL conical tubes.

- *Water Bath ($37°C$).* To thaw frozen cells and bring reagents to appropriate temperature.

- *Alarm System.* An alarm system is essential for preventing the loss of valuable frozen stocks. Set the system up to monitor and alert onsite personnel about temperature and power fluctuations, as well as low levels of LN_2 and CO_2. Offsite alerts are helpful for times when there are no personnel in the laboratory.

- *Emergency Power.* Placement of essential equipment, such as incubators and freezers, on an emergency power source, such as a backup generator, is strongly recommended in the event of a power failure.

Laboratory Layout

Placing the above listed equipment in a laboratory solely dedicated to the culturing of only hESCs and feeder cells is essential. Other major considerations for the laboratory layout include the following:

- The stem cell tissue culture laboratory should be located in an area where there will be low traffic.
- Only personnel who are working with the stem cells should be allowed in this designated workspace.
- The placement of equipment, reagents, and specific stations should be linked to their applications during the cell culture process. For instance, place the incubators near the biosafety cabinet and in an area where they are less likely to get jostled.
- Equipment should be placed in a way that allows for easy access and is ergonomic.
- A hand-washing area should be placed as distant from the biosafety cabinets and incubators as possible.
- Make sure that any equipment that may vibrate is not located near the incubators and biosafety cabinets.
- Consumables should be placed within reach of the biosafety cabinet (Fig. 4.2). Only essential items should be inside the biosafety cabinet, as this will help to

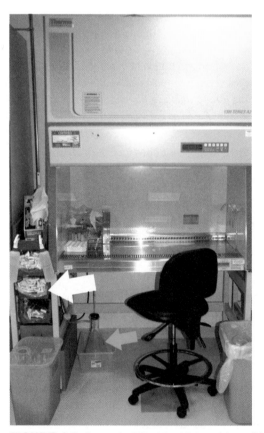

FIGURE 4.2. *Layout of the biosafety cabinet. Pictured is a standard class II, type A biosafety cabinet. The recommended placements of consumables (white arrow) for easy access and the waste container for the vacuum system (orange arrow) are noted.*

keep the work area as sterile as possible and reduce disruption of the laminar airflow inside the biosafety cabinet.

- A vacuum system for aspirating medium or filtering solutions should also be accessible from inside the biosafety cabinet and an inline filter should be placed on the vacuum line.

- Waste should be collected and decontaminated outside the biosafety cabinet, in a large waste container (vacuum flask) that holds a bleach solution. To contain any accidental waste spills, the waste container should be placed inside its own secondary containment.

- Freezers ($-20°C$, $-80°C$, and LN_2) and refrigerators ($4°C$) should be dedicated for stem cell materials and reagents only and should be placed inside or adjacent to the designated workspace.

Laboratory Maintenance

Once the laboratory has been equipped and properly arranged, it is important to follow a regular schedule of upkeep and maintenance for all equipment to ensure the optimal function of the laboratory. Maintenance includes the following:

- Keeping the stem cell culture laboratory organized and free of clutter at all times.

- Spraying all work surfaces with 70% ethanol and wiping with lint-free paper towels or Kimwipes on a daily basis to reduce the risk of contamination. After culture work is complete and the workspace of the biosafety cabinet has been wiped down, the UV lamp should be turned on for minimally 15–20 minutes in order to sterilize the biosafety cabinet. For maximal germicidal effect, UV lamps should be replaced after 8000–12,000 hours, or as recommended by the manufacturer. Appropriate filtration of the air prior to turning on the UV lamps is also recommended.

- Placing adhesive floor mats at the point of entry of each doorway and changing the mats as they become soiled to limit entry of dust and dirt into the culture lab.

- Wet mopping the floor only at the end of the day. It is imperative that dust or other particles should not be stirred up prior to performing any cell culture.

- Inspecting biosafety cabinets once they are installed and then periodically thereafter.

- Changing the HEPA filters for both the biosafety cabinets and incubators at regular intervals, such as every 6 months, or as recommended by the manufacturer.

- Disinfecting the incubators, water trays in the incubators, and water baths regularly. Incubators need to be disinfected with commercially available biocidal spray and 70% ethanol at least every 3 months. Following a regular schedule helps to keep a minor contamination from spreading. Be sure to read the manufacturers' suggestions before disinfecting equipment.

- Adding a commercially available antimicrobial agent to water trays in the incubators and water baths at regular intervals, as recommended by the manufacturer. This will reduce the chance of contamination between regularly scheduled disinfections.

- Upon detection of contamination, all work areas, water baths, incubators, and other equipment that have come in contact with contaminated cultures should be thoroughly disinfected with biocidal spray and 70% ethanol.

REFERENCES

1. Harrison R. Observations on the living developing nerve fiber. *Anat Rec Proc Soc Exp Med*. 1907:116–128, 140–143.

2. McDevitt TC, Palecek SP. Innovation in the culture and derivation of pluripotent human stem cells. *Curr Opin Biotechnol*. 2008;19:527–533.

3. Chase LG, Firpo MT. Development of serum-free culture systems for human embryonic stem cells. *Curr Opin Chem Biol*. 2007;11:367–372.

4. Pera MF, Andrade J, Houssami S, et al. Regulation of human embryonic stem cell differentiation by BMP-2 and its antagonist noggin. *J Cell Sci*. 2004;117:1269–1280.

5. Xu RH, Chen X, Li DS, et al. BMP4 initiates human embryonic stem cell differentiation to trophoblast. *Nat Biotechnol*. 2002;20:1261–1264.

6. Xu RH, Peck RM, Li DS, et al. Basic FGF and suppression of BMP signaling sustain undifferentiated proliferation of human ES cells. *Nat Methods*. 2005;2:185–190.

7. Bendall SC, Stewart MH, Menendez P, et al. IGF and FGF cooperatively establish the regulatory stem cell niche of pluripotent human cells in vitro. *Nature*. 2007;448:1015–1021.

SUGGESTED READING

Centers for Disease Control and Prevention. Biosafety in Microbiological and Biomedical Laboratories. Washington DC: US Government Printing Office; 2007. Available at http://www.cdc.gov/od/ohs/biosfty/bmbl5/BMBL_5th_Edition.pdf.

Freshney RI. *Culture of Animal Cells: A Manual of Basic Technique*. 5th ed. Hoboken, NJ: John Wiley & Sons; 2005: 73–85.

Schantz T, Ng K W. *A Manual for Primary Human Cell Culture*. 2nd ed. Singapore: World Scientific Pub Co.; 2009.

REAGENT PREPARATION

5

Alicia Allaire, Mai X. Luong, and Kelly P. Smith

Several reagents must be prepared prior to maintaining human embryonic stem cells (hESCs) in culture. It is important to have a firm grasp of the characteristics of each reagent, including components, appropriate preparation, storage, and shelf life. This chapter details the preparation of the following reagents necessary for hESC culture:

SECTION I.   Preparation of 0.1% Gelatin Solution

SECTION II.   Preparation of Collagenase Solution

SECTION III.   Preparation of Basic Fibroblast Growth Factor Stock Solution

SECTION IV.   Preparation of Inactivated Mouse Embryonic Fibroblast Culture Medium

SECTION V.   Preparation of Human Embryonic Stem Cell Culture Medium

SECTION VI.   Preparation of Differentiation/Embryoid Body Medium

General Notes

- Many reagents, especially Knockout Serum Replacer (KOSR), bFGF, and collagenase can have significant variation between lots. When working with a new lot, it is important to compare it directly to the lot that is currently in use, to ensure that the current culture quality and viability are maintained.

Abbreviations

bFGF: basic fibroblast growth factor

BSA: bovine serum albumin

$CaCl_2$: calcium chloride

DMEM/F12: Dulbecco's modified Eagle's medium, nutrient mixture F-12

ESCs: embryonic stem cells

g: gram

hESCs: human embryonic stem cells

HI-FBS: heat-inactivated fetal bovine serum

H_2O: water

KOSR: knockout serum replacer

M: molar

MEF: mouse embryonic fibroblast

MEM: modified Eagle's medium

$MgCl_2$: magnesium chloride

Human Stem Cell Technology and Biology, edited by Stein, Borowski, Luong, Shi, Smith, and Vazquez
Copyright © 2011 Wiley-Blackwell.

μg: microgram

μL: microliter

μm: micrometer

mg: milligram

mL: milliliter

mM: millimolar

NEAA: nonessential amino acid

ng: Nanogram

PBS: phosphate buffered saline

I. PREPARATION OF 0.1% GELATIN SOLUTION

Many adherent mammalian cell types require a substrate or scaffold on which they can attach for growth in culture vessels. Although some cell types, such as many transformed or cancer cells, can grow directly on the surface of a plastic culture vessel, primary cells usually require a biological matrix for attachment. A commonly used matrix, gelatin, is a substance produced by the partial degradation of animal collagen. The mixture of proteins in gelatin forms a bond, or gel, in a temperature-dependent manner to form a substrate for cell attachment. While gelatin does not directly support the undifferentiated growth of hESCs, gelatin coating of the culture vessel enhances attachment of the mouse embryonic fibroblast (MEF) cells that support hESC growth.

Overview

Weigh gelatin.

Place in clean bottle.

Add filtered H_2O.

Autoclave 30 minutes.

Cool to room temperature.

Store at 4°C.

General Notes

- Use filtered, endotoxin-free water.
- Wash glassware for cell culture with only water, as hESCs are sensitive to residual soap.
- Use gelatin within 2 months of preparation.

Equipment

- 500-mL sterile glass bottle
- Autoclave
- Sterile biosafety cabinet
- Weighing balance

Materials

- Alcohol-proof marker
- Weigh boat
- Lab tape

Reagents

- Endotoxin-free H_2O
- Gelatin, porcine

Preparation

1. Place the following items near the biosafety cabinet:
   - 70% Ethanol spray
   - Absorbent paper towels (or Kimwipes®)
   - Disposable gloves
2. Ensure that the following item is in the biosafety cabinet:
   - 500-mL sterile glass bottle

NOTES

NOTES

_____

_____

_____

_____

_____

_____

_____

_____

_____

_____

_____

_____

Procedure

1. Follow preparations according to Chapter 3, Aseptic Technique on page 19.

2. Spray a 500-mL glass bottle with 70% ethanol and place it in the biosafety cabinet.

3. Apply tape to the bottle and label with the following information:
   - "0.1% Gelatin"
   - Date
   - Initials

4. Weigh 0.5 g of gelatin powder in a weigh boat, and then add the 0.5 g of gelatin to the labeled bottle in the biosafety cabinet.

5. Add 500 mL of filtered, endotoxin-free water to the bottle, swirling the bottle to mix contents (at this stage, the gelatin is not soluble).

   Note: *Adjust the volume of water and gelatin powder proportionally if a volume other than 500 mL is to be prepared.*

6. Within 2 hours after mixing, autoclave the gelatin solution for 30 minutes.

7. Cool the 0.1% gelatin solution to room temperature on the bench, then store the 0.1% gelatin solution bottle at 4−8°C until use. Use this solution within 2 months of preparation.

II. PREPARATION OF COLLAGENASE SOLUTION

Passaging, or splitting, cultured cells involves the transfer of cells from their current culture vessel to one or more new culture containers for continued growth. Upon contact with the surface of a vessel, adherent cells secrete proteins that bind to the surface and form a layer of extracellular matrix. The extracellular matrix is composed of numerous proteins, including collagen, which help to anchor cells to the vessel by interacting with specific cell surface proteins. However, certain cell types require additional collagen for optimal adhesion and prefer vessels coated with gelatin, which is partially degraded animal collagen. Removal of adherent cells from a culture vessel involves detaching them from the underlying extracellular matrix, which is usually accomplished by treating the cells with a protease enzyme. These enzymes break down proteins by cleaving amino acid chains at specific points. Degradation of cell surface and extracellular matrix proteins aids in the release of adherent cells from the vessel surface.

The most commonly used protease in cell culture is trypsin, which cleaves all proteins at lysine and arginine residues. However, due to the sensitivity of hESCs, trypsin is considered to be too harsh to be used when passaging these cells since this protease also degrades cell surface proteins that are not involved in cell adhesion. For this reason, collagenase, a protease that specifically cleaves triple helix motifs commonly found in collagen proteins, is preferred for hESC culture. There are different preparations of collagenase (usually referred to as types I—IV) that contain varying amounts of other proteases. Each preparation has characteristics that make it suitable for culturing different cell types. For hESCs, collagenase type IV is most commonly used due to its low level of trypsin activity.

Overview

Weigh collagenase.

Add DMEM/F12 to filter cup of filter system.

Dissolve collagenase in DMEM/F12.

Vacuum filter collagenase solution.

Store collagenase solution at 4°C.

General Notes

- Collagenase solution should be used within 2 weeks of preparation as it loses activity over time.

Equipment

- Pipette aid
- Sterile biosafety cabinet
- Weighing balance

Materials

- 25-mL Sterile glass or plastic pipettes
- 150-mL Filter system
- 70% Ethanol spray
- Absorbent paper towels (or Kimwipes)
- Alcohol-proof marker
- Disposable gloves
- Weigh boat

NOTES

NOTES

_____

_____

_____

_____

_____

_____

_____

_____

_____

_____

_____

_____

_____

_____

_____

_____

_____

Reagents

- Collagenase type IV
- DMEM/F12 medium

Preparation

1. Place the following items near the biosafety cabinet:
   - 25-mL Sterile pipettes
   - 70% Ethanol spray
   - Absorbent paper towels (or Kimwipes)
   - Disposable gloves
2. Ensure that the following items are in the biosafety cabinet:
   - 50-mL Sterile conical tubes
   - 150-mL Filter system (Fig. 5.1)
   - Alcohol proof marker
   - DMEM/F12 medium sufficient for preparation of collagenase solution

Procedure

Note: *The following procedure describes how to prepare 100 mL of collagenase solution at a concentration of 1 mg/mL. If preparing another quantity, adjust the amount of collagenase type IV powder and DMEM/F12 medium proportionately.*

1. Follow preparations according to Chapter 3, Aseptic Technique on page 19.
2. Weigh 100 mg of collagenase type IV powder in a weigh boat and place the weigh boat containing collagenase in the biosafety cabinet.
3. To a 150-ml filter cup, pour 80 mL of room temperature DMEM/F12 medium.

FIGURE 5.1. *Sterile filter system. These filters are used to produce larger volumes of sterile liquid for cell culture. The apparatus contains a cup that has the filter at the bottom. The pore size of the filter is 0.22 μm, which removes most bacteria. The cup is attached to a sterile plastic bottle that receives the filtered solution. The connecting cap has a stem that attaches to a vacuum system which draws the solution through the filter. A separate cap is used to seal the bottle for storage once filtration is complete. Filter units come in a variety of sizes.*

4. Using a 25-mL pipette, transfer 20 mL DMEM/F12 medium to the weigh boat containing the collagenase.

5. Pipette DMEM/F12 medium up and down in the boat to dissolve the collagenase type IV powder. The collagenase should dissolve almost instantly.

6. When the collagenase is completely dissolved, transfer the 20 mL of collagenase solution back to the 150-mL filter cup.

7. Take 20 mL of the collagenase solution from the filter cup, rinse the boat to collect residual collagenase, and transfer the solution back to the filter system.

8. Attach filter system (Fig. 5.1) to a vacuum, and filter the solution to sterilize, and then tightly cap the bottle.

9. Record the following on the bottle:
   - 1 mg/mL Collagenase
   - Initials
   - Date of preparation
   - Expiration date (14 days after preparation)

10. Store the collagenase solution at 4°C.

NOTES

III. PREPARATION OF BASIC FIBROBLAST GROWTH FACTOR STOCK SOLUTION

Growth factors are needed for successful culture of human embryonic stem cells (hESCs). Basic fibroblast growth factor (bFGF) promotes undifferentiated growth of hESCs. bFGF has been shown to protect hESCs from stress-induced cell death and to increase hESC adhesion. Exposure of hESCs to bFGF stimulates expression of stem cell genes and represses cell death genes. In addition, bFGF cooperates with other factors to repress hESC differentiation. It is likely that suppression of differentiation and stimulation of self-renewal, as well as improved cell survival and cell adhesion, interact to promote undifferentiated growth of hESCs.

This protocol describes how to prepare bFGF stock solution for the culture of hESCs.

Overview

Prepare 30% BSA solution.

Reconstitute lyophilized bFGF.

Sterile filter bFGF solution.

Aliquot bFGF solution.

Store at -80°C.

Equipment

- Micropipettors P200 (20 to 200 µL), P1000 (200 to 1000 µL)
- Pipette aid
- Sterile biosafety cabinet

Materials

- 1.5-mL or 2-mL Sterile microcentrifuge tubes
- 10-mL and 25-mL Sterile glass or plastic pipettes
- 50-mL Sterile conical tubes
- 70% Ethanol spray
- Absorbent paper towels (or Kimwipes)
- Alcohol-proof marker
- Cryovial holder
- Disposable gloves
- Sterile tips to fit P200 and P1000 micropipettors
- Storage boxes for -80°C freezer

Reagents

- 30% BSA (bovine serum albumin, sterile-filtered, cell culture tested)
- Basic fibroblast growth factor (bFGF) (lyophilized)
- Phosphate buffered saline (PBS) with 0.01% calcium chloride ($CaCl_2$) and 0.01% magnesium chloride ($MgCl_2$)

Preparation

1. Label cryovials with the following information:
   - bFGF stock solution batch number (e.g., FGF-MM/DD/YY)
   - "4 µg/mL"
   - Expiration date (6 months from the date of preparation)

2. Place the following items near the biosafety cabinet:
- 5-mL, 10-mL, and 25-mL Sterile pipettes
- 70% Ethanol spray
- Absorbent paper towels (or Kimwipes)
- Disposable gloves
- Storage boxes for $-80°$C freezer

3. Ensure that the following items are in the biosafety cabinet:
- 50-mL Sterile conical tubes
- Alcohol-proof marker
- Cryovial holder
- Sterile 200-μL, 1000-μL pipette tips

NOTES

_____

_____

_____

_____

_____

_____

_____

_____

_____

Procedure

1. Follow preparations according to Chapter 3, Aseptic Technique on page 19.

2. Prepare 0.1% bovine serum albumin (BSA) solution:

2.1. Place the 30% BSA bottle in the biosafety cabinet after thoroughly cleaning bottle with 70% ethanol.

2.2. Label a 50-mL sterile tube as "0.1% BSA."

2.3. Add PBS to bring volume to 30 mL.

2.4. Add 0.1 mL of the 30% BSA to the tube using a 200-μL Pipetman®.

2.5. Cap the tube and mix the solution gently, inverting the tube four times. It is now a 0.1% BSA solution.

3. Prepare 4 μg/mL bFGF stock solution:

3.1. Place one vial containing 100 μg lyophilized bFGF in a 50-mL tube and briefly centrifuge at 200g to bring the lyophilized bFGF to the bottom of the vial.

3.2. Use a pair of forceps to carefully remove the vial from the 50-mL tube. Clean the external surface of the vial with 70% ethanol and place it in the biosafety cabinet.

3.3. Label a sterile 50-ml tube in the biosafety cabinet as "FGF."

3.4. Add 25 mL of the 0.1% BSA solution to the "FGF" tube.

3.5. From the "FGF" tube remove 500 μL of 0.1% BSA solution using a 1000-μL Pipetman and add it to the bFGF vial. Do not touch the FGF pellet at the bottom of the vial.

3.6. Gently pipette up and down a few times with the 1000-μL Pipetman until FGF is completely dissolved. Transfer the bFGF solution to the 50-mL "FGF" tube.

3.7. Rinse the bFGF vial with 500 μL of solution from the "FGF" tube to collect any residual bFGF protein in the vial. Transfer the solution back to the 50-mL tube. Rinse the vial three more times.

4. Aliquot bFGF (4 μg/mL) stock solution:

4.1. In the biosafety cabinet, place a set of 50 sterile 1.5-mL or 2-mL vials in a cryovial holder, and label with the following information:
- 4 μg/mL bFGF
- Date of preparation
- Expiration date
- Initials

_____

_____

_____

_____

_____

_____

_____

_____

_____

_____

_____

_____

_____

_____

_____

4.2. Loosen caps of all labeled vials.

4.3. From the 50-mL "FGF" tube, transfer 0.5 mL to each of the vials, then tighten the caps and store these vials at -80°C.

4.4. If needed, place one or two vials of FGF stock solution at 4°C to be used within 2 weeks.

IV. PREPARATION OF MOUSE EMBRYONIC FIBROBLAST CULTURE MEDIUM

Since their original derivation in 1981, mouse embryonic stem cells have been cultured with inactivated mouse embryonic fibroblasts (MEFs) as a feeder layer. This practice was subsequently applied to the derivation of human embryonic stem cells (hESCs) in 1998 and continues to be widely used today. Other alternatives are now available for hESC culture that do not use inactivated MEFs, such as human foreskin fibroblasts and feeder-free culture conditions, and more alternatives are in development. However, culturing with inactivated MEFs remains the most common method for supporting undifferentiated growth of hESCs.

A benefit of an inactivated MEF feeder layer is production of an extracellular matrix that provides a substrate for the attachment of hESCs. In addition, inactivated MEFs secrete growth factors that help maintain hESCs in an undifferentiated state.

Because MEFs have been mitotically inactivated, by irradiation or treatment with Mitomycin C, they deplete nutrients and other factors from the medium more slowly than dividing cells. Unlike medium for hESCs, which is changed daily, inactivated MEF medium can remain unchanged for up to 5 days. Inactivated MEF medium contains:

- *Dulbecco's modified Eagle's medium (DMEM)* is a serum-free base medium.
- *Heat-inactivated fetal bovine serum (HI-FBS)* provides inactivated MEFs with metabolic requirements in the form of minerals, hormones, and lipids. It also promotes cell attachment through adhesion factors and antitrypsin activity.
- *Nonessential amino acids (NEAAs)* help to prolong viability of the cell culture.

Overview

Thaw HI-FBS aliquot.

Combine inactivated MEF culture medium ingredients.

Sterile filter.

Store at 4°C.

Equipment

- 37°C Water bath
- Pipette aid
- Sterile biosafety cabinet

Materials

- 5-mL, 10-mL, and 25-mL Sterile glass or plastic pipettes
- 50-mL Sterile conical tubes
- 500-mL Filter system with a 0.22-μm filter (Fig. 5.1)
- Alcohol-proof marker
- Disposable gloves

Reagents

- 70% Ethanol
- DMEM (liquid)
- Heat inactivated fetal bovine serum (HI-FBS)
- MEM nonessential amino acids solution

Note: *Although HI-FBS is commercially available, FBS may be inactivated by heating the solution to 56°C for 30 minutes.*

NOTES

NOTES

_____
_____
_____
_____
_____
_____
_____
_____
_____
_____
_____
_____
_____
_____
_____
_____
_____
_____
_____
_____
_____
_____
_____
_____

Preparation

1. Place the following items near the biosafety cabinet:
   - 5-mL, 10-mL, and 25-mL sterile pipettes
   - 70% ethanol spray
   - Absorbent paper towels (or Kimwipes)
   - Disposable gloves

2. Ensure that the following items are in the biosafety cabinet:
   - 5-mL, 10-mL, and 25-mL sterile pipettes
   - 50-mL sterile conical tubes
   - 500-mL filter system
   - Alcohol-proof marker

3. Thaw HI-FBS (_if needed_):

   3.1. It is recommended to prepare aliquots of HI-FBS. If aliquots have already been prepared, place the frozen aliquot in a $37°C$ water bath until it has thawed and follow the Procedure section of the protocol.

   3.2. Stock bottles of HI-FBS are stored at $-20°C$ and will need to be thawed prior to making aliquots. If thawing a stock bottle in order to prepare aliquots, the stock bottle needs to be thawed following one of the options below.

     3.2.1. _Option one:_ Thaw HI-FBS overnight at $4°C$. If HI-FBS is not completely thawed the next day, finish thawing in a $37°C$ water bath.

     3.2.2. _Option two_: Thaw HI-FBS directly in a $37°C$ water bath. It will take 2–3 hours to thaw a 500-mL bottle. Swirl the bottle gently at frequent intervals to accelerate thawing.

4. Aliquot the HI-FBS if a new bottle was thawed:

   4.1. Label sterile 50-mL tubes in the biosafety cabinet with the following information:
   - HI-FBS
   - Vendor
   - Vendor's catalog and lot number
   - Expiration date
   - Initials

   4.2. Transfer the thawed HI-FBS bottle to the biosafety cabinet after thoroughly cleaning the bottle with 70% ethanol.

   4.3. Leave enough HI-FBS in the bottle to be used in the Procedure section of this protocol. Leave the bottle in the biosafety cabinet if it is to be used shortly, or store the bottle at $4°C$.

   4.4. Pour the remainder into the labeled 50-mL tubes in 40-mL aliquots and store at $-20°C$.

Procedure

1. Follow preparations according to Chapter 3, Aseptic Technique on page 19.

2. Prepare inactivated MEF culture medium:

   2.1. In the biosafety cabinet, open a 500-mL filter system. Label the bottle with the following information:
   - Inactivated MEF medium with the date of preparation
   - Expiration date (14 days after medium preparation)
   - Initials

TABLE 5.1 PREPARATION OF MEF CULTURE MEDIUM

Ingredient	Volume		
DMEM	90 mL	222.5 mL	445 mL
HI-FBS	9 mL	25 mL	50 mL
NEAAs	1 mL	2.5 mL	5 mL
Final volume	**100 mL**	**250 mL**	**500 mL**

2.2. Determine the amount of medium needed and add to the filter system the appropriate amount of the following ingredients, in the order listed in Table 5.1.

> **Note:** *The DMEM may be measured by pouring directly into the graduated filter system. The other solutions should be added using sterile pipettes.*
> *Scale up or down proportionately if another quantity of medium is needed.*

2.3. Attach filter system to a vacuum and filter the solution.

2.4. Discard the filter system. Tightly cap the bottle containing inactivated MEF medium.

2.5. Store the medium bottle at 4°C and use the medium within 14 days.

V. PREPARATION OF HUMAN EMBRYONIC STEM CELL CULTURE MEDIUM

NOTES

_____

_____

_____

_____

_____

_____

_____

_____

_____

_____

_____

_____

_____

_____

_____

_____

_____

_____

_____

_____

_____

_____

_____

_____

_____

_____

Human embryonic stem cell (hESC) medium promotes the robust growth of stem cells and maintains the undifferentiated state of hESCs during long-term culture. In contrast to other culture media, hESC medium may not contain antibiotics as they can potentially mask contamination and be problematic for downstream clinical applications. The medium has been optimized for both of these objectives.

Human embryonic stem cell medium is comprised of several key ingredients:

- *Dulbecco's modified Eagle's medium (DMEM/F12)* is a serum-free base medium.
- *KnockOut™ Serum Replacement (KOSR)* is an Invitrogen product that contains BSA, transferrin, insulin, and other protein and nonprotein components. While its composition has been published, the exact formulation is proprietary. KOSR supports prolonged hESC growth in an undifferentiated state. Although KOSR is the current standard, there are an increasing number of alternatives available.
- *Nonessential amino acids (NEAAs)* stimulate cell growth and prolong viability of the culture.
- L-*Glutamine* is an unstable essential amino acid that is most effective when supplemented fresh.
- *β-Mercaptoethanol* is an antioxidant that prevents formation of free radicals. It is also thought to assist in cysteine uptake and promote cell growth.
- *Basic fibroblast growth factor (bFGF)* supplements KOSR to support prolonged hESC growth in an undifferentiated state

Note: *DMEM/F12 contains a pH dye, phenol red, which indicates pH while in storage or in culture. When DMEM/F12 is combined with all of the other ingredients in hESC complete medium, the pH is altered and the medium becomes orange. In culture, hESC medium will change from orange to yellow-orange indicating an increase in acidity due to exposure to 5% CO_2, nutrient depletion, and/or cell death.*

In general, most media commonly being used for culturing iPS cell lines are very similar, if not identical, to the hESC medium described in this section. However, some iPS cell lines are cultured with antibiotics and/or varying concentrations of bFGF. It is suggested to follow the media recommendations from the cell line provider or supporting literature when beginning culture of a new line.

Overview

Thaw KOSR and bFGF aliquots.

Combine hESC culture medium ingredients.

Sterile filter.

Store at 4°C.

Equipment

- Micropipettors P200 (20 to 200 µL), P1000 (200 to 1000 µL)
- Pipette aid
- Sterile biosafety cabinet

Materials

- 5-mL, 10-mL, and 25-mL Sterile glass or plastic pipettes
- 20-µL and 1000-µL Sterile pipette tips
- 500-mL filter system (Fig. 5.1)

- 70% Ethanol spray
- Absorbent paper towels (or Kimwipes)
- Alcohol-proof marker
- Cryovial holder
- Disposable gloves

Reagents

- Basic fibroblast growth factor (bFGF)
- ß-Mercaptoethanol
- DMEM/F12 medium
- Knockout Serum Replacer (KOSR)
- L-Glutamine, nonanimal, cell culture tested
- MEM nonessential amino acids solution

Preparation

1. Place the following items near the biosafety cabinet:
   - 5-mL, 10-mL, and 25-mL Sterile pipettes
   - 70% Ethanol spray
   - Absorbent paper towels (or Kimwipes)
   - Disposable gloves
2. Ensure that the following items are in the biosafety cabinet:
   - 5-mL, 10-mL, and 25-mL Sterile pipettes
   - 50-mL Sterile conical tubes
   - 500-mL Filter system
   - Alcohol-proof marker
   - Sterile 20-µL and 1000-µL pipette tips

Procedure

1. Follow preparations according to Chapter 3, Aseptic Technique on page 19.
2. Thaw KOSR and bFGF reagents (*if needed*):
   2.1. It is recommended to prepare aliquots of KOSR and bFGF. If aliquots have already been prepared, place the frozen aliquots in a 37°C water bath to thaw and proceed to step 3 of this protocol.
   2.2. If there are no aliquots available, thaw a stock bottle of KOSR (stored at −20°C) using one of the two options below.
      2.2.1. *Option one:* Thaw KOSR overnight at 4°C. If KOSR is not completely thawed the next day, finish thawing in a 37°C water bath.
      2.2.2. *Option two:* Thaw KOSR directly in a 37°C water bath. It will take 2–3 hours to thaw a 500-mL bottle. Swirl the bottle gently at frequent intervals to accelerate thawing.
   2.3. If there are no frozen bFGF aliquots available, prepare bFGF aliquots according to Section III, Preparation of Basic Fibroblast Growth Factor Stock Solution.
3. Aliquot KOSR (*if a new bottle was thawed*):
   3.1. Label sterile 50-mL tubes in the biosafety cabinet as:
      - KOSR
      - Vendor's catalog and lot number
      - Expiration date
      - Initials

TABLE 5.2 PREPARATION OF HESC CULTURE MEDIUM

Ingredient	Stock Concentration	Final Concentration	Volume		
DMEM/F12	100%	N/A	386 mL	193 mL	77.2 mL
KOSR	100%	20%	100 mL	50 mL	20 mL
100 mM L-glutamine	200 mM	2 mM	5 mL	2.5 mL	1 mL
NEAAs	10 mM	0.1 mM	5 mL	2.5 mL	1 mL
ß-Mercaptoethanol	14.3 M	0.1 mM	3.5 µL	1.75 µL	0.7 µL
ß-FGF	4 µg/mL	4 ng/mL	0.5 mL	0.25 mL	0.1 mL
Final volume			**500 mL**	**250 mL**	**100 mL**

NOTES

3.2. Transfer the thawed KOSR bottle to the biosafety cabinet after thoroughly cleaning the outside of the KOSR bottle with 70% ethanol.

3.3. Leave enough KOSR in the bottle to be used at a later time. (Leave the bottle in the biosafety cabinet if it is to be used shortly, or store the bottle at 4°C.)

3.4. Pour the remainder into the labeled 50-mL tubes in 25-mL aliquots.

3.5. Store the aliquots at −20°C.

4. Prepare hESC culture medium:

Note: If the final volume of medium is different from the one listed in Table 5.2, adjust the volume of ingredients proportionately.

4.1. In the biosafety cabinet, open a 500-mL filter system. Label the bottle as:
   - hESC culture medium with the date of preparation
   - Expiration date (14 days after media preparation)
   - Initials

4.2. Determine the total amount of medium needed and add to the filter system the appropriate amount of the following ingredients, in the order listed in Table 5.2.

Note: The DMEM/F12 may be measured by pouring directly into the graduated filter cup. Other reagents should be added with sterile pipettes.

4.3. Attach filter system to a vacuum, and filter the solution.

4.4. Discard the filter unit. Tightly cap the bottle containing hESC culture medium and store the medium bottle at 4°C, and use the medium within 14 days.

VI. PREPARATION OF DIFFERENTIATION/EMBRYOID BODY MEDIUM

One of the methods used to assess pluripotency of hESCs is the production of embryoid bodies (EBs) in culture. Unlike hESC media, the medium used for production of EBs has ingredients required for the differentiation of hESCs. The EB medium contains fetal bovine serum that promotes spontaneous differentiation of stem cells. EB medium is comprised of several key ingredients:

- *Dulbecco's modified Eagle's medium (DMEM/F12)* is a serum-free base medium.
- *Nonessential amino acids (NEAAs)* stimulate cell growth and prolong the viability of culture.
- L-*Glutamine* is an unstable essential amino acid that should be supplemented fresh.
- *β-Mercaptoethanol* is an antioxidant that prevents formation of free radicals.
- *Heat-inactivated fetal bovine serum (HI-FBS)* provides metabolic requirements in the form of minerals, hormones, and lipids that enhance spontaneous differentiation of hESCs. HI-FBS also promotes cell attachment through adhesion factors and antitrypsin activity.

Overview

Thaw HI-FBS aliquots.

Combine differentiation/EB culture medium ingredients.

Sterile filter.

Store at 4°C.

Equipment

- 37°C Water bath
- Pipette aid
- Micropipettors P200 (20 to 200 μL), P1000 (200 to 1000 μL)
- Sterile biosafety cabinet

Materials

- 5-mL, 10-mL, and 25-mL Sterile glass or plastic pipettes
- 20-μL and 1000-μL sterile pipette tips
- 50-mL Sterile conical tubes
- 500-mL Filter system with a 0.22-μm filter (Fig. 5.1)
- 70% Ethanol
- Alcohol-proof marker
- Disposable gloves

Reagents

- ß-Mercaptoethanol
- DMEM/F12
- Heat-inactivated fetal bovine serum (HI-FBS)
- L-Glutamine, nonanimal, cell culture tested
- MEM nonessential amino acids solution

Preparation

1. Place the following items near the biosafety cabinet:
   - 5-mL, 10-mL, and 25-mL Sterile pipettes

TABLE 5.3 PREPARATION OF EMBRYOID BODY CULTURE MEDIUM

Ingredient	Stock Concentration	Final Concentration	Volume		
DMEM/F12	100%	N/A	390 mL	195 mL	78 mL
HI-FBS	100%	20%	100 mL	50 mL	20 mL
100 mM L-Glutamine	200 mM	2 mM	5 mL	2.5 mL	1 mL
NEAAs	10 mM	0.1 mM	5 mL	2.5 mL	1 mL
ß-Mercaptoethanol	14.3 M	0.1 mM	3.5 µL	1.75 µL	0.7 µL
Final volume			**500 mL**	**250 mL**	**100 mL**

NOTES

- 70% Ethanol spray
- Absorbent paper towels (or Kimwipes)
- Disposable gloves
2. Ensure that the following items are in the biosafety cabinet:
   - 5-mL, 10-mL, and 25-mL Sterile pipettes
   - 50-mL Sterile conical tubes
   - 500-mL Filter system
   - Alcohol-proof marker
3. Thaw either a new bottle or an appropriate aliquot of HI-FBS.

Procedure

Note: *If the final volume of medium is different from those listed in Table 5.3, adjust the volume of ingredients proportionately.*

1. Follow preparations according to Chapter 3, Aseptic Technique on page 19.
2. Open a 500-mL filter system in the biosafety cabinet.
3. Determine the amount of medium needed and add to the filter system the appropriate amount of the ingredients, in the order listed in Table 5.3.

   Note: *The DMEM/F12 may be measured by pouring directly into the graduated filter cup. Other reagents should be added using sterile pipettes.*

4. Attach filter system to a vacuum and filter the solution.
5. Discard the filter. Tightly cap the bottle containing EB culture medium.
6. Store the medium bottle at 4°C and use the medium within 14 days.

SUGGESTED READING

Freshney RI. *Culture of Animal Cells: A Manual of Basic Technique.* 5th ed. Hoboken, NJ: John Wiley & Sons; 2005:73–85.

Massachusetts Human Stem Cell Bank. 2009. Standard Operating Procedures. Available at http://www.umassmed.edu/mhscb.

PREPARATION OF MOUSE EMBRYONIC FIBROBLASTS FOR CULTURE OF HUMAN EMBRYONIC STEM CELLS

6

Meng-Jiao Shi, Maria Borowski, and Kimberly Stencel

Mouse embryonic fibroblast (MEF) cells are commonly used as a feeder layer for human embryonic stem cell (hESC) culture. Feeder cells support growth by providing nutrients essential for proliferation as well as preventing differentiation of the hESC culture.

Two stocks of mice are commonly used for the derivation of MEF cells for hESC culture: CF-1 and CD-1. Both stocks are outbred, easy to breed, and have a fairly large litter size (11–12 pups). Outbred mice are used due to their genetic variability, which renders them less prone to genetic disease.

Pregnant female mice are euthanized and embryos are harvested at day 12.5–13.5 of gestation. MEFs are isolated by dissecting the uterus and embryonic sac to release the embryo, discarding the visceral tissue, and mincing and culturing the remaining tissue. The process for deriving MEFs includes many steps, all of which should be performed in a sterile environment. Testing for both mycoplasma and other microbial contaminants should be completed for all new MEF cultures (see Chapter 3, Section 3.2, Quality Control of Cell Cultures).

Before MEF cells are plated as a feeder layer for hESCs, they must be inactivated so they no longer divide and compete for space and nutrients with the hESCs. Inactivation can be achieved by treatment with mitomycin C or gamma irradiation, both of which are discussed in this protocol. Following treatment, MEF cells must be further tested by culturing to confirm that they will not replicate before use as feeder cells.

Mouse embryonic fibroblasts are also commercially available. The MEFs arrive already irradiated, ready to be thawed and used in culture. Many laboratories prefer to purchase inactivated MEFs due to the time and expense involved in mouse husbandry as well as MEF derivation and expansion. It is also important to keep in mind that use of live animals is federally and locally regulated. All animal procedures must be preapproved by an Institutional Animal Care and Use Committee (IACUC) before experiments begin. Those who derive their own MEF cells do so to ensure the reproducibility and quality of MEF cells between different lots.

Human Stem Cell Technology and Biology, edited by Stein, Borowski, Luong, Shi, Smith, and Vazquez
Copyright © 2011 Wiley-Blackwell.

General Notes

- This chapter consists of five sections as listed in the Overview. Each section can be performed independently.
- It is good laboratory practice to have important relevant information recorded; therefore sample log sheets are provided at the end of this chapter.
- Always prepare fresh freezing medium.
- It is important to transfer cells to -80°C as soon as possible once they are in freezing medium.
- The MEFs derived in this protocol can be stored at -80°C for several months, or in a -150°C cryofreezer for up to several years.

Abbreviations

CaCl$_2$: calcium chloride

DMEM: Dulbecco's modified Eagle's medium

DMSO: dimethyl sulfoxide

EDTA: ethylenediaminetetraacetic acid

hESCs: human embryonic stem cells

HI-FBS: heat-inactivated fetal bovine serum

LN$_2$: liquid nitrogen

MEF: mouse embryonic fibroblast

MgCl$_2$: magnesium chloride

μg: milligram

mL: milliliter

μm: micrometer

mm: millimeter

P: passage

PBS: phosphate buffered saline

I. PREPARATION OF MEF DERIVATION AND CULTURE MEDIUM

Equipment
- Mechanical pipetting device
- Sterile biosafety cabinet

Materials
- 5-mL and 10-mL Sterile glass or plastic pipettes
- 50-mL Sterile conical tubes
- 250-mL Sterile bottles
- 500-mL Bottle with a 0.22-μm filter system
- 70% Ethanol
- Absorbent paper towels (or Kimwipes®)
- Disposable gloves

Reagents
- DMEM (liquid)
- MEM nonessential amino acids solution
- HI-FBS
- Penicillin–streptomycin 100× solution
- DMSO

Preparation
1. Place the following items near the biosafety cabinet:
   - 5-mL and 10-mL Sterile pipettes
   - 50-mL Sterile conical tubes
   - 70% Ethanol
   - Absorbet paper towels (or Kimwipes®)
   - Disposable gloves
2. Ensure that the following items are in the biosafety cabinet:
   - 50-mL Sterile conical tubes
   - 250-mL Sterile bottles
   - 500-mL Bottle with a 0.22-μm filter system
3. Preparation of MEF derivation medium:

   Note: *If the total volume of medium is different from those listed in the tables, adjust the volumes of ingredients proportionately.*

   3.1. Open a 500-mL filter system in a biosafety cabinet and label the bottle as:
   - MEF.CM
   - Expiration date (14 days after medium preparation)
   3.2. According to the total volume, add appropriate amount of ingredients to the 500-mL filter system (Table 6.1).

   Note: *DMEM may be measured by pouring directly into the filter cup. Other solutions should be added with pipettes.*

NOTES

TABLE 6.1 MEF CULTURE MEDIUM

Ingredient	500 mL Total	250 mL Total
DMEM medium	440 mL	220 mL
HI-FBS	50 mL	25 mL
Nonessential amino acids solution	5 mL	2.5 mL
Penicillin–streptomycin 100X solution	5 mL	2.5 mL

3.3. Filter the medium through the 0.22-μm filter, then discard the filter cup. Tightly cap the bottle containing MEF medium.

3.4. Store the bottle at 2–8°C and use the medium within 2 weeks.

4. Prepare the MEF culture medium (see Chapter 5).

II. MEF.P1 DERIVATION AND CRYOPRESERVATION OF MOUSE EMBRYONIC FIBROBLASTS

Equipment

- −80°C Freezer
- 37°C Cell culture incubator
- Benchtop centrifuge with swing rotors
- CO_2 tank and chamber to anesthetize the mouse
- Hemacytometer
- Ice bucket
- Inverted light microscope
- Liquid nitrogen sample storage tank
- Mr. Frosty™ freezing container
- Pipette aid
- Sterile biosafety cabinet or chemical hood for mouse dissection
- Sterile biosafety cabinet for embryo dissection and MEF plating
- Vacuum system (recommended)

Materials

- Dissection tools:
  - Two pairs pointed forceps, wrapped and autoclaved
  - Two pairs dissection forceps, wrapped and autoclaved
  - One pair dissection scissors, wrapped and autoclaved
  - Three pairs iris (fine) scissors, wrapped and autoclaved
  - Disposable sterile scalpel
- 1-mL Syringe sterile (optional)
- 1.5–2.0-mL Cryovials
- 18 gauge × 1.5-inch Needle sterile (optional)
- Two autoclaved sterile 250-mL glass beakers
- 5-mL, 10-mL, and 25-mL Sterile glass or plastic pipettes
- Six thumbtacks
- 25-mL Sterile Pasteur pipettes
- 40-μm Cell strainer
- Fifty T75 tissue culture flasks
- 70% Ethanol
- Absorbent paper towels (or Kimwipes)
- Alcohol-proof marker
- 15-mL and 50-mL Sterile conical tubes
- Cryovial holder
- Mice, CF-1 strain at 12.5–13.5 days gestation

Note: Timed pregnant mice can be purchased from a vendor or bred in a laboratory animal facility. Natural matings of one male and one female are recommended. Mouse gestational day is determined by observation of a copulation plug in the vaginal opening of the female mouse on the morning following a successful mating. The plug observed date is designated day 0.5 of gestation. The mating pair should be

NOTES

NOTES

separated once the copulation plug is seen to avoid a secondary mating. It is recommended that the female be weighed periodically to confirm weight gain due to pregnancy although pregnancy can usually be confirmed by external observation by day 12.5.

- Several sterile, disposable 100-mm Petri dishes
- Styrofoam board (e.g., 50-mL centrifuge tube rack)

Reagents

- 0.05% Trypsin-EDTA
- DMEM
- DMSO
- HI-FBS
- Isopropanol
- PBS without $CaCl_2$ and $MgCl_2$

Preparation

1. Prepare MEF derivation medium according to Section I of this chapter. This can be prepared a few days in advance.

2. Place the following items near the biosafety cabinet:
   - 10-mL Sterile pipettes
   - 70% Ethanol spray
   - Absorbent paper towels (or Kimwipe)
   - Disposable gloves

3. Ensure that the following items are in the biosafety cabinet (or chemical hood) **in preparation for mouse dissection**:
   - Two autoclaved beakers. Mark one as PBS, the other as ethanol. Pour PBS and 70% ethanol into respective beakers.
   - Autoclaved and wrapped: two pairs watchmakers forceps, two pairs dissection forceps, one pair dissection scissors, and three pairs iris scissors, autoclaved, several disposable scalpels.
   - An ethanol saturated styrofoam board. Cover it with an ethanol saturated paper towel.
   - Six ethanol-sprayed thumbtacks.
   - Two or three sterile 100-mm Petri dishes.

4. Ensure that the following items are in the biosafety cabinet **in preparation for embryo dissection and MEF plating**:
   - 1-mL Sterile syringe (optional)
   - 15-mL Sterile conical tubes
   - 18 gauge × 1.5-inch Sterile needle (optional)
   - Alcohol-proof marker
   - Disposable scalpels
   - Iris scissors
   - Pointed forceps
   - Sterile Pasteur pipettes

Procedure

1. Follow preparations according to Chapter 3, Aseptic Technique on page 19.

2. Euthanize the pregnant female mouse according to preapproved Institutional Animal Care and Use Committee (IACUC) protocol.

3. Dissect the mouse and remove uterine horns in the biosafety cabinet:

    3.1. Saturate the euthanized mouse with 70% ethanol by immersing the mouse in an ethanol bath or spraying the mouse thoroughly with ethanol.

    3.2. Transfer the ethanol-soaked mouse to a biosafety tissue culture biosafety cabinet and place it belly up on the paper towel on the styrofoam board.

    3.3. Pin or tape each leg to the styrofoam board. Do not stretch the limbs too tight as this will make it difficult to open the abdominal cavity.

    3.4. Add 10-mL sterile PBS to a 100-mm sterile Petri dish.

    3.5. Open the abdominal cavity using a pair of dissecting forceps and a pair of dissecting scissors and expose the cavity contents.

        Note: Soak all used scissors and forceps in the beaker containing ethanol and rinse in the beaker with PBS before using these instruments on the next mouse.

    3.6. Using a new pair of dissecting forceps and a pair of iris scissors, take out the mouse uterine horns and transfer them into the 100-mm dish containing PBS.

    3.7. Wash the uterine horns with 10 mL sterile PBS three times. Tilt the dish to separate the tissue from the PBS, then carefully aspirate the PBS.

    3.8. Transfer the uterine horns to a new 100-mm dish using forceps.

    3.9. Discard any mouse tissue and any disposable containers touched by the mouse tissue into a biohazard bag.

        Note: Transport the uterine horns inside the covered 100-mm dish to a different biosafety cabinet and spray with ethanol.

4. Remove embryos from the uterine horns:

    Note: Embryos are fragile; therefore treat them with care to ensure that they remain intact prior to removal of visceral tissue.

    4.1. There are three layers of membrane that need to be removed to release each embryo: the uterine wall, yolk sac, and very thin amniotic sac. Using two fresh pairs of pointed forceps and/or a pair of iris scissors, gently cut one end of the uterine horn, taking care to only cut the uterine layer (Fig. 6.1a,b). The embryo in its yolk sac should emerge easily from the uterus. Sometimes the yolk sac will already be removed. Repeat for the next embryo in the uterus.

    4.2. Release the embryos from the yolk sac by gently cutting the sac with scissors. Use caution not to break the embryo (Fig. 6.1c,d). Repeat for all embryos.

    4.3. Transfer the embryos to a new sterile dish using forceps and rinse them with 10 mL of PBS three times. Tilt the dish to separate the tissue from the PBS, then carefully aspirate the PBS.

    4.4. Discard the dish containing the uterine and yolk sac tissues as biohazard waste.

    4.5. Count the number of embryos and record the information on the log sheet.

    4.6. Using forceps and a sterile scalpel to cut out the dark red visceral tissue (Fig. 6.1e) from the embryos.

    4.7. When all embryos have been eviscerated (Fig. 6.1f), transfer them to a new sterile dish.

NOTES

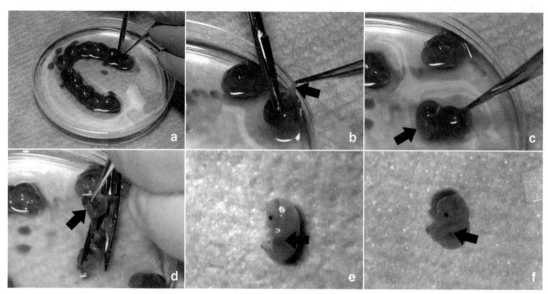

FIGURE 6.1. *Embryos are attached to the uterine wall and there are three layers of membrane that need to be removed to release the embryo: the uterine wall, yolk sac, and amniotic sac. (a) An embryo is released from the uterine horn by careful cutting of the uterine wall. (b) Removal of the uterine wall (arrow) using forceps and scissors, (c,d) Release of the embryo from the yolk sac (arrow), (e) Red viscera tissue (arrow) can be removed using scalpels or forceps and scissors, (f) Eviscerated embryo is ready for mincing and plating.*

NOTES

4.8. Discard the Petri dish containing embryo visceral tissue as biohazard waste.

5. Mince the embryos:

    5.1. Wash the embryos with 10 mL PBS until the embryonic tissue is free of blood and remnant visceral tissue is no longer present (Fig. 6.1f). This will assure a relatively pure fibroblast culture. Tilt the dish to separate the tissue from the PBS, then carefully aspirate the PBS.

    5.2. Using fresh iris scissors, mince the embryonic tissue into tiny pieces. It may take about 5 minutes. Add 5 mL trypsin-EDTA to the dish and thoroughly mix the minced tissue with the solution.

    5.3. Label the dish with initials and the number of embryos. Place the dish in a 37°C incubator and incubate for 10 minutes.

    5.4. Return the dish to the biosafety cabinet after the incubation is complete.

    5.5. Add another 5 mL trypsin-EDTA to the dish.

    5.6. Pipette up and down several times to further break up any remaining tissue.

    5.7. Return the dish to the 37°C incubator for another 10–20 minutes.

    Note: An alternative to steps 5.2 through 5.7 is as follows:

    5.8. *Cut one embryo into several pieces and carefully transfer to a sterile 1-cc syringe with plunger removed (Fig. 6.2a). Place syringe in a 15-mL tube.*

    5.9. *Add 1 mL trypsin to the syringe and use plunger to push embryos and trypsin through 18 gauge × 1.5-inch needle and into a 15-mL conical tube (Fig. 6.2b).*

    5.10. *Shear the tissue by pulling back on plunger (thus suctioning tissue mixture back into syringe) and pushing through again.*

    5.11. *Repeat for a total of three times.*

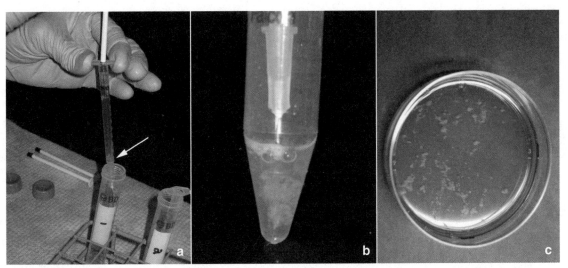

FIGURE 6.2. *Embryos can be minced one at a time using a syringe and needle. (a) An embryo was cut into several pieces (arrow) and transferred to a 1-cc syringe with needle attached, (b) Minced tissue from one embryo in a 15-mL tube, (c) Minced tissue from one embryo in a 100-mm plate.*

5.12. *Transfer minced embryo to a 100-mm dish (Fig. 6.2c).*

5.13. *Repeat steps 5.8 to 5.10 for remaining embryos, using the same syringe and conical tube, and transferring the minced embryos into the same dish.*

5.14. *Place the dish in a 37°C incubator and incubate for 10 minutes.*

6. During the incubation period, determine the number of T75 flasks needed by dividing the embryo number by 3. Round-up fractions. For example, use five T75 flasks for cells derived from 14 embryos.

6.1. Record the number of T75 flasks needed for MEF plating.

6.2. Place the necessary number of T75 flasks in the biosafety cabinet. Label each as:
   • P1.MEF
   • Date

6.3. Add 10 mL MEF derivation medium to each T75 flask.

6.4. When the incubation time is over, return the dish containing the minced embryo tissue to the biosafety cabinet.

6.5. Pipette vigorously up and down using a sterile disposable 10-mL pipette until the mixture is uniformly in a fine suspension (a few large clusters may still remain). This process may take up to 5 minutes.

6.6. Add 20 mL MEF derivation medium to the dish, then mix and transfer to a 50-mL tube.

6.7. Rinse the dish with 5 mL MEF derivation medium to collect any remaining cells. Transfer the rinse to the 50-mL tube containing the cell suspension.

6.8. Mix the cell suspension well and transfer it in equal aliquots to the T75 flasks containing 10 mL MEF derivation medium. The final volume should be 20–25 mL per T75 flask.

6.9. Swirl the plate to mix the cells and derivation medium in the flask and place the flasks in the 37°C incubator.

6.10. In the incubator, move the flasks three times side-to-side, back and forth (in a cross motion) to disperse the cells evenly across the flask surface.

6.11. Incubate the cells overnight.

NOTES

TABLE 6.2 MEF FREEZING MEDIUM

Ingredient	50 mL Total	100 mL Total
DMEM derivation medium	30 mL	60 mL
DMSO	10 mL	20 mL
HI-FBS	10 mL	20 mL

NOTES

_____

_____

_____

_____

_____

_____

_____

_____

_____

_____

_____

_____

_____

_____

_____

_____

_____

_____

_____

_____

_____

_____

7. **Next day:** Observe fibroblast cultures and refresh medium if needed:

   7.1. Remove one flask from the incubator and view the entire culture surface under a microscope. Repeat this for all flasks.

   7.2. Discard any flasks with microbial contamination (e.g., bacterial, yeast, or fungal, mouse bone fragments, or beating cells).

   7.3. If the flask surface is more than 90% confluent, harvest cells and proceed to step 8 for harvest and cryopreservation.

   7.4. If the flask surface is less than 90% confluent, aspirate the medium and replace it with fresh MEF derivation medium.

   7.5. Return the flasks to the incubator for an additional day of culture.

   7.6. Monitor the culture daily under the microscope. Harvest and freeze cells according to step 8 when the flask surface is more than 90% confluent.

8. Harvest and freeze MEF.P1:

   Note: Carefully review the entire section before beginning procedure.

   8.1. Prepare MEF cryopreservation medium:

   Note: Cryopreservation medium must be prepared fresh and used within one day. Therefore prepare only enough for use on that day.

   8.1.1. In a biosafety cabinet, place a sterile 50-mL tube or 250-mL bottle. Label it as: "2× MEF freezing medium (MEF.FM)."

   8.1.2. According to the final total volume (Table 6.2), add appropriate amount of reagents into the tube or bottle using sterile pipettes.

   8.1.3. Filter the medium through a 0.22-μm filter system; discard the filter cup. Tightly cap the bottle containing the medium.

   8.1.4. Label the tube or bottle as 2× MEF.FM.

   8.1.5. Store the medium at 2–8°C and use it within one day.

   8.2. Label sterile cryovials:

   8.2.1. Place six cryovials per T75 culture flask in a cryovial holder in the biosafety cabinet.

   8.2.2. Use a permanent alcohol proof marker to label the cryovials as:
   • MEF.P1
   • Date (e.g., MM/DD/YY)
   • Preparer's initials

   8.3. Harvest MEF.P1 cells:

   8.3.1. Remove all flasks from the incubator and place them in the biosafety cabinet.

   8.3.2. Aspirate the spent medium from each flask.

   8.3.3. Wash the cells in each flask with 10–12-mL of PBS.

   8.3.4. Add 3 mL trypsin-EDTA to each flask. Swirl the flask to make sure the trypsin-EDTA covers the whole culture surface.

   8.3.5. Incubate for 3–5 minutes in a 37°C incubator.

8.3.6. Cap the flasks tightly and dislodge the cells from the flask surface by gently tapping the side of the flask 3–5 times.

8.3.7. Check under microscope to determine whether all cells have detached from surface. If not, incubate in the incubator for another 1–2 minutes, followed by tapping at the side of the flask.

8.3.8. Add 6 mL MEF derivation medium to each T75 flask and mix to neutralize the trypsin.

8.3.9. Pipette up and down with a 10-mL sterile, disposable pipette to break up the cell clusters. Try to minimize foaming during the pipetting.

8.3.10. Pool the cell suspension from all flasks into one flask.

8.3.11. Place a 40-μm cell strainer on top of a sterile 50-mL tube.

8.3.12. Add MEF cell suspension from the flask to the cell strainer.

8.3.13. Allow cells to pass thorough the cell strainer by gravity flow.

8.3.14. Collect cells in more than one 50-mL tube if there is more than 40-mL of cell suspension.

8.3.15. Rinse the cell strainer with another 10 mL MEF derivation medium and discard the cell strainer.

8.3.16. Centrifuge the cell suspension in the 50-mL tubes at 200*g* (∼1000 rpm) for 5 minutes.

8.3.17. Aspirate the medium from each 50-mL tube and resuspend each pellet in 5 mL of MEF derivation medium.

8.3.18. Transfer all resuspended cells into one 50-mL tube.

8.4. Count cells:

8.4.1. Label a sterile 1.5–2-mL vial as "cell" to be used for cell counting.

8.4.2. Mix the MEF cells thoroughly and transfer 0.4 mL of cells to the "cell" vial.

8.4.3. Dilute the cells in the "cell" vial by adding 0.88 mL PBS (20× dilution).

8.4.4. Count diluted cells using a hemacytometer (see Chapter 3, Section 3.3, Cell Counting Using a Hemacytometer).

Note: If the individual counts differ by more than 5%, it is best to repeat the count.

8.4.5. Average two cell counts that are within 5% of each other.

8.4.6. Calculate the total number of cells by multiplying cell concentration (cells/mL) by the total volume.

8.5. Record cell number in the log sheet and calculate the total MEF derivation medium required for resuspending cells at 20×10^6 cells/ mL.

Note: The cells will be frozen at 10×10^6 cells/mL. In order to be frozen in 2× freezing medium, the cells must be at 20×10^6 cells/mL of MEF medium.

8.6. If the current cell suspension is less than 20×10^6 cells/mL, centrifuge again and resuspend the cells with the appropriate MEF medium volume.

8.7. If the cells are too concentrated, add an appropriate amount of MEF derivation medium to the cell suspension.

8.8. Freeze MEF.P1 cells:

NOTES

_____

_____

_____

_____

_____

_____

_____

_____

_____

_____

_____

_____

Note: *Cells should be frozen as soon as possible once the freezing medium is added to the cells. Several lab members can coordinate to ensure it is finished as quickly as possible. To minimize variation, distribution of the cells to cryovials must be performed by one person only.*

8.8.1. Place sets of 18 labeled cryovials in the biosafety cabinet. Loosen all caps.

8.8.2. Slowly and drop-wise, add an equal volume of $2\times$ cryopreservation medium (freezing medium) into the tube or bottle containing cells at 20×10^6 cells/mL. Mix by gently swirling the bottle/cell suspension, while adding the appropriate amount of freezing medium.

8.8.3. After thoroughly mixing the cells, aliquot 1 mL into each of the cryovials using a 10-mL pipette.

8.8.4. Once cells are transferred to 18 vials, tighten the caps and place them in an isopropanol freezing container (Mr. Frosty™ and immediately place the container into a -70°C to -80°C freezer. If help is available, this step can be done by another person to speed the freezing process.

8.8.5. Repeat steps 8.8.1–8.8.4 until all MEF cells are placed into cryovials and stored in the Mr. Frosty freezing container.

8.8.6. Transfer all the vials into a liquid nitrogen freezer (-150°C) the next day.

III. THAWING AND EXPANSION OF MEF.P1 TO MEF.P4

Equipment
- 37°C Cell culture incubator
- 37°C Water bath
- Benchtop centrifuge with swing rotors
- Hemacytometer
- Inverted microscope
- Pipette aid
- Sterile biosafety cabinet
- Vacuum system (recommended)

Materials
- 5-mL, 10-mL, and 25-mL Sterile glass or plastic pipettes
- 15-mL and 50-mL Sterile conical tubes
- Fifty T75 tissue culture flasks
- 70% Ethanol
- Absorbent paper towels (or Kimwipes)
- Alcohol-proof marker
- Cryovial holder
- Sterile disposable Petri dishes

Reagents
- DMEM
- HI-FBS
- Isopropanol
- PBS without $CaCl_2$ and $MgCl_2$

Preparation
1. Prepare reagent:
    1.1. If there is no MEF culture medium available, prepare the medium according to Section I of this chapter.
    1.2. If there is MEF culture medium available, place the medium in a 37°C water bath for about 30 minutes.
2. Place the following items near the biosafety cabinet:
    - 5-mL, 10-mL, and 25-mL Sterile glass or plastic pipettes
    - Fifty T75 tissue culture flasks
    - 70% Ethanol
    - Absorbent paper towels (or Kimwipes)
    - Cryovial holder
    - Sterile disposable Petri dishes
    - Sterile Pasteur pipettes
3. Ensure that the following items are in the biosafety cabinet:
    - 15-mL and 50-mL Sterile conical tubes
    - Alcohol-proof marker

Procedure

1. Follow preparations according to Chapter 3, Aseptic Technique on page 19.
2. Thaw P1.MEF Cells:

   Note: *Thaw no more than two vials at a time in order to minimize the length of time that cells are exposed to cryopreservation medium after thawing.*

   2.1. Place the bottle of warmed MEF culture medium in the biosafety cabinet.
   2.2. Transfer 10 mL medium per vial of MEF to be thawed into a 50-mL tube. Label it as "cell."

   Note: *Wear lab coat, gloves, and eye protection when removing vials from liquid nitrogen storage, as vials may accidentally explode upon warming.*

   2.3. Remove one or two vials of MEFs from the liquid nitrogen tank.
   2.4. Immerse the vial(s) in a 37°C water bath without submerging the cap. Swirl the vial(s) gently.
   2.5. When only small pieces of ice crystals remain, remove the vial(s) from the water bath.
   2.6. Spray the vial(s) completely wet with 70–75% ethanol to sterilize the outside of the tube. Place the vial(s) in the cryovial holder in the biosafety cabinet.

3. Remove cryoprotectant and resuspend MEFs in MEF culture medium:

   Note: *The cells in the culture are at Passage 2 and therefore called P2.MEF cells. Each vial of frozen MEFs should be distributed into two T75 flasks.*

   3.1. Using a 5-mL pipette, remove cells from the cryovial and add them dropwise to the 10 mL MEF medium in the 50-mL tube while gently moving the tube back and forth to mix the MEF cells. This reduces osmotic shock to the MEF cells.
   3.2. Rinse the cryovial(s) with MEF medium from the 50-mL tube and transfer the medium back to the tube.
   3.3. If thawing more than one vial, repeat steps 3.1 and 3.2.
   3.4. Centrifuge cells at 200*g* (about 1000 rpm in the benchtop centrifuge) for 5 minutes.
   3.5. Label two T75 flasks for each vial of cryopreserved MEF cells as:
       • P2.MEF
       • Date
       • Initials
   3.6. To each T75 flask, add 10 mL MEF culture medium.
   3.7. When the centrifugation is complete, remove the supernatant from the tube.
   3.8. Resuspend the cell pellet in 5 mL/T75 MEF culture medium. For example, resuspend the cells in 10 mL if two T75 flasks are to be used.
   3.9. Transfer equal portions of cells to each of the flasks. Rinse the tube with fresh MEF medium and evenly distribute it into the flasks.
   3.10. Place the flasks into the 37°C incubator and move the flasks three or four times in a cross motion to disperse the cells evenly across the flask surface.
   3.11. Incubate the cells overnight in a 37°C incubator.

4. **Next day:** Observe MEF cell culture and add fresh medium:
   4.1. View the culture under a microscope. The cells should grow in a monolayer and will remain attached to the surface.

4.2. Discard any flasks with contamination (e.g., bacterial, yeast, or fungus, mouse bones, beating cells).

4.3. Change MEF medium for remaining flasks (15 mL/T75).

> **Note:** *Observe MEF cells under the microscope daily for the next 3–4 days without changing medium. Generally, cells are ready to be harvested 3–4 days post-thawing, although this can vary by lot.*

5. Harvest P2.MEFs and passage them to P3.MEFs at a ratio of a 1:3 (or 1:4).

   Note: *Cells can be split at a ratio of 1:3 or 1:4 depending on when the cells are harvested. If possible, avoid harvesting cells on the weekend.*
   Cells may be cultured in either T75 or T175 flasks. The surface area of one T175 flask is equivalent to approximately 2.5 T75 flasks.

6. Observe MEF cells in T75 flasks under a microscope. Split cells when the culture surface is approximately 90% confluent by following the steps:

   6.1. Place all flasks in the biosafety cabinet.

   6.2. Aspirate MEF medium completely from the flasks. Add 10 mL of room temperature PBS to each flask. Swirl the PBS over the MEF cells and aspirate completely.

   6.3. Add 3 mL of room temperature trypsin-EDTA to each T75 flask.

   6.4. Place in the 37°C incubator for 3–5 minutes.

   6.5. While MEF cells are incubated with trypsin, label appropriate number of new T75 flasks as:
   - P3.MEF
   - Date
   - Initials

   6.6. Add 20 mL of MEF culture medium into each of the labeled new T175 flasks.

   6.7. When the trypsin incubation is complete, remove the T75 flasks from the incubator and tap the side of each flask three to five times to dislodge the MEF cells from the flask.

   6.8. Check under the microscope to determine whether all cells have detached from the surface. If not, incubate cells in the incubator for another 1–2 minutes, followed by tapping at the side of the flask.

   6.9. Add 6 mL MEF culture medium to each T75 flask and mix to neutralize the trypsin enzyme.

   6.10. Pipette up and down along the cell culture surface using a 10-mL sterile pipette to further dislodge any attached MEF cells and to break up any cell clusters. Try to minimize foaming during the pipetting.

   6.11. Pool the cell suspension from all flasks into one T75 flask.

   6.12. Rinse all flasks with 10-mL of fresh MEF culture medium. Add the medium to the flask containing the MEF cells.

   6.13. Discard the used empty flasks.

   6.14. Mix MEF cells thoroughly in the T75 flask by gently swirling.

   6.15. Transfer cells evenly to the new labeled flasks containing culture medium. Rinse the old T75 flask with fresh MEF medium and evenly distribute it into the new flasks.

   6.16. Place the new flasks in the 37°C incubator and move the flasks three to four times in a cross motion to disperse the cells evenly across the flask surface.

   6.17. Incubate the cells overnight in a 37°C incubator.

7. Harvest P3.MEFs and passage them to P4.MEFCs at a 1:3 (or 1:4) ratio by repeating step 5.

8. Observe cells daily under a microscope and record relevant information.

Note: Passage 4 MEFs can be inactivated either by irradiation (see Section IV) or with Mitomycin C treatment (see Section V). MEFs are cryopreserved in a LN₂ freezer following their inactivation. Irradiation is the recommended inactivation method, but mitomycin C treatment may be used if a gamma irradiator is not available.

IV. IRRADIATION OF MOUSE EMBRYONIC FIBROBLAST CELLS

Equipment

- -80°C Freezer
- 37°C Cell culture incubator
- 37°C Water bath
- Benchtop centrifuge with swinging-bucket rotor
- Inverted light microscope
- Irradiator
- LN_2 freezer (-130°C to -150°C)
- Mr. Frosty freezing containers
- Pipette aid
- Sterile biosafety cabinet
- Timer
- Weighing balance

Materials

- 5-mL and 10-mL Sterile glass or plastic pipettes
- Six-well culture vessels
- 15-mL and 50-mL Sterile conical tubes
- 70% Ethanol
- Alcohol-proof marker
- Cryovials
- Cryovial holder
- Disposable gloves
- Sterile Pasteur pipettes

Reagents

- 0.05% Trypsin-EDTA
- DMEM/F12
- DMSO
- HI-FBS
- Isopropanol
- MEF culture medium (see Section I of this chapter)
- PBS without $CaCl_2$ and $MgCl_2$

Preparation

1. Place the following items near the biosafety cabinet:
   - 5- mL and 10-mL Sterile glass or plastic pipettes
   - Six-well culture vessels
   - 15-mL and 50-mL Sterile conical tubes
   - 70% Ethanol
   - Cryovials
   - Cryovial holder
   - Disposable gloves
   - Sterile Pasteur pipettes

NOTES

TABLE 6.3 MEF FREEZING MEDIUM

Ingredient	50 mL Total	100 mL Total
DMEM medium	30 mL	60 mL
DMSO	10 mL	20 mL
HI-FBS	10 mL	20 mL

NOTES

_____

_____

_____

_____

_____

_____

_____

_____

_____

_____

_____

_____

_____

_____

_____

_____

_____

_____

_____

_____

_____

_____

_____

2. Ensure that the following items are in the biosafety cabinet:
   - Alcohol-proof marker
   - Sterile Pasteur pipettes
3. Preparation of MEF cryopreservation medium:
   3.1. In a biosafety cabinet place a sterile 50-mL tube or 250-mL bottle.
   3.2. Label medium name on bottle (e.g., 2× MEF.FM).
   3.3. According to the total final volume, add appropriate amount of reagents (Table 6.3) using sterile pipettes to a sterile tube or bottle. Scale up or down accordingly.
   3.4. Filter sterilize the cryopreservation medium using a 0.22-μm filter system and tightly cap the tube or bottle containing the medium.
   3.5. Label the tube or bottle, for example, 2× MEF.FM.
   3.6. Store at 2–8°C and use it within one day.

   Note: Cryopreservation medium must be prepared fresh and used within one day; therefore prepare only enough for use that day.

Procedure

1. Follow preparations according to Chapter 3, Aseptic Technique on page 19.
2. Harvest MEF.P4 for irradiation:
   2.1. Observe MEFs under a microscope. Harvest cells when the culture surface is approximately 90% confluent by following the steps.
   2.2. Place all flasks in the biosafety cabinet.
   2.3. Aspirate MEF medium completely from the flasks.
   2.4. Add 30 mL of room temperature PBS to each flask. Swirl the PBS over the MEF cells and aspirate completely.
   2.5. Add 10 mL of room temperature trypsin-EDTA to each flask.
   2.6. Place in the 37°C incubator for 3–5 minutes.
   2.7. When the trypsin incubation is complete, remove the flasks from the incubator and tap the side of each flask three to five times to dislodge the MEF cells from the flask.
   2.8. Check under microscope to determine whether all cells have detached from surface. If not, incubate for another 1–2 minutes, followed by tapping at the side of the flask.
   2.9. Add 20 mL MEF culture medium to each flask. Mix to neutralize the trypsin enzyme.
   2.10. Pipette up and down along the cell culture surface using a 10-mL sterile pipette to further dislodge any attached MEF cells and to break up any cell clusters. Try to minimize foaming during the pipetting.
   2.11. Pool the cell suspension from all flasks into one flask.
   2.12. Rinse all flasks with 20 mL of fresh MEF culture medium. Add the medium to the flask containing the MEF cells.

2.13. Discard the used empty flasks.

2.14. Mix MEF cells thoroughly by swirling the flask.

2.15. Remove 100 µL of the cell suspension from the center of the suspension; dilute in 400 µL MEF medium (1:5 dilution).

2.16. Count the sample using a hemacytometer.

2.17. Transfer cells from the flask to sterile 50-mL tubes.

2.18. Spin each 50-mL conical tube containing MEFs for 5 minutes at 200 g.

2.19. Aspirate the supernatant, then resuspend the cells at a concentration of 2.0×10^6 to 10.0×10^6 cells/mL in MEF cell culture medium.

2.20. Transfer MEFs into 15-mL conical tubes for irradiation.

3. Irradiation:

   3.1. Irradiate MEF cells at 8000 rads using a gamma irradiator. When placing the tubes into the irradiator, mix the cells slightly by tapping and rocking the tube gently.

   3.2. After irradiating, thoroughly resuspend the MEF cells with a pipette.

   3.3. Keep the irradiated MEFs in the biosafety cabinet at room temperature to decrease cell clumping.

   3.4. Remove two 100-µL samples and count each sample.

   Note: If the individual counts differ by more than 5%, it is best to repeat the count.

   3.5. Average two cell counts that are within 5% of each other.

   3.6. Calculate the total number of cells by multiplying cell concentration (cells/ml) by the total volume.

   Note: Cells can be frozen at (1–10) $\times$ 10^6 cells/mL. Approximately (1–1.5) $\times$ 10^6 cells are needed to seed one six-well plate. The steps below are calculated for freezing cells at 10×10^6 cells/mL.

   3.7. Resuspend cells at 20×10^6 cells/mL in MEF culture medium.

   Note: Cells need to be resuspended at twice the final concentration dilution with 2× freezing medium.

   3.8. If the current cell suspension is less than 20×10^6 cells/mL, centrifuge again and resuspend the cells with the appropriate MEF medium volume.

   3.9. If the cells are too concentrated, add the appropriate amount of MEF derivation medium to the cell suspension to achieve a concentration of 20×10^6 cells/mL.

4. Freeze irradiated MEFs:

   Note: Cells should be frozen as soon as possible once the freezing medium is added to the cells.

   4.1. Use a permanent alcohol proof marker to label sufficient number of cryovials as:
   - "Irradiated MEFs"
   - Total cell number in the vial (e.g., 10×10^6 cells)
   - Date (e.g., MM/DD/YY)

   4.2. Place sets of 18 labeled cryovials in the biosafety cabinet. Loosen all caps.

   4.3. Slowly and drop-wise, add equal volume of 2× cryopreservation medium (freezing medium) into the tube or bottle containing cells at

NOTES

20×10^6 cells/mL. Mix cells by gently tapping the tube or bottle while adding the freezing medium.

4.4. After thoroughly mixing the cells, aliquot 1 mL to each of the cryovials using a 10-mL pipette.

4.5. Once cells are transferred to 18 vials, transfer the vials to the next person to tighten the caps and place them in an isopropanol freezing container and immediately place the container in a -70°C to -80°C freezer.

4.6. Repeat the above until all MEF cells are placed into cryovials and stored in the -80°C freezer.

4.7. Transfer all the vials into liquid nitrogen racks the next day.

4.8. Record the number of vials frozen and the LN_2 tank storage location on the log sheet.

4.9. These frozen MEFs can now be used as a feeder layer for hESC culture. Thaw and seed MEFs for hESC culture as described in Chapter 7.

V. MITOMYCIN C INACTIVATION OF MOUSE EMBRYONIC FIBROBLAST CELLS

Equipment

- -80°C Freezer
- 37°C Cell culture incubator
- 37°C Water bath
- Benchtop centrifuge with swinging-bucket rotor
- Inverted light microscope
- Irradiator
- LN_2 freezer (-130°C to -150°C)
- Mr. Frosty freezing containers
- Pipette aid
- Sterile biosafety cabinet
- Timer
- Weighing balance

Materials

- 5-mL and 10-mL Sterile glass or plastic pipettes
- Six-well culture vessels
- 15-mL and 50-mL Sterile conical tubes
- 70% Ethanol
- Alcohol-proof marker
- Cryovials
- Cryovial holder
- Disposable gloves
- Sterile Pasteur pipettes

Reagents

- DMEM
- HI-FBS
- Isopropanol
- MEF culture medium (Section I of this chapter)
- PBS without $CaCl_2$ and $MgCl_2$

Preparation

1. Place the following items near the biosafety cabinet:
   - 5-mL and 10-mL Sterile glass or plastic pipettes
   - Six-well culture vessels
   - 70% Ethanol
   - Cyrovials
   - Cryovial holder
   - Disposable gloves
2. Ensure that the following items are in the biosafety cabinet:
   - 15-mL and 50-mL Sterile conical tubes
   - Alcohol-proof marker
   - Steril Pasteur pipettes

NOTES

TABLE 6.4 MEF FREEZING MEDIUM

Ingredient	50 mL Total	100 mL Total
DMEM medium	30 mL	60 mL
DMSO	10 mL	20 mL
HI-FBS	10 mL	20 mL

3. Preparation of MEF cryopreservation medium:

   3.1. In a biosafety cabinet place a sterile 50-mL tube or 250-mL bottle.

   3.2. Label medium name on bottle (e.g., 2× MEF.FM).

   3.3. According to the total final volume, add appropriate amount of ingredients (Table 6.4) using sterile serological pipettes to a sterile tube or bottle. Scale up or down accordingly.

   3.4. Filter sterilize the cryopreservation medium using a 0.22-μm filter system and tightly cap the tube or bottle containing the medium.

   3.5. Label the tube or bottle; for example, 2× MEF.FM.

   3.6. Store at 2–8°C and use within one day.

   Note: Cryopreservation medium must be prepared fresh and used within one day. Therefore prepare only enough for use that day.

Procedure

1. Follow preparations according to Chapter 3, Aseptic Technique on page 19.

2. Inactivation of MEF.P4 with mitomycin C:

   2.1. Observe MEFs under a microscope. Inactivate cells when the culture surface is approximately 90% confluent by following the steps.

   2.2. Place all flasks in the biosafety cabinet.

   2.3. Aspirate MEF medium completely from the flasks.

   2.4. Add 12 mL of MEF medium containing 10 μg/mL mitomycin C to each T75 flask. If using a different size culture vessel, scale amount of medium up or down accordingly.

   2.5. Place all culture vessels back in the 37°C incubator for 2–3 hours.

   2.6. After incubation, aspirate medium with mitomycin C, then wash with 10 mL of PBS three times.

   2.7. Add 10 mL of room temperature trypsin-EDTA to each flask.

   2.8. Place in the 37°C incubator for 3–5 minutes.

   2.9. When the trypsin incubation is complete, remove the flasks from the incubator and tap the side of each flask three to five times to dislodge the MEF cells from the flask.

   2.10. Check under microscope to determine whether all cells have detached from surface. If not, incubate for another 1–2 minutes, followed by tapping at the side of the flask.

   2.11. Add 20 mL MEF culture medium to each flask. Mix to neutralize the trypsin enzyme.

   2.12. Pipette up and down along the cell culture surface using a 10-mL sterile pipette to further dislodge any attached MEF cells and to break up any cell clusters. Try to minimize foaming during the pipetting.

   2.13. Pool the cell suspension from all flasks into one flask.

   2.14. Rinse all flasks with 20 mL of fresh MEF culture medium. Add the medium to the flask containing the MEF cells.

2.15. Discard the used empty flasks.

2.16. Mix MEF cells thoroughly by swirling the flask.

2.17. Remove 100 μL of the cell suspension from the center of the suspension; dilute in 400 μL MEF medium (1:5 dilution).

2.18. Count the sample using a hemacytometer.

   Note: If the individual counts differ by more than 5%, it is best to repeat the count.

2.19. Average two cell counts that are within 5% of each other.

2.20. Calculate the cell concentration (cells/mL).

   Note: The cells will be frozen at 10×10^6 cells/mL. However, they need to be first at 20×10^6 cells/mL in MEF medium before being diluted by 2× freezing medium.

2.21. If the current cell suspension is less than 20×10^6 cells/mL, centrifuge MEFs for 5 minutes at 200g and resuspend the cells with the appropriate MEF medium volume.

2.22. If the cells are too concentrated, add the appropriate amount of MEF derivation medium to the cell suspension to achieve the concentration of 20×10^6 cells/mL.

3. Freeze mytomycin C inactivated MEFs:

Note: Cells should be frozen as soon as possible once the freezing medium is added to the cells.

3.1. Use a permanent alcohol proof marker to label a sufficient number of cryovials as:
   • ''Mitomycin C treated MEFs''
   • Total cell number in the vial (e.g., 10×10^6 cells)
   • Date (e.g., MM/DD/YY)

3.2. Place sets of 18 labeled cryovials in the biosafety cabinet. Loosen all caps.

3.3. Slowly and drop-wise, add equal volume of 2× cryopreservation medium (freezing medium) into the tube or bottle containing cells at 20×10^6 cells/mL. Mix cells by gently tapping the tube or bottle while adding the freezing medium.

3.4. After thoroughly mixing the cells, aliquot 1 mL to each of the cryovials using a 10-mL pipette.

3.5. Once cells are transferred to 18 vials, transfer the vials to the next person to tighten the caps and place them in an isopropanol freezing container and immediately place the container in a $-70°$C to $-80°$C freezer.

3.6. Repeat the above until all MEF cells are placed into cryovials and stored in the $-80°$C freezer.

3.7. Transfer all the vials into liquid nitrogen racks the next day.

3.8. Record the number of vials frozen and the LN_2 tank storage location on the log sheet.

3.9. These frozen MEFs can now be used as a feeder layer for hESC culture.

SUGGESTED READING

American Veterinary Medical Association 2007. AVMA Guidelines on Euthanasia. Available at http://www.avma.org/issues/animal_welfare/euthanasia.pdf.

Conner DA. Mouse Embryo Fibroblast (MEF Feeder Cell Preparation). Current Protocols in Molecular Biology 2000; Unit 23.2.

Introduction to Human Embryonic Stem Cell Culture Methods Part I Derivation of Mouse Embryonic Fibroblasts (MEF) and Part II Mouse Embryonic Fibroblast (MEF) Culture and Propagation, WiCell Research Institute, Inc. Available at http://www.wicell.org/index.php?option=com_docman&task=doc_download&gid=126 and http://www.wicell.org/index.php?option=com_docman&task=doc_download&gid=127.

P1.MEF DERIVATION LOG SHEET

1. MEF Derivation from mouse
   - Date:_____ Recorded by:_____
   - Mouse gestation length:_____days
   - Number of embryos harvested:_____
   - Number of T75 flasks seeded for P1.MEF:_____

2. The next day observation:
   - Date: _____ Recorded by : _____
   - Number of T75 flasks contaminated by mouse bone:_____beating cells:_____or other (specify):_____
   - Culture confluence level: _____
   - Medium changed (circle one):   yes   no

3. P1.MEF harvest and cryopreservation:
   - Date: _____ Recorded by: _____
   - Total cell volume for counting: _____ mL
   - Cell counting: Chamber one cell number: _____ Chamber two cell number: _____
   - Cell concentration (average of the two chambers): _____ $\times 10^6$ cells/mL
   - Total cell number: _____ $\times 10^6$ cells
   - Total MEF medium volume at 20×10^6 cells/mL: _____ mL
   - Number of vials to be frozen at 10×10^6 cells/mL: _____
   - Location stored in LN_2 tank: _____

MEF EXPANSION RECORDING FORM

1. Thaw P1.MEF cells:

   - Start date: _____   Recorded by: _____
   - Number of cryovials thawed: _____
   - Number of T75 flasks seeded for P2.MEF: _____

2. The next day observation:

   - Date: _____   Recorded by: _____
   - Contaminated (indicate what type): _____
   - Culture confluence level: _____
   - Medium changed (circle one):   yes   no
   - Number of T75 flasks continued for culture: _____

3. Passage P2.cells to P3.MEF cells:

   - Date: _____   Recorded by: _____
   - Number of T75 flasks from which cells are harvested: _____
   - New flasks for seeding (circle one):   T75   T175
   - Number of new T75 or T175 flasks for seeding: _____

4. Passage cells to P4.MEF cells:

   - Date: _____   Recorded by: _____
   - Number of T75 or T175 flasks from which cells are harvested: _____
   - New flasks for seeding (circle one):   T75   T175
   - Number of new T75 or T175 flasks for seeding: _____
   - Date cells are harvested for irradiation: _____

RECORDING FORM FOR IRRADIATION AND CRYOPRESERVATION OF MEFS

Date: _____   Recorded by: _____

1. Total P4.MEFs before and after irradiation:
   - Before irradiation, total cell number : _____ $\times 10^6$ cells
   - After irradiation, total cell number : _____ $\times 10^6$ cells.

2. Irradiated MEFs cryopreserved:
   - Number of vials to be frozen at 10×10^6 cells/mL: _____
   - Location stored in LN_2 tank: _____

THAWING AND SEEDING OF FROZEN MOUSE EMBRYONIC FIBROBLASTS

7

Meng-Jiao Shi, Maria Borowski, and Kimberly Stencel

Human embryonic stem cells (hESCs) are cultured with a feeder layer of inactivated mouse embryonic fibroblasts (MEFs). The cells of this feeder layer provide an extracellular matrix upon which the hESCs can grow. In addition, feeder cells help the hESCs to maintain their pluripotency by secreting growth factors into the medium.

Mouse embryonic fibroblast cells are derived from 13–14-day mouse embryos (see Chapter 6). The MEFs used as a feeder layer for hESC culture are inactivated, inhibiting division of the cells so they are no longer capable of growth. MEFs can be inactivated by a chemical treatment of mitomycin C or exposure to radiation; alternatively, inactivated MEFs can be purchased commercially.

Although culturing hESC lines using inactivated MEF as feeder cells is common practice, other cell types have also been used, such as human fetal muscle and skin, adult fallopian tube epithelial cells, and human foreskin fibroblasts. For some research or applications, it is important to have a pure culture of hESCs without the presence of feeder cells. For these applications, substrates such as BD Matrigel™ or laminin-coated dishes along with commercially available defined media preparations like mTeSR®1 can substitute for feeder layers (see Chapter 11).

Overview
Prepare gelatin plates.

Calculate and aliquot MEF culture medium.

Calculate number of vials of inactivated MEFs to be thawed.

Thaw inactivated MEFs.

Seed inactivated MEFs in six-well gelatin-coated plate or other culture vessel.

General Notes
- MEFs should be seeded 1–4 days prior to the plating of hESCs.
- Either freshly harvested or frozen MEFs can be used for this purpose.
- Avoid thawing more cells than you can process efficiently.
- Discard unused MEF plates after 5–7 days.

Human Stem Cell Technology and Biology, edited by Stein, Borowski, Luong, Shi, Smith, and Vazquez
Copyright © 2011 Wiley-Blackwell.

- It is good laboratory practice to have important relevant information recorded; therefore a log sheet is provided as an example at the end of this chapter.

Abbreviations

hESCs: human embryonic stem cells

LN_2: liquid nitrogen

MEF: mouse embryonic fibroblast

mL: milliliter

Equipment

- 37°C Cell culture incubator
- 37°C Water bath
- Benchtop centrifuge with swinging-bucket rotors
- Hemacytometer or other cell counter
- Insulated gloves
- Pipette aid
- Safety glasses
- Sterile biosafety cabinet
- Vacuum system (recommended)

Materials

- 5-mL, 10-mL, and 25-mL Sterile glass or plastic pipettes
- 15-mL and 50-mL Sterile conical tubes
- 70% Ethanol spray
- Absorbent paper towels (or Kimwipes®)
- Alcohol-proof marker
- Cryovial holder
- Disposable gloves
- Sterile Pasteur pipettes

Reagent Preparation

- Frozen MEFs
- 0.1% Gelatin solution (see Chapter 5)
- MEF culture medium (see Chapter 5)

NOTES

_____

_____

_____

_____

_____

_____

_____

_____

_____

Preparation

1. Ensure that the following items are in the biosafety cabinet:
   - 15-mL, and 50-mL Sterile conical tubes
   - 5-mL and 10-mL Sterile pipettes
   - Alcohol-proof marker
   - Cryovial holder
   - Sterile Pasteur pipettes

2. Note the number of MEFs that are in each frozen vial.

3. To seed six wells in a six-well plate, $(1.2-1.5) \times 10^6$ MEFs are required (assuming a 90% recovery of MEFs from one freezing and thawing cycle).

   Note: Scale up or down proportionately if other cell culture vessels are used.

4. Calculate the amount of MEF culture medium required:

   4.1. Total culture medium per MEF vial will be 9 mL for thawing plus the volume needed to bring MEF concentration to $(0.1-0.13) \times 10^6$

MEFs/mL. For example, for thawing a vial of 1×10^6 cells, use 9 mL for thawing + 10 mL for resuspending cells, totaling 19 mL.

   4.2. Total culture medium for all MEF vials = culture medium per MEF vial × number of MEF vials to be thawed + 5 mL to compensate for pipetting errors.

5. Record necessary information about cells and culture medium.

Procedure

1. Follow preparations according to Chapter 3, Aseptic Technique on page 19. (*This protocol may be viewed on the accompanying DVD or on the web at* http://www.wiley.com/go/stein/human.)

2. Gelatin coating for cell culture plates:

   2.1. Prepare 0.1% gelatin solution.

   2.2. Before using the solution, warm bottle in 37°C water bath for 15–30 minutes.

   2.3. Place the plates that are to be coated in the biosafety cabinet. It is recommended to label the cover of the plate (not over the wells).

   2.4. For six-well plates, add 2 mL of 0.1% gelatin solution to each well.

   2.5. Tilt the plates in several directions so that the liquid covers the entire surface area (Fig. 7.1).

   2.6. Place the plates in a 37°C incubator.
      Note: Plates can be used after 4 hours and up to 4 days.

3. Prepare gelatin plate(s) in the biosafety cabinet:

   3.1. Remove gelatin plate(s) from the incubator and place inside the biosafety cabinet.

   3.2. Aspirate the gelatin solution from the plate(s) completely with a Pasteur pipette.

   3.3. Return the plate(s) to the incubator for later use.

4. Aliquot MEF culture medium in the biosafety cabinet:

   4.1. Take the MEF culture medium bottle from the refrigerator and place the bottle in the biosafety cabinet after cleaning the bottle with 70% ethanol (Fig. 7.2).

NOTES

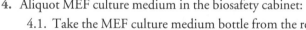

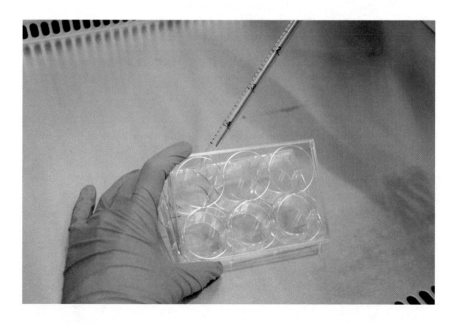

FIGURE 7.1. *After adding gelatin, tip the plate so that the liquid covers the entire surface of the bottom of the well.*

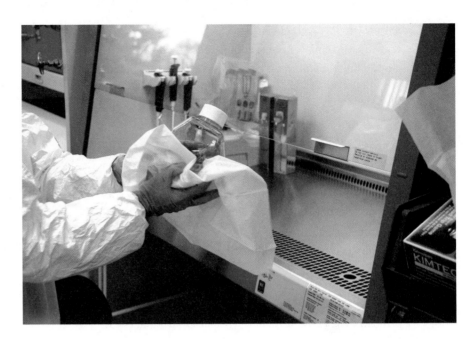

FIGURE 7.2. *Take the MEF culture medium bottle from the refrigerator and place the bottle in the hood after cleaning it with 70% ethanol.*

NOTES

_____

_____

_____

_____

_____

_____

_____

_____

_____

_____

_____

_____

_____

_____

_____

_____

_____

_____

_____

_____

4.2. According to Preparation 4.2, calculate how much MEF culture medium is required and label an appropriate-size sterile tube or bottle (e.g., "MEF.M").

4.3. Transfer an appropriate amount of MEF culture medium to the labeled tube or bottle and warm it in a 37°C water bath for 15–30 minutes.

5. Thaw MEFs:

Note: Thaw no more cells than you can process efficiently.

5.1. Place the warmed MEF culture medium in the biosafety cabinet after thoroughly cleaning its surface with 70% ethanol.

5.2. Label a sterile 50-mL tube (both on the cap and side) in the biosafety cabinet (e.g., "MEF").

Note: Label one 50-mL tube for each vial of MEFs to be thawed.

5.3. Transfer 9 mL MEF culture medium into the tube.

5.4. While wearing protective glasses and insulated gloves, obtain the correct vial(s) of MEFs and warm the vial(s) slightly between gloved hands.

5.5. In the 37°C water bath, immerse the vial in the water without submerging the cap. Swirl the vial gently (Fig. 7.3).

Note: After about 30 seconds, check frequently to see if the frozen solution is beginning to melt. This process may take up to 1 minute. When there is only a small piece of ice floating in the vial, remove the cryovials from the water bath.

5.6. Spray the vial(s) with 70% ethanol until saturated and place the vial(s) in the biosafety cabinet.

5.7. Pipette the cells up and down gently a few (three to five) times, and then transfer the cell suspension drop-wise to the MEF tube while gently swirling the tube. This will help to reduce osmotic shock to the cells.

5.8. Centrifuge for 5 minutes at 200g.

5.9. Return the 50-mL tube to the biosafety cabinet after spraying it with 70% ethanol.

FIGURE 7.3. *When thawing the MEFs, immerse the vial in the water without submerging the cap. Swirl the vial gently.*

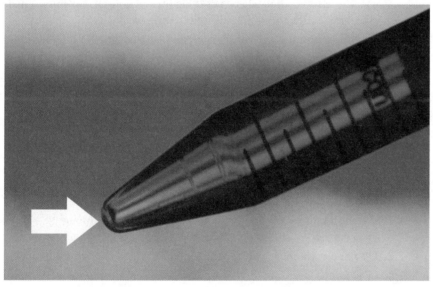

FIGURE 7.4. *After centrifuging, be careful not to disrupt the cell pellet (shown in the bottom of the tube).*

5.10. Remove the supernatant from the tube. Be careful not to touch the cell pellet (Fig. 7.4).

5.11. Gently flick the bottom of the tube to loosen the cell pellet.

5.12. Add 10 mL of MEF culture medium to the cell pellet. Resuspend and mix the cells by pipetting gently up and down a few times.

5.13. Count MEFs (see Chapter 3, Cell Counting with a Hemacytometer on page xx).

5.14. Add MEF culture medium to the MEF tube to make the final cell concentration, at $(0.1-0.13) \times 10^6$ MEFs/mL.

6. Seed MEFs in six-well plate:

6.1. Take the gelatin plate(s) (no more than three plates at a time) from the incubator and place it/them in the biosafety cabinet.

6.2. Label the plates with the following information:
   - MEF
   - Date
   - Initials

NOTES

_____

_____

_____

_____

_____

_____

_____

_____

_____

_____

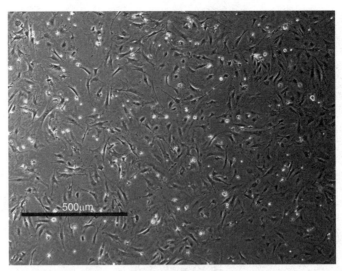

FIGURE 7.5. *MEFs 1 day after thawing.*

NOTES

6.3. Mix the MEFs completely by pipetting up and down a few times. Then take 6 mL MEFs and add 2 mL/well to three wells.

6.4. Seed the remaining wells of all plates in the biosafety cabinet by repeating step 6.3.

6.5. When all plates in the biosafety cabinet are seeded, slide the plates back and forth and side-to-side (cross motion) three to five times inside the biosafety cabinet to evenly distribute cells.

6.6. Transfer the plates to the incubator and make another cross motion three to five times with the plates.

6.7. Close the incubator door and repeat steps 6.1–6.6 if there are more MEFs to be seeded.

7. Post-seeding:

7.1. When all cells are plated, clean the biosafety cabinet thoroughly with 70% ethanol.

7.2. Do not disturb the freshly seeded plates for at least the next 12 hours.

8. Monitor MEFs in culture:

8.1. Check each plate of cells under the microscope the following day for MEF quality and for signs of contamination (Fig. 7.5).

8.2. Record observations.

8.3. Repeat step 8 each day until the plates are used for hESC culture.

SUGGESTED READING

Massachusetts Human Stem Cell Bank Standard Operating Procedures, 2009. Available at http://www.umassmed.edu/mhscb.

Stubban C, Wesselschmidt R L. Mouse embryonic fibroblast feeder cells. In: Loring J, Wesselschmidt R L, Schwartz P H, eds. New York: Academic Press; *Human Stem Cell Manual: A Laboratory Guide*, 2007: 34–46.

WiCell Research Institute. Introduction to Human Embryonic Stem Cell Culture Methods, September 2007.

GELATIN COATING, MEF SEEDING, AND CULTURE LOG SHEET

1. Coat plates with gelatin solution: Date (MM/DD/YY):_____

   - 0.1% gelatin solution batch #:_____ Prepared by:_____
   - Number of six-well plates coated:_____
   - Number of wells/plates coated:_____ Prepared by:_____

     Date:_____

2. Thaw and seed MEFs: Date (MM/DD/YY) :_____

   2.1. MEFs:
   - MEF cell lot number:_____
   - Vendor:_____
   - Number of vials thawed:_____
   - MEF cell inactivation method:_____

     Irradiation:_____

     Mitomycin C treatment:_____

   2.2. MEF culture medium:
   - MEF medium batch #:_____
   - Prepared by:_____

   2.3. MEF seeded plates:_____
   - Number of six-well plates seeded:_____
   - Number of wells/plate seeded:_____
   - Prepared by:_____Date:_____

3. Monitor MEF in culture:

Days in Culture	Date (MM/DD/YY)	Confluence %	Cell Quality	Initials	Comments
1st day	_____	_____	_____	_____	_____
2nd day	_____	_____	_____	_____	_____
3rd day	_____	_____	_____	_____	_____
4th day	_____	_____	_____	_____	_____

THAWING, SEEDING, AND CHANGING THE MEDIUM OF HUMAN EMBRYONIC STEM CELLS

8

Meng-Jiao Shi, Kimberly Stencel, and Maria Borowski

Successful recovery after cryopreservation is important for efficient propagation of human embryonic stem cells (hESCs) in culture. The average percentage range of recovery after thawing of hESCs is quite low, often only 0.1–1.0%. To ensure optimal recovery, the thawing procedure must be performed as quickly as possible.

It is important to monitor and record cell growth every day to gain understanding of the cell growth pattern. As the cells grow and divide, they deplete the medium of additives that are essential to their growth; therefore it is essential that the hESC medium be changed daily. Daily monitoring and feeding of the culture maintains healthy cells by providing fresh nutrients.

The medium used in hESC culture has been optimized to retain growth of undifferentiated cells. Human embryonic stem cell medium with all the essential nutrients added is light orange in color due to the slightly lower pH. However, hESC medium often looks yellow after 24 hours in the 5% CO_2 incubator and should be changed to replenish the essential nutrients.

Most hESCs are grown without the use of antibiotics. Therefore practice of proper aseptic technique is critical for successful hESC culture (see Chapter 3, Aseptic Technique, page 19.). (*This protocol may also be viewed on the accompanying DVD or on the web at* http://www.wiley.com/go/stein/human.)

This chapter describes the procedures for thawing and seeding as well as for changing the medium of human embryonic stem cells. These topics are discussed as Section I and II, respectively, in this chapter.

General Notes

- Read the entire chapter before initiating the activities described in Procedure.
- It is good laboratory practice to record important information, therefore a log sheet is provided as an example at the end of this chapter.
- There are significant variations in hESC growth rates after thawing. In some cases, small hESC colonies can be seen on day 1 after thawing; with others, it may take up to 5 days to see a few colonies.

Abbreviations

DMEM/F12: Dulbecco's modified Eagle's medium (F12)

Human Stem Cell Technology and Biology, edited by Stein, Borowski, Luong, Shi, Smith, and Vazquez
Copyright © 2011 Wiley-Blackwell.

hESC.M: human embryonic stem cell medium

hESCs: human embryonic stem cells

LN$_2$: liquid nitrogen

MEF: mouse embryonic fibroblast

μL: microliter

mL: milliliter

Overview for Section I, Thawing and Seeding of hESCs

Calculate volume of relevant medium required.

Aliquot medium and warm aliquots.

Thaw hESCs.

Prepare inactivated MEF plates.

Seed hESCs.

Incubate plates.

Overview for Section II, Replacement of hESC Culture Medium

Calculate volume of relevant medium required.

Aliquot medium and warm aliquots.

Change medium.

Incubate plates.

Equipment

- 37°C Cell culture incubator
- 37°C Water bath
- Benchtop centrifuge with swinging-bucket rotors
- Insulated gloves
- Inverted microscope
- Pipette aid
- Safety glasses
- Sterile biosafety cabinet
- Vacuum system (recommended)

Materials

- 5-mL, 10-mL, and 25-mL Sterile glass or plastic pipettes
- 15-mL, and 50-mL Sterile conical tubes
- 70% Ethanol spray
- Absorbent paper towels (or Kimwipes®)
- Alcohol-proof marker
- Cryovial holder
- Disposable gloves
- Sterile Pasteur pipettes

Reagents

- DMEM/F12
- hESC Culture medium (see Chapter 5)
- Inactivated MEF plate(s) (see Chapter 5)

I. THAWING AND SEEDING OF HESCS

Preparation

NOTES

1. Documentation and calculation:

   Note: *To conserve costly hESC culture medium, use DMEM/F12 to rinse completely FBS from inactivated MEFs. Use hESC culture medium for thawing and seeding of hESCs. See following steps for detailed calculation.*

   1.1. Using the formulas shown below, calculate how much DMEM/F12 and hESC culture medium are needed for this experiment. These formulas are also provided on the log sheet at the end of this chapter.

   Number of mL DMEM/F12 medium
   $$= 2 \text{ mL/well} \times \text{Number of Wells} + 2 \text{ mL extra}$$
   Number of mL hESC culture medium
   $$= \text{Volume for thawing (normally 9 mL/vial)}$$
   $$+ \text{ Volume for seeding (2.5 mL/well)} + 2\text{--}5 \text{ mL extra}$$

2. Monitoring inactivated MEFs under microscope:

   2.1. Remove inactivated MEF plates needed one at a time and evaluate them under the microscope. Return to the incubator if there is no contamination.

   2.2. Repeat process until all plates are evaluated.

   2.3. If there is contamination in any culture plate, disinfect the whole plate with bleach and discard the culture.

3. Place the following items near the biosafety cabinet:

   - 5-mL and 10-mL Sterile pipettes
   - 70% Ethanol spray
   - Absorbent paper towels (or Kimwipes)
   - Disposable gloves

4. Ensure that the following items are in the biosafety cabinet:

   - 5-mL, 10-mL, and 25-mL Sterile pipettes
   - 15-mL and 50-mL Sterile conical tubes
   - Alcohol-proof marker
   - Cryovial holder
   - Sterile Pasteur pipettes

Procedure

1. Follow preparations according to Chapter 3, Aseptic Technique on page 19.

2. Aliquot DMEM/F12 and hESC culture medium in the biosafety cabinet:

   2.1. Take a bottle of DMEM/F12 medium and a bottle of the appropriate hESC culture medium from the refrigerator.

   2.2. Spray the bottles with 70% ethanol, dry with a Kimwipe and place them in the biosafety cabinet.

   2.3. According to the volume calculated:

   2.3.1. Label an appropriate-size sterile tube as "DMEM/F12."

   2.3.2. Transfer the calculated amount of DMEM/F12 to the tube.

   2.3.3. Label an appropriate-size sterile tube or bottle as "hESC.M."

   2.4. Transfer the calculated amount of hESC culture medium to the labeled "hESC.M" tube or bottle.

   2.5. Warm the aliquots of medium in a 37°C water bath for 15–30 minutes.

   2.6. Return the bottles of stock medium to the refrigerator.

3. Thaw hESCs:

Note: *Wear protective safety glasses and insulated gloves when removing cryovials from the liquid nitrogen freezer.*

Never thaw cells simultaneously from different stem cell lines. If they are from the same line and same lot, thaw no more than two vials at a time in order to minimize prolonged exposure of thawed cells to cryopreservation medium.

3.1. Place the prewarmed DMEM/F12 and aliquots in the biosafety cabinet after thoroughly cleaning containers with 70% ethanol.

3.2. Label a sterile 15-mL tube (both on the cap and side) in the biosafety cabinet as "cell."

3.3. Transfer 9 mL of culture medium per cryovial of hESCs into the "cell" tube.

3.4. While wearing protective glasses and insulated gloves, remove the hESC vial(s) from the LN_2 freezer and verify the label on the cryovial(s).

3.5. Immerse the vial(s) in 37°C water bath without submerging the cap in the water and swirl gently.

3.6. After 30 seconds, check to see whether frozen medium has started to melt. Keep checking intermittently. This may take up to 1 minute.

3.7. When there is only a small piece of ice floating in the cryovial(s), remove the vials(s) from the water bath.

3.8. Spray the vial(s) with 70% ethanol until thoroughly saturated.

3.9. After drying with a Kimwipe, place the vial(s) in the cryovial holder that is already in the biosafety cabinet.

3.10. Pipette the cells gently up and down a few times and transfer the cell suspension drop-wise to the culture medium in the labeled "cell" tube while gently swirling the medium to reduce osmotic shock to the cells.

3.11. Centrifuge for 5 minutes at 200*g*.

3.12. During the centrifugation period, take the inactivated MEF plates from the incubator and place them in the biosafety cabinet.

3.13. Label the plate(s) in the biosafety cabinet: for example,
   • The date (MM/DD/YY)
   • The words "thaw" and the cell line ID
   • The hESC passage number from the cryovial plus one
   • Initials

3.14. Return the plates to the incubator.

3.15. When centrifugation is complete, return the tube to the biosafety cabinet after cleaning the tube with 70% ethanol.

3.16. Remove the medium from the tube (be careful not to touch the cell pellet).

3.17. Add the appropriate volume of culture medium as calculated for seeding and pipette cells gently several times to mix.

4. Seed hESCs in six-well plates:

Note: *Normally, one vial of thawed hESCs will be seeded in one well of a six-well plate.*

To avoid mislabeling cells and cross-contamination, seed cells from different lines in different inactivated MEF plates.

4.1. Take the labeled inactivated MEF plates from the incubator and place them in the biosafety cabinet.

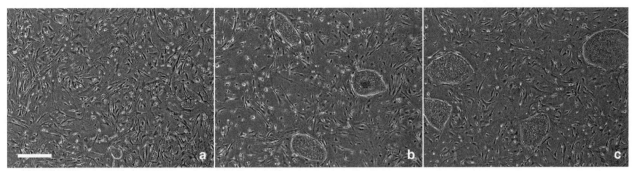

FIGURE 8.1. *It is important to monitor hESCs growth after thawing: (a) first day after thaw; (b) second day after thaw; (c) fifth day after thaw.*

4.2. Aspirate the inactivated MEF medium completely from the wells with a Pasteur pipette.

4.3. Rinse the wells with 2-mL/well of DMEM/F12 and aspirate.

4.4. Mix the hESCs in the tube by pipetting gently up and down a few times, and drop-wise add cells at 2.5 mL/well to six-well plate.

4.5. When all plates are seeded, slide the plates back and forth and side-to-side (cross motion) three to five times inside the biosafety cabinet.

4.6. Transfer the plates to the incubator.

4.7. With the plates in the incubator, make another three to five cross motions.

4.8. Close the incubator door.

4.9. Clean the biosafety cabinet thoroughly with 70% ethanol.

> **Note:** *Do not move the freshly seeded plates for at least the next 12 hours so that the cells have time to adhere.*
>
> *From here on, monitor cell growth and change medium daily (Fig. 8.1) (see Section II of this chapter for details on how to change medium). Colonies may not be visible for a few days.*

NOTES

_____

_____

_____

_____

_____

_____

_____

_____

_____

II. REPLACEMENT OF hESC CULTURE MEDIUM

Preparation

1. Documentation and calculation:

   Note: Human embryonic stem cells grow slowly during the first week after thaw. Therefore two log sheets are suggested: log sheet #1 for monitoring cell growth kinetics after thawing and log sheet #2 for subsequent passaging

   1.1. Using the formulas shown below, calculate how much of each culture medium is needed. These formulas are also provided on the log sheets at the end of this chapter.

   Culture medium = Number of wells_____ × 2.5 mL/well + 2 mL extra
   = _____ mL

2. Place the following items near the biosafety cabinet:
   - 5-mL and 10-mL Sterile pipettes.
   - 70% Ethanol spray
   - Absorbent paper towels (or Kimwipes®)
   - Disposable gloves

3. Ensure that the following items are in the biosafety cabinet:
   - 5-mL, 10-mL, and 25-mL Sterile pipettes
   - 15-mL and 50-mL Sterile conical tubes
   - Alcohol-proof marker
   - Cryovial holder
   - Sterile Pasteur pipettes

   Note: Most hESC culture medium does not contain antibiotics; therefore stringent practice of proper aseptic technique is critical for preventing contamination.

Procedure

1. Follow preparations according to Chapter 3, Section 3.1, Aseptic Technique.

2. Aliquot hESC culture medium in the biosafety cabinet:

   2.1. Place stock bottle of hESC culture medium in the biosafety cabinet after cleaning the bottle with 70% ethanol.

   2.2. According to the volume calculated in Section I, do the following:

   2.2.1. Label an appropriately sized sterile tube or bottle as "M" for medium.

   2.2.2. Transfer the calculated amount of hESC culture medium to the labeled "M" tube or bottle.

   2.2.3. Place the "M" tube or bottle in a 37°C water bath for 15–30 minutes.

3. Monitor hESC growth:

   3.1. Take the hESC plates from the incubator one-by-one and evaluate cell growth under the microscope (Fig. 8.1).

   3.2. Record observations in the recommended hESC Culture Log Sheet.

   3.3. Return the hESC culture plates to the incubator.

   Note: If there is contamination in any culture plate, perform the following:

   3.4. If no further culturing will be done with the plate, discard all contaminated hESC culture wells/plates with at least 250 μL bleach (per well) for at least 30 minutes.

3.5. If uncontaminated hESC cells in a plate with contamination are to be cultured further, then perform the following:

3.6. Move the plate to a biosafety cabinet and disinfect each contaminated well with 250 μL bleach for 5 minutes.

3.7. After removing the disinfected content from the well completely, change medium for other wells and place the plate in a specifically designated incubator for observation during the next few days.

3.8. It is recommended that you record the contamination on the hESC culture log sheet provided.

3.9. In the next few days, closely observe this plate for signs of contamination. If contamination spreads to other wells, disinfect the wells as done previously.

3.10. Once the hESC culture is free of contamination for 5 days, the plate can be returned to regular incubator for further culture.

4. Change medium for hESCs culture:

   Note: *The following protocol is for cells cultured in a six-well plate. Scale up proportionately if cells are cultured in other culture vessels according to the surface area.*
   If there are multiple identical plates of the same line requiring feeding, it is acceptable to change medium for up to three plates simultaneously in the biosafety cabinet. However, different cell lines should not be in the biosafety cabinet at the same time to prevent cross-contamination or misidentification.

   4.1. Place the warm "M" tube or bottle of hESC culture medium in the biosafety cabinet after thoroughly cleaning the container with 70% ethanol.

   4.2. Take the hESC plates from the incubator and place them in the biosafety cabinet.

   4.3. Aspirate the spent medium completely using a Pasteur pipette.

   4.4. Add 2.5 mL fresh hESC culture medium to each well in a six-well plate.

   4.5. Return the hESC plates to the incubator.

   4.6. Change culture medium daily until the cells are ready for passaging or splitting to new inactivated MEF plates.

NOTES

SUGGESTED READING

Freshney I.R. *Culture of Animal Cells: A Manual of Basic Technique.* 5th ed. Hoboken, NJ: John Wiley & Sons; 2005:205–206.

Gonzalez R, Wesselschmidt R L. Schwartz P H, Loring J. Human embryonic stem cell culture. In: Loring J, Wesselschmidt RL, Schwartz PH, eds. *Human Stem Cell Manual: A Laboratory Guide.* New York: Academic Press; 2007:3–33.

Massachusetts Human Stem Cell Bank Standard Operating Procedures, 2009. Available at http://www.umassmed.edu/mhscb.

WiCell Research Institute. Introduction to Human Embryonic Stem Cell Culture Methods, September 2007.

hESC THAWING AND SEEDING LOG SHEET

1. Cell line information:
   - Cell line name:_____   Passage #:_____

2. Thawing specifications:
   - Number of vials to be thawed_____
   - Location in LN$_2$ freezer: Room #:_____Freezer ID:_____
     Box :_____

3. Inactivated MEF plates for seeding:
   - Gelatin coating
   - The date gelatin was coated (MM/DD/YY):_____
     Coated by:_____

4. Inactivated MEF cells:_____
   - The date iMEF cells was seeded (MM/DD/YY):_____
     Seeded by:_____
   - Inactivated MEF culture observation today (write "Normal" or other observations):_____
   - Total number of wells in the six-well plate(s) to be seeded:_____

5. Volume of DMEM/F12 medium required for rinsing the wells:
   - Volume = Number of Wells_____× 2.0 mL/well + 2 ml extra
     =_____mL +_____mL =_____mL total

6. hESC culture medium specifications:
   - Culture medium batch number:_____
   - Volume of **culture medium** for thawing cells = Number of vials_____×9 mL/vial =_____mL (X)
   - Volume of **culture medium** for seeding = Number of wells_____×2.5 mL/well =_____mL
   - Total **culture medium** (mL) = Thawing + Seeding + 2 mL extra =_____mL +_____mL + 2 mL =_____mL

7. Signatures:
   - Thawed and seeded by:_____   Date (MM/DD/YY) :_____
   - Verified by:_____Date (MM/DD/YY) :_____

hESC CULTURE LOG SHEET #1

Passage #1 after thaw:_____

Fill out Section 1 when you change medium on the 1st day after thaw

Cumulative passage# of hESC in culture:_____

1. **Thaw and seed hESC cells:**   Passage# of hESCs thawed:_____

   - Performed by:_____   Date (MM/DD/YY):_____

   - Gelatin plate coated by:_____ Date (MM/DD/YY) :_____

   - Inactivated MEF plates prepared by:_____Date (MM/DD/YY):_____

2. **Change medium for hESC culture (culture medium: hESC culture medium):**

Days	Date: MM/DD/YY	Any Visible Colonies	Colony Quality	Morphology, Appearance	**Culture Medium** batch#	Initials	Comments
1st day	___	___	___	___	___	___	___
2nd day	___	___	___	___	___	___	___
3rd day	___	___	___	___	___	___	___
4th day	___	___	___	___	___	___	___
5th day	___	___	___	___	___	___	___
6th day	___	___	___	___	___	___	___
7th day	___	___	___	___	___	___	___

hESC CULTURE LOG SHEET #2

Passage # after thaw:_____

Use this form starting from the 2nd passage after thawing

Cumulative passage # of hESC in culture_____

1. hESC seeding information:
   - Gelatin coating:
   - The date gelatin was coated (MM/DD/YY):_____
     by:_____

2. Inactivated iMEF plates:
   - The date inactivated MEF cells were seeded (MM/DD/YY):_____
     by:_____

3. How to passage hESCs:
   - Method used to passage (circle one): (a) PTD–Enzyme (b) PTK–Manual
   - hESC splitting ratio_____

4. Change medium for hESC culture (culture medium: hESC culture medium):

Days	Date MM/DD/YY	Confluence %	Colony Quality	Morphology, Appearance	Differentiation %	CM Batch #	Initials	Comments
1st day	_____	_____	_____	_____	_____	_____	_____	_____
2nd day	_____	_____	_____	_____	_____	_____	_____	_____
3rd day	_____	_____	_____	_____	_____	_____	_____	_____
4th day	_____	_____	_____	_____	_____	_____	_____	_____
5th day	_____	_____	_____	_____	_____	_____	_____	_____
6th day	_____	_____	_____	_____	_____	_____	_____	_____
7th day	_____	_____	_____	_____	_____	_____	_____	_____

PASSAGING OF HUMAN EMBRYONIC STEM CELLS ON INACTIVATED MOUSE EMBRYONIC FIBROBLAST PLATES

9

Meng-Jiao Shi, Kimberly Stencel, and Maria Borowski

Successful culture of human embryonic stem cells (hESCs) requires regular passage to maintain an undifferentiated state and pluripotency. Passaging includes careful observation of the morphological changes of colonies in culture, as it is important to keep differentiation of the stem cells to a minimum. Different cell lines possess unique growth kinetics; therefore it is necessary to develop splitting schedules for each cell line. It is known that hESCs grow slowly during the first 2 weeks after being thawed and plated. Human embryonic stem cell growth then accelerates until it reaches a plateau. The cell growth rate will remain at that plateau for many passages if cultured properly. It is essential to observe cell cultures daily to become familiar with the growth kinetics and colony morphology of the cell line being cultured. These observations will help determine the frequency and method used for successful passage.

Traditional enzymatic passage uses enzymes to lift the cells for transfer to new culture vessels. For hESC culture, collagenase is the most widely used enzyme. When culturing hESCs, collagenase is used to detach the cell colonies from the cell culture vessel, not to produce a single cell suspension as is the case in other types of cell culture. Though easier to perform, enzymatic passage does have its drawbacks. It is important to note that the enzymes used, including collagenase, are toxic to the cells. Longer exposure to these enzymes increases the chance that the cells will be damaged. In addition, this toxicity can lead to significant cellular stress, which is associated with increased rates of genomic instability. All hESC cultures exhibit some degree of spontaneous differentiation. Traditional enzymatic passage transfers both differentiated and undifferentiated populations of cells into the new culture vessel. Thus, over time, the percentage of differentiated cells in the culture can increase with this method.

To maintain undifferentiated cultures, alternatives to traditional enzymatic passaging have been developed. In an adaptation of enzymatic passaging, referred to here as "Pick to Discard–Enzyme," differentiated areas of the culture are identified, are scraped off the plate, and are removed prior to enzymatic passaging of the remaining culture. This is the method of choice for cultures with lower levels of differentiation. For cultures with high levels of differentiation, "Pick to Keep," a nonenzymatic passaging method that involves manually detaching undifferentiated

Human Stem Cell Technology and Biology, edited by Stein, Borowski, Luong, Shi, Smith, and Vazquez
Copyright © 2011 Wiley-Blackwell.

cell areas via scraping and transferring to a new culture vessel, may be used. Although they are more laborious than traditional enzymatic passaging, these two methods are better alternatives for long-term maintenance of undifferentiated hESC cultures and are strongly recommended for hESC cultures with more than 15% differentiation.

Overview

Prepare sterile bent Pasteur glass pipettes.

Determine when and how to passage hESCs.

Passage hESCs by Pick to Discard (PTD)–Enzyme method.

Passage hESCs by Pick to Keep (PTK) method.

General Notes

- Read this entire chapter before initiating cell culture activities described in Procedure.
- It is good laboratory practice to have important relevant information recorded; therefore a log sheet is provided as an example at the end of this chapter.
- If possible, it is recommended to use a sterile biosafety cabinet equipped with a stereomicroscope for this protocol. However, if one is not available, these passaging methods can be performed in a regular biosafety cabinet.

Abbreviations

hESC.M: human embryonic stem cell medium

hESCs: human embryonic stem cells

MEF: mouse embryonic fibroblast

mL: milliliter

PTD: Pick to Discard–Enzyme

PTK: Pick to Keep

Equipment

- 37°C Cell culture incubator
- 37°C Water bath
- Alcohol burner
- Benchtop centrifuge with swinging-bucket rotors
- Colony marker (for cell culture microscope)
- Inverted microscope
- Pipette aid
- Sterile biosafety cabinet
- Timer
- Weighing balance
- Vacuum system (recommended)

Materials

- 5-mL and 10-mL Sterile glass or plastic pipettes
- 15-mL and 50-mL Sterile conical tubes
- Alcohol-proof marker
- Disposable gloves
- Nonsterile 9-inch Pasteur pipettes
- Sterile 9-inch Pasteur pipettes

Reagents

- Six-well inactivated MEF plates (see Chapter 5)
- Collagenase solution (see Chapter 5)
- DMEM/F12 medium
- hESC culture medium (see Chapter 5)

Preparation

1. Prepare sterile bent Pasteur glass pipettes:

   Note: *Bent Pasteur pipettes are required to either remove the differentiated cells or transfer undifferentiated hESCs to new inactivated MEF plates.*

   1.1. Turn on an alcohol burner.

   1.2. Take a 9-inch Pasteur glass pipette and heat the area about 2 inches from the tip briefly over the open flame (about 3–5 seconds) (Fig. 9.1a)

   1.3. Once the heated area is pliable, remove the pipette from the flame and gently pull the tip at a 45° angle until it separates from the base of the pipette (Fig. 9.1b). Place the tip briefly over the open flame to form a small ball. Ideally, the bent pipette should now have a small 0.5-inch hook that ends with a small ball at the tip (Fig. 9.1c).

   1.4. Place the bent Pasteur pipette gently in a Pasteur pipette container.

   1.5. Repeat steps 1.2–1.4 to make additional bent Pasteur pipettes and place them in the same container.

   1.6. Autoclave the bent Pasteur pipettes.

2. Evaluate the confluence and quality (differentiation level) of the hESC culture:

   2.1. *Identifying differentiated cells.* Healthy, undifferentiated hESC colonies generally have well-defined borders and the individual cells within the colony appear to be similar to each other (Fig. 9.2a). The exact colony morphology will differ with different cell lines and culture conditions (i.e., whether MEFs, mTeSR®1, or conditioned medium with Matrigel™ are used for hESC culture). Spontaneously differentiated cells may resemble cobblestones (Fig. 9.2b). The differentiated cells may reside at the colony perimeter (Fig. 9.2b) or in the center (Fig. 9.2c,d) of the colony. There can also be differentiated colonies containing large clumps of swirled tissue, or little balls of cells, like embryoid bodies (Fig. 9.2d). Different culture conditions can produce diverse types of differentiated cells (Fig. 9.2b,d,e).

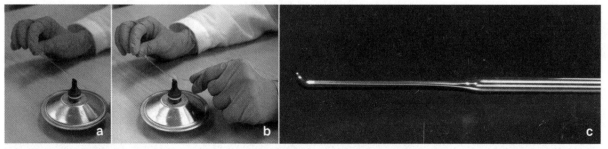

FIGURE 9.1. *Pasteur pipettes are required to either remove the differentiated cells or transfer undifferentiated hESCs to new inactivated MEF plates. (a) Warm a 9-inch Pasteur glass pipette about 2 inches from the tip briefly over the open flame. (b) Once the area is pliable, remove the pipette from the flame and gently pull the tip at a 45° angle until it separates from the base of the pipette. Place the tip briefly over the open flame to form a small ball. (c) Ideally, the bent pipette should now have a small 0.5-inch hook that ends with a small ball at the tip.*

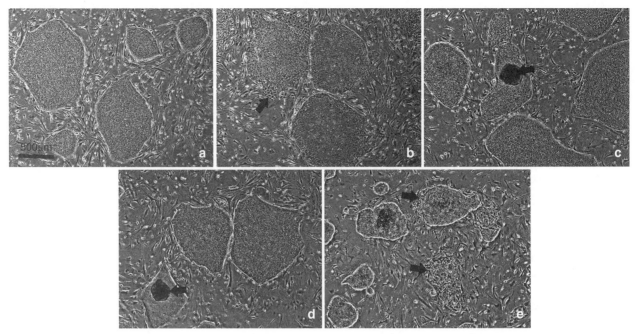

FIGURE 9.2. *Healthy and differentiated colonies: (a) healthy, undifferentiated hESC colonies generally have well-defined borders; (b, arrow) spontaneously differentiated cells may resemble cobblestones and differentiated cells may reside at the colony perimeter or (c, d, arrows) in the center of the colony; (e, arrow) different culture conditions can produce diverse types of differentiated cells.*

2.2. *Determine when to passage (or split) hESCs.* Examine the hESC cultures daily under a microscope to assess the colony size and extent of differentiation. In general, passage cells when any of the following occur:
- Inactivated MEF feeder layer is 2 weeks old
- Most of the hESC colonies are greater than 700 mm
- hESC colonies are too dense (at approximately 70% confluence, Fig. 9.3b)
- Colonies exhibit increased differentiation (Fig. 9.2b–e).

2.3. *Determine how to passage (or split) hESCs.* The method of choice depends on the differentiation level in the cultures (Table 9.1) and how many colonies are to be passaged.

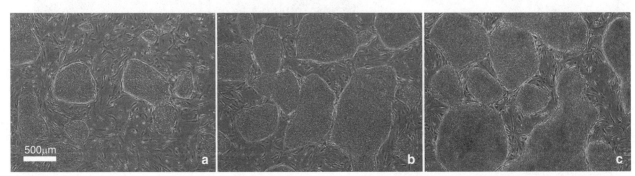

FIGURE 9.3. *The culture confluence determines when hESC colonies should be passaged: (a) colonies are not ready to be passaged at 50% confluence; (b) cultures should be passaged at approximately 70% confluence; (c) before most of the colonies have darkened areas. Bar represents 500 μm.*

TABLE 9.1 STRATEGIES TO DETERMINE PASSAGE METHOD

% Differentiation	How to Passage	Method	Split Ratio
0–20%	Remove differentiated areas, then enzymatically passage remaining hESCs	PTD–Enzyme	1:1 to 1:3
>70%	Pick only undifferentiated areas of cells	PTK	1:1
20–70%	Either above method is fine		1:1 to 1:3

- If differentiation level is **low** (less than 20%, Fig. 9.4a), the method of PTD–Enzyme is recommended.
- If differentiation level is **high** (more than 70%, Fig. 9.4d), the method of PTK is recommended.
- If differentiation level is **intermediate** (20–70% Fig. 9.4b,c), either method can be chosen depending on how many good hESC colonies are available and how many colonies are to be passaged:
  - If there are many undifferentiated hESC colonies available and only limited colonies need to be passaged, PTK is recommended.
  - If there are many undifferentiated hESC colonies available and as many good colonies as possible need to be passaged, PTD–Enzyme is recommended.
  - If there are a limited number of undifferentiated hESC colonies available and as many good colonies as possible need to be passaged, either method is appropriate.

Note: *Regardless of the method used, always set aside one well of hESCs from the same culture used for passaging for an additional 1–2 days in case of errors or contamination during passaging.*

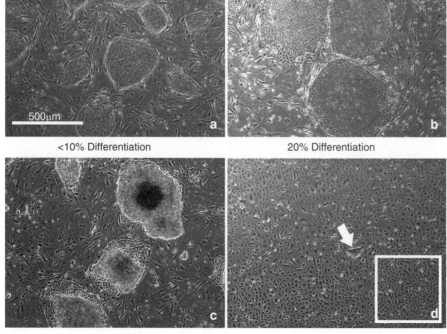

<10% Differentiation

20% Differentiation

50% Differentiation

70% Differentiation

FIGURE 9.4. *The method of passaging (PTK or PTD) depends on the differentiation level in the cultures. (a) Less than 10% differentiation, Pick to Discard (PTD) is recommended. Bar = 500 μm. (b, c) For 20–50% differentiation, either method can be used, (d) For cultures with greater than 70% differentiation, Pick to Keep (PTK) is recommended. Arrow indicates undifferentiated colony to pick. Box shows area with differentiated cells.*

2.4. *Splitting ratio during passaging.* For up to three previous sequential passages, compare the cell splitting ratios and colony confluence levels on harvest days. Splitting ratios may be adjusted up or down if growth rate changes. For instance, hESCs grow slowly during the first 1–2 passages after thawing, and they may be passaged to new inactivated MEF plates at a 1:1 ratio (i.e., one well to one well). At later passages, as cell growth accelerates, hESCs may be split using a 1:3 to 1:6 ratio.

3. Evaluate inactivated MEFs using a microscope:

   3.1. Remove one inactivated MEF plate from the incubator and evaluate it under a microscope for contamination.

   3.2. Return the plate to the incubator if there is no contamination.

   3.3. If there is contamination in any well, disinfect the whole plate with bleach and discard the plate.

   3.4. Continue evaluating inactivated MEF plates one at a time until all plates for this passage have been assessed.

4. Calculate the amount of reagents required.

5. Place the following items near the biosafety cabinet:
   - 5-mL, 10-mL, and 25-mL Sterile pipettes
   - 70% Ethanol spray
   - Absorbent paper towels (or Kimwipes®)
   - Disposable gloves

6. Ensure that the following items are in the biosafety cabinet:
   - 5-mL and 10-mL Sterile pipettes
   - 15-mL and 50-mL Sterile conical tubes
   - Alcohol-proof marker
   - A box of sterile bent Pasteur pipettes
   - Regular sterile Pasteur pipettes

I. PICK TO DISCARD (PTD)–ENZYME METHOD

Procedure

1. Follow preparations according to Chapter 3, Aseptic Technique on page 19. (*This protocol may be viewed on the accompanying DVD or on the web at* http://www.wiley.com/go/stein/human.)

2. Aliquot hESC medium and collagenase solutions in the biosafety cabinet.

    2.1. Depending on the volume calculated in the recommended log sheet, label:
    - One or two sterile 50-mL tubes as "DMEM/F12."
    - One or two sterile 50-mL tubes or an appropriate sterile bottle as embryonic stem cell medium (e.g., "hESC.M").
    - One or two sterile 50-mL tubes as "Collagenase."

    2.2. Remove from the refrigerator the appropriate lot (as recorded on the log sheet) of hESC medium and collagenase stock solution bottles, and a bottle of DMEM/F12 medium. Place them in the biosafety cabinet after thoroughly spraying the bottles with 70% ethanol.

    2.3. Transfer appropriate volume of DMEM/F12 medium to the"DMEM/F12" tube(s).

    2.4. Transfer appropriate volume of hESC medium to the "hESC.M" tube(s) or bottle.

    2.5. Transfer appropriate volume of collagenase solution to the "Collagenase" tube(s).

    2.6. Return the hESC, DMEM/F12 medium, and collagenase stock solution bottles to the refrigerator.

    2.7. Place the aliquots of DMEM/F12 and hESC medium in a 37°C water bath for 15–30 minutes.

    2.8. Leave the aliquot of collagenase solution in the biosafety cabinet for 15–30 minutes to reach room temperature.

3. Prepare inactivated MEF plate(s) (up to three plates may be processed at a time):

    3.1. Place the prewarmed DMEM/F12 and hESC medium aliquots in the biosafety cabinet after thoroughly cleaning the containers with 70% ethanol.

    3.2. Place the inactivated MEF plates that have just been inspected for contamination in the biosafety cabinet.

    3.3. Label these MEF plates with:
    - hESC line name
    - New passage number
    - The date and your initials, (e.g., "H1.p55, 10-25-08, XY").

    3.4. Aspirate the medium completely from the inactivated MEF plate(s) with a Pasteur pipette using the vacuum aspiration system.

    3.5. Add 2 mL/well of fresh DMEM/F12 medium and gently swirl the plate to rinse the well walls.

    3.6. Discard the DMEM/F12 medium completely from the plate(s).

    3.7. Add 1.5 mL of fresh hESC medium to each well.

    3.8. Return the plate(s) to the incubator.

4. HESC Passaging (PTD):

    Note: Harvest one plate at a time.

    4.1. Mark the differentiated areas of cells under a microscope (Fig. 9.5).

NOTES

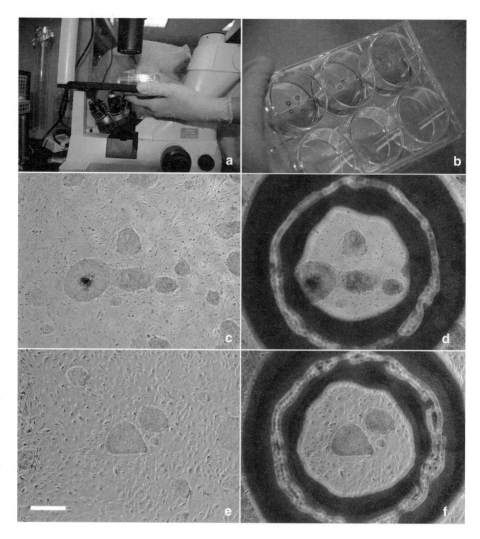

FIGURE 9.5. *Marking plates: (a) before passaging cells, evaluate hESC colonies under an inverted microscope equipped with a colony marker; (b) plate showing marked areas in wells; (c,d) mark the differentiated area if the PTD method is to be used for passaging; (e, f) mark undifferentiated areas if the PTK method is to be used. Scale bar in (e) = 500 μm.*

NOTES

- Remove a hESC plate from the incubator and place it under an inverted microscope equipped with a colony marker.
- Mark the differentiated area under the hESC plate bottom using the colony marker attached to the microscope.

4.2. Transfer the hESC plate to the biosafety cabinet.

4.3. Using a straight Pasteur pipette, aspirate all marked (differentiated) cells in the plate. The plate may need to be scraped with the pipette to detach the colonies.

4.4. Aspirate the spent medium along with the floating differentiated cells or cell debris using the same straight Pasteur pipette.

4.5. Add 1 mL/well of collagenase solution to the plate.

4.6. Return the plate to the incubator.

4.7. Start a timer set for 5 minutes.

4.8. Label a sterile 15-mL or 50-mL tube as "hESC" for collecting hESCs later.

4.9. After 3–5 minutes of incubation, check the plate(s) under a microscope to determine whether the colonies are ready to be harvested. If they are ready, the perimeter of the colony should curl up (Fig. 9.6b), separated from the inactivated MEFs. If they are not ready, return the plate back to the 37°C incubator for another 2–5 minutes of incubation.

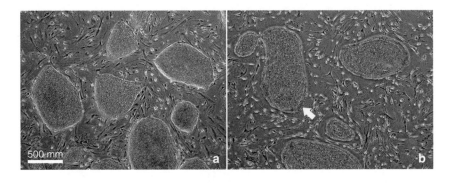

FIGURE 9.6. *Colonies (a) before and (b) after collagenase treatment. Colonies are ready to be harvested if edges of colonies are clearly curled and lifted off feeder layer (arrow, b). Bar represents 500 mm.*

Note: *Incubation length depends on the freshness of the collagenase solution. Although an older solution requires a longer incubation time, do not exceed incubation time for more than 15 minutes.*

4.10. When collagenase incubation is complete, place the hESC plate(s) in the biosafety cabinet.

4.11. If a significant number of colonies have detached:

    4.11.1. Add 2-mL/well of hESC medium.

    4.11.2. Using a 5-mL pipette, detach the remaining attached hESC colonies in each well.

    4.11.3. Transfer detached colonies from all wells into a labeled 15-mL/50-mL tube.

    4.11.4. Centrifuge at 200*g* for 5 minutes.

    4.11.5. Carefully discard the solution without touching the cell pellet and resuspend the cell pellet in 5-mL of hESC medium.

4.12. If a significant detachment of hESC colonies has not occurred, discard collagenase solution from each well. Be careful not to disturb the colonies.

    4.12.1. Add 2 mL of hESC medium to each well in the plate(s).

    4.12.2. Using a sterile 5-mL pipette, take up most of the medium from one well. Use the pipette to scrape the inactivated MEFs/hESC colonies in the well while slowly releasing the medium into the well to detach the cells/colonies from the surface. Pipette up and down gently to minimize bubbles. Repeat these steps until most of the cells/colonies are detached from the surface. Leave the contents in the well until cells/colonies in all of the wells are detached.

    4.12.3. When cells in all wells of one plate are detached, transfer the cell/colony solution from all wells into the labeled sterile "hESC" tube.

    4.12.4. Use 3 mL of fresh hESC medium to rinse all the wells in the plate. Transfer the cell/colony solution to the same "hESC" tube.

4.13. If there are more plates, repeat the above steps to harvest hESCs into the same "hESC" tube.

4.14. Pellet the cells in the "hESC" tube by centrifugation at 200*g* for 5 minutes.

4.15. Aspirate the supernatant from the tube, without touching the cell pellet.

4.16. Resuspend the cell pellet in an appropriate volume of hESC medium (1 mL per well of cells to be seeded, plus 0.5 mL extra for pipetting error).

NOTES

NOTES

4.17. Using a 5-mL pipette, disrupt the cell clusters by gently pipetting cells up and down a few times. Pipette carefully to avoid bubbles by keeping the pipette tip in the solution at all times.

Note: In the above step, initially pipette the cell suspension up and down quickly to break up big cell clusters. Check frequently to see if the big cell clusters are disrupted. After this step, it is important to pipette slowly so as not to break down clusters any more than needed. When most of the colonies are small enough to float in the solution without sinking immediately to the bottom of the tube, the hESCs are ready to be plated. It is important not to disrupt the colonies to a single cell suspension.

4.18. Add hESCs to inactivated MEF plates (processing no more than three plates at a time):

4.18.1. Transfer inactivated MEF plates (containing 1.5 mL/well of hESC medium) from the incubator and place in the biosafety cabinet.

4.18.2. Using a 5-mL pipette, gently pipette cells up and down a few times to distribute cell clusters evenly in the medium. Immediately take 3 mL of the cell suspension and, drop-wise (in a circular motion), add 1 mL/well to each of the three wells.

4.18.3. Repeat the above two steps until cells are distributed to all wells in the inactivated MEF plate.

4.18.4. When complete, slide the plates back and forth and side-to-side (cross motion) three to four times along the inside surface of the biosafety cabinet to evenly distribute the hESC colonies in the wells.

4.18.5. Transfer the plate to the incubator.

4.18.6. Slide the plates in a cross motion three to four times along the inside surface of the incubator; then close the incubator door.

4.18.7. Repeat steps above if there are more cells to plate.

4.18.8. When all cells are plated, clean biosafety cabinet thoroughly with 70% ethanol.

4.18.9. Complete recording on the recommended log sheet and place this sheet in an appropriate notebook or folder.

Note: Do not move the freshly seeded plates for at least the next 12 hours. Monitor cell growth and change medium daily (see Chapter 8, Section II, Replacement of hESC Culture Medium on page 100. (This protocol may be viewed on the accompanying DVD or on the web at http://www.wiley.com/go/stein/human.)

II. PICK TO KEEP (PTK) PROTOCOL

Procedure

Note: This section describes PTK at a ratio of 1:1 (one well to one well). If there are few viable colonies, a ratio of 2:1 or higher may be needed.

1. Follow preparations according to Chapter 3, Aseptic Technique on page 19.

2. Aliquot hESC medium in the biosafety cabinet:

2.1. Depending on the volume calculated in the recommended log sheet, label:
  - One or two sterile 50-mL tubes as "DMEM/F12."
  - One or two sterile 50-mL tubes or an appropriate sterile bottle as embryonic stem cell medium (e.g, "hESC.M").

2.2. Remove from the refrigerator the appropriate lot (as recorded on the log sheet) of hESC medium bottle, and a bottle of DMEM/F12 medium. Place them in the biosafety cabinet after thoroughly cleaning the bottles with 70% ethanol.

2.3. Transfer appropriate volume of DMEM/F12 medium to the"DMEM/F12" tube(s).

2.4. Transfer appropriate volume of hESC medium to the "hESC.M" tube(s) or bottle.

2.5. Return the hESC, DMEM/F12 medium, and collagenase stock solution bottles to the refrigerator.

2.6. Place the aliquots of DMEM/F12 and hESC medium in a 37°C water bath for 15–30 minutes.

3. Prepare inactivated MEF plate(s) (up to three plates may be processed at a time):

3.1. Place the prewarmed DMEM/F12 and hESC medium aliquots in the biosafety cabinet after thoroughly cleaning the containers with 70% ethanol.

3.2. Place the inactivated MEF plates that have just been inspected for contamination in the biosafety cabinet.

3.3. Label these MEF plates with:
  - hESC line name
  - New passage number
  - The date and your initials (e.g., "H1.p55, 10-25-08, XY").

3.4. Aspirate the medium completely from the inactivated MEF plate(s) with a Pasteur pipette using the vacuum system.

3.5. Add 2 mL/well of fresh DMEM/F12 medium and gently swirl the plate to rinse the well walls.

3.6. Discard the DMEM/F12 medium completely from the plate(s).

3.7. Add 1.5 mL of fresh hESC medium to each well.

3.8. Return the plate(s) to the incubator.

4. HESC passaging (PTK):

4.1. Place the hESC plate under a microscope.

4.2. Mark the undifferentiated areas in the hESC wells on the plate bottom using the colony marker attached to the microscope. If possible, mark at least 50 undifferentiated areas for transfer (Fig. 9.5).

4.3. Transfer the hESC plate to the biosafety cabinet.

4.4. Aspirate the spent medium from the wells with a straight sterile Pasteur pipette.

4.5. Add 1.5 mL of fresh hESC culture medium to each well containing hESCs.

4.6. Using a bent Pasteur pipette, thoroughly scrape the marked undifferentiated areas.

Note: The quality of the bent Pasteur pipettes is essential to the efficient removal of undifferentiated areas of hESCs. Refer to Figure 9.1 for an example of correctly prepared bent Pasteur pipette.

4.7. Under the microscope, verify that the marked areas have been completely scraped off the plate and return the plate to the biosafety cabinet. If necessary, repeat the previous step.

4.8. Transfer the prepared inactivated MEF plate containing 1.5 mL of hESC medium (as prepared in Procedure step 3) to the biosafety cabinet.

4.9. From the scraped hESC plate, gently pipette the hESC solution up and down a few times to disrupt the scraped colonies.

4.10. Transfer the medium (about 1.5 mL) along with the hESCs in suspension into the prepared inactivated MEF plate at 1:1 ratio (one well to one well).

4.11. Place the newly plated hESCs in the incubator. Slide the plate back and forth and side-to-side (cross motion) on the shelf of the incubator three to five times.

Note: Replace hESC culture medium daily.

SUGGESTED READING

Freshney IR. *Culture of Animal Cells: A Manual of Basic Technique.* 5th ed. Hoboken, NJ: John Wiley & Sons, 2005:192–195.

Gonzalez R, Wesselschmidt R L. Schwartz P H. Loring J. Human embryonic stem cell culture. In Loring J, Wesselschmidt RL, Schwartz PH, eds. *Human Stem Cell Manual: A Laboratory Guide*, New York: Academic Press, 2007: 5–7.

Massachusetts Human Stem Cell Bank Standard Operating Procedures, 2009. Available at http://www.umassmed.edu/mhscb.

National Stem Cell Bank Protocols: Available at http://www.wicell.org/index.php?option=com_content&task=category&sectionid=7&id=246&Itemid=248, SOP-CC-025A, and at https://www.wicell.org/index.php?option=com_docman&task=doc_download&gid=1069, Selection and Maintenance of hES cells.

WiCell Research Institute. Introduction to Human Embryonic Stem Cell Culture Methods, September 2007 (www.wicell.org).

hESC ENZYMATIC PASSAGING LOG SHEET

Passage # after thawing: _____ Passage # of hESCs harvested: _____

1. Gelatin coating and inactivated MEF cells in culture:
   - Date gelatin plates were coated (MM/DD/YY): _____ By: _____
   - The date inactivated MEF cells were seeded (MM/DD/YY): _____ By: _____
   - Inactivated MEF Morphology: observation of inactivated MEF cultures on the day hESCs are seeded (write "Normal" or other observations): _____

2. How to passage hESCs:
   - Method for passaging: _____
   - Total number of wells from which hESC are harvested — six-well plate: _____
   - Total number of wells in which hESC are seeded — six-well plate: _____
   - hESC splitting ratio: _____

3. Reagents and medium required:
   - DMEM/F12 lot #: _____   Expiration: _____
   - Collagenase lot #; (if any) : _____   Expiration: _____
                                         Prepared by: _____
   - hESC medium lot # (if any) : _____   Expiration: _____
                                         Prepared by: _____
   - Sign: _____
   - Passaged by: _____   (MM/DD/YY) _____

HARVESTING HUMAN EMBRYONIC STEM CELLS FOR CRYOPRESERVATION

10

Meng-Jiao Shi, Kimberly Stencel, and Maria Borowski

Cryopreservation is the process of slowly cooling cells or cell aggregates to subzero temperatures for storage. At these temperatures, all biological processes are arrested, allowing the cells to be preserved for extended periods of time. However, if proper care is not taken during this process or if appropriate cryopreservation medium is not used, cells can be damaged, leading to cell death. Cell death can also occur from dehydration of the cells and by extracellular and intracellular ice formation.

Even with careful cryopreservation and thawing, the percentage of cell death is very high, due to the fragile nature of human embryonic stem cells (hESCs). On average, only 0.1–1% of the cells are viable after thawing. Since hESCs grow in colonies during culture, when frozen and thawed as aggregates, a higher percentage of hESCs survive the cryopreservation process and they recover more rapidly. However, if not performed properly, this approach may also lead to higher incidence of cell death rates since the protective cryopreservation medium cannot surround each cell in an aggregate as well as when cells are frozen in a single cell suspension. Thus many cells in the aggregates die due to ice crystal formation.

To reduce the chance of ice crystal formation during the freezing procedure, the cells must be gradually frozen to temperatures between $-130°C$ and $-150°C$ for long-term storage. This gradual cooling can be achieved by placing cryovials containing the cells in the specially designed vessel containing isopropanol called Mr. Frosty™ (Nalgene® Labware Cat# 5100-0001) (Fig. 10.1). This container is placed at $-80°C$ overnight. During the first freeze day, the temperature of the cells is brought down at a rate of $-1°C$/minute. After approximately 24 hours, the cells can be taken out of the Mr. Frosty container and placed into a liquid nitrogen freezer ($-130°C$ and $-150°C$) for long-term storage.

To ensure that the hESCs remain structurally intact during the entire cryopreservation process, the freezing medium contains essential nutrients such as fetal bovine serum (FBS) and dimethyl sulfoxide (DMSO). The FBS contains proteins and growth factors to support the structural integrity of the cells and the DMSO protects the cells by partially solubilizing the cell membrane, making it more pliable and less prone to puncture, disrupting the lattice of the ice so that fewer crystals form.

Human Stem Cell Technology and Biology, edited by Stein, Borowski, Luong, Shi, Smith, and Vazquez
Copyright © 2011 Wiley-Blackwell.

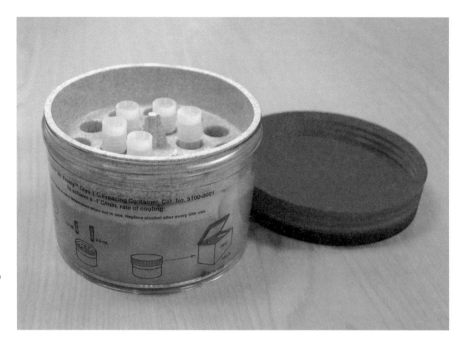

FIGURE 10.1. *The Mr. Frosty freezing container will allow the cells to slowly freeze at a rate of −1°C/minute. Note that the Mr. Frosty container can hold 18 cryovials.*

Overview

Calculate volume of relevant medium and reagents to be used.

Evaluate the confluence and quality of the hESC culture.

Prepare medium/reagent aliquots and warm to 37°C.

Harvest hESCs.

Freeze hESCs at −80°C.

Transfer cells from −80°C to long-term storage in a liquid nitrogen (LN$_2$ freezer, −130°C to −150°C).

General Notes

- It is important to carefully read and follow the protocol in this chapter for optimal recovery of viable cells after thawing as this process should be completed as quickly as possible and there may be limited time to stop and reference the protocol.
- It is good laboratory practice to record important information; therefore a log sheet is provided as an example at the end of this chapter.
- Always prepare fresh freezing medium.
- It is important to transfer cells to −80°C as soon as possible once they are in freezing medium.

Abbreviations

DMEM: Dulbecco's modified Eagle's medium

DMSO: dimethyl sulfoxide

hESCs: human embryonic stem cells

HI-FBS: heat-inactivated fetal bovine serum

LN$_2$: liquid nitrogen

MEF: mouse embryonic fibroblast

mL: milliliter

μm: micrometer

Equipment

- 37°C Cell culture incubator
- 37°C Water bath
- −80°C Freezer
- Alcohol-proof marker
- Benchtop centrifuge with swinging-bucket rotor
- Inverted microscope
- LN$_2$ freezer (−130°C to −150°C)
- Mr. Frosty freezing containers
- Pipette aid
- Timer
- Sterile biosafety cabinet
- Vacuum system (recommended)
- Weighing balance

Materials

- 5-mL and 10-mL Sterile glass or plastic pipettes
- 15-mL and 50-mL Sterile conical tubes
- Cryovials
- Cryovial holder
- Disposable gloves
- Sterile Pasteur pipettes

Reagents

- Collagenase solution at 1 mg/mL (See Chapter 5)
- DMEM/F12
- DMSO
- hESC culture medium (See Chapter 5)
- HI-FBS
- Isopropanol

Preparation

NOTES

1. Calculate the amount of medium and collagenase solution needed, according to Table 10.1.
2. In a 37°C water bath, thaw an appropriate amount of HI-FBS needed for 2× freezing medium according to Table 10.2.
3. Determine when to harvest hESCs:
    3.1. Under a microscope, carefully evaluate hESC growth and quality. Harvesting should be performed between 4 and 6 days after plating; during this time, cells will be in a period of rapid growth.
    3.2. The following observations will be used to determine when hESCs are ready for harvest. Harvest hESCs when:
    - There is less than 10% differentiation based on visual estimation.
    - Colonies of hESCs are approximately 70% confluent (Fig. 10.2).
4. Prepare cryovial:
    4.1. On the cryovial, record the following information:
    - Cell line name and passage number

NOTES

_____

_____

_____

• Date (when cells were harvested)
• Initials of person performing the procedure

Note: It is important to use a permanent alcohol-proof laboratory grade marker when writing on the cryovial(s).

TABLE 10.1 MEDIUM AND SOLUTION NEEDED FOR HARVESTING AND FREEZING HESC CELLS

Collagenase solution required: 1 mL/well × Number of wells + 2 mL extra   = 1×_____ wells +2 = _____ mL
hESC medium for harvesting: 2 mL × Number of wells + 2 mL × Number of plates   = 2×_____ wells + 2 × plate # = mL
hESC medium for freezing: 0.5 mL/well × Number of wells = 0.5 × wells =_____ mL
Total medium required: _____mL for harvesting + _____ mL for freezing   +5 mL extra =_____ mL
Preparation of 2 × freezing media: 2 × Freezing medium required = 0.5 mL× Number of cryovials + 4–5 mL   extra for pipette and filter error

TABLE 10.2 2X FREEZING MEDIUM

Ingredient	Ratio in the 2x Freezing Medium	10 mL Total	Actual Volume =_____mL
HI-FBS	60%	6 mL	_____mL
hESC culture medium	20%	2 mL	_____mL
Sterile DMSO	20%	2 mL	_____mL

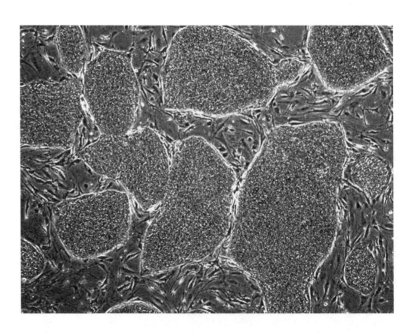

FIGURE 10.2. *Carefully evaluate hESC colony growth and quality to determine when to harvest, approximately 70% confluent.*

Procedure

1. Follow preparations according to Chapter 3; Aseptic Technique on page 19. (*This protocol may be viewed on the accompanying DVD and on the web at* http://www.wiley.com/go/stein/human.)

2. Place the following items near the biosafety cabinet:

   - 5-mL, 10-mL, and 25-mL Sterile pipettes
   - 70% Ethanol spray
   - Absorbent paper towels (or Kimwipes®)
   - Disposable gloves
   - Sufficient number of Mr. Frosty freezing containers for freezing hESCs (each Mr. Frosty container can hold a maximum of 18 cryovials).

   Note: Make sure all freezing containers have a sufficient amount of fresh iso-propanol in them. The isopropanol needs to be changed after every fifth use.

3. Ensure that the following items are in the biosafety cabinet:

   - 2-mL, 5-mL, and 10-mL Sterile pipettes
   - 15-mL and 50-mL Sterile conical tubes
   - Alcohol-proof marker
   - Cryovial holders
   - Sterile Pasteur pipettes

4. Prepare 2 × freezing medium in the biosafety cabinet:

   Note: Prepare 2× freezing medium just before use.

   4.1. Clean the hESC culture medium, sterile DMSO, and thawed HI-FBS bottles (or thawed aliquots) with 70% ethanol and transfer them to the biosafety cabinet.

   4.2. Label one sterile 50-mL conical tube or a larger size sterile bottle as 2× freezing medium (on both the cap and side).

   4.3. Add the reagents and their proper volumes listed in Table 10.2, to the labeled freezing medium container.

   4.4. Mix and filter through a 0.22-μm filter system.

   4.5. Keep the above solution on ice or at 4°C and use it within 24 hours.

   Note: This 2× freezer medium will be added to hESC suspension at a 1:1 volume ratio, resulting in 1× freezing medium with a final concentration of 10% DMSO and 30% FBS.

5. Aliquot medium and collagenase solution in the biosafety cabinet:

   5.1. According to the volume calculated in Preparation step 1, label both the cap and side of:
      - One or two sterile 50-mL tubes or an appropriate sterile bottle as "hESC.M" for harvesting and resuspending hESCs.
      - One sterile 50-mL tube as "Collag" for collagenase.

   5.2. Transfer appropriate volume of hESC medium to the "hESC.M" tube(s) or bottle from the stock bottle in the biosafety cabinet.

   5.3. Place the aliquot of ESC medium in a 37°C water bath for 15 minutes.

   5.4. Transfer appropriate volume of collagenase solution to the "Collag" tube(s).

   5.5. Leave the aliquot of collagenase solution in the biosafety cabinet for 15–30 minutes to reach room temperature.

NOTES

6. Place the labeled cryovials in the biosafety cabinet.

   Note: *Cells are normally frozen at 1 vial per well if the confluence is at or greater than 60%.*

7. Harvest hESCs (no more than two plates at a time):

   Note: *Working with more than two plates at a time during this procedure is not recommended. The cells are very sensitive to the collagenase solution and will perish if they are left in it for too long.*

   7.1. Place the warm hESC medium aliquot in the biosafety cabinet after thoroughly cleaning the bottle with 70% alcohol.

   7.2. Label a sterile container as "ESC" for collecting hESCs at a later time.

   7.3. Set timer(s) for 5 minutes and 10 minutes, respectively.

   7.4. Remove one or two hESC plate(s) from the incubator and place it/them in the biosafety cabinet.

   7.5. Aspirate the medium completely from the hESC plate(s) using a Pasteur pipette.

   7.6. Add 1 mL collagenase/well.

   7.7. Return the plate(s) to the incubator.

   7.8. Start the timer(s).

   7.9. After 5 minutes of incubation, check the plate(s) under a microscope to determine whether the cell colonies are ready to be harvested. When they are ready, the edges of the colony should appear curled up (Fig. 10.3) (See Chapter 9; *this protocol may be viewed on the accompanying DVD and on the web at* http://www.wiley.com/go/stein/human), separating from the iMEFs and lifting from the vessel surface. If they are not ready, return the plate to the 37°C incubator for another 3–5 minutes of incubation.

   Note: *Incubation time depends on the freshness of the collagenase solution. The older it is, the longer the incubation will take. However, do not exceed 15 minutes.*

   7.10. When collagenase incubation is complete, place the hESC plate in the biosafety cabinet.

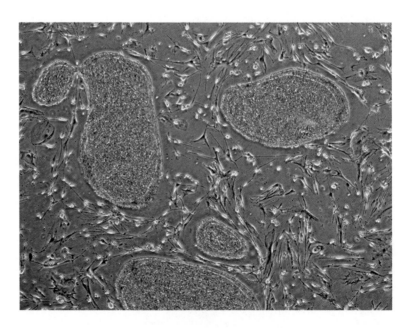

FIGURE 10.3. *Colonies should appear curled up.*

7.11. Aspirate collagenase from the wells taking care not to disturb the hESC colony layer.

7.12. Add 2 mL of hESC medium to each well.

7.13. Using a 5-mL pipette, draw up most of the medium from one well and hold it in the pipette. Scrape the MEFs and hESC colonies in the well using the pipette while slowly releasing the medium into the well to wash them off the surface. Gently pipette the medium up and down during the wash to disrupt cell clusters. Keep the pipette tip in the medium during this process to minimize bubbles.

7.14. When almost all hESCs are detached, leave the contents in the well and repeat step 7.13 to detach hESCs from remaining wells.

7.15. Combine hESCs from all wells into the labeled sterile "hESC" conical tube.

7.16. Take another 3 mL of fresh hESC medium per plate and rinse all the wells, then transfer to the same sterile "hESC" container.

7.17. Repeat the above protocol if needed

8. Freeze hESCs:

Note: Once in DMSO freezing medium, cells should be frozen as quickly as possible (within 5 minutes). This will help to ensure greater cell viability upon thawing.

When freezing a large lot, different personnel may complete different parts of the freezing process in order to reduce the time that hESCs are in the freezing medium at room temperature. For example, while one person distributes the cell aliquots to the cryovials, another person may tighten the caps of cryovials and another may transfer the cryovials to the freezing containers and place them in a −80°C freezer.

8.1. Centrifuge hESCs at 200 g (about 1000 rpm) for 5 minutes.

8.2. During the centrifugation period, loosen all cryovial caps in the biosafety cabinet.

8.3. When centrifugation is complete, return "hESC" container to the biosafety cabinet after cleaning with 70% ethanol.

8.4. Aspirate the supernatant without touching the hESC pellet.

8.5. Resuspend the cell pellet by gently pipetting up and down two to three times in an appropriate volume of hESC medium (0.5 mL/well plus 0.5 mL extra of the total volume).

8.6. Place the cold 2× freezing medium in the biosafety cabinet after cleaning with 70% ethanol.

8.7. Label a new sterile 50-mL tube as "F.ESC."

8.8. Gently pipette up and down twice to mix the cell suspension in the "ESC" container and transfer 9.2 mL to the "F.ESC" tube.

8.9. From the freezing medium tube, take an equal volume of 2 × freezing medium (9.2 mL) and add it drop-wise to the cell suspension in the "F.ESC" tube while gently tapping the tube.

8.10. Gently pipette up and down one or two times to mix the cells, then transfer the cells to cryovials at 1 mL/vial.

8.11. Tighten cryovial cap(s), then transfer to a Mr. Frosty freezing container.

8.12. Place the freezing container in a −80°C freezer overnight.

Note: If there are more than 18 cryovials to be frozen, repeat steps 8.7–8.13.

8.13. The next day, transfer vials to a LN$_2$ freezer for storage (Fig. 10.4).

FIGURE 10.4. *After placing the cryovials in a −80°C freezer overnight, transfer them to a LN₂ freezer for storage.*

SUGGESTED READING

Katkov II, Kim MS, Bajpai R, et al. Cryopreservation by slow cooling with DMSO diminished production of Oct-4 pluripotency marker in human embryonic stem cells. *Cryobiology* 2006;53:194–205.

Massachusetts Human Stem Cell Bank Standard Operating Procedures, 2009. Available at http://www.umassmed.edu/mhscb.

Ware CB, Nelson AM, Blau CA. Controlled-rate freezing of human ES cells. *Biotechniques*. 2005;38:879–880, 882–873.

WiCell Research Institute. Introduction to Human Embryonic Stem Cell Culture Methods, September 2007. Available at http://www.wicell.org/index.php?option=com_content&task=category&sectionid=7&id=246&Itemid=248.

hESC CRYOPRESERVATION LOG SHEET

Date (MM/DD/YY) _____   Cryopreserved by: _____

1. Cell line information:

   - Cell line name: _____   Cumulative hESC passage number: _____

2. Cells at freezing:

   - Colony quality: _____   Colony confluence (%): _____
   - Total number of wells to be harvested — six-well plate: _____
   - Number of vials to be frozen: _____

3. hESCs frozen (select one): _____ (a) vial/well; (b) other: _____

   - Comments:

HUMAN EMBRYONIC STEM CELL CULTURE ON BD MATRIGEL™ WITH mTeSR®1 MEDIUM

11

Meng-Jiao Shi, Kimberly Stencel, and Maria Borowski

As described in previous chapters, human embryonic stem cells (hESCs) are most often grown with inactivated mouse embryonic fibroblast (MEF) feeder cells, which help maintain hESC self-renewal and pluripotency. However, for applications such as human cell therapies, drug screening, and functional genomics, use of animal cells and products in hESC culture is undesirable, as these products may be contaminated with retroviruses and other pathogens that may alter the hESC biology and genome, as well as risk transmission of pathogens to the patient. For this reason, methods of feeder-independent and/or serum-free culture continue to be developed.

In a feeder-independent culture environment, hESCs must retain the same characteristics and growth potential as they would when grown in the presence of inactivated MEFs. The cells must continue to express factors associated with the undifferentiated state and maintain pluripotency in vitro and in vivo, as well as retain a normal karyotype and growth rate.

Several culture conditions for hESCs that are, for the most part, serum-free and/or feeder-independent are available. This protocol uses BD Matrigel™ (referred to in this text as "Matrigel"), available from BD Biosciences, as a culture matrix and inactivated MEF conditioned medium that allows hESCs to be cultured without direct contact with feeders;[1] however, this culture system is not free from nonhuman components. The medium contains animal products as it is "conditioned" by inactivated MEF cells before being placed on the hESCs; therefore it contains products secreted from the mouse cells. In addition, the Matrigel substrate is produced from mouse sarcoma cells. Feeder independent cultures are performed with extracellular matrix and a defined medium containing a combination of growth factors such as TGFb and bFGF.[2,3] Each method uses an animal component containing serum replacement in the medium. Richards et al.[4] reported in 2002 the first culture system using human fetal or adult fallopian tube epithelial cells in place of the inactivated MEF cells. Since then, bone marrow cells, foreskin fibroblasts, and other human cell types have also been used.[5,6] However, in some of these cases culture medium is supplemented with human serum and subsequently has batch-to-batch or lot-to-lot variability.

Human Stem Cell Technology and Biology, edited by Stein, Borowski, Luong, Shi, Smith, and Vazquez
Copyright © 2011 Wiley-Blackwell.

TeSR®1 medium was originally developed for the maintenance of hESCs in a serum-free, animal-product-free environment that supports long-term growth in feeder-independent cultures. Developed at the WiCell Research Institute (www.wicell.org), this medium contains high levels of bFGF with transforming growth factor TGFb, λ-aminobutyric acid (GABA), pipecolic acid, and lithium chloride. TeSR has been used with a cell support matrix composed of four human components: collagen IV, fibronectin, laminin, and vitronectin.[7] To offset this costly culture method, a modified formula, mTeSR1, was created.[8] Although mTeSR1 includes some animal sourced proteins, it is less expensive, fully defined, and serum-free.

Overview

Aliquot Matrigel stock.

Coat six-well plates with Matrigel.

Thaw or passage hESCs and seed on Matrigel-coated plates.

Change medium daily.

Harvest hESCs and seed them to new Matrigel-coated plates or freeze them using mFreSR™ medium.

General Notes

- Read the entire chapter before initiating cell culture activities described in Procedure.
- It is good laboratory practice to have important relevant information recorded; therefore a log sheet is provided as an example at the end of this chapter.
- Check lot dilution factor before aliquoting Matrigel.

Abbreviations

bFGF: basic fibroblast growth factor

CO_2: carbon dioxide

DMEM/F12: Dulbecco's modified Eagle's medium/F12

hESCs: human embryonic stem cells

LN_2: liquid nitrogen

mg: milligram

μL: microliter

mL: milliliter

TGFb: transforming growth factor b

Equipment

- 37°C Water bath
- −80°C Freezer
- Benchtop centrifuge with swinging-bucket rotors
- Cell culture incubator
- Inverted microscope
- Pipette aid
- Sterile biosafety cabinet
- Timer

Materials

- 1.5-mL or 2-mL Sterile vials

- 5-mL and 10-mL Sterile glass or plastic pipettes
- Six-well plates
- 15-mL and 50-mL Sterile conical tubes
- Alcohol-proof marker
- Disposable gloves
- Sterile Pasteur pipettes

Reagents
- DMEM/F12
- Dispase solution at 1 mg/mL (StemCell Technologies Cat # 07923)
- Matrigel™ hESC-qualified matrix (BD Biosciences Cat # 354277)
- mFreSR™ cryopreservation medium (StemCell Technologies Cat # 05854)
- mTeSR®1 maintenance medium (StemCell Technologies Cat # 05850)

Preparation

NOTES

1. Preparation of dispase aliquots from 100 mL stock:

   Note: Dispase is a protease recommended for passaging hESCs cultured on Matrigel. Generally 1 mL/well is used for six-well culture plate. Aliquots should be thawed only once. Thawed aliquots are stable for 2 weeks if stored at 2–8°C.

   1.1. Thaw the stock bottle of 1 mg/mL dispase solution overnight at 2–8°C.

   1.2. For one bottle of 100 mL dispase stock, label 15 sterile 15-mL tubes (e.g., including the name of "Dispase," lot #, manufacturer's name, and expiration date).

   1.3. When the stock dispase solution is thawed, transfer 6.5 mL to each labeled tube.

   1.4. Store the aliquots at −20°C.

2. **Day one**: Preparation of aliquots of Matrigel hESC-qualified matrix:

   Note: While working with Matrigel it is critical to keep it cold (e.g., 2–8°C) because it will begin to gel as it warms to room temperature. This can be achieved by keeping the aliquots on ice prior to plating.
   Matrigel aliquots cannot be thawed and refrozen; therefore it is important that aliquots are stored in an appropriate volume. Different lots may have different dilution factor. It is very important to aliquot the Matrigel according to the dilution factor provided by BD Biosciences for each lot. Generally, 135–175 µL is suggested for coating two six-well plates at 0.5 mg/plate.

   2.1. Confirm the dilution factor of the Matrigel lot. Calculate the amount of Matrigel needed to coat two six-well plates (i.e., 12 wells).

   2.2. Label a set of sterile 15-mL tubes (e.g., including Matrigel concentration, lot #, expiration date)

   2.3. Place the labeled tubes in a −20°C freezer.

   2.4. Place an unopened box of the appropriate sized pipette tips in the −20°C freezer.

   2.5. To thaw the Matrigel stock, place the bottle in a small beaker, then add the beaker to a filled ice bucket (Fig. 11.1). Cover the ice bucket and leave it overnight in a 4°C refrigerator.

3. **Day two**: Preparation of aliquots of Matrigel hESC-qualified matrix:

   3.1. Position thawed Matrigel stock bottle directly on ice in a small container, and place the container in the biosafety cabinet after cleaning the container with 70% ethanol.

FIGURE 11.1. *To thaw Matrigel, place the bottle in a small beaker and place the beaker on ice in an ice bucket. Cover the ice bucket and place the bucket overnight in a 4°C bibrigerator.*

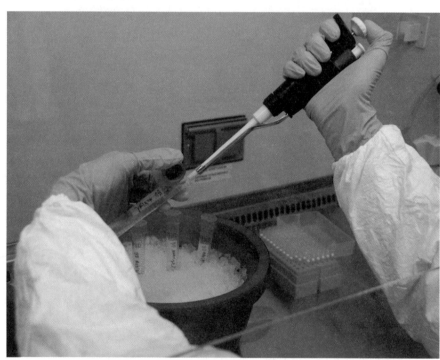

FIGURE 11.2. *For each aliquot, dispense the amount of Matrigel that is needed for two six-well plates into each 15-mL tube. It is important to work as quickly as possible so the Matrigel stays cold.*

NOTES

_____

_____

_____

_____

_____

_____

_____

_____

3.2. Arrange labeled sterile 15-mL tubes on ice in another small container, and place the container in the biosafety cabinet after cleaning the container with 70% ethanol.

3.3. Place chilled pipette tips in the biosafety cabinet.

3.4. Dispense enough Matrigel for two six-well plates (12 wells) into each labeled 15-mL tube as quickly as possible (Fig. 11.2).

> ***Note:*** *To calculate amount needed, contact BD Bioscience for the concentration of that lot, as this information is currently not included on the product sheet.*

3.5. Quickly switch tips every 5–7 tubes to ensure sterility and that the tips are chilled.

3.6. Store the aliquots at −70°C for up to 6 months.

NOTES

4. Place the following items near the biosafety cabinet:
   - 5-mL, 10-mL, and 25-mL Sterile pipettes.
   - 70% Ethanol spray
   - Absorbent paper towels (or Kimwipes®)
   - Disposable gloves
5. Ensure that the following items are in the biosafety cabinet:
   - 5-mL and 10-mL Sterile pipettes
   - 15-mL and 50-mL Sterile conical tubes
   - Alcohol-proof marker
   - Sterile Pasteur pipettes

Procedure

1. Follow preparations according to Chapter 3, Aseptic Technique on page 19. (*This protocol may be viewed on the accompanying DVD and on the web at* http://www.wiley.com/go/stein/human.)

2. Coating plates with aliquots of Matrigel hESC-qualified matrix:

   NOTES

   2.1. Chill sterile 5-mL and 10-mL pipettes in a −20°C freezer.
   2.2. Based on the number of hESC plates needed, thaw an appropriate number of Matrigel aliquots on ice in a 4°C refrigerator for 1–2 hours.

   Note: *To expedite the thawing process, transfer cold DMEM/F12 and Matrigel to the biosafety cabinet and add DMEM/F12 directly to the frozen aliquot. Pipette gently until the Matrigel aliquot is totally thawed.*

   2.3. After Matrigel is thawed, place cold DMEM/F12, chilled tubes, and pipettes in the biosafety cabinet.
   2.4. Add 12.5 mL cold DMEM/F12 to the 15-mL tube(s) containing thawed *Matrigel aliquot*. Mix thoroughly by pipetting. Avoid foaming or bubbles.
   2.5. Add 1 mL diluted Matrigel to each well in a six-well plate (Fig. 11.3).
   2.6. Swirl the plate(s) to spread the Matrigel solution evenly across the entire surface.

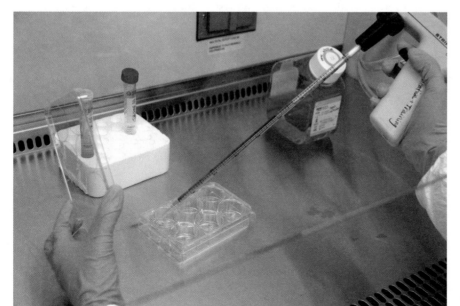

FIGURE 11.3. *Add 1 mL diluted Matrigel to each well in a six-well plate.*

NOTES

2.7. Note the date when Matrigel was plated on the plate lid.

2.8. Leave the plate(s) in the biosafety cabinet for 1–2 hours.

2.9. After 1–2 hours, aspirate the Matrigel entirely and add 1.5 mL mTeSR1 medium to each well.

2.10. Place the plate(s) in the incubator while thawing/seeding or harvesting hESCs.

> ***Note:*** *If the plates are not to be used immediately, add an additional 1 mL of DMEM/F12 to each well and seal the edges of the plate (e.g., with Parafilm). Store the plates at 2–8°C for up to 7 days. The plates must be warmed to room temperature before use. Place the plates in the biosafety cabinet for at least an hour prior to plating cells.*

3. Thaw and seed hESCs on Matrigel plates:

   3.1. Thaw hESCs according to the protocol in Chapter 8 (on page 97). (*This protocol may be viewed on the accompanying DVD and on the web at* http://www.wiley.com/go/stein/human), except resuspend cells in mTeSR1 medium instead of hESC culture medium and seed onto Matrigel plates instead of MEF plates.

4. Passage hESCs in feeder-independent plate with dispase:

   4.1. Prewarm the DMEM/F12 medium, dispase aliquots, and mTeSR1 medium to room temperature.

   4.2. Prepare Matrigel-coated plates:

      4.2.1. *If the Matrigel-coated plates were stored at 2–8°C*, place them in the biosafety cabinet to warm them to room temperature for at least 1 hour prior to plating hESCs. When the Matrigel-coated plates are at room temperature, aspirate the Matrigel, and add 1.5 mL mTeSR1 per well. Label the new plates with the cell name and the passage number. Return the plates to the incubator.

      4.2.2. *If the Matrigel-coated plates were freshly prepared*, remove the plates from the incubator and label them with the cell name and the passage number. Return the plates to the incubator.

   4.3. Using a microscope, confirm hESC colonies are ready for passaging (Fig. 11.4).

   4.4. Add 1 mL dispase solution to each well of hESCs, and incubate for 5 minutes at 37°C.

   4.5. Observe cells under a microscope. When the colony edges begin to curl up (Fig. 11.5), proceed to step 4.6. Otherwise, place in incubator for another 3–5 minutes.

   4.6. When dispase treatment is complete, aspirate the dispase solution from each well and gently wash the cells three times with 1 mL per well of DMEM/F12.

   4.7. Add 2 mL mTeSR1 medium to each well.

   4.8. Using a sterile 5-mL pipette, scrape to detach hESC colonies. Leave the contents in the well while detaching cell colonies in other wells.

   4.9. When cells in all wells of one plate are detached, transfer the colony solution from all wells into a sterile 15-mL or 50-mL tube.

   4.10. Use 2–3 mL of fresh mTeSR1 medium to rinse all the wells in the plate. Transfer the colony solution to the same tube.

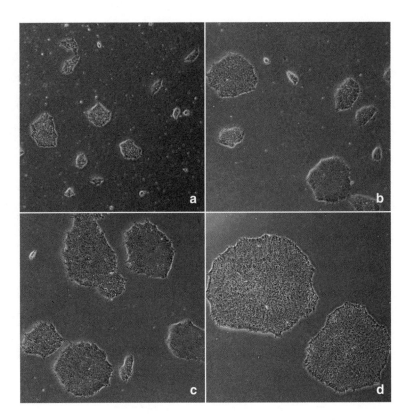

FIGURE 11.4. *Monitor the hESC colonies to confirm they are ready for passaging: (a) H1P85 day 1 after seeding on Matrigel; (b) H1P85 day 2 after seeding on Matrigel; (c) H1P85 day 3 after seeding on Matrigel; and (d) H1P85 day 4 after seeding on Matrigel. Cells are ready to passage.*

4.11. Using a sterile 5-mL pipette, disrupt the cell clusters by gently pipetting cells up and down a few times. Pipette carefully to avoid bubbles by keeping the pipette tip in the solution at all times.

4.12. According to the splitting ratio, add an appropriate amount of mTeSR1 to the tube (1 mL/well of cells to be seeded, plus 0.5 mL extra for pipetting error).

4.13. Remove the prepared Matrigel-coated plate(s) from the incubator and place them in the biosafety cabinet.

4.14. Add 1 mL/well hESCs drop-wise (in a circular motion) to each of the Matrigel-coated wells.

4.15. Transfer the plate(s) to the incubator.

4.16. Slide the plate(s) back and forth and side-to-side (cross motion) three to four times along the inside surface of the incubator to evenly distribute hESCs in the wells.

4.17. Close the incubator door.

> **Note:** *Do not move the freshly seeded plates for the next 12 hours. Monitor cell growth and change medium daily (see Chapter 8).*

5. Freezing hESCs using mFreSR medium:

5.1. Harvest cells according to the procedures described in steps 4.4–4.10.

5.2. Pellet the cells in the tube and aspirate.

5.3. After centrifugation, resuspend cells in mFreSR medium at 1 mL/well harvested.

5.4. Aliquot 1 mL/cryovial and resuspend cells into labeled cryovials at 1 mL/cryovial, then transfer the cells to a Mr. Frosty™ freezing container.

5.5. Place the freezing container in a −80°C freezer overnight.

5.6. The next day, transfer vials to a LN₂ freezer for storage.

NOTES

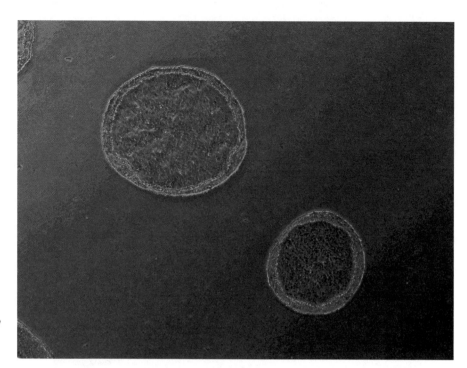

FIGURE 11.5. *After adding 1 mL of dispase solution to each well, observe cells. When edges of colonies curl up, the cells are ready for harvesting.*

REFERENCES

1. Xu C, Inokuma MS, Denham J, et al. Feeder-free growth of undifferentiated human embryonic stem cells. *Nat Biotechnol*. 2001;19:971–974.

2. Amit M, Shariki C, Margulets V, et al. Feeder layer- and serum-free culture of human embryonic stem cells. *Biol Reprod*. 2004;70:837–845.

3. Levenstein ME, Ludwig TE, Xu RH, et al. Basic fibroblast growth factor support of human embryonic stem cell self-renewal. *Stem Cells*. 2006;24:568–574.

4. Richards M, Fong CY, Chan WK, et al. Human feeders support prolonged undifferentiated growth of human inner cell masses and embryonic stem cells. *Nat Biotechnol*. 2002;20:933–936.

5. Cheng L, Hammond H, Ye Z, et al. Human adult marrow cells support prolonged expansion of human embryonic stem cells in culture. *Stem Cells*. 2003;21:131–142.

6. Hovatta O, Mikkola M, Gertow K, et al. A culture system using human foreskin fibroblasts as feeder cells allows production of human embryonic stem cells. *Hum Reprod*. 2003;18:1404–1409.

7. Ludwig TE, Levenstein ME, Jones JM, et al. Derivation of human embryonic stem cells in defined conditions. *Nat Biotechnol*. 2006;24:185–187.

8. Ludwig TE, Bergendahl V, Levenstein ME, et al. Feeder-independent culture of human embryonic stem cells. *Nat Methods*. 2006;3:637–646.

SUGGESTED READING

Lanza, R, Gearhart, J, Hogan, B, et al. *Essentials of Stem Cell Biology*. New York: Elsevier Academic Press; 2006.

Lu J, Hou R, Booth CJ, Yang SH, Snyder M. Defined culture conditions of human embryonic stem cells. *Proc Natl Acad Sci USA*. 2006;103:5688–5693.

Stem Cell Technologies, Maintenance of Human Embryonic Stem Cells in mTeSR®1, Technical Manual version 1.1.0.

Vallier L, Alexander M, Pedersen RA. Activin/nodal and FGF pathways cooperate to maintain pluripotency of human embryonic stem cells. *Cell Sci*. 2005;118:4495–4509.

Yao S, Chen S, Clark J, et al. Long-term self-renewal and directed differentiation of human embryonic stem cells in chemically defined conditions. *Proc Natl Acad Sci* USA. 2006;103:6907–6912.

hESC CULTURE ON MATRIGEL™ WITH mTESR®1 MEDIUM

1. Passage # of hESCs harvested: _____

2. Matrigel-coated plates:
   - The date Matrigel was coated (MM/DD/YY): _____ By: _____

3. How to passage hESCs
   - Number of wells to be harvested and seeded: _____
   - Total number of wells harvested: six-well plate: _____ Total number of wells seeded: six-well plate: _____
   - hESC splitting ratio: _____

4. Reagents and medium required:
   - DMEM/F12 lot #:_____   Expiration:_____
   - Dispase lot #:_____   Expiration:_____
   - mTeSR1 medium lot #: _____   Expiration: _____

5. hESC culture:

Days	Date: MM/DD/YY	Any Visible Colonies	Colony Quality	Medium Lot	Initials	Comments
1st day	_____	_____	_____	_____	_____	_____
2nd day	_____	_____	_____	_____	_____	_____
3rd day	_____	_____	_____	_____	_____	_____
4th day	_____	_____	_____	_____	_____	_____
5th day	_____	_____	_____	_____	_____	_____
6th day	_____	_____	_____	_____	_____	_____
7th day	_____	_____	_____	_____	_____	_____

Laboratory Guide for Human Stem Cell Culture: Characterization of Pluripotent Stem Cells

SECTION III

Human Stem Cell Technology and Biology, edited by Stein, Borowski, Luong, Shi, Smith, and Vazquez
Copyright © 2011 Wiley-Blackwell.

DEFINING PLURIPOTENCY

12

Kelly P. Smith and Mai X. Luong

INTRODUCTION

In developmental biology, *pluripotency* is defined as the potential of a particular cell to develop into all cell types found in the embryonic and adult organism. During mammalian development, *totipotency*, which is the capacity to form all cell lineages of an organism including extraembryonic tissues such as the placenta, is a characteristic only of the zygote (fertilized egg) and continues through the four-cell stage blastomere.[1] In contrast to totipotent cells, pluripotent cells do not form extraembryonic tissues. Pluripotency has traditionally been viewed as a characteristic of cells in the inner cell mass of the blastocyst from which embryonic stem (ES) cells are derived, but more recent studies have found that pluripotency is a complex state. In the mouse, it has been demonstrated that pluripotency extends through to cells of the post-implantation epiblast.[2,3] However, cells derived from early blastocysts (ES cells) and post-implantation epiblasts (EpiSCs) differ in their gene expression and developmental capacities.[3,4] These differences have led to the concept that, in the mouse, pluripotency exists as two different states, where ICM derived ES cells are in a "ground" state and EpiSCs are in a "primed" state that may not have the same developmental potential as the ground state.[5] Human ES cells appear to be more similar to mouse EpiSCs, which leads to questions regarding the true status of hESC pluripotency.

Demonstration of pluripotency is a key factor in the description of ES cell lines upon derivation and must continue as ES cells are expanded to ensure that the capacity for differentiation is preserved while spontaneous differentiation is held in check. With the advent of new cellular reprogramming techniques such as the derivation of induced pluripotent stem (iPS) cells,[6,7] determination of pluripotency becomes even more important. Reprogramming technology holds enormous promise as it has spurred the derivation of disease-specific pluripotent cells,[8–11] which will provide new tools for studying these diseases and may pave the way for more personalized patient-specific cellular therapies.[12,13] As new technologies are developed and more ES-like cells are produced, assays that rigorously define and measure pluripotency and the reprogramming process are critical to their characterization.

MEASURING PLURIPOTENCY

The definitive test for pluripotency would demonstrate that the ES or reprogrammed cells can functionally replace all cell types, including germline cells, in the developing

---

Human Stem Cell Technology and Biology, edited by Stein, Borowski, Luong, Shi, Smith, and Vazquez
Copyright © 2011 Wiley-Blackwell.

organism. This has been accomplished with mouse ES and iPS cells by introducing these cells into the blastocyst and producing chimeric mice that are germline competent.[14] In contrast to mouse ES and iPS cells, mouse EpiSCs do not have the capacity to produce chimeras.[3,4]

Since implanted stem cells only contribute to a subset of cells within any given chimeric mouse, it is possible that host-derived cells may complement nonautonomous defects of implanted cells. This issue can be addressed by tetraploid complementation, in which ES cells are introduced into 4n blastocysts that have been produced by cell fusion. As the 4n cells cannot contribute to the inner cell mass, the resulting organism is comprised completely of cells derived from the introduced ES cells.[15]

While this standard has been attained numerous times with mouse ES cells, attempts to reproduce this result with murine reprogrammed cells have been more problematic. Initial attempts resulted in mid- to late-stage embryos that do not develop to full term.[16] Only recently has it been demonstrated that iPS cells can support full-term development of tetraploid blastocyst-complemented embryos.[17–19] The efficiency of generating iPS mice using different iPS cell lines ranged from 0.3% to 13%, which is similar to the efficiencies for ES cells.[15]

For human ES or reprogrammed cells, it is clearly not ethically feasible to produce live organisms either as chimeras or organisms entirely derived from these cells by 4n complementation; therefore researchers must resort to less definitive alternative standards to assess pluripotency. The foundations for many of these standards were established using embryonal carcinoma (EC) cells. Teratocarcinoma formation by EC cells has uncovered many of the factors recognized today as markers for pluripotency.[20] Although defining pluripotency in reprogrammed cells is sufficiently difficult, an additional consideration regarding pluripotency is fidelity of regulatory cascades that support differentiation and responsiveness of these cells to environmental cues.

ASSAYING PLURIPOTENCY AND THE UNDIFFERENTIATED STATE

Morphology and Cell Cycle

Over the last decade, studies of undifferentiated human ES cells have revealed many characteristics that distinguish the pluripotent state. Many of these morphological and molecular characteristics are currently used to help define pluripotency in hES and reprogrammed cells. Undifferentiated pluripotent stem cells have a high ratio of nucleus to cytoplasm and prominent nucleoli, and grow in multilayered colonies with defined edges. In addition, these cells have a distinctive nuclear architecture (lamina, nuclear speckles, and heterochromatin domains) as well as chromatin structure.[21,22] Pluripotent cells also have rapid growth characterized by an abbreviated G1 phase of the cell cycle.[23–25]

Gene Expression and Epigenetics

Human ES cells express a distinct set of cell surface antigens such as SSEA3, SSEA4, TRA-1-60, and TRA-1-81, which are most often used as markers for the undifferentiated state. Other commonly used markers include telomerase and alkaline phosphatase activity, as well as three genes that play significant roles in pluripotency: OCT4, SOX2, and NANOG.[26–28] Expression of these markers is most often detected by immunocytochemical staining with fluorescence microscopy or flow cytometry, or by quantitative reverse transcription polymerase chain reaction (qRT-PCR). However, using qRT-PCR to detect OCT4 expression may be problematic due to the large number of retropseudogenes that exist, which has resulted in detection of false-positive signals in somatic stem cells.[29]

While specific pathways such as STAT3 and ERK as well as transcription factors (OCT4, NANOG, SOX2) have been clearly associated with pluripotency in mouse and humans,[30] expression studies have also shown that pluripotency is not a single state that can be defined by expression of a specific set of genes. For instance, NANOG expression is not required for ES cell pluripotency in vitro.[31] Within the mouse embryo itself, there is heterogeneity in expression of specific factors.[31–33] Similarly, expression studies have identified subpopulations of cells in pluripotent mouse[34] and human[35] ES cell cultures. In addition, the expression profiles of human ICM cells and ES cells are distinct.[36]

Studies of epigenetic genome modifications revealed that ES cells also have distinct chromatin signatures, consisting of bivalent domains that contain both repressive and activating histone modifications.[37,38] It has been suggested that these bivalent domains play a role in maintaining pluripotency. However, other studies have shown that these domains are not unique to pluripotent stem cells.[39] Despite this, bisulfite sequencing of the OCT4 and NANOG gene promoters to verify demethylation may be a reliable and rigorous way of demonstrating their activation and the cells' pluripotent state.[40] In addition, this method is objective and the results are not affected by factors such as genetic background and culture conditions.

Lastly, undifferentiated human and mouse ES cells express a unique set of microRNAs (miRNAs), which are noncoding single-stranded transcripts that regulate gene expression by inhibiting translation of mRNAs into proteins. Some of these miRNAs are conserved between the two species[41,42] and may serve as new molecular markers for pluripotency. Several lines of evidence support the role of miRNA in regulating ES cell gene expression. For example, ES cells deficient in miRNA processing enzymes demonstrate defects in self-renewal, differentiation, and perhaps viability.[43–45] In addition, specific miRNAs have been shown to regulate lineage specification as well as mammalian developmental patterning.[46–48]

ASSAYING DIFFERENTIATION POTENTIAL

In Vitro Assays

The ability to form cell lineages that represent the three germ layers (endoderm, mesoderm, and ectoderm) through embryoid body (EB) formation[49] has also been used as a measure of pluripotency. Embryoid bodies are created by the spontaneous formation of spherical clusters of ES cells in suspension. Expression of genes specific to each germ layer is typically assayed by qRT-PCR and immunofluorescence microscopy. Although EBs do not completely reproduce the pattern formation or structural organogenesis observed in the embryo, they are often used to assay pluripotency as they are relatively simple to produce and do not require the use of animal models.

The capacity to form cell types representing each germ layer can also be determined in vitro using protocols for directed differentiation.[50] In these protocols, cells are exposed to specific culture conditions and growth factors to induce differentiation along a particular pathway. Although expression of differentiation markers is generally used to confirm cell differentiation, it is not a test for functionality and furthermore marker expression is influenced by cellular stress response.[51]

In Vivo Assays: Teratomas

Short of producing chimeric organisms, the most stringent method for measuring pluripotency is to demonstrate the ability of a cell line to form cell lineages of all three germ layers through teratoma formation. Teratomas are produced by the introduction of the cells into immunodeficient mice via subdermal, intratesticular, intramuscular, or kidney-capsule injection. Once introduced into the host, pluripotent cells will develop into teratomas that possess cells of ectodermal, endodermal, and mesodermal lineages. Practical limitations to using teratomas to assess pluripotency include the following: (1) costly animal studies require a separate skill-set; (2) significant histological expertise is required to determine cell lineages and structures in teratomas, (3) frequency[52] of teratoma formation as well as their composition[53] are influenced by cell inoculation site and mouse strain, (4) this assay is qualitative and may be somewhat subjective, and (5) teratoma formation can take much longer (several weeks) than other methods. More importantly, teratoma assays do not demonstrate the ability of cells to contribute to all cell types in the adult organism. Nonetheless, teratoma formation is the most stringent functional assay currently available for assessment of pluripotency in human cells.

CLASSIFYING PLURIPOTENCY

Human embryonic stem cells can be characterized at four levels with regard to pluripotency: cellular, molecular, functional, and developmental levels (Table 12.1). Since the assays used at each level have unique advantages and limitations, cells should ideally be characterized at as many levels as possible. Currently, there is no single standard for pluripotency that can be applied across different cell types to allow accurate comparison of different cell types and the cell lines within each type. Thus all available tests should be employed when performing cell characterization to ensure that cells meet the maximum number of criteria for pluripotency.

PLURIPOTENCY AND CELLULAR REPROGRAMMING

Determining pluripotency is an especially relevant yet difficult issue for iPS cells since the mechanisms of reprogramming are not well understood and the existence of partially reprogrammed cells is widely recognized. Multiple reprogramming methods and reagents (e.g., transgenes, vectors, and small molecules) as well as different source cells likely result in different paths of reprogramming.[51,54] Each of these approaches has the potential by genetic and/or epigenetic mechanisms to result in cells that are not fully reprogrammed[39] (Fig. 12.1). It has been shown that some partially reprogrammed mouse cells resemble completely reprogrammed cells, but these cells are not functionally pluripotent in that they cannot produce chimeras.[55] It has been demonstrated that fully reprogrammed cells can be distinguished from those that have only undergone partial reprogramming by the silencing of the integrated proviral transgenes and the expression of the pluripotency markers TRA-1-60, DNMT3B, and REX1.[56] Although it is unclear whether this assay can reproducibly define completely reprogrammed cells, studies such as these, which define the specific sets of assays needed to help determine pluripotency, are crucial to the understanding of the reprogramming processes as well as the mechanisms underlying pluripotency.

UNDERSTANDING PLURIPOTENCY

As understanding of the nature of pluripotency and pluripotent cells continues to evolve, there are several important concepts to consider:

- Pluripotency is a dynamic state or set of states. These different states are highly sensitive to cell culture conditions and may influence the capacity of the cells to differentiate into functional tissues.

TABLE 12.1 LEVELS OF CHARACTERIZATION

Level	Tests	Results	Advantages	Limitations
Cellular	Cell and colony morphology Marker staining Flow cytometry Cell cycle analysis	Indicate that the cells share characteristics with undifferentiated ES cells	Most are rapid and provide an overview of entire culture	Tests are fairly nonspecific. Markers are often expressed in cells that are not pluripotent
Molecular	RT-PCR Microarray analysis Epigenetic analysis	Determine that cells are undifferentiated	Most tests are relatively fast and simple to perform on multiple cell lines	Tests do not accurately define pluripotency Does not determine capacity for differentiation
Functional	Embryoid body Teratoma Directed differentiation	Assays ability of cells to form cell types representative of the three germ layers	Demonstrates broad differentiation capacity and ability to develop into specific cell types	EBs do not demonstrate tissue or structure formation Teratomas are time consuming and subject to variability Tests do not necessarily demonstrate developmental pluripotency
Developmental	Chimeras Tetraploid complementation	Tests ability of cells to contribute to all cell types in the adult organism, including germ line cells	Most stringent test of pluripotency	Not ethically feasible to perform with human cells (surrogate criteria required)

- Variations in reprogramming reagents and methods can produce cells that are in different states of reprogramming. It is unclear at this time what these partially reprogrammed states represent.

- Many reprogrammed cells may not be equivalent to ES cells. For example, it has been reported that ES and reprogrammed cells have distinct epigenetic signatures.[57]

- Current methods used for characterizing human ES cells, such as embryoid body and teratoma formation, suggest that these cells may be pluripotent during normal development. However, studies of murine epiblast stem cells (EpiSCs)[2,3] suggest that this may not necessarily be true. Although EpiSCs can

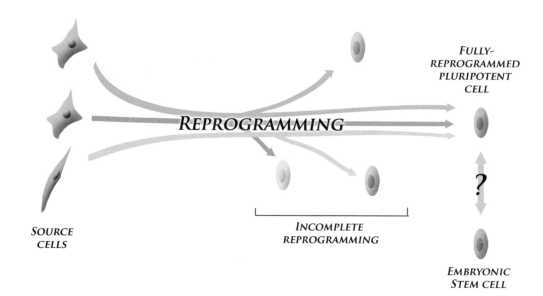

FIGURE 12.1. *Many paths to pluripotency. Differences in source cells, techniques, and reagents may result in variations in the course of reprogramming. Some of these variations may lead to incomplete reprogramming or to reprogrammed lines that may not be equivalent to embryonic stem cells.*

develop into three germ layers via EB and teratoma formation, they are not capable of contributing to the formation of a chimera upon introduction into a blastocyst. Thus the pluripotency assays currently in use do not necessarily indicate the potential to form all cell lineages during normal development.

The definition and assessment of pluripotency are vital to stem cell research.[58] However, it is important to emphasize that there are limitations inherent in the numerous available assays. In addition, the criteria for pluripotency are moving targets since the dynamic nature of pluripotency will likely require further refinement of the tests used for its determination as progress is made to better define this state. It has been suggested that, due to cost and time constraints in the therapeutic environment, many iPS lines will be banked and the lines will be tissue matched to individual patients.[59] The highest measurable level of pluripotency would be needed for these banked lines and would require adherence to a set of guidelines such as those suggested above. However, the capacity for some multipotent cell lines to readily produce a particular cell type will also likely determine their value in cellular therapy. As the field evolves, regardless of whether stem cells are used in the research or clinical environment, a key challenge will be the definition and evaluation of pluripotency.

REFERENCES

1. Van de Velde H, Cauffman G, Tournaye H, et al. The four blastomeres of a 4-cell stage human embryo are able to develop individually into blastocysts with inner cell mass and trophectoderm. *Hum Reprod*. 2008;23:1742–1747.

2. Brons IG, Smithers LE, Trotter MW, et al. Derivation of pluripotent epiblast stem cells from mammalian embryos. *Nature*. 2007;448:191–195.

3. Tesar PJ, Chenoweth JG, Brook FA, et al. New cell lines from mouse epiblast share defining features with human embryonic stem cells. *Nature*. 2007;448:196–199.

4. Guo G, Yang J, Nichols J, et al. Klf4 reverts developmentally programmed restriction of ground state pluripotency. *Development*. 2009;136:1063–1069.

5. Nichols J, Smith A. Naive and primed pluripotent states. *Cell Stem Cell*. 2009;4:487–492.

6. Takahashi K, Tanabe K, Ohnuki M, et al. Induction of pluripotent stem cells from adult human fibroblasts by defined factors. *Cell*. 2007;131:861–872.

7. Yu J, Vodyanik MA, Smuga-Otto K, et al. Induced pluripotent stem cell lines derived from human somatic cells. *Science*. 2007;318:1917–1920.

8. Park IH, Arora N, Huo H, et al. Disease-specific induced pluripotent stem cells. *Cell*. 2008;134:877–886.

9. Dimos JT, Rodolfa KT, Niakan KK, et al. Induced pluripotent stem cells generated from patients with ALS can be differentiated into motor neurons. *Science*. 2008;321:1218–1221.

10. Soldner F, Hockemeyer D, Beard C, et al. Parkinson's disease patient-derived induced pluripotent stem cells free of viral reprogramming factors. *Cell*. 2009;136:964–977.

11. Ebert AD, Yu J, Rose FF Jr, et al. Induced pluripotent stem cells from a spinal muscular atrophy patient. *Nature*. 2009;457:277–280.

12. Yamanaka S. Strategies and new developments in the generation of patient-specific pluripotent stem cells. *Cell Stem Cell*. 2007;1:39–49.

13. Kiskinis E, Eggan K. Progress toward the clinical application of patient-specific pluripotent stem cells. *J Clin Invest*. 2010;120:51–59.

14. Okita K, Ichisaka T, Yamanaka S. Generation of germline-competent induced pluripotent stem cells. *Nature*. 2007;448:313–317.

15. Nagy A, Rossant J, Nagy R, et al. Derivation of completely cell culture-derived mice from early-passage embryonic stem cells. *Proc Natl Acad Sci USA*. 1993;90:8424–8428.

16. Hanna J, Markoulaki S, Schorderet P, et al. Direct reprogramming of terminally differentiated mature B lymphocytes to pluripotency. *Cell*. 2008;133:250–264.

17. Kang L, Wang J, Zhang Y, et al. iPS cells can support full-term development of tetraploid blastocyst-complemented embryos. *Cell Stem Cell*. 2009;5:135–138.

18. Boland MJ, Hazen JL, Nazor KL, et al. Adult mice generated from induced pluripotent stem cells. *Nature*. 2009;461:91–94.

19. Zhao XY, Li W, Lv Z, et al. iPS cells produce viable mice through tetraploid complementation. *Nature*. 2009; 461:86–90.

20. Andrews PW, Matin MM, Bahrami AR, et al. Embryonic stem (ES) cells and embryonal carcinoma (EC) cells: opposite sides of the same coin. *Biochem Soc Trans*. 2005;33:1526–1530.

21. Meshorer E, Misteli T. Chromatin in pluripotent embryonic stem cells and differentiation. *Nat Rev Mol Cell Biol*. 2006;7:540–546.

22. Butler JT, Hall LL, Smith KP, et al. Changing nuclear landscape and unique PML structures during early epigenetic transitions of human embryonic stem cells. *J Cell Biochem*. 2009;107:609–621.

23. Becker KA, Ghule PN, Therrien JA, et al. Self-renewal of human embryonic stem cells is supported by a shortened G1 cell cycle phase. *J Cell Physiol*. 2006;209:883–893.

24. Ghule PN, Becker KA, Harper JW, et al. Cell cycle dependent phosphorylation and subnuclear organization of the histone gene regulator p220(NPAT) in human embryonic stem cells. *J Cell Physiol*. 2007;213:9–17.

25. Ghule PN, Dominski Z, Yang XC, et al. Staged assembly of histone gene expression machinery at subnuclear foci in the abbreviated cell cycle of human embryonic stem cells. *Proc Natl Acad Sci USA*. 2008;105:16964–16969.

26. Nichols J, Zevnik B, Anastassiadis K, et al. Formation of pluripotent stem cells in the mammalian embryo depends on the POU transcription factor Oct4. *Cell*. 1998;95:379–391.

27. Chambers I, Colby D, Robertson M, et al. Functional expression cloning of Nanog, a pluripotency sustaining factor in embryonic stem cells. *Cell*. 2003;113:643–655.

28. Mitsui K, Tokuzawa Y, Itoh H, et al. The homeoprotein Nanog is required for maintenance of pluripotency in mouse epiblast and ES cells. *Cell*. 2003;113:631–642.

29. Lengner CJ, Welstead GG, Jaenisch R. The pluripotency regulator Oct4: a role in somatic stem cells? *Cell Cycle*. 2008;7:725–728.

30. Silva J, Smith A. Capturing pluripotency. *Cell*. 2008; 132:532–536.

31. Chambers I, Silva J, Colby D, et al. Nanog safeguards pluripotency and mediates germline development. *Nature*. 2007;450:1230–1234.

32. Chazaud C, Yamanaka Y, Pawson T, et al. Early lineage segregation between epiblast and primitive endoderm in mouse blastocysts through the Grb2-MAPK pathway. *Dev Cell*. 2006;10:615–624.

33. Enver T, Pera M, Peterson C, et al. Stem cell states, fates, and the rules of attraction. *Cell Stem Cell*. 2009;4:387–397.

34. Toyooka Y, Shimosato D, Murakami K, et al. Identification and characterization of subpopulations in undifferentiated ES cell culture. *Development*. 2008;135:909–918.

35. Hough SR, Laslett AL, Grimmond SB, et al. A continuum of cell states spans pluripotency and lineage commitment in human embryonic stem cells. *PLoS One*. 2009;4:e7708.

36. Reijo Pera RA, DeJonge C, Bossert N, et al. Gene expression profiles of human inner cell mass cells and embryonic stem cells. *Differentiation*. 2009;78:18–23.

37. Bernstein BE, Mikkelsen TS, Xie X, et al. A bivalent chromatin structure marks key developmental genes in embryonic stem cells. *Cell*. 2006;125:315–326.

38. Azuara V, Perry P, Sauer S, et al. Chromatin signatures of pluripotent cell lines. *Nat Cell Biol*. 2006;8:532–538.

39. Mikkelsen TS, Hanna J, Zhang X, et al. Dissecting direct reprogramming through integrative genomic analysis. *Nature*. 2008;454:49–55.

40. Wernig M, Meissner A, Foreman R, et al. In vitro reprogramming of fibroblasts into a pluripotent ES-cell-like state. *Nature*. 2007;448:318–324.

41. Suh MR, Lee Y, Kim JY, et al. Human embryonic stem cells express a unique set of microRNAs. *Dev Biol*. 2004; 270:488–498.

42. Houbaviy HB, Murray MF, Sharp PA. Embryonic stem cell-specific MicroRNAs. *Dev Cell*. 2003;5:351–358.

43. Kanellopoulou C, Muljo SA, Kung AL, et al. Dicer-deficient mouse embryonic stem cells are defective in differentiation and centromeric silencing. *Genes Dev*. 2005;19:489–501.

44. Murchison EP, Partridge JF, Tam OH, et al. Characterization of Dicer-deficient murine embryonic stem cells. *Proc Natl Acad Sci USA*. 2005;102:12135–12140.

45. Wang Y, Medvid R, Melton C, et al. DGCR8 is essential for microRNA biogenesis and silencing of embryonic stem cell self-renewal. *Nat Genet*. 2007;39:380–385.

46. Chen CZ, Li L, Lodish HF, et al. MicroRNAs modulate hematopoietic lineage differentiation. *Science*. 2004;303:83–86.

47. Hornstein E, Mansfield JH, Yekta S, et al. The microRNA miR-196 acts upstream of Hoxb8 and Shh in limb development. *Nature*. 2005;438:671–674.

48. Mansfield JH, Harfe BD, Nissen R, et al. MicroRNA-responsive "sensor" transgenes uncover Hox-like and other developmentally regulated patterns of vertebrate microRNA expression. *Nat Genet*. 2004;36:1079–1083.

49. Itskovitz-Eldor J, Schuldiner M, Karsenti D, et al. Differentiation of human embryonic stem cells into embryoid bodies compromising the three embryonic germ layers. *Mol Med*. 2000;6:88–95.

50. Trounson A. The production and directed differentiation of human embryonic stem cells. *Endocr Rev*. 2006;27:208–219.

51. Jaenisch R, Young R. Stem cells, the molecular circuitry of pluripotency and nuclear reprogramming. *Cell*. 2008;132:567–582.

52. Prokhorova TA, Harkness LM, Frandsen U, et al. Teratoma formation by human embryonic stem cells is site-dependent and enhanced by the presence of Matrigel. *Stem Cells Dev*. 2009;18:47–54.

53. Cooke MJ, Stojkovic M, Przyborski SA. Growth of teratomas derived from human pluripotent stem cells is influenced by the graft site. *Stem Cells Dev*. 2006;15:254–259.

54. Lowry WE, Richter L, Yachechko R, et al. Generation of human induced pluripotent stem cells from dermal fibroblasts. *Proc Natl Acad Sci USA*. 2008;105:2883–2888.

55. Takahashi K, Yamanaka S. Induction of pluripotent stem cells from mouse embryonic and adult fibroblast cultures by defined factors. *Cell*. 2006;126:663–676.

56. Chan EM, Ratanasirintrawoot S, Park IH, et al. Live cell imaging distinguishes bona fide human iPS cells from partially reprogrammed cells. *Nat Biotechnol*. 2009; 27:1033–1037.

57. Doi A, Park IH, Wen B, et al. Differential methylation of tissue- and cancer-specific CpG island shores distinguishes human induced pluripotent stem cells, embryonic stem cells and fibroblasts. *Nat Genet*. 2009;41:1350–1353.

58. Daley GQ, Lensch MW, Jaenisch R, et al. Broader implications of defining standards for the pluripotency of iPSCs. *Cell Stem Cell*. 2009;4:200–201; author reply 202.

59. Nakatsuji N, Nakajima F, Tokunaga K. HLA-haplotype banking and iPS cells. *Nat Biotechnol*. 2008;26:739–740.

CHARACTERIZATION OF HUMAN EMBRYONIC STEM CELLS BY IMMUNOFLUORESCENCE MICROSCOPY

13

Shirwin M. Pockwinse and Prachi N. Ghule

It is of vital importance to monitor and characterize human embryonic stem cells (hESCs) periodically to confirm that cells remain pluripotent and undifferentiated. *In situ* immunofluoresence (IF) microscopy is a method used to characterize and confirm hESC pluripotency by the detection and localization of specific proteins that are expressed in pluripotent cells. In addition, detection of proteins expressed in the early stages of differentiation can aid in determining the degree to which an hESC culture may be differentiating.

In IF microscopy, stem cell antigens are detected using specific primary antibodies that are then coupled with fluorochrome conjugated secondary antibodies and visualized on an epifluorescence microscope. This microscope combines the power of high-resolution optics and the ability to image molecules based on fluorescence emission. Multiple antigens can be visualized simultaneously using more than one antigen-specific primary antibody, which in turn can be detected by secondary antibodies tagged with different fluorochromes (e.g., FITC, Rhodamine).

The advantage of IF microscopy is that expression and localization of specific factors can be assayed within the context of an individual cell as well as the overall culture. For instance, with this technique, one can determine if low-level expression of a specific protein is due to its expression in a subset of cells or low expression throughout the culture. In addition, localization of proteins within the cell can be informative; for example, the pluripotency associated factor Oct4 begins to migrate from the nucleus to the cytoplasm in early differentiation.

Alternative methods used in concert or individually with IF to confirm that hESC cultures have retained their pluripotency are:

- Flow cytometry, a measure of a total population of cells to determine the percentage expressing hESC-specific pluripotency markers (See Chapter 14).
- Reverse transcription polymerase chain reaction (RT-PCR) analysis for gene expression of hESCs at the RNA level (See Chapter 15).

Overview

Preparation of hESCs for IF

Fixation and processing of hESCs for IF

Human Stem Cell Technology and Biology, edited by Stein, Borowski, Luong, Shi, Smith, and Vazquez
Copyright © 2011 Wiley-Blackwell.

Staining of hESCs for IF

Visualization and data analysis

General Notes

- Read the entire chapter before initiating cell culture activities described in Procedure.

- It is good laboratory practice to have important relevant information recorded; therefore a log sheet is provided as an example at the end of this chapter.

Abbreviations

BSA: bovine serum albumin

hESC: human embryonic stem cell

MEFs: mouse embryonic fibroblasts

μg: microgram

mg: milligram

μL: microliter

mL: milliliter

μm: micrometer

PBS: phosphate buffered saline

PBSA: bovine serum albumin in phosphate buffered saline

RT-PCR: reverse transcription polymerase chain reaction

v/v: volume/volume

w/v: weight/volume

Equipment

- Fluorescence microscope with digital camera and associated image acquisition and data analysis software

- Humidity chamber (plastic container with dampened paper towels)

- Inverted microscope

- Microcentrifuge

- Oven set at 37°C

- Timer

Materials

- 22-mm × 22-mm No. 1 Sterile glass coverslips

- 25-mm × 74-mm × 1.0-mm Microscope slides

- Six-well plates of hESCs, 3–5 days after plating with one coverslip per well (See Chapter 9). (*This protocol may be viewed on the accompanying DVD and on the web at* http://www.wiley.com/go/stein/human.)

- Absorbent paper towels (or Kimwipes®)

- Aluminum foil

- Disposable gloves

- Ice bucket

- Microcentrifuge tubes (1.5 mL)

- Micropipettors P2 (0.2 to 2 μL), P20 (0.5 to 20 μL), P200 (20 to 200 μL), P1000 (200 to 1000 μL)

- Nail polish

- Pointed dissection forceps

- Parafilm
- Sterile, micropipettor filtered tips to fit P2, P20, P200, and P1000 micropipettors

Reagents

- 0.1% Gelatin solution (See Chapter 5)
- 37% Formaldehyde (Fisher Sci # F79-500)
- 4′,6-Diamidino-2-phenylindole, dihydrochloride (DAPI) (Sigma # D9542)
- BSA (Sigma # A3294-100G)
- Mounting medium/antifade reagent (Invitrogen Prolong Gold® # P36930)
- PBS
- Primary antibodies
- Secondary antibodies
- Triton® X-100 (Sigma # X100-500 mL)

Reagent Preparations

- Gelatin coated coverslips (see Preparation step 1).
- *Fixative solution:* Dilute 37% formaldehyde 1:10 in PBS (v/v) to make a final concentration of 37% formaldehyde.
- *Triton® X-100:* Make a stock solution of 10% in distilled water (v/v) and adjust to final dilution.
- *Permeabilization solution:* 0.25% Triton X-100 in Dilute 10% Triton X-100 1:40 in PBS (v/v) to make a concentration of 0.25% Triton X-100.
- *Phosphate buffered saline with bovine serum albumin (PBSA):* Dissolve BSA 0.5 g/100 mL in PBS (w/v) and filter through a 0.2-μm bottletop filter before use.
- *Primary and secondary antibodies:* Depending on the purity and specificity of the antibodies and for optimum dilutions, test ranges of dilutions between 1:100 and 1:300 for primary antibodies and 1:500 to 1:800 for secondary antibodies.
- *DAPI stain:* Dilute 1 mg/mL DAPI stock 1:20,000 in PBSA. Final concentration equals 0.5 μg/mL (w/v).

Preparation

NOTES

1. Prepare gelatin-coated coverslips:

   Note: *Use gelatin-coated glass coverslips as glass does not autofluoresce and gelatin aids the adherence of cells to coverslips. Gelatin-coated coverslips can be stored in a biosafety cabinet for up to 1 week.*

   1.1. Pour 0.1% gelatin solution into a 250-mL beaker, add sufficient number of coverslips, cover beaker with aluminum foil, and autoclave (See Chapter 5).

   1.2. In a biosafety cabinet, pick up a coverslip using pointed dissecting forceps and place it in one well of a gelatin-coated six-well plate (See Chapter 7). (*This protocol may be viewed on the accompanying DVD and on the web at* http://www.wiley.com/go/stein/human.) Repeat as needed.

2. Culturing of hESCs on coverslips:

   2.1. Once coverslips have been gelatin coated and placed in appropriate wells, plate with inactivated MEFs (See Chapter 7).

NOTES

   2.2. Plate hESCs on coverslips with inactivated MEFs (See Chapter 9). (*This protocol may be viewed on the accompanying DVD and on the web at* http://www.wiley.com/go/stein/human.)

   2.3. Allow hESCs to grow for 3–5 days (~50% confluence) before fixing and staining for IF.

> ***Note:*** *It is important to use hESCs before they exceed 50% confluence or they tend to peel off the coverslip during IF staining procedure.*
>
> *Successful fixation preserves cell structure and antigen epitope recognition, which in turn reveals protein distribution in situ.*

3. Prepare reagents and primary antibody dilutions:

   3.1. Prepare a humidity chamber for antibody incubation by placing dampened paper towels on the bottom of a plastic container.

   3.2. Label six-well culture plate lids with sample information—cell type, permeabilized or nonpermeabilized, antibody, and antibody dilutions: for example, H9p30, permeabilized (up to 20 minutes) Oct4 1:100, donkey anti-goat 1:800.

   3.3. Cover cell culture plate lids with Parafilm.

   3.4. Keep all reagents and primary antibody dilutions on ice.

   3.5. Prepare fixative solution: dilute 37% formaldehyde 1:10 in PBS to make a final concentration of 3.7% formaldehyde.

> ***Note:*** *Fixative solution must be prepared fresh for each IF staining.*

   3.6. Prepare permeabilizing solution: 0.25% Triton X-100 in PBS.

   3.7. Prepare PBSA solution: 0.5% bovine serum albumin (BSA) in PBS and filter through a 0.2-μm bottletop filter, before use.

   3.8. Prepare optimal primary antibody dilutions in PBSA for each antibody and keep on ice.

   3.9. Refer to Table 13.1 for a list of suggested primary antibodies of different pluripotency markers that have been successfully used in many laboratories. Different lots of antibodies and the same antibody from different vendors should be optimized for a specific staining procedure.

TABLE 13.1 SUGGESTED ANTIBODY REAGENTS FOR IMMUNOFLUORESCENCE STAINING

Primary Antibodies	Pluripotency Marker	Company/ Catalog Number	Dilution Used	Cellular Localization	Permeabilization Conditions
Oct4	Yes	Santa Cruz SC-8628	1:100	Nuclear	Yes, 20 minutes
SSEA3	Yes	Chemicon MAB4303	1:100	Surface marker	No
SSEA4	Yes	Chemicon MAB4304	1:100	Surface marker	No
TRA-1–60	Yes	Chemicon MABA4360	1:100	Surface marker	Yes, 5 minutes

Secondary Antibodies	Company/Catalog Number Invitrogen AlexaFluor ® (Green)		Dilution Used for Primary Antibodies
Donkey anti-goat	A11055	1:800	Oct4
Goat anti-mouse	A11001	1:800	SSEA1, SSEA4, and TRA-1–60
Goat anti-rat	A21212	1:800	SSEA3

Procedure

NOTES

1. Immunocytochemical staining of hESCs:

    Note: *Steps 1.1–1.6 are performed on ice.*

    1.1. Add 2 mL of cold PBS per well to wash cells. Aspirate after each wash.

    Note: *Gently add and aspirate solutions from side of well to minimize cell disturbance.*

    1.2. *Fixation:* Add 2 mL/well of 3.7% formaldehyde in PBS. Fix for 10 minutes. Aspirate the fixative.

    1.3. *Wash:* Wash cells once with 2 mL PBS. Aspirate the PBS.

    1.4. *Permeabilization:* Go to step 1.6 if cells are not being permeabilized. For samples that need permeabilization, add 2 mL of permeabilization solution per well, for up to 20 minutes. Time can vary with antibody (e.g., permeabilize Oct4 for 20 minutes). Aspirate permeabilization solution.

    Note: *Permeabilization, which perforates the cell membrane, allowing antibodies access to the inside of the cell, is required for detection of antigens that are inside the cell, such as transcription factors. However, for detection of cell surface antigens, do not permeabilize the hESCs as permeabilization can remove cell surface epitopes.*

    1.5. Wash two times with 2 mL PBS.

    1.6. Add 2 mL PBSA to each well for 20 minutes to block and/or reduce nonspecific binding of the primary antibody to decrease background signal. This incubation can last up to 2 hours.

    1.7. In preparation for antibody staining, carefully remove each coverslip from its well using forceps (Fig. 13.1a). Place each coverslip, cell side up,

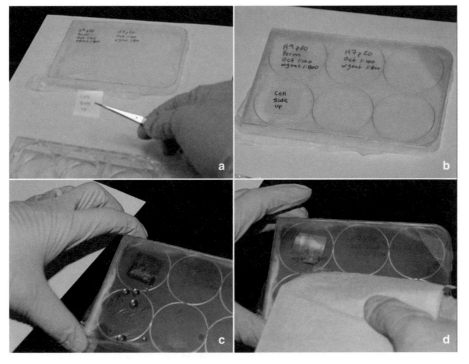

FIGURE 13.1. *Handling and washing coverslips. (a) Using pointed forceps, (b) place coverslip with cell side up on labeled plate lid; (c) to wash coverslips, tilt lid so that solution flows off them, and (d) remove residual solution after last wash by touching edge with Kimwipe.*

NOTES

_____
_____
_____
_____
_____
_____
_____
_____
_____
_____
_____
_____
_____
_____
_____
_____
_____
_____
_____
_____
_____
_____
_____
_____

on its labeled well on top of six-well plate lid that has been covered with Parafilm (Fig. 13.1b).

Note: *Now that coverslip is on plate lid, carefully add primary antibody solutions to the side of coverslip and allow solutions to flow across the surface of the coverslip. The solution will adhere to the coverslip by capillary action.*

1.8. *Primary antibody:* Add 200 µL of diluted primary antibody to the coverslip. The negative control will be 200 µL of PBSA without antibody.

1.9. Incubate for 1 hour in a humidity chamber inside a 37°C oven.

1.10. While primary antibodies are incubating, prepare secondary antibody dilutions in PBSA. Prepare 200 µL of antibody dilution per coverslip. Keep on ice and cover tubes with aluminum foil.

1.11. *Wash:* Return six-well plate lid with coverslip to ice. Gradually tilt lid so that primary antibody solution flows off coverslips and onto the lid (Fig. 13.1c). Add 200–300 µL PBSA to each coverslip and tilt lid to remove PBSA. Repeat this three more times. After the last wash, carefully remove remaining PBSA by touching the edge of the coverslip with a Kimwipe® (Fig. 13.1d).

1.12. *Secondary antibody:* Add 200 µL of the diluted appropriate secondary antibody to each coverslip and transfer the plate lid to the humidity chamber. Cover the chamber with aluminum foil and incubate for 1 hour just inside a 37°C oven.

Note: *When using secondary antibodies, always be conscious of light levels and cover the sample as much as possible. Exposure to light will cause the fluorochrome on the secondary antibody to fade.*

1.13. While secondary antibodies are incubating, prepare slides, wiping both sides of each slide with ethanol. Label each slide with the same information listed on each well of the plate as well as primary and secondary antibody and fluorochrome information. Date and initial each slide.

1.14. *Wash:* Return six-well plate lid with coverslips to ice. Wash with PBSA four times as described in step 1.11 to remove unbound secondary antibody.

1.15. *Nucleus staining:* Dilute DAPI 1:20,000 to 0.5 µg/mL in PBSA. (DAPI stock is 1 mg/mL.) Stain with 200 µL/coverslip of DAPI in PBSA. Incubate for 5 minutes in the dark. Wash once with PBSA, and then wash twice with PBS.

Note: *DAPI is a fluorescent stain that binds to DNA. When a cell is stained with DAPI, the cell nucleus fluoresces blue under UV light.*

1.16. *Mount coverslips:* Place a drop of mounting medium/antifade reagent or other fluorescence mounting medium in the middle of the correctly labeled slide. With forceps, hold coverslip, cell surface down, at an angle to the slide over the mounting medium; drop and lower gently onto the slide. Gently blot excess mounting medium with a Kimwipe.

1.17. Place three dots of nail polish equidistant on coverslip edge to adhere the coverslip to the slide and dry in the dark for 15 minutes.

Note: *It is recommended to use either colorless or light colored nail polish as dark colors (e.g., red) may produce autofluorescence under the epifluorescence microscope.*

1.18. *Seal:* Paint around the circumference of the coverslip with nail polish. Dry the sealed coverslip in the dark for 1 hour. This prevents mounting medium from mixing with the immersion oil while observing on the microscope and prevents drying out of mounting medium.

1.19. Store slides in a −20°C freezer.

2. Slide analysis using an epifluorescence microscope:

    Note: Epifluorescence microscope must be equipped with the appropriate filter sets to detect the wavelengths of light emitted. Record all necessary information for each experiment or slide preparation.

    2.1. Guidelines for identifying unique characteristics of hESC cultures by IF. The following must be noted: density of mouse feeder cells—sparse, dense, or adequate. If mitotic cells are observed in feeder layer, discard culture. Scan the entire slide to look for uniformity and intensity of the staining for any particular marker. Also note the morphology of hESC colonies and document the following characteristics:
        - The size of the colonies: small, intermediate, or large.
        - Overall percentage of undifferentiated versus differentiated colonies.
        - Apoptosis status: typically, apoptotic cells will show extensive blebbing of nuclear lamina. If extensive apoptosis is observed, discard culture.

    2.2. Record the percentage and localization of differentiation in hESC colonies, which can be observed either at the periphery or in the center of the colony. Differentiation can be identified as indicated by negative staining of any of the pluripotency markers such as Oct4 (Fig. 13.2a), SSEA3/4 (Fig. 13.2b), or TRA-1-60. If differentiation is more than 15–20%, manually passage the culture to remove the differentiated cells (See Chapter 9).

    2.3. Differentiation is also identified using markers (e.g., SSEA1) that specifically stain differentiated hESC cells (Fig. 13.2c).

    2.4. When analyzing the staining for pluripotency markers, record the following characteristics for each protein/antibody:
        - Cellular distribution of the protein if nuclear/cytoplasmic/cytoskeletal.
        - Intensity of staining. Intensity can be measured as—(negative), + (faint/weak), ++ (strong), or +++(very strong).
        - Percentage of positively stained cells per colony and percentage of positively stained colonies per slide/coverslip.

NOTES

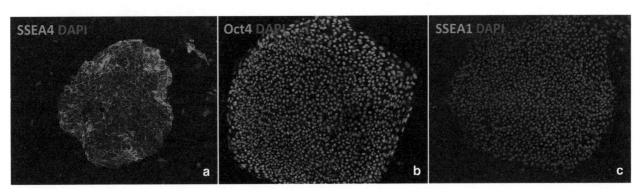

FIGURE 13.2. *Different markers for characterization of undifferentiated state of human embryonic stem cells are analyzed using immunofluorescence. H9 cells are stained for (a) SSEA4 (green), (b) Oct4 (green), and (c) SSEA1 (green). All cells are counterstained by DAPI (blue).*

2.5. Proper recording of above-mentioned guidelines will aid in successful long-term maintenance of hESC culture.

SUGGESTED READING

Harlow EW, Lane D. *Using Antibodies, A Laboratory Manual*. Cold Spring Harbor, NY: Cold Spring Harbor Laboratory Press; 1999.

International Stem Cell Forum; International Stem Cell Initiative (ISCI) project. Available at http://www.stemcellforum.org/isci_project.cfm.

Javois LC *Immunocytochemical Methods and Protocols. Methods in Molecular Biology*, Volume 115. Totowa NJ: Humana Press; 1999.

Massachusetts Human Stem Cell Bank Standard Operating Procedures, 2009. Available at http://www.umassmed.edu/iscr.

Stein GS, Montecino M, McNeil S, Pockwinse S, Van Wijnen A, Stein JL, Lian JB. Isolation and visualization of the principal components of nuclear architecture. In: Studzinski GP, ed. *Cell Growth, Differentiation and Senescence: A Practical Approach*. Oxford, England: Oxford University Press; 1999:177–208.

WiCell Research Institute. Introduction to Human Embryonic Stem Cell Culture Methods, September 2007.

IMMUNOFLUORESCENCE LOG SHEET

Cell line name and passage number: _____

Coverslips autoclaved in 0.1% gelatin on: _____ By: _____

Inactivated MEFs seeded on: _____ By: _____

hESCs seeded on: _____ By: _____

Immunofluorescence microscopy performed on: _____ By: _____

Images taken on date: _____ By: _____

Saved to: _____

Antibody	Target	Localization	IF %
Oct4	(Nuclear) Pluripotency	_____	_____
SSEA3	(Surface) Pluripotency	_____	_____
SSEA4	(Surface) Pluripotency	_____	_____
TRA-1-60	(Surface) Pluripotency	_____	_____
SSEA1	(Surface) Differentiated stem cells	_____	_____

Overall analysis: _____

_____

Slides stored in: _____

Initials: _____ Date: _____

PREPARATION OF HUMAN EMBRYONIC STEM CELL SAMPLES FOR FLOW CYTOMETRY

14

Meng-Jiao Shi, Kimberly Stencel, and Maria Borowski

Characterization is an important component of the long-term maintenance of human embryonic stem cells (hESCs) in culture. For many downstream applications, it is crucial that hESCs remain undifferentiated. In this protocol, flow cytometry is used to measure the percentage of differentiated and undifferentiated cells in a population of hESCs. With this technique, hESC colonies are first disaggregated into single cell populations. The cell suspension is then incubated with directly conjugated fluorescent antibodies that bind to specific antigens of interest located on the surface or inside the cells. Once the antibodies have labeled the cell suspension, it is passed through the flow cytometer past a beam of light from either a lamp or laser. Two detectors are aimed at the point where the stream of cells passes through the light beam: the forward scatter (FSC) detector and the side scatter (SSC) detector. The FSC detector measures cell volume and is aligned with the light beam while the SSC detector measures the inner portion of the cell and is perpendicular to the light beam. Both of these detectors recognize each of the antibodies used for labeling the sample. Each fluorescently labeled cell passes through the beam and emits light, contributing to data collected; while cells that have not been labeled will not emit light and therefore no information can be gathered. A combination of scattered and fluorescent light is recognized by the detectors, then analyzed by the fluctuations in brightness. This information is then plotted by individual physical and/or chemical structures on or in each cell.

Typically, several different antigens of interest are chosen per sample. It is suggested to choose antigens specific to undifferentiated hESCs but also antigens indicating differentiating hESCs in order to accurately assess the population. Antigens presenting on undifferentiated hESCs include the surface markers SSEA3 and SSEA4, and intracellular transcription factors, Oct4 and Nanog. Human embryonic stem cells that have differentiated express the surface marker SSEA1. In addition, an antibody against CD29, a surface antigen on fibroblasts, is included to accommodate hESCs that have been cultured on inactivated MEF feeder cells. The limitation of antigen combinations depends on the number of fluorochromes that the flow cytometer can detect per sample.

Flow cytometry can analyze several thousand cells per second and thus is a relatively quick and efficient way to evaluate an entire population of hESCs. Another

Human Stem Cell Technology and Biology, edited by Stein, Borowski, Luong, Shi, Smith, and Vazquez
Copyright © 2011 Wiley-Blackwell.

method of characterization of hESC populations is immunofluorescence microscopy (IF). Immunofluorescence microscopy can provide information about the number and location of cells in a colony that are labeled and the localization of the antigen in the cell. However, this technique cannot provide accurate quantitation of the number of labeled cells in an entire population.

Overview

Harvest cells.

Stain and fix cells.

Do flow analysis.

Do data analysis.

General Notes

- Read the entire chapter before initiating activities described in Procedure.
- It is good laboratory practice to have important relevant information recorded; therefore a sample log sheet is provided at the end of the chapter.
- Antibody sources provided here serve as an example. Other sources may be used.

Abbreviations

APC: allophycocyanin

Ca: calcium

CD29: integrin b1 chain

CSF: carboxyfluorescein

EDTA: ethylenediaminetetraacetic acid

FB: fixation buffer

FBS: fetal bovine serum

hESCs: human embryonic stem cells

HI-FBS: heat-inactivated fetal bovine serum

IF: immunofluorescence microscopy

MEF: mouse embryonic fibroblast

Mg: magnesium

μm: micrometer

μL: microliter

mL: milliliter

mm: millimeter

nm: nanometer

PBS: phosphate buffered saline

PE: phycoerythrin

SSEA1: stage-specific embryonic antigen-1

SSEA4: stage-specific embryonic antigen-4

WB: wash buffer

Equipment

- Flow cytometer with 488-nm and 633-nm detectors, or access to a flow cytometry facility
- Hemacytometer or other cell counter
- Inverted light microscope

- Tabletop centrifuge
- Vortex

Materials

- 5-mL and 10-mL Sterile glass or plastic pipettes
- 12-mm × 75-mm Round bottom tubes
- 15-mL and 50-mL Sterile conical tubes
- 20-μL and 200-μL Pipettors and pipette tips
- 40-μm Cell strainer
- Alcohol-proof marker
- Pasteur pipettes

Reagents

- 1× PBS (Dulbecco's phosphate buffered saline, Ca/Mg-free)
- HI-FBS
- 1× Trypsin-EDTA (0.05%)
- Inactivated MEF medium or any medium with 10% FBS (see Chapter 5)
- HESC culture medium (see Chapter 5)
- 16% Methanol-free formaldehyde stock solution (Thermo Fisher Scientific, Cat #28908)
- Antibodies (Table 14.1)

Preparation

1. 1× Wash buffer (1× WB):

    1.1. Prepare 100 mL of 1× WB by adding 2 mL of fetal calf serum to 98 mL of 1× PBS.

    1.2. Store the 1× WB at 4°C. It is stable up to 1 month at 4°C.

2. 1× Fixation buffer (1× FB):

    2.1. To make final 2% formaldehyde solution, add 2 mL of methanol-free formaldehyde (16%) to 14 mL of PBS. Make fresh and keep at room temperature.

NOTES

_____

_____

_____

_____

_____

TABLE 14.1 EXAMPLE ANTIBODY SOURCES

Fluorochrome Conjugated Antibody	Isotype	Target Population	Vendor	Catalog Number
CD29-APC	Armenian hamster IgG	Inactivated MEF	eBioscience BD	17-0291
Isotype control IgG-APC	Armenian hamster IgG	Inactivated MEF	eBioscience BD	17-4888
SSEA4-CSF	Mouse IgG$_3$	Pluripotent stem cells	R&D Systems	FAB1435F
Isotype control, IgG$_3$-CSF	Mouse IgG$_3$	Pluripotent stem cells	R&D Systems	IC007F
SSEA1-PE	Mouse IgM	Differentiating cells	eBioscience BD	12-8813-73
Isotype control IgM-PE	Mouse IgM	Differentiating cells	EBioscience BD	12-4752-73

NOTES

_____
_____
_____
_____
_____
_____
_____
_____
_____
_____
_____
_____
_____
_____
_____
_____
_____
_____
_____
_____
_____
_____
_____
_____
_____
_____
_____

2.2. Discard 1× FB after assay.

3. Place the following items near the biosafety cabinet:
   - 5-mL and 10-mL Sterile pipettes.
   - 70% Ethanol spray
   - Absorbent paper towels (or Kimwipes®)
   - Disposable gloves

4. Ensure that the following items are in the biosafety cabinet:
   - 5-mL Sterile pipettes
   - 10-mL and 25-mL Sterile pipettes
   - 15-mL and 50-mL Sterile conical tubes
   - Alcohol-proof marker
   - Sterile Pasteur pipettes

Procedure

1. Follow preparations according to Chapter 3, Aseptic Technique on page 19. (*This protocol may be viewed on the accompanying DVD and on the web at* http://www.wiley.com/go/stein/human.)

2. Detach and count cells:

   Note: *The steps below apply to six-well culture plates. If cells are cultured in other culture vessels, adjust the reagent volumes proportionally.*

   It is possible to harvest (1–2) × 10^6 hESCs from one well of a six-well plate when colony confluence is between 60% and 70%. A minimum of 3 × 10^6 cells is required for this procedure.

   2.1. Aspirate the medium from culture wells.

   2.2. Rinse wells twice with 3 mL/well of 1× PBS.

   2.3. Aspirate PBS from wells.

   2.4. Add 1 mL of 1× trypsin-EDTA to each well.

   2.5. Incubate at 37°C for 2–4 minutes. Confirm by microscopy that cells are detached and beginning to separate from each other into smaller aggregates.

   2.6. Using a 5-mL pipette, gently scrape cells from culture surface and pipette up and down to break up the cell clumps.

   2.7. To inactivate the trypsin enzyme, add 2 mL MEF culture medium to each well within 5–6 minutes after adding the trypsin-EDTA solution. Pool the detached cells into one 15-mL or 50-mL tube.

   2.8. If needed, add more hESC culture medium to bring the total liquid volume to 20 mL.

   2.9. Pass the cells through the center of a 40-μm cell strainer by gravity flow.

   2.10. Rinse the top of the cell strainer reservoir with 5 mL of hESC culture medium.

   2.11. Centrifuge the tube for 5–7 minutes at 200g.

   2.12. Resuspend the cell pellet in approximately 1 mL of 1× wash buffer per well.

   2.13. Count cells (see Chapter 3, Cell Counting Using a Hemacytometer on page 26).

   2.14. Using 1× wash buffer, adjust the volume to achieve a concentration of 3 × 10^6 cells/mL.

   Note: *If cells are too dilute, centrifuge and resuspend in an appropriate volume of 1× wash buffer.*

3. Stain and fix cells:

   Note: At a minimum, each assay should include two negative controls, as well as one antibody specific for inactivated MEFs, one antibody to detect differentiated hESCs, and one for undifferentiated hESCs (see Table 14.3).

   3.1. Label six 12-mm × 75-mm round bottom tubes as 1 to 6 sequentially.

   3.2. To each tube except #1, add 100 μL of the cell suspension (about 3×10^5 cells/tube).

   3.3. Add the rest of the cells to tube #1.

   3.4. To each tube except #1, add 10 μL of antibody according to Table 14.2.

   3.5. Incubate for 30 minutes at 4–8°C in the dark (e.g., wrap the tubes in foil).

   3.6. Add 2 mL of 1× wash buffer to each tube, and centrifuge for 5 minutes at 200g.

   3.7. Decant the supernatant and resuspend each cell pellet in 100 μL of 1× FB. Vortex and incubate the cells at room temperature for 30 minutes in the dark.

   3.8. Add 2 mL of 1× PBS to each tube, and centrifuge for 5 minutes at 200g.

   3.9. Decant the supernatant and resuspend each cell pellet in 300 μL of 1× wash buffer.

   3.10. Submit the samples to a flow cytometry facility or store the samples at 4–8°C until ready for analysis.

   Note: Analysis should be completed within 1 week.

4. Data acquisition and analysis:

   4.1. From the flow readout, summarize the results and record the percentage of cells appropriate to each of the subpopulations representing MEF (CD29$^+$SSEA4$^-$SSEA1$^-$), pluripotent stem cells (CD29$^-$SSEA4$^+$SSEA1$^-$), and differentiating stem cells (CD29$^-$SSEA4$^-$SSEA1$^+$), respectively, as shown in the Table 14.3.

   Note: A good quality undifferentiated culture should have less than 5% differentiated cells.

NOTES

TABLE 14.2 DATA AQUISITION FOR FLOW CYTOMETRY

Tube	Purpose	Antibodies Added
1	Baseline (autofluorescence of cells); negative control	
2	Nonspecific binding antibody; negative	IgG-APC + IgG-CSF + IgM-PE
3	Label MEF; CD29	CD29-APC + IgG-CSF + IgM-PE
4	Label hESC; SSEA4	IgG-APC + SSEA4-CSF + IgM-PE
5	Differentiating cells; SSEA1	IgG-APC + IgG-CSF + SSEA1-PE
6	Multiple labeling; all three markers	D29-APC + SSEA4-CSF + SSEA1-PE

TABLE 14.3 DATA ANALYSIS FOR FLOW CYTOMETRY

Marker	Target Population	Actual Positive (%)
CD29$^+$	Inactivated MEF	
SSEA4$^+$	Pluripotent stem cells	
SSEA1$^+$	Differentiating cells	

SUGGESTED READING

Brandenberger R, Wei H, Zhang S, et al. Transcriptome characterization elucidates signaling networks that control human ES cell growth and differentiation. *Nat Biotechnol*. 2004;22:707–716.

Draper JS, Pigott C, Thomson JA, Andrews PW. Surface antigens of human embryonic stem cells: changes upon differentiation in culture. *J Anat*. 2002;200:249–258.

Hoffman LM, Carpenter MK. Characterization and culture of human embryonic stem cells. *Nat Biotechnol*. 2005;23:699–708.

Hyslop L, Stojkovik M, Armstrong L, et al. Downregulation of NANOG induces differentiation of human embryonic stem cells to extraembryonic lineages. *Stem Cells*. 2005;23:1035–1043.

Kannagi R, Cochran NA, Ishigami F, et al. Stage-specific embryonic antigens (SSEA-3 and -4) are epitopes of a unique globoseries ganglioside isolated from human teratocarcinoam cells. *EMBO J*. 1983;2:2355–2361.

Laslett AL, Filipczyk AA, Pera MF. Characterization and culture of human embryonic stem cells. *Trends Cardiovasc Med*. 2003;13:295–301.

Niwa H, Miyazaki J, Smith AG. Quatitative expression of Oct-3/4 defines differentiation, dedifferentiation or self-renewal of ES cells. *Nat Genet*. 2000;24:372–376.

Pera MF, Filipczyk AA, Hawes SM, Laslett AL. Isolation, characterization, and differentiation of human embryonic stem cells. *Methods Enzymol*. 2003;365:429–446.

Sato N, Sanjuan IM, Heke M, Uchida M, Naef F, Brivanlou AH. Molecular signature of human embryonic stem cells and its comparison with the mouse. *Dev Biol*. 2003;260:404–413.

Sidhu KS, Tuch BE. Derivation of three clones from human embryonic stem cell lines by FACS sorting and their characterization. *Stem Cells Dev*. 2006;15:61–69.

Sperger JM, Chen X, Draper JS, et al. Gene expression patterns in human embryonic stem cells and human pluripotent germ cell tumors. *Proc Natl Acad Sci USA*. 2003;100:13350–13355.

FLOW DATA LOG SHEET

- Cell ID: _____   Passage number: _____

- Staining done by: _____   Date: _____

- Analysis done by: _____   Date: _____

- Test results:

Marker	Target Population	Positive (%)
CD29$^+$	MEF	_____
SSEA4$^+$	Pluripotent stem cells	_____
SSEA1$^+$	Differentiating cells	_____

Comments and decisions made regarding the test sample:

_____

_____

_____

_____

_____

_____

CHARACTERIZATION OF HUMAN EMBRYONIC STEM CELL GENE EXPRESSION USING REVERSE TRANSCRIPTION POLYMERASE CHAIN REACTION

15

Alicia Allaire, Kelly P. Smith, and Mai X. Luong

In each cell nucleus, genes encoded by deoxyribonucleic acid (DNA) provide a blueprint for the entire organism. Gene expression involves the transcription of genes into ribonucleic acids (RNAs), which are subsequently translated into proteins. Each cell type expresses only a subset of these proteins, which are required to carry out the functions specific to that cell type.

Human embryonic stem cells (hESCs) in culture must be continuously monitored to ensure that they remain largely undifferentiated and retain pluripotency. This can be accomplished by assaying the expression levels of numerous genes that have been linked to the undifferentiated state. Gene expression in hESCs can be monitored by examining the levels of proteins or RNAs produced by these genes. Proteins can be detected by Western blot analysis or immunofluorescence staining using protein-specific antibodies. These tagged proteins can then be assayed by fluorescence microscopy (see Chapter 13) or flow cytometry (see Chapter 14). While quantitative assessment of protein levels is a measure of gene expression, factors such as the rate of protein turnover can affect the results. Thus in conjunction with protein analysis, measurement of messenger RNA (mRNA) is often used to determine pluripotency. However, it should be noted that analysis of mRNA levels may not accurately reflect gene transcription since factors such as mRNA processing and degradation can influence mRNA levels in the cell.

In the past, mRNA levels were most commonly measured by Northern blot analysis. In this procedure, RNA molecules are extracted from cells and separated by size using gel electrophoresis. The RNA on the gel is transferred to a membrane filter and specific RNAs are detected using labeled RNA or DNA probes. However, the development of the reverse transcription polymerase chain reaction (RT-PCR) assay revolutionized the field by making detection faster, easier, and more sensitive, while avoiding the radioactive probes used in Northern blot analysis. In RT-PCR, short oligonucleotides (random hexamers or oligo dT primers) bind to a particular region of a specific RNA molecule and prime it for processing by reverse transcriptase, an enzyme that copies RNA to DNA (Fig. 15.1). The resulting complementary DNA (cDNA) is then amplified using another set of gene-specific primers. Amplification

Human Stem Cell Technology and Biology, edited by Stein, Borowski, Luong, Shi, Smith, and Vazquez
Copyright © 2011 Wiley-Blackwell.

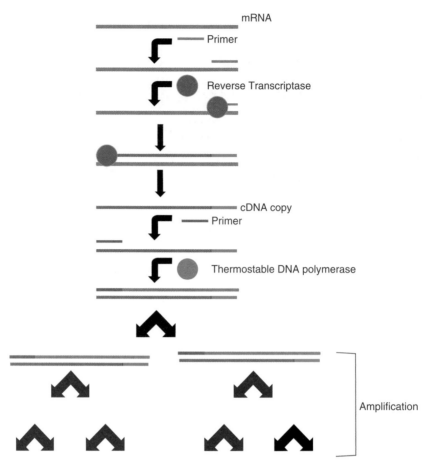

FIGURE 15.1. *This diagram shows the reverse transcription of an RNA sample into a cDNA copy and subsequent amplification of the cDNA by polymerase chain reaction. Either nonspecific hexamer primers or oligo dT primers (red bars) are used to prime the production of the cDNA by reverse transcriptase. The cDNA copy is then amplified using a set of primers (green bars) specific to the mRNA being assayed. These primers bind to target sequences in the cDNA at specific temperatures and begin synthesis of double-stranded copies of the cDNA by a thermostable DNA polymerase. The PCR cycle of DNA strand separation, primer annealing, and polymerase elongation are carried out multiple times (usually 30–40 cycles) to amplify many copies of cDNA.*

is accomplished by heating and cooling the sample in cycles. Heating the sample denatures double-stranded RNA:DNA or DNA:DNA molecules. As the sample is cooled, the specific primers bind the cDNA and a heat resistant DNA polymerase produces a complementary strand starting at the primer. As the cycles are repeated, the cDNA copies are copied again, amplifying the target to produce millions of copies. These products are then run on an agarose gel. Nucleic acid bands are commonly detected using nonradioactive agents (such as ethidium bromide) that are incorporated into the DNA in a sequence-nonspecific manner. This PCR technique is far more sensitive than Northern blot analysis and is technically capable of detecting a single RNA molecule. However, accurate quantification of RNA levels using traditional end-point RT-PCR reactions can be difficult.

PCR amplification occurs in three phases: exponential, linear, and plateau. In the exponential phase, primers and other reagents are in excess, so there is precise doubling of products with each cycle. Linear phase marks the point at which reagents are being consumed and the reaction slows. This leads to significant variability of results. In the plateau phase, the reactions have stopped and some product degradation can occur. Thus for accurate comparison of relative RNA levels in different samples, it is necessary to collect quantitative data at a point in which every sample is in the exponential phase (when amplification is extremely reproducible). This requires a validation experiment to determine the ideal cycle for each gene of interest.

A modification of RT-PCR, termed quantitative reverse transcription polymerase chain reaction (qRT-PCR) has made quantification more accurate and faster. In qRT-PCR, fluorescent probes are used to measure the amount of product after each cycle, thus allowing measurements of the exponential phase of the reactions. Quantitative RT-PCR requires a multifunctional instrument to perform the thermal cycling, fluorescence detection, and data analysis of the products.

Overview

Section I.   Preparing RNA Using RNeasy® Plus Mini Kit

Section II.   Quantifying RNA using NanoDrop

Section III. cDNA Synthesis Using Reverse Transcriptase

Section IV.   PCR

Section V.   Analysis of PCR Products by Agarose Gel Electrophoresis

Section VI.   qRT-PCR

Abbreviations

cDNA: complementary DNA

DNA: deoxyribonucleic acid

EDTA: ethylenediaminetetraacetic acid

gDNA: genomic deoxyribonucleic acid

mRNA: messenger RNA

μg: microgram

μL: microliter

mL: milliliter

M: molar

mM: millimolar

ng: nanogram

OD: optical density

PCR: polymerase chain reaction

qRT-PCR: quantitative reverse transcription polymerase chain reaction

RNA: ribonucleic acid

RT: reverse transcription

RT-PCR: reverse transcription polymerase chain reaction

TE buffer: tris(hydroxymethyl)aminomethane EDTA buffer

General Notes

- **Cells should be harvested at approximately 70% confluence.**
- **All protocols are performed at the bench unless stated otherwise**.
- RNA is very unstable, so the following precautions must be observed when working with it.
  - Keep samples out of direct sunlight.
  - Never vortex RNA.
  - Always wear gloves while handling reagents and RNA samples.

- Change gloves frequently and spray gloves, pipettes, and surfaces with 70% ethanol and RNaseZap® regularly.
- Work quickly and keep RNA on ice to reduce the chance of degradation by ribonucleases (RNases).
- Use proper aseptic technique. Hands and dust particles may carry bacteria and RNases.
- Handle tubes as little as possible and avoid touching the inside of the lids.
- Keep tubes closed when not in use.
- Work with sterile disposable polypropylene tubes. Buy tubes that are certified to be RNase-free. Use caution in keeping them sterile once the package is opened.
- Keep amplified PCR products and plasmids away from RNA samples.
- RNases are very stable enzymes that are difficult to inactivate and small amounts of RNases are sufficient to significantly degrade an RNA sample. Products such as RNaseZap® (Applied Biosystems) or RNase Away® (Molecular Bioproducts), which remove RNase contamination, are helpful in prepping equipment and workspace. Diethyl pyrocarbonate (DEPC) is an efficient inhibitor of RNases that may be used to treat water, plastics, glassware, and other solutions needed for working with RNA.

I. PREPARING RNA USING RNEASY® PLUS MINI KIT

RNA was first isolated and purified using phenol and chloroform. Although inexpensive, performing phenol chloroform extractions is a time-consuming procedure that involves the use of hazardous materials. In addition, it takes practice to yield quality RNA. In contrast, commercially available RNA purification kits provide a faster method and cleaner yield, even for those with less experience. When choosing a kit, it is important to consider not only price, but also the time required for extraction, and the quantity and purity of the RNA. Kits include all of the buffers and reagents needed, as well as silica-based nucleic acid affinity columns. In these kits, cells are lysed, homogenized, and loaded onto the column. RNA is retained on the column while contaminants are washed through and discarded. Lastly, the bound purified RNA is eluted into a clean tube. Typically, centrifugation is used at each step to accelerate the process.

When choosing a kit, it is best to select one that will produce RNA with the least amount of contaminants or inhibitors. Find a kit that works best for isolating and purifying RNA from hESCs in culture. Also, remember to consider the limitations of a kit such as the amount of RNA that a column can bind and the size of the RNA molecules that can be bound.

Overview

Lyse cells.

Homogenize cell lysate.

Eliminate genomic DNA from lysate.

Bind RNA in lysate to affinity column.

Wash column several times.

Elute RNA.

Equipment

- 37°C Water bath
- Chemical hood
- Microcentrifuge (room temperature)
- Micropipettors: P2 (0.2 to 2 μL), P20 (0.5 to 20 μL), P200 (20 to 200 μL), P1000 (200 to 1000 μL)
- Timer

Materials

- 15-mL Sterile conical tubes
- Absorbent paper towels (or Kimwipes®)
- Disposable gloves
- RNase-free 1.5-mL microcentrifuge tubes
- Small cell scraper
- Sterile micropipettor filtered tips to fit P2, P20, P200, and P1000 micropipettors

Reagents

- RNeasy® Plus Mini Kit (Qiagen 74134). This kit contains:
  - 50 gDNA eliminator mini spin columns (uncolored), each in a 2-mL collection tube
  - 50 RNeasy Mini spin columns (pink)
  - (50) 1.5-mL collection tubes

NOTES

- (50) 2-mL collection tubes
- 45 mL Buffer RLT Plus
- 45 mL Buffer RW1
- 11 mL Buffer RPE
- 10 mL RNase-free water
- Handbook
- RNaseZap® solution
- 4.3 M ß-Mercaptoethanol
- 96–100% Ethanol
- 70% Ethanol
- QIAshredder™ Homogenizer (Qiagen 79654)
- Spray bottle of 70% ethanol
- Six-well plates with hESCs at 70% confluence (for preparation of hESCs see Chapter 7, page 88). (*This protocol may be viewed on the accompanying DVD and on the web at* http://www.wiley.com/go/stein/human.)

Preparation

1. Prepare laboratory bench and equipment:

   1.1. Wash hands thoroughly with soap; rinse them completely with tap water; dry hands with paper towel, then put on appropriate-sized gloves.

   1.2. Spray workspace on laboratory bench thoroughly with 70% ethanol and wipe dry with a Kimwipe. Spray workspace with RNaseZap solution and wipe thoroughly with a Kimwipe. Rinse with 70% ethanol again, and then dry with a Kimwipe.

   1.3. Spray a Kimwipe with RNaseZap solution and wipe any pipettes that may be used. Wet a Kimwipe with ethanol and wipe pipettes again. Allow drying.

   1.4. Spray gloves with RNaseZap and spread solution all over gloves, and then spray gloves with 70% ethanol. Allow drying.

   1.5. Confirm that microcentrifuge is at room temperature (no cooler than 20°C).

2. Prepare RNeasy Plus Mini Kit:

   2.1. Prepare Buffer RLT in a chemical hood by adding 10 μL of ß-mercaptoethanol per 1 mL of Buffer RLT.

   Note: *Since Buffer RLT Plus is stable at room temperature for 1 month after addition of ß-mercaptoethanol, it is generally recommended that 5 mL of Buffer RLT be prepared at a time. However, this volume may be changed as needed.*

   Before the addition of ß-mercaptoethanol ensure that Buffer RLT Plus has not formed a precipitate during storage. If present, dissolve precipitate by warming in a 37°C water bath and then place at room temperature.

   2.2. Prepare Buffer RPE by adding 44 mL of 96–100% ethanol to Buffer RPE and swirl to mix.

3. Label and prepare tubes:

   Note: *In this procedure, one column will be used to extract cells from one well of a six-well plate. A typical yield for one well of hESCs that is at 70% confluence is 12 μg, an amount that is generally sufficient for analysis by RT-PCR.*

   Prepare one set of tubes per sample. Directions below are for one sample. Prepare as many sets as needed.

3.1. Label:

    3.1.1. One sterile RNase-free microcentrifuge tube per well of cells.

    3.1.2. One 2-mL collection tube (provided with QIAshredder column) and place a QIAshredder spin column inside.

    3.1.3. One 2-mL collection tube with gDNA spin column per sample (from RNeasy Plus Mini Kit).

    3.1.4. One 2-mL collection tube with an RNeasy mini spin column per sample (from RNeasy® Plus Mini Kit).

    3.1.5. A second 2-mL collection tube per sample (from RNeasy Plus Mini Kit).

    3.1.6. A 1.5-mL collection tube (from RNeasy Plus Mini Kit).

    3.1.7. A second 1.5-mL RNase-free microcentrifuge tube per sample.

3.2. Close all tube lids.

Procedure

Note: *This procedure should be carried out as quickly as possible.*

A frozen hES cell pellet may be used in this procedure. Thaw pellet on ice, flick tube to mix, add 600 µL of Buffer RLT Plus with ß-mercaptoethanol and begin protocol, at step 6.

There are places where this protocol can be stopped if necessary:

- *After the cells are collected in Buffer RLT Plus in step 6, this cell lysate may be stored in a freezer at −80°C.*

- *Homogenized cell lysates from step 8 may be stored in a freezer at −80°C.*

To resume this protocol from either of the stopping points, thaw the samples in a 37°C water bath until completely thawed and salts are dissolved. Avoid any prolonged incubation at 37°C, which may compromise RNA integrity. If any insoluble material is visible in the sample, centrifuge for 5 minutes at 5000 g. Transfer supernatant to a new RNase-free microcentrifuge tube.

1. On the bench, carefully aspirate all media from wells.

2. To each well add 600 µL of Buffer RLT Plus with ß-mercaptoethanol.

3. Swirl plate gently to evenly distribute solution.

4. Use a new cell scraper to scrape each well of cells.

5. Collect lysate from each well using a clean 1000-µL tip. Pipette into the appropriate sterile RNase-free microcentrifuge tube. Close the tube cap.

6. Vortex each sample for 1 minute.

7. Pipette lysate directly onto the appropriately labeled QIAshredder spin column placed over a 2-mL collection tube. Centrifuge for 2 minutes at maximum speed.

8. Transfer the homogenized lysate to a gDNA eliminator spin column placed in a 2-mL collection tube. Centrifuge for 30 seconds at 8000*g*. Discard the column and save the flow-through.

9. Add 600 µL of 70% ethanol to the flow-through and mix well by pipetting. Do not centrifuge.

10. Transfer 700 µL of the sample, including any precipitate that may have formed, to an RNeasy spin column placed in a 2-mL collection tube. Close the lid gently and centrifuge for 15 seconds at 8000*g*. Discard the flow-through.

11. Add the remainder of sample to the column (approximately 600 µL). Close the lid gently and centrifuge for 15 seconds at 8000*g*. Discard the flow-through.

NOTES

NOTES

_____

_____

_____

_____

_____

_____

_____

_____

_____

_____

_____

_____

12. Add 700 µL Buffer RW1 to the RNeasy spin column. Close the tube lid gently and centrifuge for 15 seconds at 8000g to wash the spin column membrane. Discard the flow-through.

13. Add 500 µL Buffer RPE to the RNeasy spin column. Close the tube lid gently and centrifuge for 15 seconds at 8000g to wash the spin column membrane. Discard the flow-through.

14. Again, add 500 µL Buffer RPE to the RNeasy spin column, close tube lid gently, and centrifuge for 2 minutes at 8000g to wash the spin column membrane.

15. Place the RNeasy spin column over a new 2-mL collection tube and discard the old collection tube with the flow-through. Centrifuge at full speed for 1 minute.

16. Place the RNeasy spin column in a new 1.5-mL collection tube. Add 30 µL of RNase-free water directly to the spin column membrane. Close the lid gently, and centrifuge for 1 minute at 8000g to elute the RNA. **Important: RNA is in the eluate**.

17. To increase yield, add 15 µL of RNase-free water to the RNeasy spin column. Close the tube lid gently, and centrifuge for 1 minute at 8000g.

18. Pipette (approximately 45 µL) of eluate from tube into a clean RNase-free microcentrifuge tube.

19. Proceed directly to Section II, Quantifying RNA Using NanoDrop or freeze immediately in a −80°C freezer.

II. QUANTIFYING RNA USING NANODROP

Once extracted, it is necessary to determine the RNA concentration and yield prior to proceeding with cDNA synthesis. Traditionally, RNA was quantified by measuring UV absorbance using a spectrophotometer. By measuring the optical density (OD) of the RNA solution at wavelengths of 260 and 280 nanometers, it is possible to determine the concentration of the solution as well as the presence of contaminants such as DNA, proteins, or salts in the sample. An OD 260 reading of 1.0 is approximately equivalent to an RNA concentration of 40 μg/mL. The ratio of absorbance reading between 260 nm and 280 nm can be used as a measure of the sample's purity. A pure RNA sample will have a 260/280 ratio range of 1.80 to 2.00. The absorbance of some contaminants like proteins is pH dependent; therefore the RNA sample should be eluted in TE buffer (10 mM Tris, 1 mM EDTA) that has a pH of 8.0. Although TE is optimal, water is often used. A decreased 260/280 ratio may indicate DNA contamination. Many kits provide an additional step to eliminate genomic DNA.

Conventional spectrophotometers can be used to measure absorbance, but the large cuvette size requires the use of a large fraction of the sample if the RNA is dilute. Currently, products such as the NanoDrop™ spectrophotometer (Thermo Scientific) have been developed to measure sample volumes as small as 0.5 μL, making quantification much more efficient. This section describes measurement of RNA concentration using the NanoDrop Spectrophotometer.

Equipment
- Ice bucket
- Micropipettors: P2 (0.2 to 2 μL), P20 (0.5 to 20 μL), P200 (20 to 200 μL), P1000 (200 to 1000 μL)
- NanoDrop 1000 Spectrophotometer (ThermoFisher Scientific ND-1000)

Materials
- Absorbent paper towels (or Kimwipes®)
- Disposable gloves
- Microcentrifuge tube rack
- RNase-free Eppendorf tubes (1.5 mL)
- Sterile micropipettor filtered tips to fit P2, P20, P200, and P1000 micropipettors

Reagents
- Spray bottle of 70% ethanol
- Sterile RNase-free water
- RNaseZap® solution

Preparation
1. Wash hands thoroughly with soap; rinse them completely with tap water; dry hands with paper towel, then put on appropriate-sized gloves.
2. Spray workspace on laboratory bench thoroughly with 70% ethanol and wipe dry with a Kimwipe. Spray workspace with RNaseZap solution and wipe thoroughly with a Kimwipe. Rinse with 70% ethanol again, and then dry with a Kimwipe.
3. Spray a Kimwipe with RNaseZap solution and wipe any pipettes that you may be using. Wet a Kimwipe with ethanol and wipe pipettes again. Allow drying.

NOTES

NOTES

_____

_____

_____

_____

_____

_____

_____

_____

_____

_____

_____

_____

_____

_____

_____

_____

_____

_____

_____

_____

_____

_____

_____

4. Spray gloves with RNaseZap and spread solution all over gloves, and then spray gloves with 70% ethanol. Allow drying.

Procedure

1. Resuspend the sample:

   Note: _These steps take place at the bench._

   1.1. _If RNA has been prepared from live cells_ (see Section I, Preparing RNA Using RNeasy® Plus Mini Kit), place samples on ice. Resuspend the sample gently with a pipette and briefly centrifuge. (Do not vortex.) If sample quantification has been done, proceed directly to Section III, cDNA Synthesis Using Reverse Transcriptase.

   1.2. _If RNA samples are frozen, thaw_. Resuspend the sample gently with a pipette. Briefly centrifuge. (Do not vortex.)

2. Measure RNA concentration:

   Note: _Steps 2.1–2.13 take place at NanoDrop Station._ **NanoDrop 1000 Optimum Range**_: The optimum range of the NanoDrop 1000 Spectrophotometer is 2–3700 ng/μL. If the sample does not fall in this range, concentrate or dilute it appropriately._

   2.1. Turn on NanoDrop computer and open software.

   2.2. Wipe NanoDrop pedestal with a clean Kimwipe.

   2.3. Pipette 1.5 μL of RNeasy elution buffer onto the NanoDrop pedestal.

   2.4. Select "OK" to initialize instrument.

   2.5. In NanoDrop software, select RNA.

   2.6. Wipe pedestal with a Kimwipe and pipette 1.5 μL RNase-free water (or buffer that was used to elute RNA).

   2.7. Select "Blank."

   2.8. Wait for Blank to be read (about 10 seconds).

   2.9. Wipe pedestal with a Kimwipe.

   2.10. Pipette 1.5 μL of first sample.

   2.11. Enter first sample and name.

   2.12. Select "Measure."

   2.13. Wipe pedestal clean with a Kimwipe.

   2.14. Apply 1.5 μL of RNase-free water to pedestal.

   2.15. Wipe pedestal clean with a Kimwipe.

   2.16. Repeat steps 2.10–2.14 for each sample.

   2.17. Reapply 1.5 μL of water to pedestal and wipe pedestal clean with a Kimwipe.

3. Documentation of RNA quantity:

   3.1. Record the following information on each tube:
   - Type of sample (RNA), concentration of RNA
   - Treatment if any
   - Initial of preparer
   - Date

   3.2. If proceeding immediately to Section III, cDNA Synthesis Using Reverse Transcriptase, place RNA samples on ice and out of direct sunlight. Immediately store any RNA not to be used right away in a −80℃ freezer.

III. cDNA SYNTHESIS USING REVERSE TRANSCRIPTASE

When performing RT-PCR using RNA as a template, the RNA must be reverse transcribed into complementary DNA (cDNA) through a reverse transcription (RT) reaction. This RT and PCR can be completed in two sequential steps or in one step. There are many kits commercially available for both methods.

One-step RT-PCR contains all of the enzymes needed for both the RT and PCR step in one reaction. The reagents have been optimized so that once the RT reaction is completed, the enzymes needed for PCR are activated. In traditional two-step RT-PCR, the initial RT reaction occurs in one tube, while the subsequent PCR reaction (which uses the RT product as template) occurs in a new tube with its own reagents and enzymes. While one-step RT-PCR is faster and minimizes the possibility for contamination and pipetting errors, it is often more expensive and the resulting cDNA cannot be used again. This chapter provides protocols for RT-PCR as two separate sequential steps.

Positive and negative controls should be included in each experiment. Reverse transcription of each RNA sample should be carried out in parallel with two negative controls. The three reactions are: +RT, which contains both reverse transcriptase enzyme and RNA; −RT, which contains RNA but no reverse transcriptase enzyme; and No RNA, which contains reverse transcriptase enzyme but RNA is substituted with water. An outline of a typical PCR experiment and the rationale for the experimental design are provided (Fig. 15.2 and Table 15.1).

Equipment

- Microcentrifuge (with adapters for 0.2-mL tubes)
- Micropipettors P2 (0.2 to 2 µL), P20 (0.5 to 20 µL), P200 (20 to 200 µL), P1000 (200 to 1000 µL)
- PCR tube rack
- Thermal cycler

NOTES

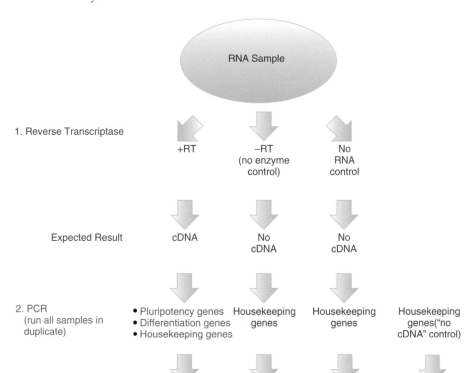

FIGURE 15.2. *An outline of a typical PCR experiment and the rationale for the experimental design (See Table 15.1).*

TABLE 15.1 RT-PCR CONTROLS

Positive Control	Content of Reaction	Expected PCR Result
+RT	RT enzyme and RNA	Amplification of housekeeping genes and target genes if these are expressed. Housekeeping genes (e.g., GAPDH, b-Actin) are involved in basic functions needed for cell maintenance. Housekeeping genes are used as an internal standard in this protocol to account for tube-to-tube differences caused by variability in RNA quality or reverse transcription efficiency or inaccurate quantitation or pipetting.

Negative Control	Content of Reaction	Expected PCR Result
−RT	RNA	No cDNA is produced in the absence of RT enzyme. Thus unless the RNA samples are contaminated with (genomic) DNA, the contents of this tube should not provide DNA templates for amplification in the subsequent PCR reaction.
No RNA	RT enzyme	No cDNA is produced in the absence of RNA. Thus unless the reagents are contaminated with RNA and/or DNA, the contents of this tube should not provide DNA templates for amplification in the subsequent PCR reaction.
No cDNA	Use water in PCR reaction in place of RT product	No PCR product is produced in the absence of cDNA, unless the reagents used in the PRC reaction are contaminated with DNA.

NOTES

Materials
- Absorbent paper towels (Kimwipes®)
- Disposable gloves
- RNase-free 1.5-mL microcentrifuge tubes
- Sterile 0.2-mL PCR tubes
- Sterile, micropipettor filtered tips to fit P2, P20, P200, and P1000 micropipettors

Reagents
- SuperScript™ III First-Strand (Invitrogen 1080-051) Synthesis System for RT-PCR. This kit contains:
  - 50 µL Oligo (dT) (50 µM)
  - 250 µL Random hexamers (50 ng/µL)
  - 1 mL 10x RT buffer
  - 500 µL 25 mM $MgCl_2$
  - 250 µL 0.1 M DTT
  - 500 µL 25 mM dNTP mix
  - 50 µL SuperScript™ III RT (200 U/µL)
  - 100 µl RNaseOUT™ (40 U/µl)
  - 50 µL *Escherichia coli* RNase H (2 U/µL)
  - 1.2 mL DEPC-treated water
  - 20 µL Total HeLa RNA (10 ng/µL)
  - 25 µL Sense Control Primer (10 µM)
  - 25 µL Antisense Control Primer (10 µM)
- RNaseZap® solution
- Squirt bottle of 70% ethanol
- RNase-free water

Preparation

1. Prepare laboratory bench and equipment:

   1.1. Wash hands thoroughly with soap; rinse them completely with tap water, dry hands with paper towel, then put on appropriate-sized gloves.

   1.2. Spray workspace on laboratory bench thoroughly with 70% ethanol and wipe dry with a Kimwipe. Spray workspace with RNaseZap and wipe thoroughly with a Kimwipe. Rinse with 70% ethanol again, and then dry with a Kimwipe.

   1.3. Spray a Kimwipe with RNaseZap and wipe any micropipettors that you may be using. Wet a Kimwipe with ethanol and wipe pipettes again. Allow drying.

   1.4. Spray gloves with RNaseZap and spread solution all over gloves, and then spray gloves with 70% ethanol. Allow drying.

   1.5. Confirm that microcentrifuge is at room temperature (no cooler than 20°C).

2. Thaw reagents and program cycler:

   2.1. Remove reagents to be used from the SuperScript III First-Strand Synthesis System and thaw on ice.

   2.2. Program cycler as follows:
   - **RT1**
     - 65°C for 5 minutes
   - **RT2**
     - 25°C for 10 minutes
     - 50°C for 50 minutes
     - 85°C for 5 minutes
   - **RT3**
     - 37°C for 20 minutes

3. Label tubes:

   3.1. Label two tubes for each RNA sample:
   - *First tube:* +RT and the sample name (includes RNA and reverse transcriptase enzyme).
   - *Second tube:* −RT (includes RNA but no reverse transcriptase enzyme).

   3.2. Label 2 tubes as: No RNA (includes reverse transcriptase but no RNA) as a control to confirm that reagents are not contaminated with DNA or RNA.

Procedure

1. For each RNA sample use 1 μg of RNA per reaction. Using the concentration that was calculated in Section II, Quantifying RNA Using NanoDrop, dilute the samples in RNase-free water into a labeled and sterile 0.2-mL PCR tube, so that each samples' final concentration is 1 μg/8 μL. Prepare at least 3 μg of diluted RNA per sample in order to have sufficient RNA for two reactions and to account for pipetting errors.

   Note: Errors in pipetting of RNA samples or reagents can cause significant variation in PCR reactions. To help control these variations, prepare a master mix which contains reagents that are common to all reactions. This will reduce the number of pipetting steps. In addition, avoid pipetting very small volumes; the template RNA should be diluted so that between 3 and 10 μL are used for each reaction. Variability can also be reduced by using a dedicated set of micropipettes.

2. Dilute RNA to a final concentration of 1 μg/8 μL with RNase-free water. If samples are more dilute than 1 μg/8 μL, use less RNA in each reaction,

but all reactions must have the same amount of RNA. SuperScript III First-Strand Synthesis System for RT-PCR is optimized for 1–5 μg of total RNA.

3. For each sample, prepare the RNA/primer mixtures in sterile 0.2-mL PCR tubes in the order listed in Table 15.2.

4. Mix briefly by pipetting up and down after all components of the reaction have been added.

5. Briefly centrifuge tubes.

6. Confirm that all tube lids are tightly closed. Place tubes in the cycler and close lid.

7. Select the RT1 program (65°C for 5 minutes) and select "Start Run."

8. While the RT1 program is running, prepare cDNA Synthesis Mix for each sample (refer to Fig. 15.2 to determine how many reactions are needed). Remember to add an additional sample to account for pipetting error.

9. Prepare cDNA synthesis mix by adding the reagents to an RNase-free micro-centrifuge tube labeled "cDNA Synthesis Mix" in the order they are listed in Table 15.3 for +RT reactions and 2 No RNA reactions. For –RT reactions prepare the materials in Table 15.4.

TABLE 15.2 RNA/PRIMER MIXTURES FOR RT-PCR

Components	+RT Reaction	–RT Reaction	No RNA (Negative Control)
Primers (random hexamers or oligo dT)	1 μL	1 μL	1 μL
10 mM dNTP mix	1 μL	1 μL	1 μL
RNase-free water	0 μL	0 μL	8 μL
Diluted total RNA (1 μg/8 μL)	8 μL	8 μL	0 μL
Each reaction total:	**10 μL**	**10 μL**	**10 μL**

TABLE 15.3 PREPARATION OF cDNA SYNTHESIS MIX

Component	One Reaction	Ten Reactions
10 × RT buffer	2 μL	20 μL
25 mM MgCl$_2$	4 μL	40 μL
0.1 M DTT	2 μL	20 μL
RNaseOUT™ (40 U/μL)	1 μL	10 μL
SuperScript™ III RT (200 U/μL)	1 μL	10 μL

TABLE 15.4 PREPARATION OF –RT CONTROL MIX

Component	One Reaction	Ten Reactions
10 × RT buffer	2 μL	20 μL
25 mM MgCl$_2$	4 μL	40 μL
0.1 M DTT	2 μL	20 μL
RNaseOUT™ (40 U/μL)	1 μL	10 μL
Water	1 μL	1 μL

10. Once the run is complete, remove each tube carefully from the cycler. Place samples on ice for at least 1 minute.

11. Add 10 µL of cDNA Synthesis Mix to each sample. Mix gently by pipetting up and down.

12. Briefly centrifuge tubes.

13. Confirm that all tube lids are tightly closed. Place tubes in cycler and close lid.

14. Select the RT2 program (25°C for 10 minutes, 50°C for 50 minutes, 85°C for 5 minutes) and select "Start Run."

15. Once the run is complete, remove each tube carefully from the cycler. Place samples on ice.

16. Collect the reactions with a brief centrifugation. Add 1 µL of *E. coli* RNase H (2 U/µL) to each tube. Mix gently by pipetting up and down.

17. Confirm that all tube lids are tightly closed. Place tubes in cycler and close lid.

18. Select the RT3 program (37°C for 20 minutes) and select "Start Run."

19. Once the run is complete, remove each tube carefully from the cycler. Place samples on ice.

Note: RT products can be kept on ice if used immediately or they can be stored in −20°C freezer.

NOTES

IV. PCR

One common approach for assessing the quality of hESC cultures is by monitoring expression of genes associated with pluripotency and differentiation using RT-PCR. Although there are multiple genes associated with these different states (Table 15.5), OCT4, SOX2, and NANOG are most frequently used as pluripotency markers and alpha fetoprotein (AFP) as a differentiation marker. Sample primers are provided in Table 15.6 for these four genes, as well as the housekeeping gene beta-actin.

Primer Quality. Primer sets need to be carefully chosen for RT-PCR. Many programs are available online to aid in primer selection. The following elements are important in the design of optimal primers:

- The melting temperature of primer sets should be within 5°C of each other.
- Primers should not have significant complementarity to avoid dimerization.
- Primers should be selected that bind to unique regions of specific mRNAs.
- Having one primer that binds to an exon–exon junction can avoid amplification of contaminating genomic DNA.

Equipment

- Microcentrifuge (with adapters for 0.2-mL tubes)
- Micropipettors: P20 (0.5 to 20 µL), P200 (20 to 200 µL)
- PCR tube rack
- Thermal cycler

TABLE 15.5 GENES ASSOCIATED WITH PLURIPOTENCY AND DIFFERENTIATION

Gene	Cell State
DNMT3B	Undifferentiated
Nanog	Pluripotent
SOX2	Pluripotent
POU5F1 (Oct4)	Pluripotent
ZFP42 (Rex1)	Pluripotent
TDGF1	Undifferentiated
TERT	Undifferentiated
FGF4	Undifferentiated
CD9	Undifferentiated
FoxD3	Pluripotent
PAX6	Ectoderm pancreatic/neuron
AC133	Ectoderm early epithelial
GATA4	Mesoderm early cardiac
AFP	Endoderm
18S	Housekeeping
ACTB	Housekeeping
GAPDH	Housekeeping
EEF1A1	Housekeeping
CTNNB1	Housekeeping

TABLE 15.6 SAMPLE PRIMERS FOR RT-PCR

Gene	Primers[a]	Product Size (bp)
Pluripotency associated genes		
Oct4/POU5F1	CGACCATCTGCCGCTTTGAG CCCCCTGTCCCCCATTCCTA	573
NANOG	GACAGCCCTGATTCTTCCAC CAGGTTGCATGTTCATGGAG	718
SOX2	CCCCCGGCGGCAATAGCA CTCGGCGCCGGGGAGATA	449
Differentiation associated gene		
AFP	CAGAACCTGTCACAAGCTGTG GACAGCAAGCTGAGGATGTC	677
Housekeeping gene		
β-actin	CCACACCTTCTACAATGAGC CGTCATACTCCTGCTTGCTG	830

[a]These primers were designed to be intron-spanning to preclude the amplification from genomic DNA.

Materials

- Disposable gloves
- Sterile, micropipettor filtered tips to fit P2, P20, P200, and P1000 micropipettors

Reagents

- AccuPrime™ *Taq* DNA Polymerase High Fidelity (Invitrogen 12346-086)
- Primers
- Autoclaved, double-distilled water

Preparation

1. Prepare laboratory bench and equipment:
   1.1. Wash hands thoroughly with soap; rinse them completely with tap water, dry hands with paper towel, then put on appropriate-sized gloves.
   1.2. Spray workspace on laboratory bench thoroughly with 70% ethanol and wipe dry with a Kimwipe.
   1.3. Spray gloves with 70% ethanol.
2. Remove reagents to be used from the AccuPrime PCR system and thaw on ice.
3. Label tubes:
   - For PCR reactions containing primers for housekeeping genes, label one tube for the "no RNA" control, as well as for each "+RT" and "−RT" product.
   - For PCR reactions containing primers for pluripotency and differentiation genes, label one tube for each "+RT" product.

Procedure

Note: Cycler annealing temperatures and times depend on the sequences of the primers and the product size.

1. Program the thermocycler as follows if using primers given in Table 15.6:
   - Initial denaturation step: 95°C for 3 minutes

NOTES

_____
_____
_____
_____
_____
_____
_____
_____
_____
_____
_____
_____
_____
_____
_____
_____
_____
_____
_____
_____
_____
_____

TABLE 15.7 PREPARATION OF PCR MASTER MIX

Components	1x	8x	16x
10X AccuPrime™ PCR Buffer I	5 µl	40 µl	80 µl
Sense primer (10 µM)	1 µl	8 µl	16 µl
Anti-sense primer (10 µM)	1 µl	8 µl	16 µl
AccuPrime™ *Taq* High Fidelity	0.2 µl	1.6 µl	3.2 µl
Distilled water	40.8 µl	327 µl	653 µl

- Thirty cycles of:
  - Denature: 95°C for 30 seconds
  - Anneal: 58°C for 30 seconds
  - Extend: 68°C for 1 minute
  - Final extension: 68° for 10 minutes

2. Make a "master mix" (Table 15.7).

3. Add 48 µL of above master mix into each 200 µL tube.

4. Add 2 µL of RT product.

5. Cap the tube, tap the tube to mix, and centrifuge briefly to collect the contents.

6. Place samples in the thermocycler.

7. Once the run is complete, remove each tube carefully from the cycler. Place samples on ice.

Note: *PCR products can be kept on ice if used immediately or they can be stored in −20°C freezer.*

V. ANALYSIS OF PCR PRODUCTS BY AGAROSE GEL ELECTROPHORESIS

Equipment

- Container for transferring gel to lightbox
- Electrophoresis chamber
- Gel casting tray and combs
- Power supply
- UV light box with camera for imaging a stained gel

Materials

- Disposable gloves
- Pipette and tips
- 500 mL Flask

Reagents

- 6X Sample Loading Buffer (0.25% bromophenol blue, 0.25% xylene cyanol, 60% glycerol)
- Agarose
- DNA ladder standard
- Ethidium bromide stock solution (10 mg/mL)
- TAE buffer (40 mM Tris-acetate, 1 mM EDTA, pH 8.3).

Procedure

Note: Ethidium bromide is a mutagen and a carcinogen. Always wear gloves and a laboratory coat when working with this reagent.

1. Prepare agarose gel:

   1.1. Add 1.5 g agarose and 100 mL TAE buffer to a 500-mL flask.

   1.2. Melt agarose by heating in a microwave oven for 2–3 minutes, pausing two or three times to swirl and mix the solution. Heating the solution for short intervals, rather than letting it boil for long periods of time, also prevents the solution from boiling out of the flask.

   1.3. Let the solution cool to 50–55°C, swirling the flask occasionally to cool evenly.

   1.4. Add 5 µL of ethidium bromide per 100 mL gel solution and swirl to mix.

   1.5. Place the combs in the gel casting tray.

   1.6. Pour the melted agarose solution into the casting tray and remove bubbles as needed. Allow the agarose to cool until it is solid with an opaque appearance.

   1.7. Carefully pull out the combs.

   1.8. Place the gel in the electrophoresis chamber.

   1.9. Add enough TAE buffer so that there is 2–3 mm of buffer over the gel.

   > *Note: Gels can be made 4–5 days in advance, sealed in plastic wrap (without combs) and stored at 4°C to prevent dehydration.*

2. Load the gel:

   2.1. Add 12 µL of 6X Sample Loading Buffer to each 50 µL PCR reaction.

   2.2. Record the order in which each sample will be loaded on the gel, including controls and ladder.

NOTES

2.3. Carefully pipette 20 μL of each sample into separate wells in the gel.

2.4. Pipette 10 μL of the DNA ladder standard into at least one well of each row on the gel.

Note: If multiple gels are electrophoresed on the same day, avoid later confusion by loading the DNA ladder in different lanes on each gel.

3. Gel electrophoresis:

3.1. Place the lid on the gel box and connect the electrodes to the box and power supply. Make sure the negative (black) lead is on the side closest to the wells.

Note: DNA is negatively charged and thus runs from negative to positive or from the black to red side when the electrodes are properly connected.

3.2. Turn on the power supply and set the voltage according to the size of the electrophoresis chamber, as maximum allowed voltage varies with chamber size. **Voltage should not exceed 5 volts/cm between electrodes**.

3.3. Make sure the current is running through the buffer by looking for bubbles forming on each electrode.

3.4. Check to make sure that the current is running in the correct direction by observing the movement of the blue loading dye—this will take a couple of minutes. The loading dye will run in the same direction as the DNA.

3.5. Electrophorese the gel for 45 minutes, or until the blue dye has migrated two-thirds of the gel length.

3.6. Turn off the power.

3.7. Carefully transfer the gel in its casting tray to a container and carry it to an ultraviolet (UV) transilluminator.

Note: Wet agarose gels are very slippery and have a tendency to slide off the casting tray.

3.8. Slide the gel off the casting tray onto the transilluminator and photograph gel under UV illumination.

Note: Utilize protective equipment to limit exposure to UV light.
Photograph the gel as soon as possible after running, as 100–300 bp bands will blur quickly.

4. Determine the ideal number of cycles for PCR:

- Start at 20, 25, or 30 cycles and use 10 μL of sample for gel electrophoresis.

- If the bands are very strong, it is necessary to set up a new reaction with fewer cycles. If there are very weak bands or no bands, then perform PCR with 35 or 40 cycles, respectively.

- If the bands are faint, but differences are clear between the samples, load more of each sample on the gel.

VI. qRT-PCR

This section provides a brief overview of quantitative reverse transcription polymerase chain reaction (qRT-PCR), a technique used to amplify and simultaneously quantify a gene of interest. This technique differs from standard PCR as it quantifies PCR products in real time after each cycle, thus eliminating the need for a validation experiment to determine the ideal cycle for each gene of interest. Standard PCR only amplifies the products, which are then analyzed visually on an agarose gel. Real-time RT-PCR requires a specialized thermocycler that allows absolute or relative quantification of specific mRNAs in real time, depending on the standards used.

There are four different chemistries currently available for qRT-PCR. SYBR Green, TaqMan®, Molecular Beacons, and Scorpions. SYBR Green is a fluorogenic dye that has a very low fluorescence in solution but once it intercalates with double-stranded DNA, it emits a strong fluorescent signal. TaqMan probes, Molecular Beacons, and Scorpions use Forster resonance energy transfer (FRET) to generate a fluorescence signal through the pairing of a fluorogenic dye molecule and a quencher.

Irrespective of the chosen technology, positive and negative controls should be included in each experiment. Reverse transcription of each RNA sample should be carried out in parallel with two negative controls as described in Figure 15.2 and Table 15.1.

Once the proper controls are determined, genes of interest need to be selected. Expression of genes associated with the undifferentiated state as well as particular lineages (Table 15.2) is monitored using qRT-PCR to evaluate the quality of the hESC culture. In addition, at least one housekeeping gene should also be selected.

Quantitation of Results. Regardless of the probes and instrumentation used to detect real-time PCR products, the fluorescence measurements for each reaction are typically presented in an amplification plot (Fig. 15.3), which plots fluorescence signal versus cycle number. In the early cycles there is little change in fluorescence signal. This defines the baseline for the amplification plot and represents background signal. Fluorescence detected above the baseline represents accumulated PCR products. The higher the starting copy number of target cDNA, the sooner a significant increase in fluorescence is observed.

The relative cDNA concentration of different samples in the same experiment can be compared to each other by setting a fixed fluorescence threshold above the baseline. For accurate comparison, the threshold must lie within the exponential growth region of the amplification curve and where all the sample lines are parallel. The threshold cycle (C_t) is defined as the fractional cycle number at which the fluorescence crosses the threshold. The higher the concentration of target cDNA

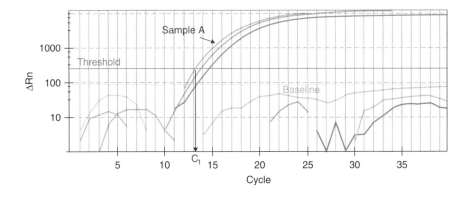

FIGURE 15.3. *Amplification plot. ΔRn is the fluorescence of the reporter normalized to the baseline, which represents background signal. Blue line, baseline; red line, fluorescence threshold; C_t, threshold cycle for Sample A.*

within a test sample, the fewer cycles (lower C_t value) it takes to reach the fixed threshold. Thus Sample A has a higher concentration of OCT4 cDNA than Samples B or C. Note that the fluorescence values are plotted on a logarithmic scale to detect small differences at earlier cycles.

Real-time RT-PCR results are quantitated using two common approaches: standard curve method and comparative threshold cycle (C_t) method. An overview of these strategies is provided below.

Standard Curve Method. The standard curve method can be used to obtain absolute quantitation (i.e., the copy number of a specific RNA transcript in a test sample), or relative quantitation, which only allows comparison of the RNA concentration of a given sample relative to that of a reference sample. The standard curve can be obtained by software in most real-time PCR instruments or manually calculated by Excel. The acceptable standard curve has a slope between approximately -3.1 and -3.6 (equivalent to a calculated 90–110% reaction efficiency), and Rsq value equal to or greater than 0.985.

The absolute standard curve method is used to quantify unknown samples by interpolating their quantity from a standard curve. The copy number of a specific mRNA transcript in a test sample can be determined by constructing an absolute standard curve, which involves construction of cDNA plasmids, in vitro transcription into the RNA standard, and accurate measurement of the RNA concentration. This method is laborious and time consuming, and can be avoided if absolute quantitation is not necessary.

To construct the standard curve for relative quantification, other nucleic acids can be used in addition to RNA. These include purified plasmid double-stranded DNA, in vitro generated single-stranded DNA, or any cDNA sample expressing the target gene. However, since DNA references will not control for variability in the reverse transcription step, these standards will only provide information on relative changes in mRNA expression.

Comparative Threshold Cycle (C_t) Method. The comparative C_t method is a relative quantitation approach that uses arithmetic formulas in place of the relative standard curve. This method is employed to analyze changes in gene expression in a given sample relative to a reference sample or calibrator, such as an untreated control sample, normal tissue, or any arbitrarily chosen sample within the experiment.

The amount of target is calculated using the formula

$$2^{-\Delta\Delta C_t}, \quad \text{where } \Delta\Delta C_t = \Delta C_{t,\text{sample}} - \Delta C_{t,\text{calibrator}}$$

Here, $\Delta C_{t,\text{sample}}$ is the C_t of the test gene normalized to the housekeeping gene.

For the $\Delta\Delta C_t$ calculation to be valid, the amplification efficiencies of the target gene and the housekeeping gene must be approximately equal. This can be ascertained by performing a validation experiment to determine how ΔC_t values change with template dilution. If the amplification efficiencies are similar, the slope is close to zero when cDNA dilutions are plotted against ΔC_t values. If there are no housekeeping genes with amplification efficiency that is similar to the target gene, then the standard curve method is recommended.

SUGGESTED READING

Ambion. RT-PCR: The Basics. Available at http://www.ambion.com/techlib/basics/rtpcr/index.html.

Ambion, Ten Most Common Real-Time qRT-PCR Pitfalls. Available at www.ambion.com/techlib/tn/102/17.html.

Boom R, Sol CJ, Salimans MM, Jansen CL, Wertheim-van Dillen PM, van der Noordaa J. Rapid and simple method for purification of nucleic acids. *J Clin Microbiol*. 1990;28(3):495–503. PMID: 1691208.

Chomczynski P, Sacchi N. Single-step method of RNA isolation by acid guanidinium thiocyanate-phenol-chloroform extraction. *Anal Biochem*. 1987;162:156–159.

Dorak MT. REAL-TIME PCR. Available at dorakmt.tripod.com/genetics/realtime.html.

http://www3.appliedbiosystems.com/cms/groups/mcb_support/documents/generaldocuments/cms_040980.pdf.

http://www3.appliedbiosystems.com/cms/groups/mcb_marketing/documents/generaldocuments/cms_042485.pdf.

http://www3.appliedbiosystems.com/cms/groups/mcb_marketing/documents/generaldocuments/cms_053906.pdf.

http://pathmicro.med.sc.edu/pcr/realtime-home.htm.

Nolan T, Hands RE, Bustin SA. Quantification of mRNA using real-time RT-PCR. *Nat Protocols*. 2006;1(3);1559–1582.

Qiagen® OneStep Rt-PCR Kit Handbook. February 2008.

RNeasy® Mini Handbook. 4th ed. April 2007. Qiagen. Available at www.qiagen.com.

VanGuilder HD, Vrana KE, Freeman WM. Twenty-five years of quantitative PCR for gene expression analysis. *Biotechniques*. 2008;44:619–626.

DETERMINING PLURIPOTENCY OF HUMAN EMBRYONIC STEM CELLS: EMBRYOID BODY FORMATION

16

Shirwin M. Pockwinse and Prachi N. Ghule

The hallmark of human embryonic stem cells (hESCs) is self-renewal and pluripotency. As discussed in several chapters in this book, hESCs can grow continuously and maintain their undifferentiated state when cultured under appropriate conditions. Pluripotency refers to the capacity of hESCs to differentiate, forming all three germ layers (endoderm, mesoderm, and ectoderm) of the embryo proper and, ultimately, all tissues and organs of the body.

Pluripotency of hESCs can be assessed by formation of intermediate structures called embryoid bodies (EBs). Human embryonic stem cells grown in suspension culture without an inactivated mouse embryonic fibroblast (MEF) feeder layer will form small aggregates that undergo spontaneous differentiation forming embryoid bodies. Fully formed EBs are composed of all three germ layers. Examples of specific markers for each germ layer are:

- Endoderm: AFP, GATA4
- Ectoderm: PAX6, SOX1
- Mesoderm: Brachyury, Flt1

The expression of these markers can be validated using reverse transcription polymerase chain reaction (RT-PCR) or immunocytochemistry.

Alternative methods for the determination of hESCs' pluripotency are (1) in vitro induced or directed differentiation, which involves differentiation of stem cells into specific cell lineages using defined factors; and (2) in vivo teratoma formation, which is the ability of stem cells to form teratomas (germ cell tumors) in immune-compromised SCID (severe combined immunodeficiency) mice (see Chapter 19).

General Notes
- Read the entire chapter before initiating cell culture activities described in Procedure.
- It is good laboratory practice to have important relevant information recorded; therefore a log sheet is provided as an example at the end of this chapter.

Human Stem Cell Technology and Biology, edited by Stein, Borowski, Luong, Shi, Smith, and Vazquez
Copyright © 2011 Wiley-Blackwell.

Overview

Plate hESCs and grow until 70–80% confluence.

Treat with dispase to detach and digest hESC colonies into small aggregates.

Transfer hESC aggregates to T25 flask with differentiation/EB medium.

If necessary, transfer the hESC aggregates to a new T25 flask.

Feed EBs every day by replacing 50% of the medium.

Grow EBs 2–3 weeks depending on desired application.

Abbreviations

DMEM: Dulbecco's modified Eagle's medium

EBs: embryoid bodies

hESCs: human embryonic stem cells

HI-FBS: heat-inactivated fetal bovine serum

mg: milligram

mL: milliliter

MEFs: mouse embryonic fibroblasts

NEAAs: nonessential amino acids

RT-PCR: reverse transcription polymerase chain reaction

Equipment

- 37°C Cell culture incubator
- 37°C Water bath
- Benchtop centrifuge with swinging-bucket rotors
- Inverted microscope with a digital camera and associated image acquisition software
- Pipette aid
- Sterile biosafety cabinet
- Vacuum system (recommended)
- Weighing balance

Materials

- 5-mL and 10-mL Sterile glass or plastic pipettes
- 15-mL Sterile conical tubes
- T25 flasks (non-tissue-culture grade or with low attachment surface)
- Disposable gloves

Reagents

- Dispase solution 1 mg/mL (Stem Cell Technologies #07923) prepared in DMEM/F12 medium
- Differentiation/EB medium (see Chapter 5)
- hESC medium (see Chapter 5)

Preparation

1. Prepare differentiation/EB medium.

 2. Grow hESCs for at least 3–5 days or until they reach 70–80% confluence before beginning EB culture (see Chapter 9 page 105). (*This protocol may be viewed on the accompanying DVD and on the web at* http://www.wiley.com/go/stein/human.)

3. Place the following items near the biosafety cabinet:
   - 5-mL and 10-mL Sterile pipettes
   - 70% Ethanol spray
   - Absorbent paper towels (or Kimwipes®)
   - Disposable gloves

4. Ensure that the following items are in the biosafety cabinet:
   - 5-mL and 10-mL Sterile pipettes
   - 15-mL Sterile conical tubes
   - T25 flasks (non-tissue-culture grade)
   - Alcohol-proof marker

Procedure

1. Follow preparations according to Chapter 3, Aseptic Technique on page 19. (*This protocol may be viewed on the accompanying DVD and on the web at* http://www.wiley.com/go/stein/human.)

2. Remove hESC plate from incubator that was passaged 3–5 days earlier and place in the sterile biosafety cabinet.

3. Warm the dispase solution to 37°C before adding to cells. Aspirate the hESC medium and add 1 mL of dispase solution (1 mg/mL) into each well of the six-well plate (Collagenase can also be used as an alternative to dispase at same concentration.)

4. Place in 37°C incubator for 5–10 minutes.

   Note: Observe the plate under an inverted microscope. Optimal enzymatic treatment has been achieved when the borders/edges of hESC colonies start to curl up or colonies start to peel off (Fig. 16.1).

5. Place the plate back in the sterile biosafety cabinet and add 1.5 mL of differentiation/EB medium per well. Gently swirl the plate to dislodge most of the loosely attached hESC colonies from inactivated MEF layer. Using a 5-mL sterile glass pipette, detach any remaining hESC colonies by gently nudging them off the plate with the pipette tip. Avoid scraping or breaking up the colonies.

   Note: Try to avoid detaching the inactivated MEF layer from the plate surface.

6. Use two wells of a six-well plate for each T25 flask of EB culture. Transfer medium containing dislodged hESC colonies from the two wells to a 15-mL conical tube. Centrifuge at 50*g* (approximately 500 rpm) for 2 minutes at room temperature. In a sterile biosafety cabinet, aspirate supernatant without touching the cells.

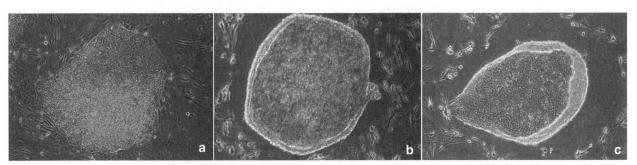

FIGURE 16.1. *Morphology of hESC colony: (a) before treatment, (b) 5 minutes after dispase treatment, and (c) 10 minutes after dispase treatment. All images were taken at 20 × magnification on an inverted phase contrast microscope.*

Note: *If the dislodged hESC colony aggregates are too small, centrifugation may disrupt them further. Hence an alternative to the centrifugation step is to allow the medium containing hESC colonies to stand in the biosafety cabinet for 5–10 minutes so that they settle at the bottom of the tube by gravity. Then carefully aspirate the supernatant without disturbing the cells and wash cells with differentiation medium. Each wash step can be done using this procedure instead of centrifugation.*

7. In a sterile biosafety cabinet, wash the hESC colonies by adding 12 mL of fresh EB/differentiation medium; centrifuge at 50*g* (approximately 500 rpm) for 2 minutes at room temperature and then aspirate supernatant.

8. Resuspend the cells by adding 12 mL of fresh EB/differentiation medium and break the hESC colonies into smaller pieces by pipetting up and down three to five times.

Note: *It is advisable not to break up the cell colonies into single cells, trying to keep the small clumps of hESCs about twice the size of that required for regular passaging.*

9. Transfer the cell suspension containing smaller hESC aggregates to a T25 flask.

Note: *Do not to use tissue culture treated flasks, as cells will attach. Use non-tissue-culture grade, low attachment surface T25 flasks for EB culture as EBs are grown in suspension.*

Human embryonic stem cells in suspension do not attach under these culture conditions unless there is considerable contamination by inactivated MEFs. During the first few days of EB culture, if inactivated MEFs are present and attaching to the bottom of the flask, the small floating EBs should be transferred to a new flask to eliminate the inactivated MEFs from the culture (Fig. 16.2).

10. Feed the EB cultures daily by replacing half the medium from the flask. Place the flask in the biosafety cabinet at a tilted angle of 45 degrees so that the small cell aggregates settle in one corner of the flask. Remove top half of the medium (5–6 mL) by pipetting (Fig. 16.3) and add the same amount of fresh EB medium to replenish the aspirated amount.

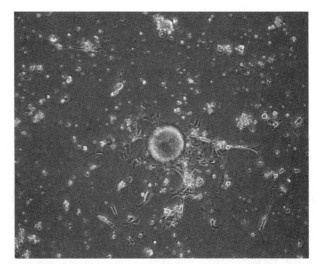

FIGURE 16.2. *EB culture after 24 hours. Note the presence of inactivated MEFs surrounding the hESC aggregate. If such inactivated MEF contamination is observed, the small hESC aggregates should be transferred to a new T25 flask. This image was taken at 10 × magnification on an inverted phase contrast microscope.*

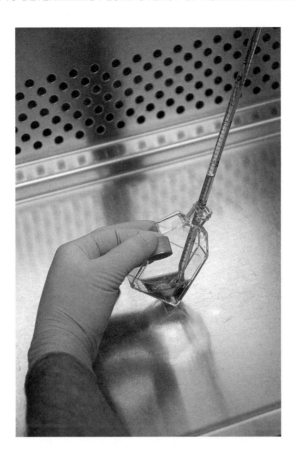

FIGURE 16.3. *Feeding of EB culture by replacing half of the EB medium.*

11. The hESC aggregates/EBs must be fed every day and can be cultured for up to 2–3 weeks depending on desired application (early or late differentiation).

12. Observe EB culture daily under an inverted microscope. Initially, for first 1–2 days, EBs appear as dense spherical aggregates of cells. After a few days of culture, they appear as hollow sphere-like structures (Fig. 16.4).

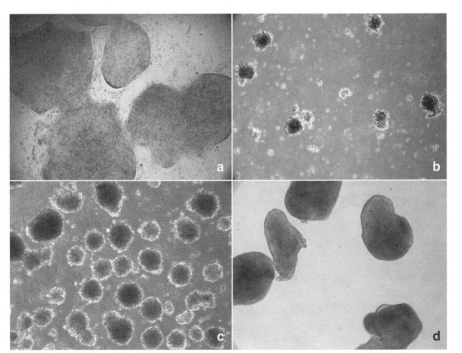

FIGURE 16.4. *Embryoid body cultures at different stages of culture: (a) day 0, hESC culture at 70–80% confluence before harvesting for EB culture; (b) day 2, EB culture; (c) day 7, EB culture; and (d) day 14, EB culture. All images were taken at 10 × magnification on an inverted phase contrast microscope.*

Note: *Be sure not to remove the small hESC aggregates from the flask in the process of removing the medium.*

13. To determine the pluripotency of the hESC culture, EBs grown as described here can be tested for expression of lineage specific markers by RT-PCR (See Chapter 15) or immunocytochemical staining assays (See Chapter 13) to confirm the presence of cell types representing all three germ layers. Examples of some of the lineage markers are mentioned in the introduction of this chapter.

14. These EB cultures can be further differentiated into specific cell types or lineages of interest by culturing them in specialized media to form cell types such as neuronal cells, cardiac muscle cells, or hematopoietic cells.

Note: *The EB/differentiation medium discussed in this protocol is used for undirected differentiation of hESCs. Depending on the desired application of the EBs, the composition of EB medium may be different. For directed differentiation toward a particular lineage, the EB medium may need different concentrations of FBS or addition of specialized supplements, such as neurobasal medium with N2 supplements for neuroepithelial differentiation. EB medium made with different basal media such as Iscove's modified Dulbecco's medium (IMDM) or DMEM may also be used. The length of time of harvest for EB culture may vary depending on the days of induction required for a particular lineage. It is very important to find the correct protocol needed for the desired application of EBs.*

SUGGESTED READING

Stubban C, Wesselschmidt RL, Katov I, Loring J. Cryopreservation of human embryonic stem cells. In Loring J, Wesselschmidt RL, Schwartz PH, eds. *Human Stem Cell Manual: A Laboratory Guide*. New York: Academic Press; 2007.

Sathananthan H and Trounson, *In vitro* differentiation of human ES cells. In Sullivan S, Cowan CA, Eggan K, eds.

Human Embryonic Stem Cells. Hoboken, NJ: Wiley, 2007, pp. 149–167.

WiCell Research Institute. Introduction to Human Embryonic Stem Cell Culture Methods, September 2007.

EMBRYOID BODY (EB) FORMATION LOG SHEET

Date (MM/DD/YY):_____   EB formation initiated by:_____

1. Cell line information:

   - Cell line name:_____   Cumulative hESC passage number:_____

2. hESC colonies prior to dispase treatment:

   - Colony quality:_____   Colony confluence level (%):_____

   - Total number of wells to be harvested:_____

   - Total number of T25 flasks needed:_____

3. Medium and solutions needed for harvesting and EB formation:

   - Dispase/collagenase solution required = 1 mL/well × Number of wells + 2 mL extra:
     1 ×_____   wells + 2 =_____mL

   - EB/differentiation medium for harvesting = 1.5 mL × Number of wells + 2 mL extra:
     1.5 ×_____   wells + 2 =_____mL

   - EB/differentiation medium = 12 mL/15-mL conical tube

   - EB/differentiation medium for feeding = 5–6 mL per T25 flask
     Dispase/collagenase lot #_____, expiration_____, prepared by_____
     EB/differentiation medium lot #_____, expiration_____, prepared by_____

4. Colony morphology documentation at scheduled intervals of EB culture:

   - Prior to starting EB culture:
     Comments:_____
     Images taken by:_____   Images stored in:_____

   - Day 1 to day 14 of EB culture at regular intervals:
     Comments:_____
     Images taken by:_____   Images stored in:_____

CHARACTERIZATION OF HUMAN EMBRYONIC STEM CELLS BY CYTOGENETICS: KARYOTYPING AND FLUORESCENCE *IN SITU* HYBRIDIZATION

17

Shirwin M. Pockwinse, Prachi N. Ghule, and Anne Higgins

Careful monitoring of human embryonic stem cells (hESCs) in culture is necessary to achieve successful long-term growth. A main characteristic of normal cells is the presence of the correct complement of chromosomes in each cell. It is known that long-term continuous passaging of hESCs can cause chromosomal instability that may subsequently lead to chromosomal abnormalities in a subset of the hESC population. These chromosomal abnormalities can render a growth advantage to the cells, allowing them to replace normal cells in the culture. Therefore it is essential to ensure that hESCs in culture maintain the right number and gross morphology of chromosomes. The most common chromosomal abnormalities are aneuploidies of chromosomes. In a normal diploid cell there are two copies of each chromosome, but in an aneuploid cell, certain chromosome pairs either gain or lose a chromosome. Most often, aneuploidies found in hESC cultures are trisomies (gain of a chromosome) of chromosomes 12 and/or 17.

Chromosomal stability of hESCs in culture is evaluated using classical or molecular cytogenetic approaches. Cytogenetics involves the microscopic evaluation of chromosomes to detect structural and/or numerical abnormalities. Karyotyping is the primary classical cytogenetic technique used to assess the chromosomal complement of cells, including the number, morphology, and size of the mitotic chromosomes. Unequivocal identification of chromosomes requires an analysis of banding patterns. In the United States, GTG-banding (G-bands by trypsin and Giemsa) is the most commonly used chromosomal banding technique. GTG-banding involves the use of trypsin (an enzyme) and Giemsa (a dye), which give a characteristic banding pattern to individual chromosomes in a cell. The banded chromosomes are then analyzed and arranged (karyotyped) in a karyogram (Fig. 17.1) based on the International System of Cytogenetic Nomenclature (ISCN) 2009. The advantage of classical cytogenetics is that it can assess the entire genome of a cell and identify any genetic abnormalities at a gross chromosome level.

Molecular cytogenetics combines molecular biology and cytogenetics. Fluorescence *in situ* hybridization (FISH) is a commonly used molecular cytogenetic

Human Stem Cell Technology and Biology, edited by Stein, Borowski, Luong, Shi, Smith, and Vazquez
Copyright © 2011 Wiley-Blackwell.

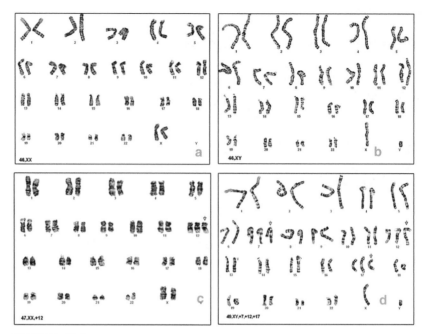

FIGURE 17.1. *Karyotype of hESC lines (a) H9 (46,XX) and (b) H1 (46,XY) showing normal diploid cells. Human embryonic stem cell lines with an abnormal karyotype showing aneuploidies of chromosomes: (c) H9 (47,XX,+12) and (d) H1 (49,XY,+7,+12,+17). Images courtesy of Cytogenetics Laboratory, University of Massachusetts Memorial Medical Center, Worcester, MA.*

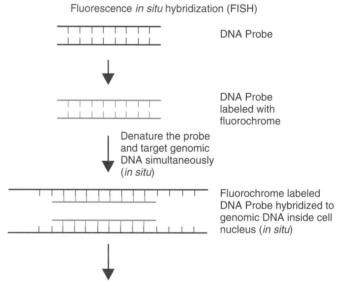

Fluorescence *in situ* hybridization (FISH)

DNA Probe

DNA Probe labeled with fluorochrome

Denature the probe and target genomic DNA simultaneously (*in situ*)

Fluorochrome labeled DNA Probe hybridized to genomic DNA inside cell nucleus (*in situ*)

Fluorochrome labeled DNA-DNA hybridized inside cell nucleus (*in situ*) is visualized by fluorescence microscopy

FIGURE 17.2. *Basic principle of FISH.*

technique. In FISH, DNA probes complementary to one or more specific genomic sequences are used to identify individual genes or sequences ranging in size from ~35 kilobase pairs to entire chromosomes (chromosomes paints). These probes are labeled with different colored fluorescent tags and then hybridized to DNA *in situ* and visualized by fluorescence microscopy (Figs. 17.2 and 17.3). The advantage of FISH is that it can be performed on interphase cells, whereas classical cytogenetics is performed on mitotic cells and is labor intensive and requires analysis by an experienced

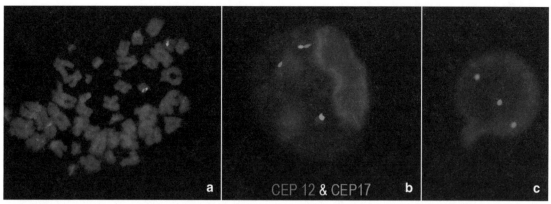

FIGURE 17.3. *FISH image showing staining for chromosomes 12 and 17 in a normal diploid hESC preparation: (a) metaphase cell and (b) interphase cell. (c) FISH image of H9 cells showing a trisomy of chromosome 12.*

cytogeneticist. The disadvantage of FISH is that, depending on the type of probe (region specific or chromosome paint), it can only identify genetic variations limited to the specific regions of the probes used and not the entire genome.

Spectral karyotyping and multicolor FISH are advanced techniques that can be used to identify chromosomal abnormalities in the entire genome. These methods use a combination of many probes, which specifically hybridize to individual chromosome pairs giving distinct color to them. Usually, a combination of at least four to five fluorochromes is used to label each chromosome pair. Disadvantages of this technique are that only metaphase cells are analyzed and it requires an extensive, dedicated equipment setup as well as interpretation and analysis by an expert.

A molecular cytogenetic method that allows investigation of the entire genome at very high resolution is genomic microarray comparative genomic hybridization (CGH). Genomic microarray analysis can detect DNA copy number gains or losses in a single experiment. DNA is extracted from the cells of interest, fluorescently labeled, and mixed in equal proportion with reference DNA labeled with a different fluorescent dye. The mixture of both DNA samples are then hybridized to the microarray slide containing the probes (most often oligonucleotides or single nucleotide polymorphisms, SNPs), and relative concentrations of test DNA and control DNA at each probe site are compared to determine if there is a loss or gain of any genomic regions. The resolution of analysis is determined by how many probes are arrayed on the slide and where the probes are found in the genome. Microarray slides with probes spanning the entire genome, slides with probes targeting a specific region of the genome, or a combination of the two approaches are used for analysis. Two limitations of microarray analysis are the inability to detect balanced chromosomal rearrangements such as translocations or inversions and the inability to detect low-level mosaicism (chromosomal or genetic abnormalities found in a small population of cells). Despite these limitations, genomic microarray analysis is a powerful tool that identifies genetic variations and copy number imbalances across the entire genome at a very high resolution.

A normal diploid human cell has a chromosomal complement of 46 chromosomes comprised of 22 pairs of autosomes and two sex chromosomes (XX or XY). Human embryonic stem cells are the only nontransformed normal diploid cells that can be maintained in long-term culture. Thus regular cytogenetic analysis is essential to evaluate their chromosome stability in culture for their long-term maintenance.

To prepare hESCs for cytogenetic analysis, cells are plated at least 3–4 days prior to treatment (2 hours before harvest) with a mitotic inhibitor such as colcemid to arrest cells in metaphase. Colcemid is related to colchicine (a natural plant product). Colcemid is preferred to colchicine as it is less toxic to the cells. The arrested cells are treated with a hypotonic solution such as potassium chloride at low molarity (0.075 M). Due to change in osmotic balance by hypotonic treatment, the cells swell up and easily burst open when dropped onto a glass slide from a height of 2 feet. The cells are then fixed using Carnoy's fixative, which is a mixture of methanol and acetic acid (3:1 ratio). The fixed mitotic cell preparation can be examined immediately for classical or molecular cytogenetic analysis or stored at −20°C until further use.

This chapter provides detailed protocols for the characterization of hESCs using classical and molecular cytogenetic techniques. However, due to the technical complexities of cytogenetic analysis, it is recommended to be done by an experienced cytogenetic laboratory.

Overview

Plate hESCs and grow for 3–5 days.

Mitotic arrest of hESCs.

Harvest to make a single cell suspension.

Treat with hypotonic solution and fix.

Prepare slides for either karyotyping or FISH analysis.

General Notes

- Read this entire chapter before initiating cell culture activities described in Procedure.
- It is good laboratory practice to have important relevant information recorded; therefore a log sheet is provided as an example at the end of this chapter.

Abbreviations

EDTA: ethylenediaminetetraacetic acid

g: gram

HCl: hydrochloric acid

hESCs: human embryonic stem cells

ISCN: International System of Cytogenetic Nomenclature

KCl: potassium chloride

µg: microgram

µL: microliter

M: molar (molarity)

mM: millimolar

N: normal (normality)

NaCl: sodium chloride

NaOH: sodium hydroxide

PBS: phosphate buffered saline

SSC: sodium saline citrate

Equipment

- 37°C Cell culture incubator
- 37°C Water bath
- Centrifuge with swinging-bucket rotors
- Fluorescence microscope with a digital camera and associated image acquisition and data analysis software

- Inverted light microscope
- Pipette aid
- Slide warmer
- Sterile biosafety cabinet
- Transmitted light microscope with a digital camera and associated karyotyping software
- Vacuum system (recommended)
- Weighing balance

Materials

- 5-mL and 10-mL sterile glass or plastic pipettes
- 22 × 22 × 60 mm No. 1 glass coverslip
- 50-mL sterile conical tubes
- Coplin jars
- Diamond-tip pen
- Disposable gloves
- Glass slides
- T25 tissue culture flask

Reagent Preparation

- *hESC medium* (See Chapter 5)
- *Phosphate buffered saline 10X (PBS):* Dissolve 14.4 grams (g) of disodium hydrogen orthophosphate (Na_2HPO_4) (100 mM), 2.4g of potassium dihydrogen phosphate (KH_2PO_4) (20 mM), 2 g of potassium chloride (KCl) (27 mM), and 80 g of sodium chloride (NaCl) (1.37 M) in 800 mL of water. Adjust the pH to 7.4 with hydrochloric acid (HCl). Adjust the volume to 1 liter with water. Filter and autoclave. Store at room temperature.
- *KaryoMAX® Colcemid™ solution*, liquid (10 µg/mL) (Gibco Cat # 15212-012), in PBS.
- *Sterile trypsin-EDTA (10X) stock solution*, for making single cell suspension: Thaw a 100-mL bottle of sterile 0.5% trypsin-EDTA (10X) (Gibco Cat # 15400-054) and aliquot into 5-mL aliquots. Label with reagent name, concentration, volume, date prepared, and expiration date of 6 months. Store at 0°C.
- *Sterile trypsin-EDTA (1X) working solution*, for making single cell suspension: Thaw a 5-mL trypsin-EDTA (10X) aliquot (see above) and mix with 45 mL of sterile PBS-(1X) (without Ca^{2+}, Mg^{2+}, Gibco Cat # 14190-144) in a sterile 50-mL conical tube. Label tube with reagent name, concentration, volume, date prepared, and expiration date of 1 month. Store at 4°C.
- *Hypotonic solution* (0.075 M KCl): Dissolve 5.55g of potassium chloride in 800 mL of water. Adjust the volume to 1 liter with water. Store at room temperature.
- *Carnoy's modified fixative:* 3 parts methanol + 1 part glacial acetic acid (3:1).
- *pH 6.8 Phosphate buffer:* Dissolve 10 tablets of Gurr's buffer pH 6.8 (Biomed Spec #331992P) into 1 liter of distilled water. Adjust the pH to 6.8 using 1 N HCl or 1 N sodium hydroxide (NaOH). Store solutions at room temperature (expiration date of 1 month).
- *pH 7.0 Phosphate buffer:* Dissolve 1 tablet of Gurr's buffer pH 7.2 (Biomed Spec #33201) and 9 g of sodium chloride (NaCl) into 1 liter of distilled water. Adjust the pH to 7 using 1 N HCl or 1 N NaOH. Store solutions at room temperature (expiration date of 1 month).

- *Giemsa stain (5% stock)–Gurr (Biomed Spec #350862) for banding chromosomes:* For large 50-mL Coplin jars, mix 2.5 mL of Giemsa with 47.5 mL of pH 6.8 buffer. Mix well and prepare fresh as needed.

- *0.25% Trypsin stock, for banding chromosomes:* Thaw a 100-mL bottle of Trypsin (Gibco Cat # 15050-065) and aliquot to 4-mL quantities using 15-mL conical tubes. Label and store at 0°C. Expiration date: 6 months.

- *0.02% Trypsin working solution*, for banding chromosomes: Thaw a 4-mL aliquot of trypsin stock (see above) and mix with 46 mL of pH 7.0 phospate buffer in a Coplin jar. Prepare fresh as needed.

- *20X Sodium saline citrate (SSC) (pH 5.3):* Dissolve 175.3 g of NaCl (3 M) and 88.2 g of trisodium citrate ($Na_3C_6H_5O_7$) (0.3 M) in 800 mL of water. Adjust the pH to 5.3 with HCl. Adjust the volume to 1 liter with distilled water. Filter and autoclave. Store at room temperature.

- *0.4X SSC:* Dilute 20 mL of 20X SSC in 1000 mL of distilled water. Adjust pH to 7.

- *2X SSC/0.1% Nonidet P-40 (NP-40):* Dilute 100 mL of 20X SSC in 1000 mL of distilled water followed by addition of 1 mL of NP-40. Adjust pH to 7.

Preparation

NOTES

1. Place the following items near the biosafety cabinet:
   - 5-mL and 10-mL Sterile pipettes
   - Absorbent paper towels (or Kimwipes®)
   - 70% Ethanol spray
   - Disposable gloves

2. Ensure that the following items are in the biosafety cabinet:
   - 5-mL Sterile pipettes
   - 50-mL Sterile conical tubes
   - 10-mL Sterile pipettes
   - Alcohol-proof marker

3. Preparation of hESCs:
   3.1. To prepare hESCs for karyotypic analysis, grow them for at least 3–5 days or until they reach 50–60% confluence (See Chapter 9 on page 105). (*This protocol may be viewed on the accompanying DVD and on the web at* http://www.wiley.com/go/stein/human.)

Procedure

1. Follow preparations according to Chapter 3, Aseptic Technique on page 19. (*This protocol may be viewed on the accompanying DVD and on the web at* http://www.wiley.com/go/stein/human.)

2. Mitotic arrest of hESCs for cytogenetic analysis:

   Note: *The calculations of colcemid solution in hESC medium given below are for a six-well plate. If using any other culture vessel, the calculations should be adjusted accordingly.*

   2.1. The stock concentration of colcemid is 0.01 µg/µL. In the biosafety cabinet, add 5 µL of colcemid per milliliter of hESC medium to achieve a concentration of 0.05 µg/mL. It is recommended to prepare enough hESC medium containing colcemid (0.05 µg/mL) for an entire six-well plate by adding 75 µL of stock colcemid to 15 mL of hESC medium.

2.2. Remove hESC plate that was passaged 3–5 days earlier from incubator and place in the biosafety cabinet. Aspirate old hESC media from the wells and replace with 2.5 mL per well of hESC medium with colcemid.

2.3. Return hESC plate to 37°C incubator for 2 hours.

> *Note: It is recommended that the colcemid concentration and incubation time be optimized for each cell line as the growth rates of individual hESC lines may differ. Incubation time may be adjusted within a range of 2–4 hours and the concentration of colcemid can be adjusted from 0.05 μg to 0.1 μg per milliliter of culture medium. Alternatively, incubate overnight with 0.01 μg of colcemid per milliliter of culture medium.*

3. Harvesting of hESCs for cytogenetic analysis:

3.1. After the 2-hour incubation with colcemid, remove hESC plate from the incubator and place in the biosafety cabinet. Collect the hESC medium from each well in a 50-mL conical tube (harvest all six-wells in one tube).

3.2. Add 1 mL of trypsin-EDTA (0.02% trypsin working stock) into each well of the six-well plate and place in a 37°C incubator for 5–7 minutes.

3.3. Place the plate back in the biosafety cabinet and if the cells have not lifted off the plate surface, then gently disrupt them by using a 5-mL pipette. Collect the cell suspension and place in the same 50-mL conical tube.

3.4. Mix well by pipetting up and down to make a single cell suspension.

3.5. Centrifuge the cell suspension at $200g$ (~1000 rpm) for 5 minutes at room temperature.

3.6. Decant supernatant and resuspend the pellet in 5–10 mL of prewarmed hypotonic solution (0.075 M KCl). Add hypotonic solution to the pellet drop-wise and mix gently.

3.7. Incubate the tube in a 37°C water bath for 20–30 minutes.

3.8. Fix the cells using chilled Carnoy's fixative (3:1, methanol: acetic acid). Prepare fresh fixative for each harvest. Add 1 mL of fixative drop by drop into the cell suspension with hypotonic solution. Let stand for 5–10 minutes.

3.9. Centrifuge for 5 minutes at $200g$ (~1000 rpm). Decant supernatant and add 10 mL of fresh fixative to the pellet and mix by pipetting up and down. Let stand for 10 minutes at room temperature.

3.10. Repeat fixation twice. Store the cells at 4°C or −20°C until ready for slide preparation.

> *Note: The fixed hESC cells can be stored at 4°C for up to 2 weeks. For long-term storage, the fixed hESCs can be transferred to a cryovial and stored in a −20°C freezer.*

4. Slide preparation for making mitotic spreads for karyotyping (classical cytogenetic technique):

4.1. Resuspend the fixed cells in 1–2 mL of fresh Carnoy's fixative.

4.2. Drop 200 μL of the cell suspension (8–9 drops) onto a wet chilled slide held at a distance of 2–3 feet (handheld) and blow with the mouth slightly across the slide surface to assist in the spreading of metaphase chromosomes and cells.

4.3. Observe chromosomes and spreading under an inverted microscope. Adjust dropping procedure (e.g., how many drops of suspension used, distance for dropping, and/or vigorousness of blowing) as needed to achieve the best quality of metaphase spreading for analysis.

NOTES

_____
_____
_____
_____
_____
_____
_____
_____
_____
_____
_____
_____
_____
_____
_____
_____
_____
_____
_____
_____
_____
_____
_____
_____
_____
_____

4.4. Dry the slides on a hot plate at 60°C and store in a box at room temperature for 2–4 days for aging. (Aging the slides improves banding patterns of chromosomes.) Alternatively, to do the Giemsa–trypsin banding immediately, the slides can be baked overnight at 65°C or 2–3 hours at 90°C. Follow the banding procedure given below.

5. Giemsa–trypsin banding of mitotic spreads for chromosome analysis:

   5.1. Prepare reagents as described in Reagents section of this chapter.

   5.2. Set up FIVE numbered Coplin jars as follows:
   #1—50 mL working 0.2% trypsin solution
   #2—50 mL of pH 7.0 phosphate buffer
   #3—50 mL of pH 7.0 phosphate buffer
   #4—50 mL of 4% Giemsa solution
   #5—50 mL distilled water

   5.3. Place slides in Coplin jar #1 for 1 minute and 30 seconds. (Adjustment to the timing may be needed, depending on the quality, age, and so on of the slides.)

   5.4. Dip the slides briefly into Coplin jar #2.

   5.5. Place the slides in Coplin jar #3 for 60 seconds.

   5.6. Place the slides in Coplin jar #4 for 3 minutes and 30 seconds.

   5.7. Dip the slides briefly into Coplin jar #5.

   5.8. Allow slides to dry thoroughly on a 60°C slide warmer.

   5.9. Place a 22-mm × 22-mm × 60-mm No. 1 glass coverslip on each slide with a drop of mounting media (Cytoseal™ XYL, Stephens Scientific 8312-4).

6. Analysis of Giemsa–trypsin banded slides:

   6.1. Observe slides under a transmitted light microscope fitted with appropriate filters and objectives (10×, 40×, and 100× magnification).

   6.2. A total of 20 well-banded metaphase cells should be analyzed and the images captured and karyotyped using associated software. The chromosomes should be compared to ISCN 2009 standards to determine if the karyotype is normal or abnormal.

   6.3. The hESC culture should be considered clonally abnormal (containing a population of cells derived from a single abnormal progenitor cell) if there is a chromosome gain noted in two or more or a loss in three or more spreads. Abnormalities occurring less frequently can be considered nonclonal or random.

7. Fluorescence _in situ_ hybridization (FISH) of hESC for aneuploidy analysis (molecular cytogenetic analysis):

   7.1. Follow Procedure as discussed in Steps 2 and 3 of this chapter to harvest cells for FISH analysis.

   7.2. Drop 200 µL of the fixed cell suspension on prechilled slides and dry the slides by keeping on a hot plate (50–60°C) for a few seconds.

   7.3. Incubate the slides in a 60°C oven for 1–2 hours.

   7.4. Viewing under a phase contrast microscope, mark the area of cell suspension on the slide with a diamond-tip pen.

8. Hybridization:

   8.1. Probes used: CEP® 17 (chromosome 17 centromere specific) Spectrum Green™ DNA Probe (Vysis, Abbott Inc.), CEP® 12 (chromosome 12 centromere specific) Spectrum Orange™ DNA probe (Vysis, Abbott

Inc.). Probe mixture: prepare according to the manufacturer's protocol (1 µL probe +7 µL hybridization buffer +2 µL deionized water).

8.2. Prewarm a hot plate to 72°C before adding probe mixture to the slide.

8.3. Dehydrate the slides by placing in Coplin jars with increasing ethanol concentrations of 70%, 85%, and 100% for 5 minutes each. Air-dry the slides.

8.4. Apply 5–10 µL of probe mixture onto the marked area on the slide and apply a coverslip. Avoid air bubbles and seal the coverslip with rubber cement.

8.5. Co-denature the cells affixed to the slide with the probe mixture by heating to 72°C for 3–5 minutes on a prewarmed hot plate.

8.6. After denaturation, transfer the slides to a humidity chamber and incubate at 37°C overnight for hybridization.

9. Posthybridization washing (according to manufacturer's protocol: Vysis, Abbott Inc.):

9.1. Prewarm 0.4X SSC (pH 7–7.5) in a Coplin jar in a 72°C water bath for least 30 minutes prior to starting the wash.

9.2. Remove the rubber cement and coverslip carefully and immediately put the slides in the Coplin jar containing 0.4X SSC (pH 7–7.5) at 72°C. Agitate the Coplin jar holding the slides for a few seconds and incubate for 2 minutes.

9.3. Remove the slides from wash bath and place in Coplin jar containing 2X SSC/0.1% NP-40 at room temperature for 1 minute, agitating for 2–3 seconds as the slides are placed in the jar.

9.4. Remove slides from the Coplin jar and allow to air dry in the dark.

9.5. Apply 10 µL of DAPI-II counterstain (Vysis, Abbott Inc.) to the marked area of the slide and apply a coverslip. Store the slides at −20°C in the dark.

9.6. Observe slides on a fluorescence microscope fitted with appropriate filters and score according to standard criteria provided by the manufacturer of DNA probes used in the assay (e.g., Vysis, Inc.). If a chromosome specific probe was used for the assay, count the copy number of that chromosome for each cell and note any abnormalities in the number. In a normal state a chromosome has two copies per cell. An aneuploid cell will either have a gain (three or more copies) or a loss (single copy) of a chromosome pair.

SUGGESTED READING

Brimble SN, Zeng X, Weiler DA, et al. Karyotypic stability, genotyping, differentiation, feeder-free maintenance, and gene expression sampling in three human embryonic stem cell lines derived prior to August 9, 2001. *Stem Cells Dev.* 2004;13(6):585–597.

International System for Human Cytogenetic Nomenclature (ISCN) 2009. Recommendations of the International Standing Committee on Human Cytogenetic Nomenclature Published in collaboration with "Cytogenetic and Genome Research." Edited by Shaffer LG, Slovak ML, and Campbell LJ.

Shen Y, Wu BL. Microarray-based genomic DNA profiling technologies in clinical molecular diagnostics. *Clin Chem.* 2009;55:659–669.

Wi-Cell Research Institute. Introduction to Human Embryonic Stem Cell Culture Methods, September 2007.

hESC CYTOGENETIC ANALYSIS LOG SHEET

Date (MM/DD/YY): _____   Cytogenetic analysis done by: _____

1. Cell line information:

   • Cell line name: _____ Cumulative hESC passage number: _____

2. Medium and solutions needed for harvesting and karyotyping:

   • Colcemid solution required = 2.5 mL/well × Number of wells +2 mL extra = 2.5 × _____ wells +2 = _____ mL

   • Giemsa lot #: _____   Expiration : _____   Prepared by: _____

   • Trypsin (0.2%) lot #: _____   Expiration : _____   Prepared by: _____

   • Giemsa (4%) lot #: _____   Expiration : _____   Prepared by: _____

3. Number of metaphase spreads analyzed microscopically: _____

4. Number of metaphase spreads karyotyped: _____

5. Chromosome analysis:

   a. Chromosome index number: _____

   b. Abnormalities (if any): _____

6. Final karyotype for the cell line: _____

   Comments: _____

hESC FISH ANALYSIS LOG SHEET

Date (MM/DD/YY): _____   Cytogenetic analysis done by: _____

1. Cell line information:

   • Cell line name: _____ Cumulative hESC passage number: _____

2. Probes used: _____

3. Number of interphase cells scored: _____

4. Number of metaphase spreads analyzed: _____

5. FISH analysis:

   a. Number of copies of each chromosome probe per cell: _____

   b. Number of cells counted: _____

   c. Abnormalities (if any): _____

6. Final FISH analysis for the cell line: _____

HIGH-RESOLUTION CHROMATIN IMMUNOPRECIPITATION ASSAY

18

Beatriz Pérez-Cadahía, Bojan Drobic, and James R. Davie

Chromatin is a highly dynamic structure composed of DNA, proteins, and RNA[1,2] that plays a major role in regulating cellular processes such as gene expression, DNA replication, and repair. The need to study the interplay between transcription factors and transcriptional machinery with the chromatin modifications has raised the necessity of developing new techniques.

The chromatin immunoprecipitation (ChIP) method was first developed by Solomon and co-workers to study the distribution of histones in *Drosophila* heat shock gene promoters[3] and further adapted for the study of a broader spectrum of proteins, including the transcription factors.[4]

Currently, ChIP is a powerful tool widely used to study protein–DNA interactions and location of proteins at specific chromosomal sites. This method has provided very reliable information on DNA-associated proteins in a wide variety of organisms from yeast to mammals.[5] The method relies on the specificity of antibody-mediated protein detection to selectively enrich chromatin that is associated with the protein of interest. The subsequent DNA isolation allows for quantitative analysis of protein–DNA interactions.[6] In order to fulfill the needs of the different biological contexts, modifications of the method have been developed. ChIP can be performed under native conditions or with the use of protein–DNA crosslinking reagents. PCR or quantitative PCR is usually the preferred method for the analysis of the immunoprecipitated DNA. Nevertheless, the implementation of microarrays (ChIP-on-chip) and sequencing (ChIP seq) has allowed for a genome-wide ChIP analysis.

A major limitation of the conventional ChIP assay is the requirement of a large number of cells (10^6–10^7 range). This was resolved by the development of Carrier ChIP[7] and micro ChIP[8]. These developments allowed the ChIP technique to be used for small cell samples such as biopsies, small stem cell populations, or embryonic cells, thereby allowing the assessment of samples using as few as 100 cells.

Here we describe a detailed ChIP protocol, optimized to increase the resolution, to minimize background signal level, and to generate a highly homogeneous pattern.

An essential feature presented in this method is the use of micrococcal nuclease (MNase) digestion instead of sonication sheering to generate the chromatin fragments. MNase cleaves unprotected DNA and if the conditions are optimized, MNase digestion can generate mononucleosomal length of the chromatin fragments (~150 base pairs), allowing for an increased resolution of the ChIP procedure. The

Human Stem Cell Technology and Biology, edited by Stein, Borowski, Luong, Shi, Smith, and Vazquez
Copyright © 2011 Wiley-Blackwell.

additional implementation of an accurate DNA quantification step, high quality PCR primer sets, and precise qPCR data analysis greatly improve ChIP resolution, reliability, and reproducibility, as ChIP has been demonstrated by its successful application to several cell types/systems.

Overview

Cell culture

Crosslinking

Nuclei isolation

MNase digestion

Nuclei lysis

Lysate preclearing and quantification

Input, IP, *Negative Control, Positive Control*

Immunoprecipitation

Elution

Reverse crosslinks

DNA isolation and quantification

QPCR input, QPCR IP, QPCR Negative Control, QPCR Positive Control

Abbreviations

$CaCl_2$: calcium chloride

ChIP: chromatin immunoprecipitation

DNA: deoxyribonucleic acid

DSP: dithiobis(succinimidyl propionate)

EDTA: ethylenediaminetetraacetic acid

KCl: potassium chloride

KOH: potassium hydroxide

LiCl: lithium chloride

MNase: micrococcal nuclease

μg: microgram

μL: microliter

μm: micrometer

mL: milliliter

M: molar

mM: millimolar

NaCl: sodium chloride

$NaHCO_3$: sodium bicarbonate

NP-40: nonyl phenoxylpolyethoxylethanol

PCR: polymerase chain reaction

PIPES: piperazine-1,4-bis(2-ethanesulfonic acid)

RNA: ribonucleic acid

SDC: sodium deoxycholate

SDS: sodium dodecyl sulfate

Tris-HCl: tris(hydroxymethyl)aminomethane hydrochloride

Equipment

- Cell culture incubator
- Centrifuge

- MagneSphere® Technology Stands
- Orbitron
- Spectrophotometer
- Water bath

Materials
- Dynabeads® Protein G
- GIAQuick PCR Purification Kit
- PicoGreen® dsDNA Quantitation Assay
- Proteinase K
- Ribonuclease A

Reagent Preparation
- Cell lysis buffer
  - 5 mM PIPES (pH with KOH to 8.0)
  - 85 mM KCl
  - 0.5% NP-40
- MNase digestion buffer
  - 10 mM Tris-HCl pH 7.5
  - 0.25 M sucrose
  - 75 mM NaCl
- RIPA buffer
  - 10 mM Tris-HCl pH 8.0
  - 1% Triton®X-100
  - 0.1% SDS
  - 0.1% SDC
- Low salt wash buffer
  - 0.1% SDS
  - 1% Triton X-100
  - 2 mM EDTA
  - 20 mM Tris-HCl pH 8.1
  - 150 mM NaCl
- High salt wash buffer
  - 0.1% SDS
  - 1% Triton X-100
  - 2 mM EDTA
  - 20 mM Tris-HCl pH 8.1
  - 500 mM NaCl
- LiCl wash buffer
  - 250 mM LiCl
  - 1% NP-40
  - 1% Deoxycholate
  - 1 mM EDTA
  - 10 mM Tris-HCl pH 8.1
- 1 × TE buffer
  - 10 mM Tris-HCl pH 7.5
  - 1 mM EDTA

- Autoclave the solution
- Elution buffer
  - 1% SDS
  - 100 mM $NaHCO_3$

Preparation

1. Grow cells to ~80–90% confluence (see Chapter 9 on page 105.) (*This protocol may be viewed on the accompanying DVD and on the web at http://www.wiley.com/go/stein/human.*)

2. Crosslinking:

   2.1. Directly add formaldehyde to the medium to a final concentration of 1% (made up in $1\times$ PBS).

      Note: Alternatively, resuspend cells in 1–5 mL of 1 × PBS with 1% formaldehyde.

      Crosslinking step is optional depending on the nature of the DNA–protein interaction of interest. When studying histone modification, it is not required, although it helps to maintain the chromatin structure.

      Double crosslinking (formaldehyde+dithiobis(succinimidyl propionate) (DSP)) is sometimes recommended when studying proteins that associate with DNA indirectly.

   2.2. Incubate cells at room temperature for 8–10 minutes.

      Note: A 10-minute incubation with formaldehyde usually provides a good efficiency. However, this step should be optimized prior to the study, as excessive crosslinking could reduce the protein recovery and may cause problems due to epitope occlusion.

   2.3. Add glycine to a final concentration of 125 mM to stop crosslinking. Incubate 3–5 minutes at room temperature.

   2.4. Remove the medium, and wash the cells twice with ice-cold PBS.

   2.5. Centrifuge at 2000*g* for 5 minutes to harvest cells.

      Note: Cells can be stored at—80° C after this step.

Procedure

1. Resuspend the cell pellet with 5–10 mL of cell lysis buffer containing phosphatase/protease inhibitors. Rotate for 5–10 minutes. To obtain the nuclei, pellet the resuspension for 5 minutes (2000*g*). Repeat this wash step one more time.

2. Resuspend the nuclear pellet with 1–2 mL of MNase digestion buffer plus phosphatase/protease inhibitors.

   2.1. Quantify DNA in solution. Measure the A_{260} of the suspension with spectrophotometer: 1 µL sample + 99 µL 5 M urea/2 M NaCl.

   $$1A_{260} = 50\ \mu g\ DNA/mL.$$

3. Add $CaCl_2$ to a final concentration of 3 mM and prewarm the samples at 37°C for 10 minutes.

4. Quickly add 2.5 U of MNase per 50 µg of DNA and incubate the samples at 37°C for 20 minutes. (The amount and the time of MNase incubation should be tested for each cell line used for ChIP. DNA should be run on agarose gel to check for fragment size, which should be 150–160 bp in length.)

NOTES

_____

_____

_____

_____

_____

_____

_____

_____

_____

_____

_____

_____

_____

_____

_____

_____

_____

_____

_____

_____

_____

_____

Note: Lysate quantification is important in order to establish a constant ratio of starting material/units of MNase.

The concentration and duration of MNase digestion should be optimized for each cell type and revised when starting a new batch of the enzyme.

5. Stop the reaction with the addition of EDTA pH 8.0 to a final concentration of 5 mM.

6. Lyse the nuclei with SDS (0.5% final concentration) by rotating at room temperature for 1 hour.

7. Spin down the insoluble material by centrifugation (2000*g* for 5 minutes).

8. Dilute the nuclear lysate (5×) to 0.1% SDS with RIPA buffer containing phosphatase/protease inhibitors.

9. Preclear the lysate with Protein A/G agarose beads (60 μL per mL of lysate) for 1 hour at 4°C.

10. Spin down the lysate to pellet the beads and transfer the supernatant to new tubes.

11. Measure the concentration of soluble DNA with a spectrophotometer (see step 2.1).

    Note: Sample quantification after preclearing is important to keep the amount of starting material constant per antibody and time point.

    Lysates can be stored at −80°C after quantification.

12. Aliquot 12 A_{260} units (600 μg of soluble DNA) of the lysate and add the antibody of interest. Incubate overnight at 4°C with rotation.

    Note: The specificity of each antibody must be tested and the efficiency of the IP must be predetermined to optimize the lysate/antibody ratio.[9]

    Negative controls can include preimmune serum, normal isotype matched antibody, or performing ChIP with no antibody.

13. The next day, add 7 μL of magnetic Dynabeads Protein A or G beads for every 50 μg of DNA and incubate for 2 hours with rotation at 4°C.

    Note: The type of beads chosen for the IP depends on the origin of the antibody.

    Magnetic beads are recommended as they minimize nonspecific binding to the precoated portion of magnetic beads.

14. Pellet the magnetic beads and remove the supernatants (use a magnetic stand or low-speed centrifugation for 30 seconds).

15. Wash the beads at room temperature for 7 minutes with rotation. Perform each wash with the indicated buffer order below at least twice. Use 1 mL of buffer:
    - Low salt wash buffer
    - High salt wash buffer
    - LiCl wash buffer
    - 1× TE buffer

16. Elute the antibody/chromatin complexes by adding 250 μL of elution buffer to the beads. Vortex and incubate at room temperature for 15–30 minutes. In a separate tube, incubate 1 A_{260} unit (50 μg, 8% of Input added to each antibody sample) in an identical manner (room temperature for 15–30 minutes with rotation followed by overnight incubation at 65°C).

    Note: The magnetic beads can be recovered for reuse after elution of antibodies.

17. DNA isolation and PCR:

    17.1. The next day, treat all the samples with RNase A (0.02 μg/mL final concentration) for 30 minutes at 37°C.

NOTES

_____

_____

_____

_____

_____

_____

_____

_____

_____

17.2. Subsequently, treat the samples with Proteinase K (0.5 μg/mL final concentration) for 1 hour at 55°C.

17.3. Pellet the magnetic beads (magnetic stands or low-speed centrifugation) and use the supernatant to isolate DNA. The DNA can be isolated using Qiagen's QIAquick PCR Purification Kit.

17.4. Quantify the resulting DNA by using the PicoGreen dsDNA Quantitation Assay.

17.5. Load equal amounts of input and ChIP DNA for quantitative RT-PCR (see Chapter 15). Primers and PCR conditions should be optimized.

Note: _Accurate quantification and loading are important for improving reliability when analyzing the effect of different experimental conditions (time points, treatment doses, etc.) on the levels of the target protein. As antibodies usually have different efficiencies, in order to avoid misleading results, comparisons should only be made among data from experiments carried out with the same antibody._

Acknowledgments

Grant sponsors: Canadian Institute of Health Research (MOP-9186), CancerCare Manitoba Foundation, Inc., Canadian Breast Cancer Foundation, Canadian Cancer Society Research Institute, Canada Research Chair to JRD, Canadian Cancer Society Research Institute Terry Fox Foundation studentship to BD, Manitoba Health Research Council Fellowship to BP.

REFERENCES

1. Bernstein E, Allis CD. RNA meets chromatin. _Genes Dev_. 2005;19:1635–1655.

2. Felsenfeld G, Groudine M. Controlling the double helix. _Nature_. 2003;421:448–453.

3. Solomon MJ, Larsen PL, Varshavsky A. Mapping protein–DNA interactions in vivo with formaldehyde: evidence that histone H4 is retained on a highly transcribed gene. _Cell_. 1988;53:937–947.

4. Orlando V, Strutt H, Paro R. Analysis of chromatin structure by in vivo formaldehyde cross-linking. _Methods_. 1997;11:205–214.

5. Haring M, Offermann S, Danker T, et al. Chromatin immunoprecipitation: optimization, quantitative analysis and data normalization. _Plant Methods_. 2007;3:1–16.

6. Turner FB, Cheung WL, Cheung P. Chromatin immunoprecipitation assay for mammalian tissues. _Methods Mol Biol_. 2006;325:261–272.

7. O'Neill LP, VerMilyea MD, Turner BM. Epigenetic characterization of the early embryo with a chromatin immunoprecipitation protocol applicable to small cell populations. _Nat Genet_. 2006;38:835–841.

8. Collas P, Dahl JA. Chop it, ChIP it, check it: the current status of chromatin immunoprecipitation. _Front Biosci_. 2008;13:929–943.

9. Spencer VA, Sun JM, Li L, et al. Chromatin immunoprecipitation: a tool for studying histone acetylation and transcription factor binding. _Methods_. 2003;31:67–75.

ASSAYING PLURIPOTENCY VIA TERATOMA FORMATION

19

M. William Lensch and Tan A. Ince

Pluripotency is a functional definition describing a cell's ability to form derivatives of all three embryonic germ layers: ectoderm, mesoderm, and endoderm. While the verification of functional pluripotency in mouse systems is best accomplished in vivo via the formation of germline competent chimeras, such assays are ethically proscribed using human cells. Instead, in vivo pluripotency is determined via experimental teratoma formation following the injection of undifferentiated pluripotent stem cells into immunodeficient murine hosts. This chapter outlines the rationale, methodology, interpretation, and caveats of the teratoma formation assay using human pluripotent stem cells.

INTRODUCTION AND BACKGROUND

Pluripotency is functionally defined by the ability of a cell to form derivatives of all three embryonic germ layers: ectoderm (e.g., skin and nerve), mesoderm (e.g., muscle, bone, and cartilage), and endoderm (e.g., liver, gut, and lung)(for review, please see Ref. 1). Such a determination is retrospective; that is, one knows a functionally pluripotent cell was present at the *beginning* of an experiment if it generates a downstream entity containing all three germ layers at the *conclusion* of that experiment.

Historically, the only such downstream entities noted in vivo other than embryonic, fetal, and live-born animals were naturally occurring tumors containing various cells, tissues, and structures originating from all three germ layers including ovarian dermoids, embryonal carcinomas, teratocarcinomas, and teratomas.[2–7] The suspected origins of such masses have long been thought to be from germ cells dysregulated at various stages of development.[8–11] The seminal findings that permitted a detailed study of pluripotency at the genetic and cellular levels were the observations that the 129 strain of inbred mouse was prone to testicular teratoma[12] and that the pluripotent, tumor-inducing cells within these masses, the so-called embryonal carcinoma or EC cell, could be isolated and cultured in vitro.[13]

Once EC cells became experimentally available, a field of study opened up that permitted detailed analyses of pluripotency both in vivo[14] and in vitro.[15–17] Knowledge gained from the culture of abnormal, tumor-derived pluripotent cells would inform studies leading to the isolation of pluripotent cells from both normal,

Human Stem Cell Technology and Biology, edited by Stein, Borowski, Luong, Shi, Smith, and Vazquez
Copyright © 2011 Wiley-Blackwell.

preimplantation mouse[18,19] and human embryos,[20] so-called embryonic stem or ES cells. In turn, analysis of the data generated from these pluripotent cells led to the functional reprogramming of mature, murine somatic cells to pluripotency, so-called induced pluripotent stem or iPS cells, via direct genetic manipulation,[21] a process that was refined to demonstrate germline competency very soon thereafter.[22–24] Lines of iPS cells have likewise been generated from human somatic tissues and while their capacity to contribute to gestational chimeras cannot be ascertained due to ethical constraints, human iPS cells are capable of forming teratomas in immunodeficient murine hosts.[25–29] Of note, human ES and iPS cells do not appear to be equivalent to mouse ES cells where the human lines are more similar to mouse epiblast-derived stem cells or Epi-SCs.[30] Epi-SCs are obtained from a later developmental stage than ES cells and as a result are unable to chimerize recipient blastocysts (reviewed in Ref. 31).

In the mouse, the truest assessment of pluripotency is had by chimerizing murine embryo hosts and implanting them into the uterus, where an input cell's contribution to all three germ layers in the resulting mouse's soma, and more significantly its germline, defines the pluripotency of the input cells. While germline competency stands as a high standard for cellular potency, tetraploid complementation presents an even higher bar.

In tetraploid complementation, the blastomeres of the early two-cell mouse embryo are electrofused to generate a single-celled, tetraploid (4n), pseudozygote.[32,33] While this tetraploid embryo continues to divide, the 4n cells are poor contributors to the inner cell mass, from whence the mouse proper derives, though they remain competent to establish the trophectodermal, extraembryonic outer layer of the early embryo. Chimerizing the 4n embryo with healthy, pluripotent (and diploid) cells will complement the 4n embryo's inability to generate an entire gestating mouse (extraembryonic and embryonic portions together). Studies show that not all pluripotent cells are capable of meeting the standard of complementing the tetraploid embryo[34] and thus those that do are considered to be of the highest grade of potency.[35–37]

In human cell systems, embryo chimerism/gestation experiments are ethically impermissible.[38] Thus a series of surrogate assays are often brought to bear on human cells in order to determine their pluripotency. The results of most of these assays are obtained entirely in vitro and include the spontaneous or forced differentiation of cell cultures as embryoid bodies followed by the demonstration of gene expression signatures from all three germ layers via RT-PCR, immunodetection of tissue proteins (e.g., Western blots, immunofluorescence, and flow cytometry analysis), or via a variety of other methods. While of great utility for certain applications,[39] in isolation, each of these assays presents a low degree of confidence that a given cell under study is actually pluripotent.[34] However, combining several of these assays into a panel of tests contributes greater confidence to any potency determination.

That said, while a combination of findings obtained in vitro may be suggestive of pluripotency, it often remains desirable to generate results from systems with an even greater degree of stringency when possible. What the teratoma assay provides above and beyond results obtainable using in vitro assays alone is the capacity to investigate cellular differentiation in an integrated host, taking full advantage of incompletely understood soluble factors, cell–cell contacts, and other differentiation-promoting cues where the resulting structures are routinely capable of forming functional, three-dimensional tissue architecture. As one example here, Figure 19.5c (see Analysis of Pluripotent Cell-Derived Masses) on page 223 depicts a highly organized focus of gut tissue with crypts and villi, a surrounding layer of muscle, and a central lumen filled with secreted, mucinous material, all of which was formed from a line of human iPS cells subjected to the teratoma formation assay. As a cautionary note, while the interpretation of structures as obvious as this example is routine, the overall evaluation of teratoma assays is complex and requires evaluation by an experienced pathologist.[40,41] The assay is entirely capable of being subjected to greater experimental rigor than is routinely used[29] and the coming days will no doubt see increased efforts in this direction (see Analysis of Pluripotent Cell-Derived Masses on page 223.).

Experimental teratomas are very similar but not identical to true teratomas, where the latter are a form of naturally occurring dysplasia bearing genetic and/or epigenetic damage[38,41] (see Analysis of Pluripotent Cell-Derived Masses on page 223.). If a human pluripotent stem cell line cannot reliably form a teratoma containing identifiable representatives of all three embryonic germ layers—ectoderm, mesoderm, and endoderm—then it may be a cell line with impaired developmental capacity; a point deserving consideration when evaluating the utility of the cell line in various assays. Furthermore, the developmental potency of stem cell lines is known to change due to the fact that abnormal genetic variants will occasionally arise over time in cell culture.[42] Thus it is important to perform routine quality control experiments, including teratoma formation, with all cell lines in use.

In summary, the teratoma assay is an important contribution to the battery of tests to which human pluripotent cells are subjected in order to define them as pluripotent prior to publication and use in laboratory and eventual clinical platforms. Furthermore, teratoma formation is a very useful and highly instructive assay alongside other tests (such as karyotyping) for routine monitoring during the maintenance of pluripotent cells following long-term culture or upon culture reestablishment after long-term cryopreservation.

METHODOLOGY: HOST STRAIN SELECTION

A wide variety of immunodeficient mice are capable of serving in human cell xenotransplantation studies,[43–46] including teratoma formation assays.[38] Such immunodeficient animals may be selected from a number of possible genetic backgrounds, each with a differing genetic basis for immunodeficiency as well as variation with regard to what components of the immune system are lacking. Common strains include:

- SCID (severe-combined immunodeficient; lacking mature B and T cells)
- SCID-Beige (similar to SCID with an added reduction in NK cells)
- NOD/SCID (non-obese diabetic-SCID; lacking mature B and T cell, reduced NK cells, and no serum complement)
- BalbC/Nude (athymic, lacking T cells)
- Rag2 (lacks mature B and T cells)
- Rag2gammaC (lacks mature B, T, and NK cells)

Strain selection may be based on any number of factors including cost effectiveness, where SCID mice are routinely available within institutionally based vivaria and inexpensive compared to strains with restricted commercial availability. Immunodeficient mice may experience a range of immune system-related morbidities including a high incidence of spontaneous lymphomas, reduced fecundity, shortened life span, and infections; the latter of which are lethal regardless of the variety and locus of the infection and often require rapid culling of infected animals demonstrating signs of illness, such as circling behavior, in order to prevent the spread of infection throughout the entire colony. Investigators are strongly encouraged to consult with their institutional animal care and use committee (IACUC) or equivalent for advice on strain selection and colony maintenance issues. Beyond this, prior experience with a specific strain is invaluable when choosing which background to use.

TERATOMA FORMATION

One of three common injection routes is routinely used for teratoma formation: subcutaneous (SC), intramuscular (IM), or intratesticular (IT). Of these three, intratesticular stromal cells are more likely to produce a milieu of supportive factors capable of fostering input pluripotent cells with the caveat that this route is the most invasive of the three and requires added expertise to perform and more extensive monitoring of experimental animals (see below). While some investigators have made use of other anatomical locations (including the kidney capsule or intraperitoneal routes), intramuscular implantation has been found to be the most efficient for a variety of reasons.[47] It is worthy to comment that some investigators note restrictions on which anatomical locations are capable of producing a mass with a given cell line; for example, a certain cell line may not form subcutaneous tumors while simultaneously demonstrating robust intratesticular teratoma capacity.

Also, a threshold number of input cells is required in order to form a teratoma, where one laboratory notes that at least a few hundred input cells are needed, a result that depends on the cell line used.[47] This finding presents some caveats including that teratoma formation is not a clonal assay and thus may not necessarily prove that a single pluripotent stem cell was present a priori. However, it is entirely possible to subclone lines of human pluripotent stem cells in advance, expand the subclone in culture, and then demonstrate robust teratoma formation, which does indicate that a single input cell has the capacity to generate a population of teratoma-forming cells. Discussion of the possible reasons and underlying pathophysiological bases for such disparities is beyond the scope of this chapter, though obviously presenting fertile ground for study.

The ability to perform control studies with well-described teratocarcinoma-derived pluripotent cells (embryonal carcinoma or EC cells) such as NTERA-2 (NT2)[48] or human ES cells such as H1 or H9,[20] may be desirable in order to establish some initial experience with the assay. We recommend planning experiments and composing animal care and use committee protocols that provide for the potential to use all three common routes noted above (SC, IM, and IT). Teratoma formation should first be attempted using subcutaneous or intramuscular routes, as these locations are the least invasive and are more easily tolerated by recipient animals. Final attempts using intratesticular injections may be attempted before concluding that a given pluripotent stem cell line may bear a developmental limitation. Intratesticular injection is the most stressful to the animal as well as technically challenging to the investigator. This is a surgical procedure performed under animal care and use committee standards and thus is perhaps best reserved as a last resort. In order to not underestimate the number of animals required (or limit the availability of certain assays) the investigator should plan for the possibility of all three routes though intending to use the least invasive route possible. As intratesticular injections may be required, the protocol should articulate the use of only male mice (if possible) in order to control for endocrinological differences between experiments.

General Guidance

Less than 5×10^6 pluripotent stem cells are subcutaneously injected in a volume not to exceed 200 μL in phosphate buffered saline (PBS), Hank's balanced salt solution (HBSS), or serum-free culture medium. For intramuscular (IM) or intratesticular (IT) injections, a more concentrated cellular bolus is required. For IM injections, a volume of no more than 100–150 μL should be used. For IT injections, the volume should not exceed 50 μL. All injections should be performed with a needle of less than 23 gauge diameter. For IM and IT injections (see below), the investigator is strongly cautioned against infusing the cellular inoculum too rapidly, where excessive pressure has the potential to injure the mouse as well as to generate excessive shear force on the cells involved, which could lead to a high degree of cellular lysis, resulting in a negative assay.

Mice no less than 6–8 weeks old should be briefly anesthetized via isoflurane inhalation (2–4%, visually evaluated or by pinch) in a certified halogen scavenging apparatus prior to subcutaneous (above rear flank) or intramuscular (quadriceps/posterior thigh) injection.

The investigator is strongly cautioned to guard against accidental needle sticks during this procedure and is encouraged to be particularly mindful of the location of all exposed and contaminated sharps throughout the entire injection procedure and to use institutionally approved, self-capping safety needles. Investigators are also strongly encouraged to consult with their institutional animal care and use committee for advice on strain injection routes, methods, and obtaining and using proper institutionally approved syringes and needles prior to initiating any work in this area. Safety should always come first when using human cells of any variety and in any setting.

Subcutaneous Tumor Formation

Subcutaneous injection is easily performed by pinching the mouse's skin between the thumb and forefinger to produce a dermal "tent" where the cellular inoculum is introduced to the subcutaneous space via puncture of the tented skin just below and between the thumb and forefinger. Successful location in the subcutaneous space is typified by experiencing very little resistance to delivery of the cells.

Intramuscular Tumor Formation

Intramuscular injection is accomplished by softly extending the mouse's rear leg and injecting the cellular inoculum at an angle but directly into the thigh muscle. Swelling of the muscle as the inoculum is introduced should be noted. The investigator is cautioned against too deeply penetrating the muscle where any contact with the femur should be strictly avoided, though at the same time, a superficial injection is also to be avoided. Some investigators employ IM assays where the cellular inoculum is injected in a dilute slurry of BD Matrigel™ basement matrix[49] made up of 25% Matrigel, 25% bovine collagen solution, and 50% pluripotent cells in an appropriate suspension solution. Hentze et al.[47] have published an evaluation of the use of BD Matrigel in teratoma formation assays.

Intratesticular Tumor Formation

Intratesticular injection is challenging and requires additional training and monitoring by animal care and use committee staff. For IT routes, mice should be more deeply anesthetized prior to injections by intraperitoneal administration of Avertin® (125–240 mg/kg) or some other institutionally approved anesthetic. When sedation has been achieved, the animal should be shaved and sterilized with betadine/70% ethanol. A 1-centimeter long incision is made in order to visualize the testes. The testis is gently pulled from its position by grasping with sterile, blunt forceps. The testis is then injected with less than 5×10^6 pluripotent stem cells in a volume not to exceed 50 μL in phosphate buffered saline (PBS), Hank's balanced salt solution (HBSS), or serum-free culture medium under low pressure. All injections must be performed with a needle of less than 23 gauge diameter. The testis is then replaced in the body cavity using blunt forceps. The incision may be closed using wound clips (consult animal care and use committee for recommended wound closure methods). These animals receive postoperative monitoring (daily) and analgesia (again, consult the animal care and use committee).

Postinjection Monitoring

All experimental animals should be monitored for signs of tumor formation for up to 6 months or until a visible tumor mass is noted, whereupon they are euthanized via carbon dioxide-mediated asphyxiation (or some other institutionally approved methodology) in order to harvest the tumors for further analysis (see below). A common rule of thumb for what might be expected is that an inoculum of 1–2 million pluripotent stem cells should form a visualizable mass within 4–6 weeks. The investigator should keep records on the date of injection and any notes thereafter as animals are visually reviewed for tumor formation over the length of the assay. Teratomas should be slowly growing masses,[38,41] where rapid tumor formation may indicate differentiation failure or transformation of the injected cells.

Experimental endpoints should be tumor masses not to exceed 1.5 grams in weight (approximately 2.0 cm in diameter) in the subcutaneous space or 6 months of incubation (whichever comes first). Smaller masses should be harvested from the IM and IT spaces where larger tumors have a significant likelihood of causing distress, impairing ambulation, and/or causing other types of complications in host animals. Here, the size threshold should be less than 1.0 cm, though the investigator is encouraged to consult with the animal care and use committee staff for advice. Any signs of unexpected morbidities including skin ulcerations at the injection site, loss of appetite, lethargy, hunched habitus, and/or impaired ambulation are indications for early euthanasia of experimental animals. Investigators are further encouraged to perform a routine necropsy of euthanized animals in order to evaluate any possible "metastases" or migration of the tumor beyond the injection site and into distant anatomical locations, although this is unexpected.

Tumor Harvest

The point of tumor harvest is an important time to note specific findings that are relevant to determining the overall differentiation state of the mass. For example, subcutaneous masses should easily be removed, most often requiring little more than gentle scraping with the side of a pair of curved forceps in order to dislodge the tumor from the underside of the skin or the surface of the muscle. If a mass is noted to have "burrowed" into the muscle following subcutaneous injection, this may suggest an invasive tumor; a feature indicative of a poorly differentiated tumor, where a well-differentiated teratoma does not invade surrounding tissues.[38,41] Intramuscular and intratesticular masses require the complete removal of those structures via surgical scissors due to the penetrating nature of the original injections. As one caveat here, IM and IT injections generate a wound at the injection site and, as a result, make definitive determinations of tumor invasiveness somewhat of a challenge.

Also, the investigator should be able to note whether the teratoma is cystic, containing small fluid-filled sacs—a sign of a well-differentiated tumor. Firm, noncystic tumors indicate a poorly differentiated teratoma (or a nonteratoma following additional analysis). If possible, it is also useful to cleave the fresh mass in two with a razor blade or via cutting it in half with surgical scissors where the presence of bone or cartilage is easily discovered by noting resistance or "crunching" during cutting. These are examples of simple observations that may be performed prior to any ability to review stained tissue sections and are well worthwhile.

Once the tumor or tumor-containing tissue is harvested, drop it into a vial containing PBS. Once in the laboratory, the tissue may be transferred to ethanol or methanol for dehydration prior to paraffin embedding and cutting to create slides. If frozen sections are preferred, place the tumor sections directly onto small squares cut from thick paper or thin cardboard and coat the mass entirely with resin (such as OCT/Tissue Tek®) prior to flash freezing by dropping into liquid nitrogen or a dry-ice/ethanol bath, taking care to avoid skin contact with the liquid in any case.

The frozen specimens may be quickly wrapped in aluminum foil for further storage prior to making frozen sections.

Note: If a serial transplantation assay is to be done, a sterile operating field, receptacles, and surgical equipment are required. Serial transplantability indicates the continued presence of pluripotent stem cells and suggests too large an initial inoculum, where a microenvironmental pocket of autocrine/paracrine conditioning may permit the proliferation of undifferentiated cells or an outright differentiation incapacity for certain cells within the initial cell culture; that is, cells with genetic and/or epigenetic damage that impairs or prevents their ability to differentiate.

ANALYSIS OF PLURIPOTENT CELL-DERIVED MASSES

Differences Between Pluripotent Cell-Derived Tissue Masses and Naturally Occurring Teratomas

As described previously, one of the in vivo methods frequently employed in determining the differentiation potential of ESCs and iPSCs is microscopic examination of the tissues following injection of cells into immunocompromised mice. In such a context, ESCs and iPSCs generally form a mixture of various tissue types that belong to endodermal, mesodermal, and endodermal lineages as a reflection of their differentiation potential.

This assay is commonly referred to as "teratoma formation," although the term is misleading to some extent.[41] In clinicopathologic terms, "teratoma" is a human germ cell tumor that is thought to arise from cells in the ovary and testis de novo. These rare tumors can develop in children and young adults of both genders, and they are often but not always composed of multiple mature somatic tissue types, with the exception of monodermal (single lineage) teratomas. For example, some monodermal teratomas consist of thyroid tissue alone and are referred to as "struma ovarii" and squamous epithelium-only masses are referred to as an "epidermoid cyst." Hence use of the clinical term teratoma does not in fact necessarily indicate the presence of differentiation into all three germ cell layers.[50] In any case, based on the morphological similarity between cultured pluripotent cell-derived tissues and some human teratomas, many investigators have defined the ES and iPS cell-derived tissue masses[38] as "teratomas." While this terminology has gained widespread acceptance, it is nevertheless misleading in at least one important respect.

When properly derived, the cultured embryonic stem cells used in these assays should be genetically and epigenetically normal, being derived from preimplantation embryos, and would be capable of contributing to the embryo proper in blastocyst chimerism assays. This is routinely demonstrated with mouse embryonic stem cells, although such an assay, as mentioned previously, is ethically proscribed with human cells.

In contrast, when human pluripotent stem cells are injected into the subcutaneous, intramuscular, or other ectopic areas of recipient mice, they form disorganized tissue masses that superficially resemble a true teratoma. However, unlike ESC and iPSC derived tissues, true teratomas are tumors that arise from cells that have accumulated genetic and epigenetic oncogenic changes, and are by definition no longer normal cells. Hence teratomas are not the same as an embryo even if they are growing in a uterine environment receptive to pregnancy. Indeed, the demonstration of teratomas coincident with a normal gestation was first described as far back as the late 17th century.[51] In this regard, the experimental, pluripotent cell-derived tumors are more akin to the *disorganized tissue development* seen in the ectopic implantation of even normal embryos.[52] Thus "in vivo tissue generation assay" is perhaps a more accurate term to use for the mouse implantation of ES and iPS cells.

Microscopic Evaluation of Pluripotent Cell-Derived Tissues

The microscopic examination of Hoechst and eosin (H&E) stained sections of formalin-fixed and paraffin-embedded pluripotent cell-derived tissues has commonly been used to determine the trilineage differentiation potential of these cells. The key feature of this analysis is identification of tissues that arise from each germ layer. It is important to point out that while many investigators routinely interpret the results of experimental teratoma formation assays themselves, establishing a working relationship with a pathologist skilled in the examination and interpretation of pathological specimens can only improve upon the results obtained.[40] Here, commonly formed tissues such as muscle and cartilage are easily described, especially as the investigator gains experience with the assay. That said, many tissue types remain difficult to interpret, often requiring the use of specific counterstains or immunohistochemical methods in order to determine their true identity. Again, evaluation by a pathologist will permit the investigator to obtain an even greater degree of confidence in the developmental maturity, invasiveness, and tissue plasticity of the cells under investigation—all highly relevant points of consideration. Such added value increases the reliability of data obtained in basic studies and is likely a sine qua non for safety evaluations of cells intended for eventual clinical use.

1. *Identification of Mesodermal Differentiation:* Most ES and iPS cell-derived masses include varying amounts of cartilage and, more rarely, bone or skeletal muscle. Locating any one of these tissues can be diagnostic. In general, identification of mesodermal tissues with light microscopic examination of H&E stained slides is straightforward and least problematic in this type of evaluation (Fig. 19.1a (bone) and 19.1b (cartilage)). However, large sections of ESC and iPS cell-derived tissues can be composed of immature mesenchyme with no specific differentiation (Fig. 19.1c).

2. *Identification of Ectodermal Differentiation:* The most reliable evidence of ectodermal differentiation is the presence of dermal structures or skin. In many human mature teratomas, skin is a predominant component (Fig. 19.2a), including fully grown hair (Fig. 19.2b) and other skin-related structures such as sweat and sebaceous glands (Fig. 19.2b,c). In contrast, we have rarely seen full skin formation

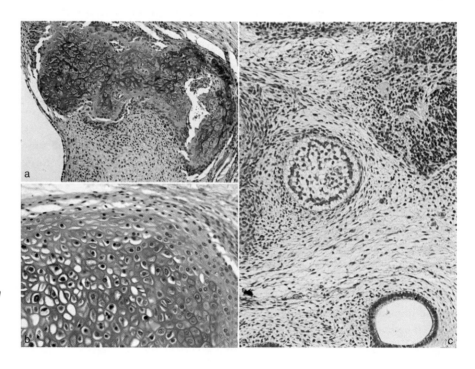

FIGURE 19.1. *Identification of mesodermal differentiation: (a) fragments of bone with numerous osteoblasts; (b) an area of cartilage formation; and (c) in most iPS and hES cell-derived teratomas, vast areas within the tissue mass are often composed of immature mesenchyme with no specific differentiation. All panels are H&E stained sections.*

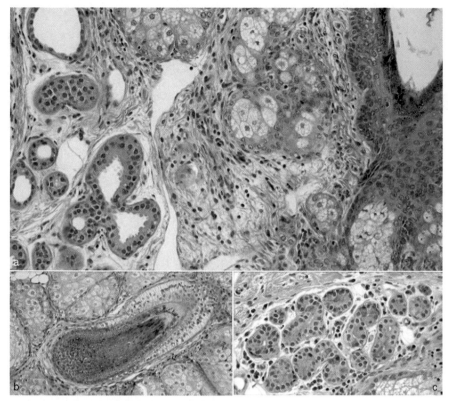

FIGURE 19.2. *Identification of ectodermal differentiation. (a) An area of skin differentiation from a true (naturally occurring) human teratoma. Adnexal structures characteristic of skin tissues such as sweat glands and sebaceous glands help distinguish this tissue from simple squamous epithelium. (b) The presence of hair is also a distinguishing characteristic of skin tissue formation in teratomas. Here, the bulb of the nascent hair is at left and the hair shaft extends toward the right. (c) A cluster of glands derived from ectoderm and skin tissue present in a naturally occurring teratoma. All panels are H&E stained sections.*

in human ES and iPS cell-derived tissues regardless of the source of the cells injected or the length of incubation in murine hosts.

In many cases the presence of squamous epithelium has been taken as evidence of ectodermal differentiation (Fig. 19.3). However, squamous epithelium can also derive from endoderm such as in esophagus. Also, several endodermal and mesodermal derived glandular epithelia such as lung, endometrium, and upper urinary tract may demonstrate what is known as "squamous metaplasia." Squamous metaplasia arises from the transdifferentiation of glandular epithelium into squamous epithelium and is frequently seen in cervix, endometrium, and lung, as well as other epithelia under physiologic and pathologic conditions. As shown in Figure 19.3, the squamous epithelium in this iPS cell-derived tissue is contiguous with glandular epithelium and may represent squamous metaplasia. Thus in the absence of other types of skin-associated tissues or hair follicles, it is difficult to consider squamous differentiation as a definitive proof of ectodermal differentiation (Fig. 19.3). Furthermore, the glandular epithelium shown in Figure 19.3 is not fully differentiated and does not have mature morphologic features that would allow a particular tissue type designation. Hence it is difficult to know whether these are primitive sweat glands (ectoderm), cervical or endometrial glands (mesoderm), or respiratory epithelium (endoderm). Consequently, without additional molecular studies that could positively identify the lineage of glandular epithelium, it would

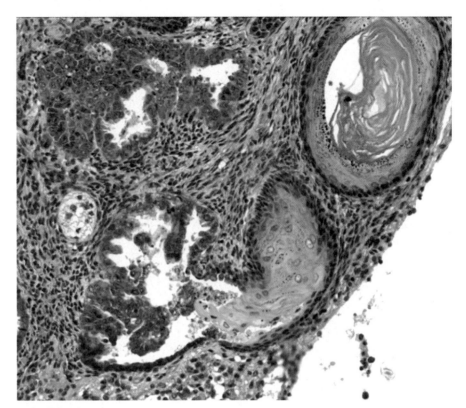

FIGURE 19.3. *Squamous differentiation in an iPS-derived teratoma. In the lower central portion of this image, an area of squamous epithelium at right merges with glandular epithelium on the left. This is typically seen in squamous metaplasia of glandular epithelium and is not definitive evidence for ectodermal (skin) differentiation, particularly in the absence of any adnexal structures as sweat glands, sebaceous glands, or hair. As such, it is difficult to definitively conclude that there is ectodermal differentiation in this teratoma based only on this area. All panels are H&E stained sections.*

be difficult to conclude that such an area provides definitive evidence for ectodermal differentiation.

Beyond this, many pluripotent cell-derived teratomas contain neural ectodermal epithelium, and in the absence of skin differentiation the presence of neuroectodermal epithelium has been considered as evidence of ectodermal differentiation potential including the identification of neural rosettes (Fig. 19.4a,b) and pigmented retinal epithelium (Fig. 19.4c). However, whether the presence of neuroectodermal differentiation is the same as full ectodermal differentiation potential is uncertain. We have noted a general trend in experimental teratomas toward fetal or immature forms of various tissues relative to fully differentiated structures, with primitive neural ectodermal epithelium representing the predominant tissue type.

3. *Identification of Endodermal Differentiation:* Definitive identification of endodermal tissues in a teratoma represents the most difficult challenge for the pathologist. Most of the glandular epithelial linings of the internal organs are derived from the endoderm. Such epithelia are composed of cuboidal or columnar cells without squamous differentiation. However, glandular epithelia can arise from ectoderm and mesoderm as well as endoderm. For example, the sweat glands of the skin, breast, and pituitary gland are lined by a simple glandular epithelium that is solely ectodermal in origin.

Furthermore, several organs that originate from the mesoderm, such as uterus and kidney, are also lined by a simple glandular epithelium. Hence the presence of

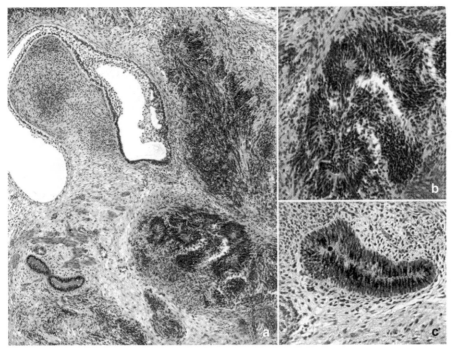

FIGURE 19.4. *Neuroectodermal differentiation in teratomas. (a) In the absence of definitive skin differentiation (such as shown in Fig. 19.2), the presence of neuroectodermal tissue (bottom right quarter of this panel) is often used as evidence for the presence of ectodermal differentiation in a teratoma. (b) An area of neural rosettes. (c) In some areas neuroectodermal tissue may contain dark pigment consistent with retinal differentiation. All panels are H&E stained sections.*

a simple glandular epithelium is not in itself a definitive proof of endodermal differentiation capacity in pluripotent cell-derived tissues. In adult tissues the glandular epithelia of different organs have specific morphologic features that allow one to distinguish them with certainty. In some cases the pluripotent cell-derived tissues are fully differentiated and are easily recognized as endodermal differentiation based on morphology alone. Identification of areas with well-formed tissue architecture and the context of the epithelium can be critical in positive identification of specific epithelia. For example, bronchial (lung) epithelium is frequently associated with supporting cartilage, and the identification of both components in the right context can aid in determining the presence of endodermal differentiation (Fig. 19.5a). The presence of colloid in the cystically dilated lumina of the glands is useful in identifying thyroid differentiation (Fig. 19.5b). A muscular layer that is surrounding a villous epithelium is typical of gastrointestinal differentiation (Fig. 19.5c). However, in many cases much of the glandular epithelium is not fully differentiated and has not gained features that are diagnostic of a specific tissue (Figure 19.6). Hence it is often not possible to discern whether the presence of glandular epithelium is the definitive indication of endodermal differentiation. In such cases, the use of molecular markers and immunohistochemistry can be very helpful. For example, lower gastrointestinal epithelium expresses cytokeratin (CK20), which can definitively identify endodermal differentiation.

Beyond this, many investigators will simply look for areas of cavitation within the teratoma, where such cystic structures are most often indicative of endodermal differentiation (e.g., Fig. 19.4). Here, one may even note functional, integrated tissue architecture where luminal depositions of secretory material such as mucus may be frequently observed. That said, it behooves the investigator to probe a bit deeper in the attempt to specifically identify such epithelial layers were brush-border

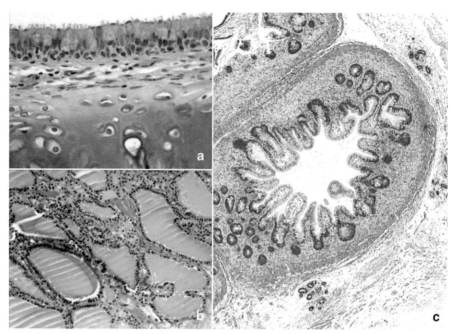

FIGURE 19.5. *Identification of endodermal differentiation in teratomas. (a) The presence of multiple, integrated components in a complex tissue is helpful in identifying the tissue origin and germ-layer identity of an epithelium. For example, in this panel a single layer of epithelium with a brush border is present adjacent to an underlying layer of cartilage, such as what is typically seen in bronchial epithelium. (b) In this panel an area of thyroid differentiation with typical proteinaceous colloid (red luminal material) in the center of microcystic spaces is seen. This architecture is nearly indistinguishable from true human thyroid tissue and helps identify the presence of definitive endodermal differentiation. (c) In this panel a cross section of intestinal tissue is seen with a central lumen surrounded by finger-like villus projections lined by glandular epithelium. Importantly, this glandular epithelium is surrounded by stromal cells defined by an outer layer of smooth muscle cells that form a muscular wall. This specific architecture is typical of gastrointestinal tissue. Note: The three examples here highlight the importance of viewing multiple components of a tissue in order to designate an origin from within a specific lineage. In the absence of other components such as cartilage, a muscular wall, or luminal secretions, it would be difficult to identify these glandular epithelia as respiratory, thyroid, and gastrointestinal based solely on their epithelial component (e.g., see Fig. 19.6). Unfortunately, in most iPS and hES cell-derived teratomas, such well-formed tissue architecture is rarely seen. Hence adjunct studies such as immunohistochemistry are generally necessary to definitively identify the differentiation lineage of epithelium in such teratomas. All panels are H&E stained sections.*

(respiratory) epithelia and other types may be noted. Such probing may contribute a significant degree of added insight (and power) to these in vivo assays where a differentiation bias toward (or away from) specific structures may provide fertile ground for additional experimentation.

In summary, many pluripotent cell-derived tissues are only partially differentiated and morphologic examination alone may not be adequate to definitively score the differentiation potential of the cells being assayed. In such cases, the use of immunohistochemistry is highly recommended.

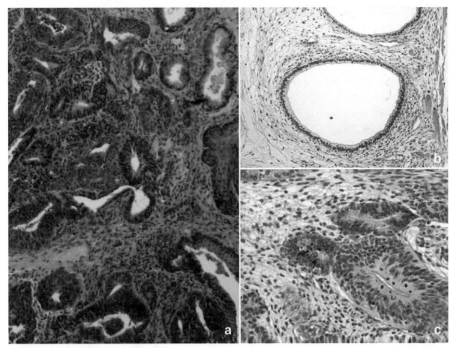

FIGURE 19.6. *Glandular epithelia with no specific differentiation in iPS and ES cell-derived teratomas. (a) The gross identification of glands superficially suggests endodermal differentiation. However, in the absence of clear tissue architecture or cellular differentiation, it is difficult to be certain that such areas are definitive evidence for endodermal differentiation potential. (b,c) While these glandular epithelia may be consistent with some types of endodermal differentiation, they could also represent partially differentiated endometrial or renal glands (both mesodermal in origin), or even sweat glands (ectodermal). All panels are H&E stained sections.*

Completeness of Differentiation

The current clinical criteria to asses teratoma differentiation were designed to distinguish patients with no future progression risk (i.e., well-differentiated or "mature" teratoma) versus patients with some risk, even if very small, that an overall differentiation impairment or overt failure may be in place (i.e., poorly differentiated or "immature" teratoma). Here, the presence of immature neuroepithelium is what distinguishes an immature teratoma from a mature one. Thus if a teratoma contains even one low-power field (4–5-mm diameter area) of immature neuroepithelium, it is considered an "immature teratoma, grade 1" in the clinic setting. The presence of even a small amount of immature tissue is clinically significant because of an associated risk of tumor recurrence and progression. If such immature tissue is present in more than four low-power fields, it is regarded as a high-grade immature teratoma. Hence the clinical "immature teratoma'" diagnosis may mean that the mature tissue can range between a few percent of the tumor bulk to nearly the entire tumor mass. As the purpose of this pathologic classification is risk assessment in the clinical setting, these criteria are only appropriate for clinical practice.

However, for advancing the science of stem cells, a field of scientific endeavor fully capable of undertaking empirical, hypothesis-driven studies, this is an overly imprecise method of assessing teratoma differentiation and may lead one to miss important nuances in different pluripotent stem cell lines. For example, the immature versus mature clinical distinction does not allow one to adequately evaluate the differences between different cell lines and examine effects of differentiation-promoting reagents that might be tested in the future. Thus there is a great need to develop a biologically and scientifically relevant molecular and histopathological classification system to supplant the clinical pathologic system currently used for evaluating pluripotent cell-derived tissues. Such work should eventually lead to a universally accepted and reproducible standard that all investigators can use. Several good studies have already initiated such work.[29,47,53,54].

In a recent study we showed that immunohistochemical staining of iPS and ES cell-derived tissues with SOX2, OCT4, Nestin, pan-keratin, and tissue-specific keratins such as K7 and K20 can be very helpful in quantitating the amount and type of differentiation (Fig. 19.7).[29] Many undifferentiated areas can be highlighted with positive SOX2 and OCT4 staining (Fig. 19.7a,b). At times it may be difficult to distinguish primitive neuroectoderm from epithelial differentiation (Fig. 19.7c). In these cases positive staining with Nestin would indicate neuroepithelium (Fig. 19.7d) and pan-keratin staining epithelial tissue (Fig. 19.7e). Interestingly, Figure 19.7f shows an area where CK20 positive epithelium (red) merges into a cytokeratin 7 positive (brown) epithelium. Cytokeratin 7 can be expressed in genitourinary,

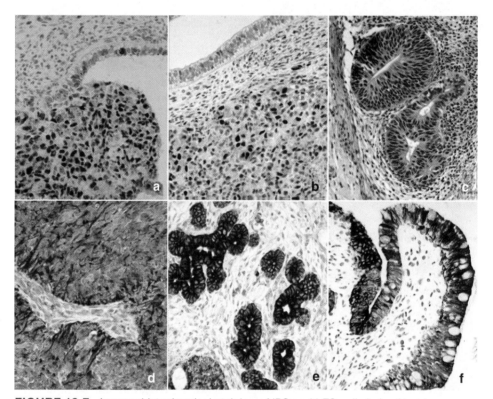

FIGURE 19.7. *Immunohistochemical staining of iPS and hES cell-derived teratomas. Molecular stains can be helpful in quantifying the amount and type of differentiation in teratomas, particularly when differentiation is partial and difficult to interpret based on H&E staining alone. Positive immunostaining against (a) SOX2 or (b) OCT4 may indicate areas of undifferentiated, residual, pluripotent stem cells. (c) An area of primitive neural ectoderm may be difficult to distinguish from other types of epithelium by H&E staining alone, although staining for Nestin (as in panel d), a marker of early neural differentiation, is an effective follow-on stain to definitively identify neural tissue. (e) Pan-keratin stains distinguish epithelial structures. (f) Subtype-specific keratin stains can be helpful in identifying specific differentiation lineages. For example, in this panel keratin 7 (K7, brown) and keratin 20 (K20, red) are contiguously expressed within the same epithelium layer. K7 may be expressed in endodermal or mesodermal epithelial layers, although K20 is expressed in the lower gastrointestinal tract (endoderm) and thus identifies this area as definitive endodermal differentiation.*

lung, breast, and upper gastrointestinal epithelium. This type of mixed epithelial differentiation would be difficult to evaluate with H&E stains alone.

Overall, a large degree of neuronal differentiation likely remains an indicator of differentiation impairment in lines of pluripotent stem cells, as do nests of undifferentiated, residual, or proliferating pluripotent cells. These nests will stain positive for OCT4, Nanog, or other common markers of pluripotent stem cells in immunohistochemical assays. Here, the presence of such cells will likely correlate with serial transplantability to secondary animals, which is a poor sign clinically and likely in the research setting as well.[38] As stated earlier, one caveat is that inoculation with too large a bolus of pluripotent cells may inadvertently generate an autocrine/paracrine microenvironment at the injection site, which has the potential to encourage the proliferation and self-renewal of pluripotent cells rather than continued differentiation. This result is easily controlled for by titrating the cell number inoculated, where a common metric is a palpable or visually recognized mass or swelling at the implant site within 1 month's time (usually on the order of 1 cm for subcutaneous injections, less for intramuscular sites).

In conclusion, it is very important to stress that a complete evaluation and reporting of the teratoma assay should do more than provide a yes/no response to the question of whether or not tissues from the three embryonic germ layers are present. A complete evaluation must also note the overall differentiation state of the mass, such as whether abundant, primitive, neural tissues or undifferentiated nests of residual pluripotent stem cells are present, whether the mass was cystic, and other key features described above. By performing a more rigorous teratoma formation assay, individual experiments as well as the field as a whole stands to exist on a much surer footing with regard to the evaluation of the in vivo differentiation capacity of human pluripotent stem cells.

Acknowledgments

The authors thank George Q. Daley and Thorsten M. Schlaeger for helpful suggestions and a review of this chapter during its preparation. M.W.L. is supported by funding from a Howard Hughes Medical Institute Investigator Award to George Q. Daley. T.A.I. is supported by ARRA grant 1RC2HL102815-01.

REFERENCES

1. Lensch MW. Cellular reprogramming and pluripotency induction. *Br Med Bull*. 2009;90:19–35.

2. Andrews PW. Human teratocarcinomas. *Biochim Biophys Acta*. 1988;948:17–36.

3. Anderson PJ, David DJ. Teratomas of the head and neck region. *J Craniomaxillofac Surg*. 2003;31:369–377.

4. Isaacs H Jr. Perinatal (fetal and neonatal) germ cell tumors. *J Pediatr Surg*. 2004;39:1003–1013.

5. Ueno T, Tanaka YO, Nagata M, et al. Spectrum of germ cell tumors: from head to toe. *Radiographics*. 2004;24:387–404.

6. Duwe BV, Sterman DH, Musani AI. Tumors of the mediastinum. *Chest*. 2005;128:2893–2909.

7. Heerema-McKenney A, Harrison MR, Bratton B, Farrell J, Zaloudek C. Congenital teratoma: a clinicopathologic study of 22 fetal and neonatal tumors. *Am J Surg Pathol*. 2005;29:29–38.

8. Askanazy M. Die Teratome nach ihrem Bau, ihrem Verlauf, ihrer Genese und im Vergleich zum experimentellen Teratoid. *Verh Dtsch Pathol Ges*. 1907;11:39–82.

9. Friedman NB, Moore RA. Tumors of the testis: a report on 922 cases. *Mil Surgeon*. 1946;99:573–593.

10. Teilum G. Classification of endodermal sinus tumour (mesoblatoma vitellinum) and so-called "embryonal carcinoma" of the ovary. *Acta Pathol Microbiol Scand*. 1965;64:407–429.

11. Stevens LC. Origin of testicular teratomas from primordial germ cells in mice. *J Natl Cancer Inst*. 1967;38:549–552.

12. Stevens LC, Little CC. Spontaneous testicular teratomas in an inbred strain of mice. *Proc Natl Acad Sci USA*. 1954;40:1080–1087.

13. Kleinsmith LJ, Pierce GB Jr. Multipotentiality of single embryonal carcinoma cells. *Cancer Res*. 1964;24:1544–1551.

14. Brinster RL. The effect of cells transferred into the mouse blastocyst on subsequent development. *J Exp Med*. 1974;140:1049–1056.

15. Martin GR. Teratocarcinomas as a model system for the study of embryogenesis and neoplasia. *Cell*. 1975;5:229–243.

16. Martin GR, Evans MJ. The morphology and growth of a pluripotent teratocarcinoma cell line and its derivatives in tissue culture. *Cell*. 1974;2:163–172.

17. Martin GR, Evans MJ. Differentiation of clonal lines of teratocarcinoma cells: formation of embryoid bodies in vitro. *Proc Natl Acad Sci USA*. 1975;72:1441–1445.

18. Evans MJ, Kaufman MH. Establishment in culture of pluripotential cells from mouse embryos. *Nature*. 1981;292:154–156.

19. Martin GR. Isolation of a pluripotent cell line from early mouse embryos cultured in medium conditioned by teratocarcinoma stem cells. *Proc Natl Acad Sci USA*. 1981;78:7634–7638.

20. Thomson JA, Itskovitz-Eldor J, Shapiro SS, et al. Embryonic stem cell lines derived from human blastocysts. *Science*. 1998;282:1145–1147.

21. Takahashi K, Yamanaka S. Induction of pluripotent stem cells from mouse embryonic and adult fibroblast cultures by defined factors. *Cell*. 2006;126:663–676.

22. Wernig M, Meissner A, Foreman R, et al. In Vitro reprogramming of fibroblasts into a pluripotent ES-cell-like state. *Nature*. 2007;448:318–324.

23. Maherali N, Sridharan R, Xie W, et al. Directly reprogrammed fibroblasts show global epigenetic remodeling and widespread tissue contribution. *Cell Stem Cell*. 2007;1:55–70.

24. Okita K, Ichisaka T, Yamanaka S. Generation of germline-competent induced pluripotent stem cells. *Nature*. 2007;448:313–317.

25. Lowry WE, Richter L, Yachechko R, et al. Generation of human induced pluripotent stem cells from dermal fibroblasts. *Proc Natl Acad Sci USA*. 2008;105:2883–2888.

26. Park IH, Zhao R, West JA, et al. Reprogramming of human somatic cells to pluripotency with defined factors. *Nature*. 2008;451:141–146.

27. Takahashi K, Tanabe K, Ohnuki M, et al. Induction of pluripotent stem cells from adult human fibroblasts by defined factors. *Cell*. 2007;131:861–872.

28. Yu J, Vodyanik MA, Smuga-Otto K, et al. Induced pluripotent stem cell lines derived from human somatic cells. *Science*. 2007;318:1917–1920.

29. Chan EM, Ratanasirintrawoot S, Park IH, et al. Live cell imaging distinguishes bona fide human iPS cells from partially reprogrammed cells. *Nat Biotechnol*. 2009;27:1033–1037.

30. Tesar PJ, Chenoweth JG, Brook FA, et al. New cell lines from mouse epiblast share defining features with human embryonic stem cells. *Nature*. 2007;448:196–199.

31. Jaenisch R, Young R. Stem cells, the molecular circuitry of pluripotency and nuclear reprogramming. *Cell*. 2008;132:567–582.

32. Nagy A, Gocza E, Diaz EM, et al. Embryonic stem cells alone are able to support fetal development in the mouse. *Development*. 1990;110:815–821.

33. Nagy A, Rossant J, Nagy R, Abramow-Newerly W, Roder JC. Derivation of completely cell culture-derived mice from early-passage embryonic stem cells. *Proc Natl Acad Sci USA*. 1993;90:8424–8428.

34. Daley GQ, Lensch MW, Jaenisch R, Meissner A, Plath K, Yamanaka S. Broader implications of defining standards for the pluripotency of iPSCs. *Cell Stem Cell*. 2009;4:200–201; author reply 202.

35. Boland MJ, Hazen JL, Nazor KL, et al. Adult mice generated from induced pluripotent stem cells. *Nature*. 2009;461:91–94.

36. Kang L, Wang J, Zhang Y, Kou Z, Gao S. iPS cells can support full-term development of tetraploid blastocyst-complemented embryos. *Cell Stem Cell*. 2009;5:135–138.

37. Zhao XY, Li W, Lv Z, et al. iPS cells produce viable mice through tetraploid complementation. *Nature*. 2009;461:86–90.

38. Lensch MW, Schlaeger TM, Zon LI, Daley GQ. Teratoma formation assays with human embryonic stem cells: a rationale for one type of human-animal chimera. *Cell Stem Cell*. 2007;1:253–258.

39. Ellis J, Bruneau BG, Keller G, et al. Alternative induced pluripotent stem cell characterization criteria for in vitro applications. *Cell Stem Cell*. 2009;4:198–199; author reply 202.

40. Ince TA, Ward JM, Valli VE, et al. Do-it-yourself (DIY) pathology. *Nat Biotechnol*. 2008;26:978–979; discussion 979.

41. Lensch MW, Ince TA. The terminology of teratocarcinomas and teratomas. *Nat Biotechnol*. 2007;25:1211.

42. Draper JS, Smith K, Gokhale P, et al. Recurrent gain of chromosomes 17q and 12 in cultured human embryonic stem cells. *Nat Biotechnol*. 2004;22:53–54.

43. Kamel-Reid S, Dick JE. Engraftment of immune-deficient mice with human hematopoietic stem cells. *Science*. 1988;242:1706–1709.

44. McCune JM, Namikawa R, Kaneshima H, Shultz LD, Lieberman M, Weissman IL. The SCID-hu mouse: murine model for the analysis of human hematolymphoid differentiation and function. *Science*. 1988;241:1632–1639.

45. Mosier DE, Gulizia RJ, Baird SM, Wilson DB. Transfer of a functional human immune system to mice with severe combined immunodeficiency. *Nature*. 1988;335:256–259.

46. Behringer RR. Human–animal chimeras in biomedical research. *Cell Stem Cell*. 2007;1:259–262.

47. Hentze H, Soong PL, Wang ST, Phillips BW, Putti TC, Dunn NR. Teratoma formation by human embryonic stem cells: evaluation of essential parameters for future safety studies. *Stem Cell Res*. 2009.

48. Andrews PW, Damjanov I, Simon D, et al. Pluripotent embryonal carcinoma clones derived from the human teratocarcinoma cell line Tera-2. Differentiation in vivo and in vitro. *Lab Invest*. 1984;50:147–162.

49. Kleinman HK, McGarvey ML, Liotta LA, Robey PG, Tryggvason K, Martin GR. Isolation and characterization of type IV procollagen, laminin, and heparan sulfate proteoglycan from the EHS sarcoma. *Biochemistry*. 1982;21:6188–6193.

50. Crum CP, Lee KR, eds. *Diagnostic Gynecologic and Obstetric Pathology*. New York: Elsevier Saunders; 2005.

51. Birch S, Tyson E. An extract of two letters from Mr. Sampson Birch, an Alderman and Apothecary at Stafford, concerning an extraordinary birth in Staffordshire, with reflections thereon by Edw. Tyson M. D. Fellow of the Coll. of Physitians, and of the R. Society. *Philos Trans R Soc*. 1683;13:281–284.

52. Stevens LC. The development of transplantable teratocarcinomas from intratesticular grafts of pre- and postimplantation mouse embryos. *Dev Biol*. 1970;21.

53. Gertow K, Przyborski S, Loring JF, et al. Isolation of human embryonic stem cell-derived teratomas for the assessment of pluripotency. *Curr Protoc Stem Cell Biol*. 2007;Chapter 1:Unit1B 4.

54. Prokhorova TA, Harkness LM, Frandsen U, et al. Teratoma formation by human embryonic stem cells is site-dependent and enhanced by the presence of Matrigel. *Stem Cells Dev*. 2009;18:47–54.

Perspectives in Human Stem Cell Technologies

SECTION IV

Human Stem Cell Technology and Biology, edited by Stein, Borowski, Luong, Shi, Smith, and Vazquez
Copyright © 2011 Wiley-Blackwell.

GENOMIC ANALYSIS OF PLURIPOTENT STEM CELLS

20

David Lapointe

INTRODUCTION

Genomic analysis of pluripotent stem cells (PSCs) seeks to understand the global mechanisms underlying the unfolding of events leading to the progression of PSCs through the differentiation program. Through genomic analysis, the key players in these processes and their relationships can be identified. Determining the interplay between differential gene expression, transcription factor binding events, miRNA species, and other genomic features will provide a more complete understanding of pluripotent stem cells. Ultimately these determinations will lead to an understanding of the genomic landscape of stem cell biology.[1] These efforts follow on other large scale initiatives that seek to annotate the genomic landscape such as ENCODE, an encyclopedia of DNA elements,[2] and FANTOM, functional annotation of the mammalian genome.[3]

In the past decade advances in genomic technology have led to an accelerated understanding of the fine details of eukaryotic genomes. Primarily these advances have allowed a greater understanding of genome-wide events through advanced DNA sequencing techniques, which have led to a phenomenal increase in the amount of generated sequence data per experiment, and through array-based technologies, which have given insights into the transcriptome and transcriptional control mechanisms on a global level. These technologies complement each other to provide a more complete view of events that define pluripotent stem cells and the transitions from these cells to derivative cell types (Fig. 20.1).

TECHNOLOGIES FOR GENOMIC ANALYSIS

Array-Based Technologies

Array-based technologies provide a global high-throughput view into genomic events (also known as DNA chips). These are used for many applications: gene expression profiling, where the levels of several thousands to tens of thousands of expressed genes can be determined concurrently, mapping of chromatin immunoprecipitated fragments to determine the binding site occupancy of specific proteins throughout the genome leading to an understanding of the cis-regulatory circuitry underlying genomic events, SNP detection, and alternate transcription detection using exon arrays or whole genome tiling arrays.

Human Stem Cell Technology and Biology, edited by Stein, Borowski, Luong, Shi, Smith, and Vazquez
Copyright © 2011 Wiley-Blackwell.

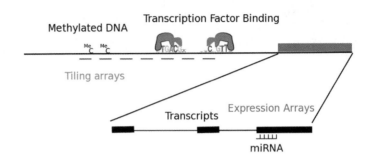

FIGURE 20.1. *Genomic features involved in the control of gene expression. Array technology as well as next-generation sequencing technology provide insight into many components of the genomic regulatory landscape.*

There are two basic types of array formats: spotted and fabricated. Spotted arrays require collecting together a library of sequences, representing hundreds to several thousand genes, and chemically attaching them to a prepared array substrate, for example, a polylysine or epoxy-silane coated glass slide, using a robotic spotting station. This type of array is useful for custom sets of genes and genetic features. The start-up requirements are quite large: isolation and production of sufficient quantity of material to be spotted, a spotting station, wet-lab station for hybridization of samples to spotted arrays, and a scanning and image processing station to quantitate binding. Spotted arrays are often referred to as two-color arrays since in general use a reference sample tagged with one fluorescent tag (Cy3) is mixed with an experimental sample tagged with a second fluorescent tag (Cy5), allowed to bind to the spotted array, and the fluorescence for the two fluors is used to estimate relative differential expression of expressed genes in the samples. Some commercial array technologies use spotted or bead-based two-color array systems.

Fabricated arrays are often formed using photolithographic techniques to build up custom oligos on a substrate, for example, Affymetrix chip technology, which lead to high array densities, 500,000 to 2,000,000 features per chip with high reproducibility of fabrication. Features are oligos corresponding to short regions of gene transcripts, regions of genes corresponding to single nucleotide polymorphisms, tiles of upstream flanking promoter regions, or tiles sampling the whole genome. Samples can be hybridized along with reference sequences in two-color analysis or as a single-color analysis with elements that control for cross-hybridization. As with all experimental data, good results come from carefully planned experimental design especially with the use of biological replicates. Although it goes without saying, only the features actually contained on the chips can be measured, which gives a disadvantage to array technology in relation to sequencing technologies.

Next-Generation Sequencing Technologies

Next-generation (NG) sequencing technology has brought phenomenal advances to the study of the fine structure of genomic control elements. Next-generation sequencers, such as 454 from Roche, Genome Analyzer from Illumina, and the SOLiD from ABI, are able to generate sequence data in quantities that are orders of magnitude greater than the previous generation of gel and capillary-based sequence technologies. NG sequencers can generate 30–50 million 36-bp sequence reads per run. Newer enhancements can provide extended sequence reads to 154 bp for the Genome Analyzer, as well as the generation of paired end tags.

Roche 454

The Roche 454 synthesizer uses a pyrosequencing technique based on sequencing by synthesis. Briefly, DNA samples are prepared and sized, and custom linkers are attached to the ends of each single stranded sequence. Prepared fragments are attached to chemically prepared beads in such a way that each bead has only one sequence attached. The sequence on each bead is then cloned and amplified to several million copies per bead. Hundreds of thousands of these beads are then sequenced simultaneously in the instrument, in a custom picotiter plate holding one bead per well, generating up to nearly a million concurrent reads of up to several hundred base pairs per read. The 454 can generate longer but fewer reads than the other technologies.

Illumina Genome Analyzer

The Illumina Genome Analyzer (GA) also uses sequencing by synthesis, where prepared and sized DNA fragments ($\sim$200 bp) have custom adapters attached to each end. Prepared DNA fragments, with adapters, are attached to linkers on the surface of the flow cell and amplified in place using bridge hybridization to densities that yield distinct cluster colonies in the tile areas of the flow cell. The GA uses a flow cell with eight independent channels permitting several unique experiments to be run in parallel, or utilizing several channels for a single experiment with increased coverage. The sequencing phase takes place in real time on the Genome Analyzer, where all sequences are extended by one base pair using a reversible dye-terminator system at each cycle of chemistry and image gathering. At each cycle, four images, one for each base dye, are collected for each region or tile in a channel. In the end, each channel of the flow cell can provide up to 5 million sequences of 36 bp and greater. The GA can utilize both single and paired end analyses. Different experiments can be multiplexed within a channel by putting custom tags, or barcodes, in the sequences to identify each experiment. Prior to genomic remapping, these short tags are used to partition or filter the sequence reads for each experiment and then removed before genomic mapping. Illumina provides software to process the generated image data into base calls from which the short read sequences are generated. Quality information, related to a phred score, is generated along with the base calling phase, and can be used to map longer sequence reads. Genome remapping against a reference genome using Eland, custom software from Ilumina, is the last stage of the sequencing pipeline. There are increasing numbers of software applications that can be used with short (and longer) sequence reads either in the place of Eland or as a continuation of analysis.

ABI SOLiD

The SOLiD system utilizes a sequencing by ligation method where dibase fluorescent tagged probes (four colors) are used to determine sequences. Prepared and sized DNA fragments are attached to silica beads using a custom adapter sequence so that each bead contains a single sequence that is amplified to populate the bead. After amplification and selection of beads with extended templates, beads are covalently attached to slides for ligation analysis. The SOLiD system uses five rounds of primer reset, where each round starts with a different primer, which hybridizes to the custom adapter sequence with a unique offset. A set of fluorescently labeled dibase probes are used to extend the primer sequence through a number of cycles that determine the read length. Since there are 16 dibase combinations and four fluorescent tags, each base is interrogated twice, once as the leading base, and again as the trailing base in different rounds of primer reset in order to resolve the color space degeneracy where each fluorescent color represents four different dibases (Fig. 20.2). In the end, each sequence is represented as colors (0123), termed color space, instead of bases (ACGT), where the colors represent dibases and hence do not correspond one-to-one with bases. For example, in Figure 20.2 the color green occurs four times and the color red occurs four times, but the color sequence green–red (13) has T as

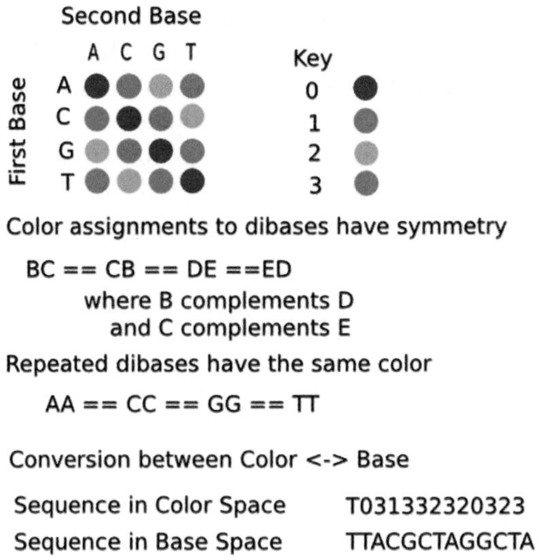

FIGURE 20.2. *Color dibase encoding for the SOLiD platform. The color space encoding matrix for dibase probes. Each nucleotide sequenced is sampled as an initial base and as a second base. Starting with a known base, T here, allows a translation from color space to base space.*

the common base. The SOLiD 3 system can generate 20 gigabases of usable data per run. The dibase system utilized inherently detects inaccuracies and is claimed to have a high overall accuracy rate. Analysis of color space data requires different software from the more commonly used sequence analysis software, although more recently software packages have added functionality for reading and processing color space data. For some analyses, it is preferable to do the analysis in color space, converting reference data into color space and doing genome remapping, SNP detection, and filtering of sequences against genomic information (RepBase, miRbase, etc.).

The ABI SOLiD has advantages over the Solexa system in some areas, but not all. SNPs are distinguishable in color space from sequencing errors that can alter the entire downstream sequence. ABI provides a number of packages for different genomic analysis using the color space data. In other areas, the Solexa has advantages in that it is more like traditional sequencing and there are a wide variety of third-party software tools for analysis and manipulation of sequences. In the end, both systems generate large quantities of sequence data for which the analysis is computationally intensive as well as storage intensive.

APPLICATIONS

Differential Gene Expression

Gene expression refers to the process by which the information contained within DNA is transcribed and translated into a functional entity, such as an enzyme, structural protein, or regulatory protein (hormone or transcription factor). The process of gene expression is complex and is controlled at a number of steps in the overall process. When genes become active or are silenced, the rate at which transcribed RNA is produced is generally considered to be gene expression, although subsequent events, such as RNA processing, translational control, as well as post-translational modifications, are involved also in the terminal function of gene expression, functional gene products. The rate at which mRNA is turned over plays a significant role in overall gene expression. In short, gene expression is a dynamic process.

A central problem in stem cell biology is understanding the dynamics, control, and progression of gene expression throughout the lineage pathways. Different measures can be comparisons among different cell types along the pathway, or a temporal comparison along a differentiation pathway.

One of the key techniques applied to this central problem is microarray analysis of differential gene expression. As mentioned above, microarrays contain probes (small oligonucleotide sequences) to tens of thousands of genes for an organism (human, mouse, etc.). Expressed nucleotide sequences from cells are extracted, amplified, labeled, and hybridized to microarray chips. Since the probes on the chip bind complementary sequences with different affinities, absolute determination or comparison of different expressed sequences is not possible. Typically, several experimental determinations with different cell conditions are compared, with sufficient biological replicates to satisfy the statistical power analysis.

If the degree of gene expression change is not large, all of the raw data from the analysis can be normalized to a representative value, background adjustments can be made, and expression values can be calculated, using a robust statistical technique such as RMA. Most commercial expression array software as well as open source packages such as R/Bioconductor contain a variety of robust statistical tools well suited to array analysis. Due to the intrinsic noise and experimental variations, solid statistical techniques are invaluable for separating signal from error. In addition, due to the large number of genes on each chip, statistical comparisons, such as t-test and ANOVA, must be adjusted for false discovery rates. For example, comparing two experiments that have 20,000 genes per chip, 1000 comparisons would be expected at random to meet a $p < 0.05$ criterion. In practice, eliminating by filtering genes that have no activity, are below some predetermined level, or have low variance across the experiments will reduce the number of statistical comparisons and yield a better resolution of changes between conditions.

There are several approaches to making comparisons using microarrays. Genes can be compared across experiments using fold-change measures and t-tests. More complicated experimental designs can use analysis of variance (ANOVA) to examine effects between conditions, which can be time, treatments, knock-out genes, and so on. By examining the way that genes change across the series of treatments, and then clustering with other genes that change in a correlated fashion, functional groupings can be assigned, which can yield insight into more complicated processes or identify substantial distinctions between cell populations along the differentiation pathways.[4] One hypothesis that can be tested is whether genes that change in a highly correlated fashion are controlled by similar mechanisms. To test this hypothesis, different approaches, which examine upstream events, are needed to determine which similar mechanisms control gene expression across functional groups of genes.

Profiling the transcriptome of cells by large scale sequencing using next-generation sequencing technology overcomes many of the limitations of chip technology.[5] In addition to physical limitations of chip technology, such as cross-hybridization effects, and sensitivity to low levels of expressed transcripts, detection and quantitation of multiple transcripts from single loci is inadequate with current chip technology. After generating a large set of short sequence reads derived from total cell transcripts using NG sequencing, remapping to a reference genome samples the available transcripts, not necessarily in totality. In addition, care must be taken to identify fragments from partially spliced transcripts, which are expected at some level and will contain intronic regions.

Chromatin Immunoprecipitation

Chromatin immunoprecipitation is a powerful technique for investigating the binding of proteins to DNA elements on a genome-wide basis, such as targets for the transcription factors NANOG and OCT[6] or chromatic remodeling events.[7] This technique can be used with array-based technologies as well as with NG sequencing technologies. With this technique, proteins that are bound to different regions of DNA are chemically crosslinked to the DNA, and the crosslinked complexes are extracted, fragmented into small pieces, and immunopurified with

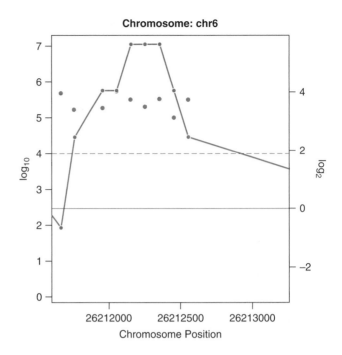

FIGURE 20.3. *ChIP-chip peaks. Peaks are located against a background of sample/total ratios. Grey symbols (right legend) represent ratios of sample to control DNA as log 2 values. Red symbols (left axis) represent probability scores from sliding window of data points. In this case all data points but the first are well above the cutoff threshold for probability (dotted line).*

monoclonal antibodies to a protein of interest. The purified fragments have the crosslinks removed and the resulting DNA fragments are amplified for analysis. Fragmented DNA, which has not been immunopurified, is often used as a reference control. It should be noted that smaller size fragments give better resolution of peak locations.

For array-based analysis, depending on the type of array chip used, a measure of occupancy is determined as a ratio of purified DNA to total DNA for each feature on the chip. Peaks are located against an often noisy background by statistical measures, for example, considering a peak as being a series of consecutive upward outliers using a chi-square test to perform significance cutoffs (Fig. 20.3).

With sequencer-based analysis, amplified immunopurified DNA, as above, is directly sequenced and mapped to a reference genome. Map locations are accumulated and counted to form histograms from which peak regions are identified. Once peak regions are identified, the data can be expressed as BED, GFF, or WIG formatted files for visualization in a genome browser, such as UCSC genome browser, or used to perform further bioinformatic analysis such as motif discovery. By combining gene expression analysis results with motif discovery using ChIP techniques, motifs can be found that bind factors that exert global control of gene expression.[8]

In a similar analysis as used with chromatin immunoprecipitation, methylated regions of DNA can be determined using MeDIP analysis by immunoprecipitating regions of DNA with a monoclonal antibody specific to 5-methylcytidine. Immunoprecipitated DNA fragments are amplified and analyzed using array-based or sequencer-based techniques as above.

Small RNA Sequencing (miRNA)

Small RNA molecules, particularly miRNA, are implicated in the fine control of large numbers of cellular post-transcriptional processes. MicroRNA (miRNA) is derived from endogenous noncoding RNA genes, processed to roughly 22 nt in length, that regulate gene expression at the post-transcriptional level. A large, and increasing, number of these molecules have been found; however, efforts to determine the functions of various miRNAs is ongoing. As of this writing, nearly

11,000 microRNAs have been identified and deposited in miRBase (Release 14 September 2009).[9]

The strategy for sequencing and identification of miRNAs is similar to performing sequencing on the GA or SOLiD platforms. In this case, total cellular RNA is gel purified and sized to isolate total small RNA. At this point appropriate adapters or linkers are ligated to the small RNAs, which are in turn amplified and attached to the sequencing matrix, GA flow cell or SOLiD beads, then further amplified, and finally sequenced. Depending on the density and number of distinct species, a small RNA library of 500,000 to several million sequence reads should be present.

At this point this sequence reads can be filtered for known elements, such as custom tag sequences. If the small RNA prep had small custom tags ligated onto the 5′ and 3′ ends, the population of those reads that contain both tags could be isolated as candidates for miRNA. Indeed, species of RNAs smaller than the read lengths (36 bp) will contain sequences that will not match to genomic sequences and must be trimmed appropriately.

Once sequence reads have been obtained, mapping to genomic sequences divides the population into unmatched and matched sequences, which then can be further resolved. For example, after locating and masking/removing custom tags or portions of adapters, reads can first be filtered for repetitive sequences, then mapped to miRBase to filter known miRNA species. Remaining reads can be mapped to genomic sequences to identify reads that map to other RNAs, such as fragments of tRNA or other species.

At this point there are a variety of directions using bioinformatics tools with which to proceed, for example, RNA structural analysis to determine characteristic stem-loop structures.

CONCLUSION

Many technologies are available for probing genomic features in pluripotent stem cells. Array-based and short read sequencing technologies, such as the SOLiD and Illumina platforms, offer a global analysis of genomic events.

REFERENCES

1. MacArthur BD, Ma'ayan A, Lemischka IR. Toward stem cell systems biology: from molecules to networks and landscapes. *Cold Spring Harb Symp Quant Biol*. 2008;73:211–215.

2. Birney E, Stamatoyannopoulos JA, Dutta A, et al. Identification and analysis of functional elements in 1% of the human genome by the ENCODE pilot project. *Nature*. 2007;447:799–816.

3. Kawaji H, Severin J, Lizio M, et al. The FANTOM web resource: from mammalian transcriptional landscape to its dynamic regulation. *Genome Biol*. 2009;10:R40.

4. Xu XQ, Soo SY, Sun W, et al. Global expression profile of highly enriched cardiomyocytes derived from human embryonic stem cells. *Stem Cells*. 2009;27:2163–2174.

5. Cloonan N, Forrest AR, Kolle G, et al. Stem cell transcriptome profiling via massive-scale mRNA sequencing. *Nat Methods*. 2008;5:613–619.

6. Mathur D, Danford TW, Boyer LA, et al. Analysis of the mouse embryonic stem cell regulatory networks obtained by ChIP-chip and ChIP-PET. *Genome Biol*. 2008;9:R126.

7. Ho L, Jothi R, Ronan JL, et al. An embryonic stem cell chromatin remodeling complex, esBAF, is essential for embryonic stem cell self-renewal and pluripotency. *Proc Natl Acad Sci USA*. 2009;106:5181–5186.

8. Kidder BL, Yang J, Palmer S. Stat3 and c-Myc genome-wide promoter occupancy in embryonic stem cells. *PLoS One*. 2008;3:e3932.

9. Griffiths-Jones S, Saini HK, van Dongen S, et al. miRBase: tools for microRNA genomics. *Nucleic Acids Res*. 2008;36:D154–D158.

PROTEOMIC ANALYSIS OF HUMAN PLURIPOTENT CELLS

21

Andy T. Y. Lau, Yan-Ming Xu, and Jen-Fu Chiu

INTRODUCTION

Embryonic stem (ES) cells are pluripotent cells derived from the inner cell mass of blastocyst-stage embryos. They are capable of self-renewal and differentiating into the three germ layers, which encompass more than 220 cell types in the human body.[1] It is because of this plasticity that ES cells may have extensive applications in regenerative medicine.[2] Because the destruction of an embryo, which is necessary for derivation of ES cells for research, has led to ethical controversies, scientists have searched for alternate techniques to obtain pluripotent cells. This now seems to be possible with the discovery by Yamanaka of reprogramming of adult cells to obtain "induced pluripotent stem" (iPS) cells using mouse and human fibroblasts in 2006 and 2007, respectively.[3,4] This breakthrough opens new research directions in stem cell investigation and paves the way for stem cell therapy—since adult cells used for reprogramming can be obtained from patients, eliminating the risk of rejection. However, this iPS cell technology is not without drawbacks; these include the difficulties of controlling the extent of reprogramming when using different iPS cell factors and culture conditions.[5] There have also been concerns about whether iPS cells and ES cells are truly equivalent in their capacity to differentiate into all cell types.[5]

Despite the pros and cons described above the ES and the iPS cell approaches are a principal focus of stem cell research because of their implications in medicine, science, and technology. It has been established that mouse and human ES/iPS cells share similar properties, but also have important differences.[6,7] Proteomic studies, which involve the identification, characterization, and functional assessment of all proteins expressed by cells, have advanced our understanding of the properties and capabilities of ES and iPS cells. Although genomic studies have provided signatures of gene expression that are characteristic of ES cells, these studies cannot determine which proteins are synthesized.[8–10] In addition, the diversity and specificity of proteins are reflected by more than 200 different post-translational modifications (PTMs), including acetylation, methylation, phosphorylation, ubiquitylation, sumoylation, ADP ribosylation, deimination, proline isomerization, glycosylation, farnesylation, myristoylation, palmitoylation, or proteolysis[11,12] (Table 21.1). These

Human Stem Cell Technology and Biology, edited by Stein, Borowski, Luong, Shi, Smith, and Vazquez
Copyright © 2011 Wiley-Blackwell.

TABLE 21.1 COMMON PTMs ON PROTEINS

Modifications	Residues Modified
Acetylation (ac)	**K**-ac
Methylation (me) (arginine)	**R**-me1 **R**-me2s **R**-me2a
Methylation (lysine)	**K**-me1 **K**-me2 **K**-me3
Phosphorylation (ph)	**S**-ph **T**-ph **Y**-ph
Ubiquitylation (ub)	**K**-ub
Sumoylation (su)	**K**-su
ADP ribosylation (ar)	**E**-ar
Deimination	**R** to citrulline
Proline isomerization	**P**-cis form to **P**-trans form

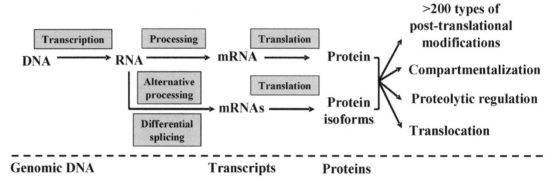

FIGURE 21.1. *The flow of genomic information to protein products. The concept of one-gene to one-protein is oversimplified since RNA can be differentially spliced and can produce various protein products. Furthermore, the protein may be affected by more than 200 different types of PTMs. Finally, as a result of compartmentalization and translocation, the same protein can be found with different properties and functions in different locations.*

protein modifications cannot be predicted solely from DNA sequences (Fig. 21.1). Thus effective integration of genomic and proteomic data is important for understanding biological control. The utilization of proteomic approaches has become an indispensable technique for understanding mechanisms that regulate self-renewal and differentiation as well as options for clinical utilization of ES cells. This chapter will discuss the application of proteomic approaches for studying the properties of human ES cells and biomarker discovery.

DEFINITION AND TYPES OF PROTEOMICS

The term "proteomics" was first used in 1995 to describe the large-scale characterization of the entire protein component of a cell type, tissue, or whole organism.[13–15] Proteomics studies the global protein expression profile rather than the behavior of a single protein. Proteomics can be classified into three areas of study.[16] The first area is protein expression, or protein profiling proteomics. It is the comprehensive and quantitative study of the proteins expressed in a given sample. The second area of study is structural proteomics. This approach maps the structure of protein complexes or the proteins present in a subcellular location or an organelle.[17] The third and last area is functional proteomics, which analyzes protein profiles at subcellular sites to understand the functional organization of cells at the molecular level. Information about the specific subcellular localization of a protein is invaluable in determining its function. The combination of protein identification by mass spectrometry with fractionation techniques that include immunoprecipitation or chromatography for enrichment of specific subcellular structures is termed "subcellular proteomics."[18]

TECHNIQUES OF PROTEOMICS

Two-Dimensional Polyacrylamide Gel Electrophoresis (2D-PAGE)

As a corollary of the limited resolving capability of 1D-PAGE, alternative methods are required to separate complex or crude samples. 2D-PAGE separates proteins based on two properties: their sizes and their individual unique isoelectric points. The most useful application of 2D-PAGE is to resolve protein isoforms of the same size, or proteins that have undergone various PTMs that change their electric charges (e.g., phosphorylation).[19,20]

There are multiple approaches to detect proteins in two-dimensional gels. The method used is very dependent on protein loading (analytical or preparative), the purpose of the gel (for protein quantitation or blotting), and the sensitivity required. The most common staining methods are silver staining and Coomassie blue staining. The 2D-PAGE has been improved recently with the introduction of difference gel electrophoresis (DIGE) technology. This technology utilizes fluorescent tagging of two protein samples with two different dyes. These dyes react with amines and possess the same molecular mass to eliminate the mass differences between tagged samples. The tagged proteins are mixed and run on the same two-dimensional gel. After image acquisition by a fluorescent scanner using different excitation wavelengths for each dye, the gel images are superimposed to identify the differences.[21]

Mass Spectrometry

The protein spots obtained by 2D gel electrophoresis are excised from the gel and digested in the gel with an enzyme (e.g., trypsin or chymotrypsin). The digest is then applied to a sample plate and coated with a matrix. The matrix is typically a small energy-absorbing molecule such as 2,5-dihydroxybenzoic acid or α-cyano-4-hydroxycinnamic acid. The analyte is spotted with the matrix on the sample plate and allowed to evaporate, resulting in the formation of crystals. The plate is then put into the matrix-assisted laser desorption/ionization–time of flight (MALDI-TOF) mass spectrometer and the laser is automatically targeted to the specific locations on the plate to obtain the peptide mass spectra. Several software packages are available for performing database matching:

- MASCOT at www.matrixscience.com[22]
- ProFound at http://prowl.rockefeller.edu/[23]
- Protein Prospector at prospector.ucsf.edu/[24]

In general, five variables are required for a peptide mass fingerprinting (PMF) search: (1) the peptide mass list; (2) the specification of the cleavage agent; (3) the error tolerance, that is, the accuracy of mass measurement; (4) the peptide modifications (e.g., phosphorylation); and (5) the species origin of the unknown protein.

Proteins are usually identified by MALDI-TOF-MS, where a PMF of all the peptides in a sample is generated for database searching.[25–29] However, it is sometimes necessary to use tandem MS (MS/MS), where mass fragmentation spectra are generated for each peptide in a sample to confirm the search result in the event of an insufficient number of proteolytic peptides available for confident matching.

New MS/MS technology platforms such as Triple Quadrupole or Ion Trap have been developed.[30] These mass spectrometers, which are integrated with electrospray ionization tandem MS (ESI-MS/MS) technology, are a marked improvement over earlier technology that employed MALDI-TOF/TOF for MS/MS. Since MS/MS can deduce the amino acid sequence of peptides from normal, post-translationally modified, or novel proteins, this approach can greatly enhance the degree of accuracy in the protein identification process and pave the way for the discovery of new proteins.

It is important to note that in the process of protein identification, proteins that should not be in the samples may be identified. This is usually due to contamination with foreign proteins that can occur at any step. The most common culprit encountered in proteomic studies is keratin, which can be introduced from dust or from shed human skin. Therefore gloves and a lab coat must be worn at all times during proteomic studies to prevent any possible keratin contamination in reagents, apparatus, or pipette tips. Also, it is often impossible to obtain some peptides from a digest for MS/MS because these protein sequences contain multiple arginines or lysines (which are readily cleaved by trypsin). Hence the selection of proteases for target protein digestion should be based on generating peptides with reasonable mass or the sequence of the focus (studying of PTMs) for MS or MS/MS diagnosis, respectively.

With the power of mass spectrometry, many problems encountered in the past are no longer obstacles when studying protein PTMs. For example, histone H3 is heavily modified by a series of PTMs. It has been shown that antibody against H3S10ph could not detect H3S10ph when H3S10 and K14 were phosphorylated and acetylated, respectively, on the same H3 N-terminal tail.[31] This revealed the potential complication of occlusion created by using site-specific antibodies against dually or heavily modified proteins. The problem was not considered at the time of antibody production by using the peptide bearing only single modification. Antibody occlusion problems are not confined to histones. To study protein PTMs unambiguously, MS is the gold standard and the preferred approach. MS can circumvent the problems associated with the use of site-specific antibodies such as specificity, cross reactivity, and epitope occlusion through interference by neighboring modifications. MS also has advantages over single modification antibody analysis as it permits the simultaneous detection of multiple modifications on the same peptide and identification of unexpected modifications.[32] Therefore the generation of dual modification-detecting antibodies is meaningful only after the modifications of the protein that occur in vivo are confirmed by MS. The antibody thus produced can facilitate detection of the protein modifications on a routine basis (e.g., Western blot, immunofluoresence assay).

Surface-Enhanced Laser Desorption Ionization (SELDI)–ProteinChip Separation and Profiling

Analogous to DNA chip technologies, ProteinChip technology coupled with SELDI-TOF-MS (surface-enhanced laser desorption/ionization–time of flight–mass spectrometry) has been developed by Ciphergen Biosystems, Inc. to facilitate the protein profiling of complex biological mixtures.[33] These ProteinChip arrays capture individual proteins from raw or complex mixtures, which are subsequently resolved by mass spectrometry using the principle that applies to MALDI-TOF. However, peaks obtained from this method preferentially represent the proteins or peptides in the low molecular size range (<10 kDa) and probably cannot represent all proteins in a cell. Additionally, SELDI–ProteinChip approaches are not quantitative and protein identification cannot be directly determined.

Protein Prefractionation and Shotgun Approach

The "shotgun" or liquid chromatography MS/MS (LC MS/MS) identifies the total cohort of proteins in the sample, using reversed-phase liquid chromatography to separate protein digests followed by online ESI-MS/MS for peptide sequencing.[34,35] The MS spectra generated from a total cellular fraction are carefully analyzed and protein identification is performed through peptide assignment and database searching.

Protein Labeling by iCAT, SILAC, or iTRAQ

Recently, novel methods for protein expression profiling have been developed that do not require separation of proteins by 2D-PAGE. These methods include isotope-Coded Affinity Tags (iCAT) and Stable Isotope Labeling with Amino acids in Cell culture (SILAC), as well as isobaric Tags for Relative and Absolute Quantitation (iTRAQ). For iCAT, protein samples from two different sources are labeled using two chemically identical reagents that differ only in mass as a result of the isotope composition.[36] Differential labeling of samples by mass allows the relative amount of proteins in two samples to be quantitated in the mass spectrometer. Limitations of this method are that it only works for cysteine-containing proteins and peptides must contain appropriately spaced protease cleavage sites flanking the cysteine residues.[37] Another approach to introducing isotopic mass shift for SILAC is to culture cells with specific isotope-containing amino acids.[38]

Recently, a new generation of isotope-coded chemical tags, iTRAQ, has been introduced for peptide relative quantitation using the MS/MS.[39] The iTRAQ tags (up to eight types) have equal mass; therefore the same peptide from each pooled sample appears at the same mass, regardless of the tag. The same peptide from all the samples is thus selected for simultaneous fragmentation. Relative quantitation is obtained via fragmentation of the iTRAQ tag, which generates a chemically identical (but isotopically distinct) reporter group of mass specific to each sample. iCAT/SILAC requires both peptides in a pair to be fragmented to ensure they are indeed identified as the same sequence in different samples. While a single fragmentation yields relative quantitation, iTRAQ labeling provides greater sensitivity and diminished machine operating time.

APPLICATIONS OF PROTEOMICS IN STEM CELL RESEARCH

Mechanisms Governing Stem Cell Self-Renewal and Differentiation

Mechanisms governing human stem cell biology, self-renewal, and differentiation are the focus of several chapters in this book. Here we will highlight the similarities and differences in properties of mESCs and hESCs, which include cell surface markers, transcription factor requirements for pluripotency, and signal transduction pathways involved for pluripotency (Table 21.2). The data derived from mESCs may not be directly applicable or comparable to mechanisms governing hESCs, illustrating the necessity of human stem cell research. Results from genomic studies provide valuable insight into gene expression in stem cells as well as changes in gene expression during stem cell differentiation. However, proteins control cell function and dictate biological activity. Proteomics is therefore indispensable to globally examine the complexity of the proteome in a systematic manner.

Studying Models for Human Pluripotent Cells

In general, two principal strategies have dominated proteomic analysis of human pluripotent cells: evaluating the cellular proteins from hESCs or the extracellular

TABLE 21.2 DIFFERENCES AND SIMILARITIES BETWEEN THE mESCs AND hESCs

Property	mESCs	hESCs
Extracellular signals for pluripotency		
LIF	+	−
BMP	+	−
Activin/Nodal	−	+
FGF	−	+
Cell surface biomarkers		
SSEA1	+	−
SSEA3	−	+
SSEA4	−	+
TRA-1-60	−	+
TRA-1-81	−	+
Alkaline phosphatase	+	+
Factors for pluripotency		
Oct4	+	+
Sox2	+	+
Nanog	+	+
Marker expression of trophectodermal differentiation	−	+
High telomerase activity	+	+
Trophoblast cell formation	−	+
Teratoma formation	+	+

Source: Data extracted from Nagano et al.[7]

proteins from feeder conditioned medium that support the growth of hESCs (Fig. 21.2). Proteomic characterization of iPS cells is emerging and contributes to understanding the biological parameters and clinical applications of epigenetic reprogramming. The significant variations in results obtained from different laboratories on the properties of pluripotent cells necessitate a stringent set of criteria for culturing ES cells. Such standardization is important for pursuit of biological control and for proteomic investigation of stem cells. An example of culture conditions that influence stem cells is O_2 concentration. The cellular responses are substantially

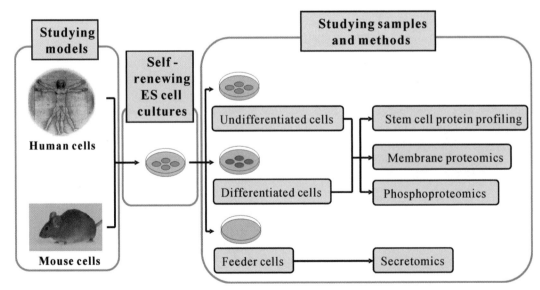

FIGURE 21.2. *Current studying models and methods employed in proteomics for biomarker discovery in stem cell research.*

different at atmospheric O_2 concentration compared with the cellular response at lower O_2 concentrations that more accurately approximate the physiologic state of the embryo. These lower O_2 concentrations are traditionally designated "hypoxia" and represent an in situ normoxia.[40] A reduced O_2 condition for stem cell culture is very significant. For instance, Yamanaka's group has demonstrated that hypoxic culture conditions enhance the production of iPS cells.[41]

Using various cutting-edge proteomic approaches, in recent years, numerous newly identified biomarkers of pluripotent cells have been reported. We will focus on stem cell (1) protein profiling, (2) membrane proteomics, (3) secretomics, and (4) phosphoproteomics.

Stem Cell Protein Profiling

Protein profiling and differential expression analysis in hESCs was initiated by Salekdeh's group in 2006 when these investigators provided the first comprehensive 2D map of hESCs. 2D-PAGE and MALDI-TOF/TOF MS/MS were utilized to construct a comprehensive and expandable map of proteins expressed in three hESC lines with emphasis on defining the inherent similarities and differences in independently derived diploid (Royan H2, Royan H5) and triploid (Royan H3) lines. A comprehensive analysis of 844 proteins, resolved by two-dimensional gel electrophoresis was carried out by MALDI-TOF/TOF. Fifty-four proteins exhibited quantitative changes and 14 proteins changed qualitatively. These differentially expressed proteins are involved in cell growth, metabolism, and signal transduction, reflecting involvement in self-renewal and pluripotency.[42]

In 2007, Zeng's group utilized another method termed the "PowerBlot large-scale Western blot assays" to examine expression of more than 1000 proteins in undifferentiated hESCs. This method detected more than 600 proteins that were grouped into 18 functional classes. Additionally, these investigators identified 42 examples of multiple bands for a single protein, which were interpreted as protein isoforms and/or post-translational modifications (PTMs). This proteomic approach, combined with analysis by commercially available antibodies, is a realistic strategy to massively identify proteins involved in hESC self-renewal from a global perspective.[43]

Using iTRAQ labeling technology together with the 2D LC-TOF/TOF-MS proteomic approach, Andrews's group in 2008 studied the global proteomics of hESCs during noggin-induced neural and bone morphogenetic protein-4 (BMP-4)-induced epidermal ectoderm differentiation. These investigators pointed out that, because hESCs are frequently maintained on irradiated mouse embryonic fibroblasts (MEFs) or in conditioned medium from MEFs, there is a risk that culture conditions may limit conclusive proteomic studies. Therefore they avoided the use of MEFs or MEF-conditioned medium for hESC culture. Instead, they maintained BG01 hESCs on gelatin-coated plates in medium supplemented with Knockout Serum Replacer and additional supplements. This experimental approach resulted in the identification of 603 unique proteins implicated in neural cell-specific pathways or epidermal ectoderm differentiation, as well as several novel markers of pluripotency and differentiation.[44]

Membrane Proteomics

Cell surface biomarkers are one of the signatures that define pluripotency and identify hESCs (Table 21.2). Currently, the number of known cell surface biomarkers is limited. There is a requirement to identify membrane proteins to facilitate development of a nomenclature for hESC lines and establish the biological principles of pluripotency that can be exploited for development of pluripotent cell-based therapies.

Human embryonal carcinoma cells (hECCs), the malignant counterparts of hESCs found in testis tumors, are constructive models to identify cell surface markers that are unique to specific tumors. In 2008, Krijgsveld's group performed plasma membrane proteomics on these two cell lines. Their strategy identified 237 and 219 specific plasma membrane proteins in the hESC line HUES-7 and the hECC line NT2/D1, respectively. Stemness-associated cell surface markers that include ALP, CD9, and CTNNB were observed. In addition, a large number of receptors, transporters, signal transducers, and cell–cell adhesion proteins were identified. Distinct HLA molecules were revealed on the surface of hESCs and hECCs despite their low abundance. This study provided a cell surface signature and distinguished normal hESCs from hECCs, which reflected the "benign" versus "malignant" phenotype.[45]

As indicated previously, SILAC is a powerful quantitative proteomic platform for comprehensive characterization of complex biological systems. In 2009, Blagoev's group exploited a SILAC-based proteomic strategy for discovery of the hESC-specific surface markers. Of the 811 identified membrane proteins, 6 displayed significantly higher expression levels in the undifferentiated state compared with the differentiated cells. These proteins included the established marker CD133/Prominin-1 as well as novel candidates for hESC surface markers: Glypican-4, Neuroligin-4, ErbB2, receptor-type tyrosine-protein phosphatase ζ (PTPRZ), and Glycoprotein M6B. These specific hESC markers can be employed in conjunction with the current marker reference list to increase understanding of mechanisms controlling hESC self-renewal and differentiation.[46]

Secretomics

hESCs are routinely co-cultured with a fibroblast feeder layer or in feeder-free culture with conditioned medium (CM) from the feeders. Feeder dependency can limit large-scale culture of undifferentiated hESCs. Consequently, the identification of factors that support the growth of undifferentiated hESCs has been pursued to establish a more defined culture environment that does not require a fibroblast feeder layer. In 2002, Lim's group concentrated the CM from the MEF feeder layers, STO cell line, and performed 2D-PAGE and MS analyses. A total of 136 unique proteins were identified, which included proteins controlling cell growth and differentiation, extracellular matrix formation and remodeling, as well as a spectrum of intracellular proteins. These studies demonstrated that feeder cells provide a complex environment for hESCs and that the factors identified could facilitate formation of a more defined culture environment for sustaining the undifferentiated growth of the hESCs.[47]

In 2007, Choo's group, using 2D-PAGE and MALDI-TOF MS/MS proteomic approaches, identified six candidate growth factors—PEDF, PAI, IGFBP-2, IGFBP-7, MCP-1, and IL-6—that were significantly down-regulated in conditioned medium from an isogenic but nonsupporting MEF line (ΔE-MEF) compared with those from the supporting primary MEF feeder layer. Replacement of the CM with the above six factors extended the viability and pluripotency of the hESCs for five passages. This study validated the concept of "feeder-free hESC culture" by functional assay and initiated the ongoing search for additional important factors that could support undifferentiated hESCs growth in feeder-free culture.[48] In 2007, Gray's group compared the proteome of conditioned media from two human fibroblast lines and one MEF line by multidimensional chromatography coupled with MS/MS. Among the three cell lines, 175 unique proteins were identified. These proteins were organized into 17 categories. The data from this study confirmed the complexity of conditioned media from fibroblast feeder layers and reinforced the expectations of identifying factors responsible for the support of undifferentiated hESCs.[49]

In 2008, Thomson's group, using column chromatography, immunoblotting, and MS-based proteomic analysis, identified multiple heparin sulfate proteoglycan (HSPG) species—Perlecan, Agrin, Glypican-4, Glypican-1, and Syndecan-4—in the conditioned medium secreted from the MEF feeder layer. These proteins stabilize basic fibroblast growth factor 2 (bFGF2) in unconditioned medium at levels comparable to those observed in the CM. Also, they could directly mediate the binding of bFGF2 to the hESC cell surface, and their removal from the CM impaired proliferation, demonstrating that HSPG species were key signaling cofactors in CM-based hESC culture.[50]

Phosphoproteomics

There is growing appreciation for the contribution of protein phosphorylation to proliferation and differentiation of hESCs. In 2009, several stem cell research groups characterized the phosphoproteome in hESCs. To avoid the influence of exogenous proteins, the hESCs were grown under feeder-independent conditions. The Coon laboratory used both MS/MS "collision-activated dissociation (CAD)" and "electron transfer dissociation (ETD)" to characterize the human ES cell phosphoproteome. This led to the identification of 11,995 unique phosphopeptides with 10,844 nonredundant phosphorylation sites. Two important pluripotency transcription factors, OCT4 and SOX2, were found to be phosphorylated (Table 21.3), indicating that phosphorylation may modulate the functions of these transcription factors.[51] Ding's group conducted phosphoproteomic analyses of hESCs and their differentiated derivatives. A total of 2546 phosphorylation sites were identified on 1602 phosphoproteins; 389 proteins contained more phosphorylation sites in undifferentiated hESCs, whereas 540 contained more phosphorylation sites in hESCs undergoing differentiation. Significantly, this approach identified the activation of additional receptor tyrosine kinase (PDGFR) in hESCs, revealing an effect of PDGF in the maintenance of pluripotency.[52] Krijgsveld's group carried out SILAC analysis of hESCs during differentiation induced by the BMP and removal of hESC growth factors. Of 5222 proteins identified, 1399 were phosphorylated at 3067 residues. The phosphorylation states of approximately 50% of these phosphosites were modified within the first hour following induction of differentiation, suggesting a complex interplay of phosphorylation networks spanning diverse signal transduction pathways and kinase activities. These investigators also observed sumoylation

TABLE 21.3 SUMMARY OF PTMs ON KEY hESC MARKERS

IPI Number[a]	Gene/Protein Name	Residues Modified
IPI00930598	OCT4 (transcription factor OCT4)	S236ph
IPI00009703	SOX2 (transcription factor SOX2)	K245su, S246ph, S249ph, S250ph, S251ph
IPI00002948	LIN28 (Lin-28 homolog A)	S3ph,[b] S200ph, T202ph
IPI00012593	DNMT3B (isoform 1 of DNA (cytosine-5)-methyltransferase 3B)	S82ph,[b] S136ph, S202ph, S209ph[b] T383ph, S387ph
IPI00026637	GAL (galanin precursor)	S116ph[b]
IPI00020668	UTF (undifferentiated embryonic cell transcription factor 1)	S18ph, T35ph,[b] S245ph[b]

[a]IPI, International Protein Index.
[b]These phosphorylations are found to be regulated during the process of differentiation.
Source: Data were extracted from Swaney et al.[51] and Van Hoof et al.[53]

of SOX2 as a result of phosphorylation (Table 21.3), providing a potential mechanism to overcome the stem cell regulatory circuitry during the initiation of cell differentiation.[53]

FUTURE OF STEM CELL PROTEOMICS

We have summarized the principal proteomic strategies that have been employed for defining the proteome of human pluripotent cells with approaches that have been pursued to establish post-translational modifications. There is a requirement to standardize proteomic methodologies that are utilized by different investigators and implement standard operating procedures to improve the comparability of proteomic data. Together with genome analysis that provides a comprehensive understanding of genes that are transcribed, we are gaining valuable insight into the biological control of pluripotency and the mechanisms that regulate the onset and progression of lineage commitment, cell specialization, and organ formation. Mechanisms that mediate biological control of development and differentiation are emerging and a platform for regenerative medicine is being established. Recently, to facilitate interactions between specialists in proteomics and stem cell biology, a new initiative designated the "Proteome Biology of Stem Cells Initiative" (PBSCI) has been established with support of the Human Proteome Organization (HUPO) and the International Society for Stem Cell Research (ISSCR). Through this initiative, specialists from both fields are collaboratively facilitating stem cell proteomic investigation.

Current proteomic technology provides a window of opportunity for further characterization of the stem cell proteome. However, the proteome of a cell or organism is analogous to a "snapshot" of activity at a single point in time. Significant components of processes that occur transiently may be underestimated or escape detection. As technology evolves, there will be further advances in functionally defining the stem cell proteome. It is realistic to anticipate that proteomic capabilities will extend to "time lapse" monitoring of the stem cell proteome where physiological responsiveness of the stem cell proteome can further support a platform for stem cell based tissue renewal and regenerative medicine.

Acknowledgments

We would like to acknowledge our debt to the authors whose works are cited in this chapter. Their works are illuminating and insightful. We would also like to express our sincere apologies to the other hundreds of authors whose works are not listed here due to time and space constraints. Last but not least, we gratefully thank Celia N. L. Lau for her proofreading of the manuscript for this chapter.

REFERENCES

1. Keller G. Embryonic stem cell differentiation: emergence of a new era in biology and medicine. *Genes Dev.* 2005;19:1129–1155.

2. Murry CE, Keller G. Differentiation of embryonic stem cells to clinically relevant populations: lessons from embryonic development. *Cell.* 2008;132:661–680.

3. Takahashi K, Yamanaka S. Induction of pluripotent stem cells from mouse embryonic and adult fibroblast cultures by defined factors. *Cell.* 2006;126:663–676.

4. Takahashi K, Tanabe K, Ohnuki M, et al. Induction of pluripotent stem cells from adult human fibroblasts by defined factors. *Cell.* 2007;131:861–872.

5. Smith KP, Luong MX, Stein GS. Pluripotency: toward a gold standard for human ES and iPS cells. *J Cell Physiol.* 2009;220:21–29.

6. Liu N, Lu M, Tian X, et al. Molecular mechanisms involved in self-renewal and pluripotency of embryonic stem cells. *J Cell Physiol.* 2007;211:279–286.

7. Nagano K, Yoshida Y, Isobe T. Cell surface biomarkers of embryonic stem cells. *Proteomics.* 2008;8:4025–4035.

8. Sun Y, Li H, Liu Y, et al. Cross-species transcriptional profiles establish a functional portrait of embryonic stem cells. *Genomics.* 2007;89:22–35.

9. Zhan M. Genomic studies to explore self-renewal and differentiation properties of embryonic stem cells. *Front Biosci.* 2008;13:276–283.

10. Cloonan N, Forrest AR, Kolle G, et al. Stem cell transcriptome profiling via massive-scale mRNA sequencing. *Nat Methods.* 2008;5:613–619.

11. Krishna RG, Wold F. Post-translational modification of proteins. *Adv Enzymol Relat Areas Mol Biol.* 1993;67:265–298.

12. Kouzarides T. Chromatin modifications and their function. *Cell.* 2007;128:693–705.

13. Wasinger VC, Cordwell SJ, Cerpa-Poljak A, et al. Progress with gene-product mapping of the Mollicutes: *Mycoplasma genitalium. Electrophoresis.* 1995;16:1090–1094.

14. Anderson NG, Anderson NL. Twenty years of two-dimensional electrophoresis: past, present and future. *Electrophoresis.* 1996;17:443–453.

15. Wilkins MR, Sanchez JC, Gooley AA, et al. Progress with proteome projects: why all proteins expressed by a genome should be identified and how to do it. *Biotechnol Genet Eng Rev.* 1996;13:19–50.

16. Lau AT, He QY, Chiu JF. Proteomic technology and its biomedical applications. *Sheng Wu Hua Xue Yu Sheng Wu Wu Li Xue Bao (Shanghai).* 2003;35:965–975.

17. Blackstock WP, Weir MP. Proteomics: quantitative and physical mapping of cellular proteins. *Trends Biotechnol.* 1999;17:121–127.

18. Lau AT, Chiu JF. Biomarkers of lung-related diseases: current knowledge by proteomic approaches. *J Cell Physiol.* 2009;221:535–543.

19. Lau AT, He QY, Chiu JF. A proteome analysis of the arsenite response in cultured lung cells: evidence for in vitro oxidative stress-induced apoptosis. *Biochem J.* 2004;382:641–650.

20. Zhou Y, Bhatia I, Cai Z, et al. Proteomic analysis of neonatal mouse brain: evidence for hypoxia- and ischemia-induced dephosphorylation of collapsin response mediator proteins. *J Proteome Res.* 2008;7:2507–2515.

21. Unlu M, Morgan ME, Minden JS. Difference gel electrophoresis: a single gel method for detecting changes in protein extracts. *Electrophoresis.* 1997;18:2071–2077.

22. Perkins DN, Pappin DJ, Creasy DM, et al. Probability-based protein identification by searching sequence databases using mass spectrometry data. *Electrophoresis.* 1999;20:3551–3567.

23. Zhang W, Chait BT. ProFound: an expert system for protein identification using mass spectrometric peptide mapping information. *Anal Chem.* 2000;72:2482–2489.

24. Clauser KR, Baker P, Burlingame AL. Role of accurate mass measurement (± 10 ppm) in protein identification strategies employing MS or MS/MS and database searching. *Anal Chem.* 1999;71:2871–2882.

25. James P, Quadroni M, Carafoli E, et al. Protein identification by mass profile fingerprinting. *Biochem Biophys Res Commun.* 1993;195:58–64.

26. Mann M, Hojrup P, Roepstorff P. Use of mass spectrometric molecular weight information to identify proteins in sequence databases. *Biol Mass Spectrom.* 1993;22:338–345.

27. Pappin DJ, Hojrup P, Bleasby AJ. Rapid identification of proteins by peptide-mass fingerprinting. *Curr Biol.* 1993;3:327–332.

28. Yates JR 3rd, Speicher S, Griffin PR, et al. Peptide mass maps: a highly informative approach to protein identification. *Anal Biochem.* 1993;214:397–408.

29. Jensen ON, Podtelejnikov AV, Mann M. Identification of the components of simple protein mixtures by high-accuracy peptide mass mapping and database searching. *Anal Chem.* 1997;69:4741–4750.

30. Canas B, Lopez-Ferrer D, Ramos-Fernandez A, et al. Mass spectrometry technologies for proteomics. *Brief Funct Genomic Proteomic.* 2006;4:295–320.

31. Clayton AL, Rose S, Barratt MJ, et al. Phosphoacetylation of histone H3 on c-fos- and c-jun-associated nucleosomes upon gene activation. *EMBO J.* 2000;19:3714–3726.

32. Garcia BA, Barber CM, Hake SB, et al. Modifications of human histone H3 variants during mitosis. *Biochemistry.* 2005;44:13202–13213.

33. Fung ET, Thulasiraman V, Weinberger SR, et al. Protein biochips for differential profiling. *Curr Opin Biotechnol.* 2001;12:65–69.

34. McCormack AL, Schieltz DM, Goode B, et al. Direct analysis and identification of proteins in mixtures by LC/MS/MS and database searching at the low-femtomole level. *Anal Chem.* 1997;69:767–776.

35. Peng J, Gygi SP. Proteomics: the move to mixtures. *J Mass Spectrom.* 2001;36:1083–1091.

36. Gygi SP, Rist B, Gerber SA, et al. Quantitative analysis of complex protein mixtures using isotope-coded affinity tags. *Nat Biotechnol.* 1999;17:994–999.

37. Haynes PA, Yates JR 3rd. Proteome profiling-pitfalls and progress. *Yeast.* 2000;17:81–87.

38. Ong SE, Blagoev B, Kratchmarova I, et al. Stable isotope labeling by amino acids in cell culture, SILAC, as a simple and accurate approach to expression proteomics. *Mol Cell Proteomics.* 2002;1:376–386.

39. Ross PL, Huang YN, Marchese JN, et al. Multiplexed protein quantitation in *Saccharomyces cerevisiae* using amine-reactive isobaric tagging reagents. *Mol Cell Proteomics.* 2004;3:1154–1169.

40. Ivanovic Z. Hypoxia or in situ normoxia: the stem cell paradigm. *J Cell Physiol.* 2009;219:271–275.

41. Yoshida Y, Takahashi K, Okita K, et al. Hypoxia enhances the generation of induced pluripotent stem cells. *Cell Stem Cell.* 2009;5:237–241.

42. Baharvand H, Hajheidari M, Ashtiani SK, et al. Proteomic signature of human embryonic stem cells. *Proteomics.* 2006;6:3544–3549.

43. Schulz TC, Swistowska AM, Liu Y, et al. A large-scale proteomic analysis of human embryonic stem cells. *BMC Genomics.* 2007;8:478.

44. Yocum AK, Gratsch TE, Leff N, et al. Coupled global and targeted proteomics of human embryonic stem

cells during induced differentiation. *Mol Cell Proteomics.* 2008;7:750–767.

45. Dormeyer W, van Hoof D, Braam SR, et al. Plasma membrane proteomics of human embryonic stem cells and human embryonal carcinoma cells. *J Proteome Res.* 2008;7:2936–2951.

46. Prokhorova TA, Rigbolt KT, Johansen PT, et al. Stable isotope labeling by amino acids in cell culture (SILAC) and quantitative comparison of the membrane proteomes of self-renewing and differentiating human embryonic stem cells. *Mol Cell Proteomics.* 2009;8:959–970.

47. Lim JW, Bodnar A. Proteome analysis of conditioned medium from mouse embryonic fibroblast feeder layers which support the growth of human embryonic stem cells. *Proteomics.* 2002;2:1187–1203.

48. Chin AC, Fong WJ, Goh LT, et al. Identification of proteins from feeder conditioned medium that support human embryonic stem cells. *J Biotechnol.* 2007;130:320–328.

49. Prowse AB, McQuade LR, Bryant KJ, et al. Identification of potential pluripotency determinants for human embryonic stem cells following proteomic analysis of human and mouse fibroblast conditioned media. *J Proteome Res.* 2007;6:3796–3807.

50. Levenstein ME, Berggren WT, Lee JE, et al. Secreted proteoglycans directly mediate human embryonic stem cell-basic fibroblast growth factor 2 interactions critical for proliferation. *Stem Cells.* 2008;26:3099–3107.

51. Swaney DL, Wenger CD, Thomson JA, et al. Human embryonic stem cell phosphoproteome revealed by electron transfer dissociation tandem mass spectrometry. *Proc Natl Acad Sci USA.* 2009;106:995–1000.

52. Brill LM, Xiong W, Lee KB, et al. Phosphoproteomic analysis of human embryonic stem cells. *Cell Stem Cell.* 2009;5:204–213.

53. Van Hoof D, Munoz J, Braam SR, et al. Phosphorylation dynamics during early differentiation of human embryonic stem cells. *Cell Stem Cell.* 2009;5:214–226.

BIOINFORMATICS STRATEGIES FOR UNDERSTANDING GENE EXPRESSION IN HUMAN PLURIPOTENT CELLS

22

Gustavo Glusman, Bruz Marzolf, Kai Wang, Ji-Hoon Cho, Burak Kutlu, and Qiang Tian

INTRODUCTION

Gene expression profiling has proved to be a powerful approach for unraveling stem cell biology. Over the past decade, technologies have evolved from expressed sequence tag (EST) sequencing and tag counting (serial analysis of gene expression—SAGE), to a variety of DNA microarrays, and the more recent high-throughput sequencing, such as Massively Parallel Signature Sequencing (MPSS) and NextGen), all of which have been successfully applied to decipher the properties of stem cells,[1–9] leading to the identification of a score of stem cell regulators (e.g., Nanog, Bmi1, Bmpr1a[9–12]). As the technology matures, it appears that DNA microarrays and next-generation (NextGen) sequencing have been widely adopted for gene profiling studies, due to their robust performance and deep coverage of the transcriptome. MicroRNAs (miRNAs), an emerging class of small RNAs, have demonstrated critical roles in modulating the biological process of stem cells[13] and can also be measured using either the array or sequencing technology. Computational data analysis strategies have also become increasingly sophisticated, morphing from the early era of cherry-picking individual genes, into the more comprehensive pathway/network-based schema,[14] yielding a more global view of biological systems and physiological processes.

Over the years, researchers at the Institute for Systems Biology (ISB) have performed tens of thousands of transcriptomic analyses by employing several different technologies. We have established a microarray data analysis pipeline that streamlines the sample and data handling processes. We have also generated and/or maintained one of the largest MPSS and NextGen transcript tag-counting datasets, including some pluripotent stem cell species. We have evaluated current approaches for miRNA measurement. ISB pioneered a systems approach to integrate and visualize complex data, which led to the development of Cytoscape,[15] the gold standard for high-throughput data visualization, and Gaggle,[16] a software suite for data integration. We have also evaluated a variety of commercial software packages for data integration and visualization. Bioinformatics strategies for analyzing data are still evolving and have not yet been standardized. To facilitate future studies

Human Stem Cell Technology and Biology, edited by Stein, Borowski, Luong, Shi, Smith, and Vazquez
Copyright © 2011 Wiley-Blackwell.

of pluripotent stem cells, we share here our experience using these technological platforms and strategies and through handling diverse and high-throughput datasets.

MICROARRAY-BASED RNA MEASUREMENT

Microarray technology is widely used to generate global mRNA and miRNA expression datasets that are essential to systems biology research, primarily due to the maturity, cost effectiveness, and availability of microarrays relative to similar global measurement technologies. Numerous DNA microarray platforms exist. They all measure relative abundances of nucleic acid species in an experimental sample, via hybridization through sequence complementarity to a predefined set of probes attached to a substrate. Computational tools for microarray-based RNA measurement range from desktop software suitable for small-scale experiments to large-scale enterprise infrastructure necessary to manage large data sets.

Platforms

Although all mRNA and miRNA expression microarrays ultimately provide an expression level measurement per gene/miRNA probed, technical differences require that varying methodologies be applied to different platforms. Commercial platforms were initially very costly in comparison with in-house spotted microarrays; however, dramatic price reductions and technical variability make the commercial microarrays the preferred choice for RNA measurement. The major commercial platforms such as Affymetrix, Agilent, Exiqon (miRNA only), and Illumina BeadChip have all been demonstrated to provide similar data quality,[17] making the choice of platform more dependent on factors such as cost, throughput requirement, and the availability of instrumentation and informatics tools.

Commercial microarray providers typically offer software for extracting probe intensities from the hybridization images as well as preprocessing of the data, including background correction and normalization. Beyond this low-level processing, researchers usually must acquire their own tools for data management and analysis.

Desktop Software

Small-scale experiments often don't require the same level of organization and data management as larger projects and may be effectively accommodated with tools that can run on a researcher's desktop computer. For the commercial platforms, extraction of raw probe intensities from images typically occurs immediately following acquisition of the hybridization images, with the instrument software generating raw intensity files that are available to the user. Normalization may also be a part of the image processing software, as with Agilent's Feature Extraction package, or may be done separately, as in the case of Affymetrix's Expression Console software.

Following normalization, various statistical, clustering, and visualization methods are generally useful in identifying trends in the data and subsets of genes that are of interest. Numerous software packages that provide this set of functionality exist, including free and open source options such as the TIGR MultiExperiment Viewer[18] as well as many commercial options. Normalized data or subsets of data generated by statistics or clustering may then be loaded into network visualization tools such as Cytoscape,[15] as described further in Section 22.5. Although using desktop software for transcriptomic/miRNA analyses can be convenient, there are notable reasons for using enterprise software, particularly as datasets grow in size.

Enterprise Software

Enterprise systems for transcriptomic microarray analysis address a number of issues raised by processing large numbers of microarrays: namely, (1) providing centralized, persistent storage of the data and results so that they're accessible even after experiments are complete, (2) allowing structured metadata capture and querying abilities to aid in the organization of large datasets, especially when they are generated by multiple researchers, (3) allowing a detailed record of various procedures applied to data, and (4) providing a streamlined workflow so that researchers don't have to worry about file formats or transporting files from one piece of software to another.

The first enterprise systems developed to provide informatics infrastructure for microarrays were large, tightly integrated, and modeled after the relational databases that house all relevant information. They typically span the entire microarray workflow, from sample entry, image extraction, and normalization through data analysis. These systems also typically provide a framework that can be expanded upon by other developers. Examples of such systems that are still actively being developed include SBEAMS-Microarray,[19] BASE,[20] and Stanford Microarray Database.[21]

Some newer enterprise systems have adopted a service-oriented architecture, where multiple and often separately developed pieces of software are loosely coupled, often using web services. A purported advantage to this type of architecture is greater flexibility in making often-unforeseen changes as research needs evolve and new technologies become available. Examples include Addama[19] and caBig (https://cabig.nci.nih.gov). The pipeline described below takes such a service-oriented approach.

Transcriptome Analysis Pipeline at ISB

The Institute for Systems Biology has a rich history of developing informatics infrastructure to support microarray analyses, ranging from algorithm development for early in-house cDNA microarrays[23] to entire architectures for data capture, management, and analysis.[22] As an example of enterprise microarray infrastructure, we detail an analysis pipeline created at ISB to enable high-throughput Affymetrix microarray processing.

Messenger RNA expression microarrays are typically designed to probe the 3′ end of transcripts, reflecting the 3′ bias of most labeling methods. The newest generation of Affymetrix Exon and Gene microarrays probe the entire length of RNA transcripts, by using a different labeling strategy that doesn't exhibit a 3′ bias. The large number of these arrays performed by ISB researchers has prompted the establishment of a pipeline to streamline their usage. This pipeline encompasses the entire sample workflow, from initial sample entry through hybridization, processing on the Affymetrix instrumentation, acquiring of the raw data, and normalization.

The first component of the pipeline involves capturing sample metadata and allowing samples to be tracked as they are prepared and hybridized. These are primarily concerns of a core facility, and we developed the web application SLIMarray[24] to serve as a LIMS for our core facility. SLIMarray allows researchers to define vocabularies to describe their samples that are specific to their research, and use those vocabularies when entering samples so that a consistent structured metadata is captured. SLIMarray then allows the facility to track samples as they are hybridized onto microarrays, and propagates sample information to instrument control software. Once hybridizations are complete and the raw data files from the instrumentation are produced, SLIMarray makes the metadata and raw data it has captured available through a web service. Providing this information via a web service decouples SLIMarray from downstream software so that either SLIMarray or the downstream software can be changed or even replaced.

The second component in the Affymetrix whole transcript expression pipeline is an Addama content repository, which provides a flexible means of storage for data and metadata. Information from SLIMarray is pulled into the content repository as new microarray data is produced. Views built on the content repository then allow users to query their microarrays based on various metadata that exists, and then submit the microarrays to an actual analysis pipeline.

The pipeline itself exists as a set of modules in GenePattern,[25] an application that allows for the creation and execution of pipelines, which are simply a series of modules. The Affymetrix whole transcript expression pipeline comprises the following series of modules:

1. Raw microarray data retrieval
2. Preprocessing/normalization using R/Bioconductor
3. Annotation of normalized data using a separate Synonym Service
4. Depositing the results back into the content repository

Once the pipeline has deposited the data back into the content repository, the researcher is notified by email and is able to retrieve the normalized data for downstream analysis.

Challenges for Microarray Technology

Although microarray technology enjoys widespread use, it is not without its drawbacks. Early problems with microarrays such as inconsistent spot morphology and uneven hybridization have largely been addressed, but issues relating to the hybridization-based nature of this technology still remain:

- *Cross-hybridization.* In addition to labeled targets of interest, probes will hybridize to other targets with similar sequences, artificially inflating the signal. Solutions such as identifying unique representative probes or using modified nucleotides to increase discrimination have been developed; however, this remains a key concern for hybridization-based RNA or DNA measurement.

- *Hybridization Condition.* Microarray technology applies a single hybridization condition across many thousands or millions of probes on an array, despite varying melting temperatures and GC content of probes. Efforts have been made in probe design to normalize these parameters; however, it is not always possible to do so while maintaining probe specificity.

- *Correlation Between Signal Intensity and RNA Level.* Fluorescent probe intensities cannot be easily converted to absolute number of transcripts and are only meaningful for relative comparisons between samples.

Some of these drawbacks are addressed by new technologies for transcriptomic measurements, as described in the next section.

FROM CHIP-BASED TRANSCRIPTOMICS TO SEQUENCING-BASED TRANSCRIPTOMICS

One of the hallmarks of chip-based gene expression assessments is that the intensity of signal observed after hybridization leads to a continuous value for the expression level of a gene. In contrast, sequencing-based transcriptomic analyses yield discrete values for gene expression levels, essentially "counting transcripts" digitally. This leads to a second difference: while microarray hybridization intensity values are only meaningful by comparison between samples (as signal ratios), digital transcript counting by sequencing aims to produce absolute expression values in each sample. A third, highly significant difference between the two approaches is that sequencing-based transcriptomics can discover and measure previously uncharacterized transcripts or variants of known transcripts, while chip-based analyses are limited to measuring what was previously known. We will briefly describe what sequencing-based techniques are available to transcript analysis, what information they produce, and how to interpret it.

Next-Generation Sequencing

Recent years have seen a revolutionary change in DNA sequencing technologies, and this change has created viable alternatives to chip-based transcriptomics. The first drafts of the human genome were produced, at great cost and labor, largely using fluorescent capillary sequencing technology. This technology yields long sequence reads, typically in the 500–800-bp range, and the usual protocols produce one or two reads from each cloned DNA fragment. A sequencing run typically takes 2 hours to complete. Recently, a variety of novel technologies have been developed that produce very large numbers of short sequence reads from a given DNA sample without a traditional cloning step. Roche's 454 FLX technology (http://www.454.com) produces over a million of 500-nucleotide reads in 10 hours. ABI's SOLiD technology (http://solid.appliedbiosystems.com) can produce 800 million reads up to 50 nucleotides long in a week. Illumina's Genome Analyzer II, also known as Solexa (http://www.illumina.com/pages.ilmn?ID=204), takes 4 days to produce a similar number of reads, which can range in length from 27 to over 75 nucleotides. Other technologies exist with unique capabilities, for example, Helicos Bioscience's HeliScope (http://www.helicosbio.com) sequences single molecules, and Complete Genomics (http://www.completegenomics.com) provides fee-for-service whole genome sequencing.

The ability to produce many sequence reads from a single sample makes some of these modern sequencing techniques directly applicable to transcriptome analysis. Even transcripts expressed at very low levels have a good chance of being observed when the protocol produces millions of reads. Within limits, it is reasonable to assume that transcripts with a higher number of copies in a sample will be observed (sequenced) more frequently than those present at only a few copies. The number of times a transcript is identified can therefore be used as a proxy for its expression level in the original sample. Since these numbers are, by definition, discrete, methods for transcriptome analysis by sequencing can be referred to as "transcript counting" techniques. This stands in contrast with the continuous hybridization signal intensity obtained when using microarrays.

Two different but complementary protocols for sequencing-based transcriptomic analysis exist: "tag counting" and "RNA-seq." Tag counting protocols are conceptually equivalent to the Serial Analysis of Gene Expression (SAGE) method,[26] while RNA-seq can be conceptualized as "shotgun sequencing the transcriptome," ideally observing the entire length of each transcript.

Tag Counting

The SAGE methodology is based on three ideas: (1) short sequence tags suffice to uniquely identify transcripts, (2) transcript tags can be physically linked, cloned, and sequenced, and (3) tag frequency is indicative of transcript expression level. The application of next-generation sequencing technologies to transcript counting essentially bypasses the linking of transcript tags prior to sequencing. Instead, each transcript is digested using a frequently employed cutting restriction enzyme (e.g., *Dpn*II, which recognizes the palindromic GATC pattern), the 3'-most fragment is selected via its poly-A tail, and a short segment is read starting at the restriction site. The observed tags are then used to computationally identify the transcripts from which they are derived, and the number of specific tag sequences is used to estimate the corresponding transcript level. A very similar protocol has been used in Massively Parallel Signature Sequencing (MPSS), in which special adapters and decoder probes were used for the sequencing stage.[27] The application of modern sequencing methods simplifies the protocol and significantly increases the depth to which samples can be studied.

The "raw" output of a tag counting experiment can be described as a very long list of statements, each claiming that a specific tag was observed, and qualifying this with a variety of scores describing base qualities, errors, and so on. Some quality scores are specific to the technique, for example, one of the several determinants of sequence reliability in the Solexa platform refers to quality of physical separation between clones: if a clone giving rise to a sequence tag partially overlaps with another clone, it is less reliable than one that can be easily distinguished from neighboring clones. The reliability of individual reads is also influenced by length and background distributions, with some errors being cumulative along the read and others being position specific. Detailed analyses of such influences can help improve usefulness of reads.[28]

After applying relevant technique-specific cutoffs for accepting or rejecting each sequence read, the results can be simplified into a long list of pairs: the first element of the pair is the sequence tag itself, the second element indicates how many times the tag was observed (and passed quality filters). Naturally, the sum of all these values (i.e., the total number of observed sequences passing quality filters) will change from experiment to experiment and from sample to sample. Thus there is a need for a normalization step, to make the results comparable across samples and experiments.

Normalization: The simplest (and most frequently used) normalization is the linear transformation to counts per million (CPM), in which each observed tag count is divided by the total tag count in the sample and multiplied by one million. The resulting values indicate, for each observed transcript, its fraction of the total material observed. These numbers are therefore directly comparable if one assumes that the total, ensemble level of expression of all genes in a cell is approximately constant. This very simple method of normalization can be applied at any stage in the analysis process.

The assumption of total constant expression need not be correct. Some genes are constitutively expressed at similar levels in most tissues, while others are tissue-specific and can vary wildly. They are usually referred to as "housekeeping genes" (HKGs) and "tissue-enriched genes" (TEGs), respectively.[29] The expression levels of HKGs are more appropriate as references for normalizing observed gene expression counts, than the sum total value observed in a given experiment. Under this model, each sample is scaled differently, in such a way that a predetermined set of HKGs display near-constant expression levels. This more elaborate method of normalization relies on the interpretation of signatures as derived from different genes. Therefore this method is usually applied only after translating observed signature counts into gene expression counts, as described next.

Interpretation: One of the basic premises of tag counting methods is that short sequence tags suffice to uniquely identify transcripts, and consequently the genes represented by those transcripts. There are two methods for translating tags to genes: *signature prediction* and *signature mapping*.

The *signature prediction* method is based on (1) the specific protocol used and (2) prior genome-wide knowledge of previously observed (or predicted) transcripts. Suppose, for example, that one used *Dpn*II to digest mRNAs from a sample, then selected for polyadenylated molecules, and finally sequenced 17-bp long signatures. Given a set of previously identified transcripts and/or gene models, it is computationally simple to identify the instance of the GATC pattern (identified by *Dpn*II) closest to the 3′ end (the polyadenylation site), and extract the 17 letters starting at that GATC site. One can therefore precompute an index of signatures expected to represent each transcript in the dataset, also known as a signature "hash table." The method can be made more elaborate by considering alternative restriction sites, taking into account incomplete digestion and/or the enzyme's "star activity," which may cut at nonstandard sites under certain conditions. However complex the signature prediction algorithm is made, it is a very efficient option for interpreting observed signatures back to transcripts and genes. The main deficiency of this method is, naturally, that not all possible transcripts have been observed. An observed signature might represent a novel transcript, which is absent from the hash table.

The *signature mapping* method compares each observed signature to the entire *reference genome*, on both strands, and provides hypotheses for the source of each signature. As such, this method is independent of the experimental protocol and is not limited by partial prior knowledge of transcripts. Several different algorithms can be used to compare a signature (a short sequence) to the reference genome (a set of very long sequences—the chromosomes). Some algorithms require a perfect match between the signature and the genome, but this requirement is not realistic due to sequencing errors and natural variation (SNPs). Most modern algorithms allow for a small number of mismatches between the signature and the reference genome and can rank hypotheses based on the quality of the match.

RNA-seq

The main weakness of the SAGE-like tag counting methods is that, at best, they measure the prevalence of a very specific position along the sequence of the transcript. This paradigm breaks down in three different scenarios: (1) the specific sequence of the gene lacks a suitable tag (e.g., no GATC when using *Dpn*II), making the gene invisible to the technique; (2) two or more genes share an identical tag (frequently the case in gene families); and (3) alternative splicing forms for the same gene share the same tag. A partial solution to these problems is to use alternative restriction enzymes, with different restriction specificities. This approach resolves some of the cases and is valuable as an internal control, but its success is limited and (naturally) it significantly raises the cost and labor involved.

A more successful and comprehensive solution has been described,[30] involving the direct, high-throughput sequencing of cDNA produced from randomly fragmented RNA. Sequencing of more than 40 million reads from a sample thus prepared can detect and quantify RNAs with very good dynamic range, and simultaneously discover novel splicing structures and possibly entirely new genes. Since a given gene can produce different transcripts, observed at different levels, the observed expression level may vary along a gene's structure. Thus there is no simple way to indicate the gene's expression in terms of CPM. Instead, a general measure of gene expression used in RNA-seq is the number of reads per kilobase of exon, per million mapped reads (RPKM).

The reads observed in RNA-seq experiments can, in principle, derive from any position along the mRNA sequence. A *signature prediction* protocol would therefore involve a very large signature index. Furthermore, such a hash table would include a very large number of ambiguous entries, with one signature supporting a number of source genes. For these reasons, the protocols of choice for interpreting RNA-seq experimental results involve *signature mapping* to the reference genome.

Since the reads are actually made from transcribed (and very frequently spliced) sequences, mapping each read directly to the genome will fail whenever a read spans a splicing junction. A simple but partial solution is to map reads also to a database of known transcripts. A more successful approach starts by mapping unspliced reads to the genome, immediately identifying transcribed exons, and then using the spliced reads to combine suitable pairs of exons (on the same strand of a chromosome, and at a reasonable distance from each other) by modeling the result of splicing them together. The reads spanning splice junctions therefore improve the coverage and quality of exon ends, and confirm the specific splicing junction. The ability to discover novel splicing patterns is one of the core strengths of the RNA-seq methodology.[31]

Data Analysis Tools

The task of mapping a very large number of short reads to the reference genome can be formidable due to sheer numbers. A simple indexing approach (hash table) is difficult for two reasons: sequencing errors and natural variability. All sequencing technologies have various probabilities of making some basic types of errors: misreading a base, skipping a base, or inserting a spurious base. In addition, the various technologies have some specific error modes (e.g., misreading a series of identical bases as one) and may have position-specific error propensities. Even a perfect sequencing technology should produce results that differ from the reference genome, when sequencing RNA from an individual. Pre-indexing all possible SNPs, copy-number variations, and other types of natural variation is simply not feasible. Nevertheless, a number of tools have been developed that are based on indexing the genome using hash tables (e.g., SOAPv1) or using the Burrows–Wheeler Transform (e.g., BWA and Bowtie).

Many tools have been developed to perform the mapping task. An extremely useful list of Next-Generation Sequencing Alignment programs has been compiled by Dr. Heng Li (http://lh3lh3.users.sourceforge.net/NGSalign.shtml), listing, for each tool, the sequencing platform(s) it supports, its availability (license), features, and limitations. Even more comprehensive lists of Short-Read Sequence Alignment programs exist in Wikipedia (http://en.wikipedia.org/wiki/Sequence_alignment_software) and in the SEQanswers online forum (http://seqanswers.com/forums/showthread.php?t=43).

Additional tools are being developed to identify splicing junctions, for example, TopHat (http://tophat.cbcb.umd.edu/) and QPALMA (http://www.fml.tuebingen.mpg.de/raetsch/suppl/qpalma), and for computing transcript abundance from RNA-seq data, for example, ERANGE (http://woldlab.caltech.edu/rnaseq/) and Cufflinks (http://cufflinks.cbcb.umd.edu/).

Application to ES Cell Analysis

Sequencing-based protocols have been successfully used to characterize the transcriptomes of ES cells. In one work,[2] three million MPSS tags led to the identification of 11,000 unique transcripts from pooled human ES cell lines, confirming the expression of previously known markers and identifying many novel genes. MPSS was also used[3] to compare RNA samples from feeder-free cultures of undifferentiated (passages 40–50) and differentiated (day 14) H1, H7, and H9 lines, identifying a large number of genes that changed expression levels upon differentiation. The Illumina sequencing technology was used[8] to generate three million tags from F1 mouse ES cells, most of which were confirmed to derive from murine transcripts. These and other analyses, revealing the complexity and regulation of the transcriptome,[32] confirm the applicability and power of sequencing-based transcriptomics.

MICRORNA PROFILING IN STEM CELLS

MicroRNA plays an important role in regulating the pluripotency of stem cells. MicroRNAs (miRNAs) are noncoding RNA molecules, 17–25 nucleotides long, that have profound effects on modulating the activities of genes at both transcription and translation levels in the cells. The existence of stable extracellular miRNA further suggests it plays a role in mediating cell–cell communication outside of the cells. miRNAs have also been suggested as key players in the maintenance of pluripotent stem cells, like the miRNAs in the mir-302 cluster in humans and mice, and the mir-290 cluster in mice.[33] miRNAs in these two clusters are highly expressed in the stem cell stage and drastically reduce their levels once the cells start to differentiate.

Recognizing the importance of miRNAs, global profiling methods to assess miRNA levels have been developed based on different technologies including microarray (such as products from Exiqon and Agilent), qPCR (such as products from Qiagen, Exiqon, and ABI), and next-generation sequencing (such as products from Illumina and Applied Biosystems). Data obtained from these platforms are in a similar format as the standard gene expression profiling results, and the general analysis approaches for mRNA expression can directly be adapted for miRNA analysis. The key is to integrate miRNA and mRNA information and identify possible miRNA-based regulatory networks. A general outline of miRNA data analysis processes used in our laboratory is listed below.

Measuring and Analyzing the miRNA Expression

As described in earlier sections, miRNA profiling data can be obtained from microarray, qPCR, or next-generation sequencing platforms; the data can be handled similarly as mRNA. Here, we are using microarray data as an example to illustrate the process of integrating miRNA and mRNA information.

- *Consolidation and Normalization.* Unlike gene array, the miRNA array usually contains replicated identical probes for each miRNA. To avoid complications in the data analysis steps, we usually consolidate the data by taking the mean or median over the identical probes. Microarray data normalization, transformation, and present/absent probe calling can be performed with commonly used methods/tools in gene array data analysis as described earlier. For normalization, a global normalization approach such as quantile normalization or normalizing with a global mean or median can be used; otherwise, a specific normalization approach such as normalizing against RNU6B, SNORDs, or a selection of commonly expressed miRNAs such as mir-103 can also be used for cross-sample comparison.[34]

- *Data Analysis.* The R-packages such as LIMMA (http://www.bioconductor. org) and EDGE (http://www.genomine.org/edge) can be used to identify static and time-course differentially expressed miRNAs, respectively. Differentially expressed miRNAs can also be analyzed, displayed, and clustered using any one of the mRNA cluster programs like MultiExperiment Viewer (MeV, http://www.Tm4.org/mev.html).

- *Obtain Putative miRNA–mRNA Relationships.* A putative relationship between individual miRNA and mRNA could be inferred, based on various miRNA target prediction programs. Commonly used target prediction algorithms include TargetScan (http://www.targetscan.org), miRTarget (http://mirdb.org), and miRanda (http://www.microrna.org). Since miRNA is known to have an inverse relationship with its targets, the putative miRNA interacting genes are most likely going to have inversely correlated expression profiles with the miRNA.

When dealing with probable interactions between miRNA and a group of genes like those from cluster analysis, the statistical significance of the association can be evaluated using Fisher's exact test with information including (1) the number of genes in the cluster that have been predicted as targets for the specific miRNA,

(2) the number of all predicted targets for the miRNA, (3) the number of genes in the cluster, and (4) the total number of genes involved in the clustering analysis. This type of analysis would lead us to construct a hypothetical miRNA–mRNA interaction network.

Challenges in Handling miRNA Data

The process of miRNA data analysis is straightforward; however, the nature of the miRNA molecules creates several unique challenges for both measurement accuracy and functional integration.

- *Short and extreme conservation of miRNA sequences make the accurate measurement of specific miRNA species difficult.* Even though the levels of individual miRNAs can be assessed by standard laboratory methods such as hybridization and polymerase chain reaction, the length of miRNAs limits the option of designing highly specific probes or primers for all the miRNA species. The other unique feature about microRNA is the sequence conservation among different miRNAs. miRNAs are usually clustered in the genome and conserved through evolution, which leads to the speculation that a number of miRNAs are derived from either ancient or recent gene duplication events. One such example is the stem cell "specific" mir-302 cluster, which contains four miRNAs—mir-302a, mir-302b, mir-302c, and mir-302d—and these miRNA sequences are conserved between humans and mice. In addition, the sequences for the four mir-302s are almost identical except two nucleotides at the 3′ end for both the human and mouse 302s (Table 22.1). This level of sequence conservation adds additional problems on the accuracy of assessing the level of specific miRNA.

- *End region sequence heterogeneity of miRNAs may affect the accuracy of measurements.* Using next-generation sequencing platforms, we and others have found significant sequence heterogeneities at both 5′ and 3′ ends for all the miRNA species. One such example is the mir-451—a microRNA highly expressed in the embryonic body stage; the species matching the sequence listed in the mirbase (http://www.mirbase.org) database is less than 50% of the total read (Table 22.2). In some other cases, the "correct" miRNA sequence represents even less of the overall isomer population. These end region sequence variations pose significant challenges to accurately assessing the level of specific miRNA species, especially for methods relying heavily on the sequence integrity of ends such as the miRNA Taqman® assay.

TABLE 22.1 ALIGNMENT OF HUMAN AND MOUSE miRNA SEQUENCES[a] IN THE mir-302 CLUSTER

hsa-miR-302a	TAAGTGCTTCCATGTTTTGG
mmu-miR-302a	TAAGTGCTTCCATGTTTTGG
hsa-miR-302b	TAAGTGCTTCCATGTTTTAG
mmu-miR-302b	TAAGTGCTTCCATGTTTTAG
hsa-miR-302c	TAAGTGCTTCCATGTTTCAG
mmu-miR-302c	AAGTGCTTCCATGTTTCAG
hsa-miR-302d	TAAGTGCTTCCATGTTTGAG
mmu-miR-302d	TAAGTGCTTCCATGTTTGAG
Consensus	TAAGTGCTTCCATGTTTTAG

[a]Human sequences are labeled with "hsa" prefix, while mouse sequences are labeled with "mmu." The nucleotide sequences that are different from the consensus are in red.

TABLE 22.2 ISOMER POPULATION OF MOUSE mir-451

Category	Sequence[a]	Number of Reads[b]		
		ES	EB	Diff
mmu-miR-451 isomirs	AAACCGTTACCATTACTG	1	32	1
	AAACCGTTACCATTACTGA	1	166	7
	AAACCGTTACCATTACTGAG	2	739	64
	AAACCGTTACCATTACTGAGT	17	4970	367
mmu-miR-451	**AAACCGTTACCATTACTGAGTT**	**32**	**9950**	**674**
mmu-miR-451 isomirs	AAACCGTTACCATTACTGAGTTT	28	5192	323
	AAACCGTTACCATTACTGAGTTTA	7	734	46
	AAACCGTTACCATTACTGAGTTTAG	7	320	20
	AAACCGTTACCATTACTGAGTTTAGT	1	7	1
	AACCGTTACCATTACTGAGT	0	4	1
	AACCGTTACCATTACTGAGTT	1	8	4
	AACCGTTACCATTACTGAGTTT	0	4	1
	CGTTACCATTACTGAGTT	0	5	4

[a]The sequence matching the sequence deposited in the database is in red.
[b]The number of reads are from three different stem cell differentiation stages: embryonic stem cell (ES), embryonic/embryoid body (EB), and the differentiated cell (Diff).

- *Very little is known about miRNA–mRNA interaction, making the data integration unreliable.* It is generally believed that the interactions between miRNA and mRNA are based on partial sequence complementation. Various prediction algorithms have been developed to predict potential miRNA interacting targets. However, there are very few validated miRNA–mRNA interactions, and most of the miRNA target genes are based on predictions in silico. Some miRNAs are predicted to target hundreds of genes spread over a number of biological pathways and networks, making it difficult to build and interpret a reliable miRNA–mRNA interaction network.

Summary of miRNAs in Stem Cells

Despite serious challenges in measuring miRNA levels and integrating miRNA into various biological networks, several interesting findings from miRNA-based biomarkers to miRNA-based therapeutics have been reported.[35] The most exciting finding for miRNAs in the stem cell field is probably the observation that stem cell "specific" miRNAs like mir-290 and mir-302 clusters can be used to modulate or reverse the fate of differentiated cells.[36] The field of miRNAs poses tremendous challenges for experimental and theoretical biologists, but at the same time provides remarkable opportunities to further understand the complexity of biological processes.

SOME EXAMPLES OF TOOLS/SOFTWARE SUITES FOR DATA INTEGRATION, NETWORK ANALYSIS, AND DATA VISUALIZATION

Network Analysis Tools and Databases at the ISB

Network analysis is a key component of a systems biology approach. At the Institute for Systems Biology, we are actively developing a wide variety of bioinformatics tools to visualize and analyze biological networks. We also use commercial tools, such as GeneGO and Pathway Studio Enterprise. As the systems biology discipline matures, visualization tools are being developed with capabilities to address the specialized needs of a wide variety of projects and users. All network analysis software developed at ISB is open source and platform independent. Below is a summary of the software and data content that are currently available to researchers.

Cytoscape

Cytoscape is an open source bioinformatics software platform for visualizing molecular interactions and complex networks. Although it started as an in-house project,[15] Cytoscape is now actively developed by an international and multi-institute network of programmers. Cytoscape core distribution provides a basic set of features for data integration and visualization. More features can be added with the help of plug-ins. The plug-ins give access to a wide variety of functions such as network and molecular profiling analyses, new layouts, additional file format support, scripting, and connection with databases. Plug-ins may be developed by anyone using the Cytoscape open API based on Java technology and plug-in community development is encouraged. Cytoscape can be downloaded at http://www.cytoscape.org/download_list.php. Cytoscape can be installed on Windows, Mac, and Linux OS.

CyGoose Within the Gaggle Framework

Gaggle is a Java-based software environment that integrates various software tools and databases, including Cytoscape.[16] Gaggle can handle all four data types typically used in systems biology research: networks, a list of names, a matrix, and an associative array. CyGoose is a Cytoscape plug-in that helps connect the program with other popular bioinformatics programs, such as MeV,[37] Data Matrix Viewer, KEGG pathways,[38] STRING,[39] Bioinformatics Resource Manager,[40] Firegoose,[41] and R/BioConductor.[42] With the implementation of simple Java methods, the programs become "Geese" and acquire the ability to broadcast at the user's request to other "Gaggled" programs. Instances of these data types are transmitted in serialized form using Java RMI within the Gaggle. A user can easily perform expression data analysis (R), clustering (MeV), correlation (DMV), pathway and GO term enrichment analysis (FireGoose), network subtraction (R), and visualization (Cytoscape) within the same environment, without having to save and upload the results after working with each tool. Gaggle and associated tools are available at http://gaggle.systemsbiology.org/docs/. Gaggle tools are also platform independent.

PIPE

Protein Information and Property Explorer (PIPE) is a web-based tool to explore functional enrichments of lists.[43] PIPE allows graphical exploration of GO term enrichments on a directed acyclic graph by a "Gaggled" instance of Cytoscape. Users can explore the terms and associated genes and broadcast gene lists to each other in an interactive and persistent manner. PIPE is available at http://pipe. systemsbiology.net.

GeneGO

GeneGO is a commercial network visualization and analysis tool that displays information from a proprietary manually curated database of human protein–protein, protein–DNA, and protein compound interactions, metabolic and signaling pathways, and the effects of bioactive molecules in gene expression.[44] MetaCore is an integrated platform to analyze expression data, CGH arrays, proteomics, metabolomics, pathway analysis, Y2H, and other custom interactions. The analytical package includes tools for data visualization, implementation of graph algorithms, and filters. Users can find more information about GeneGo on the company website (http://www.genego.com/).

Pathway Studio Enterprise (PSE)

PSE is another commercial tool that contains literature-derived interactions that are stored in the ResNet database. The interactions are extracted from publications by Natural Language Processing techniques. The tool allows building and expanding networks; relationships between genes can be explored by extracting facts from new abstracts that are in the ResNet. PSE features a free network-drawing module that can be used to create publication quality diagrams. PSE is available at http://www.ariadnegenomics.com/products/pathway-studio/.

Alternative tools exist for network analysis. Ingenuity Pathway Analysis (IPA) is a web-based, platform-independent network visualizer. IPA maintains its own expert curated data. GenMAPP (Dahlquist KD et al, PMID: 11984561) is a free, open source network analysis tool. GenMAPP is developed in Visual Basic 6.0 and can run on Windows OS. PathVisio is a closely related Java-based network analysis tool (van Iersel MP et al 2008, PMID: 18817533). PathVisio can recognize GenMAPP or GXML formatted network files.

Databases of Systems Biology Data

The data that are used for building networks are equally as important as the software tools used for network analyses. A summary of some of the public and commercial databases is given below.

Protein Interactions: Public Databases These databases are from/by academic groups that gathered significant resources to collect physical protein interaction by manual curation of scientific papers: Human Protein Reference Database (HPRD),[45] Molecular Interaction database (MINT),[46] IntAct,[47] MIPS Mammalian Protein–Protein Interaction Database,[48] and Biological General Repository for Interaction Datasets (BioGRID).[49] (See Table 22.3.)

Protein Interactions: Commercial Databases The Prolexys Human Interactome Database (Prolexys HyNet) is an broad collection of experimentally determined and verified human protein–protein interactions.[50] It is the biggest human protein–protein interaction database in the world with more than 120,000 nonredundant human protein interactions.

TABLE 22.3 COMPARISON OF PROTEIN–PROTEIN INTERACTION SOURCES[a]

Source	HPRD	MINT	IntAct	MIPS	BioGRID
HPRD	37081	10022	9940	554	19144
MINT	NA	15712	8212	152	6920
IntAct	NA	NA	25595	180	9026
MIPS	NA	NA	NA	739	439
BioGRID	NA	NA	NA	NA	25825

[a]Each cell contains the number of interactions shared between each database.

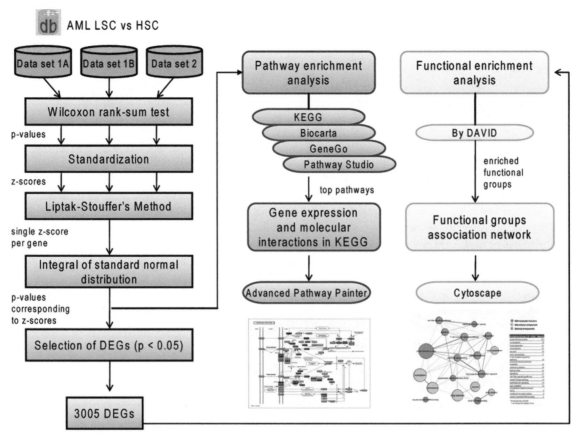

FIGURE 22.1. *Strategy for identifying differentially regulated networks between two stem cell populations. (Adapted from Majeti et al.[14])*

Custom Application of Public and Commercial Software Tools for the Analysis of Human Stem Cells. A number of the above data analysis tools can be integrated in a custom manner to gain global network level understanding of stem cells. For instance, we have developed a computational strategy to analyze one of the best characterized human stem cells, the hematopoietic stem cells (HSCs) and their malignant counterpart, the leukemia stem cells (LSCs), for dysregulated gene expression networks that differentiate the two populations.[14] We obtained gene expression data of highly purified HSC and LSC populations by using the Affymetrix GeneChip. The DNA microarray data was processed through ISB's microarray data pipeline (Section 22.2 of this chapter) for normalization. The normalized gene expression data from independent datasets were combined by using a custom designed statistical approach, with a significant score (p values, and z scores) computed for each individual gene based on the expression level (Fig. 22.1). We then employed a bifurcated approach, (1) to take all the genes in a given network and (2) to take only those differentially expressed genes (DEGs) in a given network, to identify differentially expressed networks between HSCs and LSCs. In these analyses, both public (KEGG, Biocarta) and commercial (GeneGo, Pathway Studio) databases were utilized. We have also adopted DAVID (http://david.abcc.ncifcrf.gov/) and Cytoscape programs for data integration and visualization (Fig. 22.1). Using this strategy, some key pathways, such as the adherens junction, T cell receptor signaling, and Wnt pathways, were found to be significantly dysregulated in leukemia stem cells. We anticipate that similar approaches can be extended to numerous biological systems, including pluripotent stem cells, for comparative analysis of gene expression at the heightened network level.

REFERENCES

1. Richards M, Tan SP, Tan JH, et al. The transcriptome profile of human embryonic stem cells as defined by SAGE. *Stem Cells*. 2004;22:51–64.

2. Brandenberger R, Khrebtukova I, Thies RS, et al. MPSS profiling of human embryonic stem cells. *BMC Dev Biol*. 2004;4:10.

3. Miura T, Luo Y, Khrebtukova I, et al. Monitoring early differentiation events in human embryonic stem cells by massively parallel signature sequencing. *Stem Cells Dev*. 2004;13:694–715.

4. Brandenberger R, Wei H, Zhang N S, et al. Transcriptome characterization elucidates signaling networks that control human ES cell growth and differentiation. *Nat Biotechnol*. 2004;22:707–716.

5. Akashi K, He X, Chen J, et al. Transcriptional accessibility for genes of multiple tissues and hematopoietic lineages is hierarchically controlled during early hematopoiesis. *Blood*. 2003;101:383–389.

6. Ivanova NB, Dimos JT, Schaniel C, et al. A stem cell molecular signature. *Science*. 2002;298:601–604.

7. Ramalho-Santos M, Yoon S, Matsuzaki Y, et al. "Stemness": transcriptional profiling of embryonic and adult stem cells. *Science*. 2002;298:597–600.

8. Rosenkranz R, Borodina T, Lehrach H, et al. Characterizing the mouse ES cell transcriptome with Illumina sequencing. *Genomics*. 2008;92:187–194.

9. Park IK, He Y, Lin F, et al. Differential gene expression profiling of adult murine hematopoietic stem cells. *Blood*. 2002;99:488–498.

10. Mitsui K, Tokuzawa Y, Itoh H, et al. The homeoprotein Nanog is required for maintenance of pluripotency in mouse epiblast and ES cells. *Cell*. 2003;113:631–642.

11. Park IK, Qian D, Kiel M, et al. Bmi-1 is required for maintenance of adult self-renewing haematopoietic stem cells. *Nature*. 2003;423:302–305.

12. He XC, Zhang J, Tong WG, et al. BMP signaling inhibits intestinal stem cell self-renewal through suppression of Wnt-beta-catenin signaling. *Nat Genet*. 2004;36:1117–1121.

13. Chen CZ, Li L, Lodish HF, et al. MicroRNAs modulate hematopoietic lineage differentiation. *Science*. 2004;303:83–86.

14. Majeti R, Becker MW, Tian Q, et al. Dysregulated gene expression networks in human acute myelogenous leukemia stem cells. *Proc Natl Acad Sci USA*. 2009;106:3396–3401.

15. Shannon P, Markiel A, Ozier O, et al. Cytoscape: a software environment for integrated models of biomolecular interaction networks. *Genome Res*. 2003;13:2498–2504.

16. Shannon PT, Reiss DJ, Bonneau R, et al. The Gaggle: an open-source software system for integrating bioinformatics software and data sources. *BMC Bioinformatics*. 2006;7:176.

17. Canales RD, Luo Y, Willey JC, et al. Evaluation of DNA microarray results with quantitative gene expression platforms. *Nat Biotechnol*. 2006;24:1115–1122.

18. Saeed AI, Sharov V, White J, et al. TM4: a free, open-source system for microarray data management and analysis. *BioTechniques*. 2003;34:374–378.

19. Marzolf B, Deutsch EW, Moss P, et al. SBEAMS-Microarray: database software supporting genomic expression analyses for systems biology. *BMC Bioinformatics*. 2006;7:286.

20. Saal LH, Troein C, Vallon-Christersson J, et al. BioArray Software Environment (BASE): a platform for comprehensive management and analysis of microarray data. *Genome Biol*. 2002;3:SOFTWARE0003.

21. Demeter J, Beauheim C, Gollub J, et al. The Stanford Microarray Database: implementation of new analysis tools and open source release of software. *Nucleic Acids Res*. 2007;35:D766–D770.

22. Boyle J, Breitkreutz B J, Stark C, et al. The BioGRID Interaction Database: 2008 update. *Nucleic Acids Res*. 2008;36:D637–D640.

23. Ideker T, Thorsson V, Siegel AF, et al. Testing for differentially-expressed genes by maximum-likelihood analysis of microarray data. *J Comput Biol*. 2000;7:805–817.

24. Marzolf B, Troisch P. SLIMarray: lightweight software for microarray facility management. *Source Code Biol*. 2006;1:5.

25. Reich M, Liefeld T, Gould J, et al. GenePattern 2.0. *Nat Genet*. 2006;38:500–501.

26. Velculescu VE. MINT: the Molecular INTeraction database. *Nucleic Acids Res*. 2007;35:D572–D574.

27. Brenner S, Johnson M, Bridgham J, et al. Gene expression analysis by massively parallel signature sequencing (MPSS) on microbead arrays. *Nat Biotechnol*. 2000;18:630–634.

28. Philippe N, Boureux A, Brehelin J, et al. Using reads to annotate the genome: influence of length, background distribution, and sequence errors on prediction capacity. *Nucleic Acids Res*. 2009;37:e104.

29. She X, Rohl CA, Castle JC, et al. Definition, conservation, and epigenetics of housekeeping and tissue-enriched genes. *BMC Genomics*. 2009;10:269.

30. Mortazavi A, Williams BA, McCue K, et al. Mapping and quantifying mammalian transcriptomes by RNA-Seq. *Nat Methods*. 2008;5:621–628.

31. Pan Q, Shai O, Lee LJ, et al. Deep surveying of alternative splicing complexity in the human transcriptome by high-throughput sequencing. *Nat Genet*. 2008;40:1413–1415.

32. Wang ET, Sandberg R, Luo S, et al. Alternative isoform regulation in human tissue transcriptomes. *Nature*. 2008;456:470–476.

33. Barroso-del Jesus A, Lucena-Aguilar G, Menendez P. The miR-302-367 cluster as a potential stemness regulator in ESCs. *Cell Cycle*. 2009;8:394–398.

34. Hua YJ, Tu K, Tang ZY, et al. Comparison of normalization methods with microRNA microarray. *Genomics*. 2008;92:122–128.

35. Bartels CL, Tsongalis GJ. MicroRNAs: novel biomarkers for human cancer. *Clin Chem*. 2009;55:623–631.

36. Lin SL, Chang DC, Chang-Lin S, et al. Mir-302 reprograms human skin cancer cells into a pluripotent ES-cell-like state. *RNA*. 2008;14:2115–2124.

37. Saeed AI, Bhagabati NK, Braisted JC, et al. TM4 microarray software suite. *Methods Enzymol*. 2006;411:134–193.

38. Kanehisa M, Goto S, Hattori M, et al. From genomics to chemical genomics: new developments in KEGG. *Nucleic Acids Res*. 2006;34:D354–D357.

39. von Mering C, Jensen LJ, Kuhn M, et al. STRING 7—recent developments in the integration and prediction of protein interactions. *Nucleic Acids Res*. 2007;35:D358–D362.

40. Shah AR, Singhal M, Klicker KR, et al. Enabling high-throughput data management for systems biology: the Bioinformatics Resource Manager. *Bioinformatics*. 2007;23:906–909.

41. Bare JC, Shannon PT, Schmid AK, et al. The Firegoose: two-way integration of diverse data from different bioinformatics web resources with desktop applications. *BMC Bioinformatics*. 2007;8:456.

42. Gentleman RC, Carey VJ, Bates DM, et al. Bioconductor: open software development for computational biology and bioinformatics. *Genome Biol*. 2004;5:R80.

43. Ramos H, Shannon P, Aebersold R. The protein information and property explorer: an easy-to-use, rich-client web application for the management and functional analysis of proteomic data. *Bioinformatics*. 2008;24:2110–2111.

44. Nikolsky Y, Nikolskaya T, Bugrim A. Biological networks and analysis of experimental data in drug discovery. *Drug Discov Today*. 2005;10:653–662.

45. Mishra GR, Suresh M, Kumaran K, et al. Human protein reference database—2006 update. *Nucleic Acids Res*. 2006;34:D411–D414.

46. Chatr-aryamontri A, Ceol A, Palazzi LM, et al. MINT: the Molecular INTeraction database. *Nucleic Acids Res*. 2007;35:D572–574.

47. Kerrien S, Alam-Faruque Y, Aranda B, et al. IntAct–open source resource for molecular interaction data. *Nucleic Acids Res*. 2007;35:D561–D565.

48. Pagel P, Kovac S, Oesterheld M, et al. The MIPS mammalian protein–protein interaction database. *Bioinformatics*. 2005;21:832–834.

49. Breitkreutz BJ, Stark C, Reguly T, et al. The BioGRID Interaction Database: 2008 update. *Nucleic Acids Res*. 2008;36:D637–D640.

50. LaCount DJ, Vignali M, Chettier R, et al. A protein interaction network of the malaria parasite *Plasmodium falciparum*. *Nature*. 2005;438:103–107.

EPIGENETIC ANALYSIS OF PLURIPOTENT CELLS

23

Mojgan Rastegar, Geneviève P. Delcuve, and James R. Davie

INTRODUCTION

Epigenetics refers to a variety of processes that have long-term effects on gene expression programs without changes in DNA sequence. Key players in epigenetic control are histone modifications and DNA methylation, which, in concert with chromatin remodeling complexes, nuclear architecture, and microRNAs, define the chromatin structure of a gene and its transcriptional activity. The balance between pluripotency and differentiation is regulated by epigenetic mechanisms. Although epigenetic marks are established early during development and differentiation, adaptations occur throughout life in response to intrinsic and environmental stimuli and may lead to late life disease and cancer.[1]

CHROMATIN STRUCTURE AND HISTONE MODIFICATIONS

Nuclear DNA is packaged into nucleosomes, the basic repeating structural units in chromatin. The nucleosome consists of a histone octamer, arranged as a $(H3-H4)_2$ tetramer and two H2A-H2B dimers, around which DNA is wrapped. Histone H1 binds to the linker DNA, which joins nucleosomes together. The core histones (H2A, H2B, H3, H4) have a similar structure with a basic N-terminal domain, a globular domain organized by the histone fold, and a C-terminal tail. The N-terminal tails emanate from the nucleosome in all directions and are available to interact with linker DNA, nearby nucleosomes, or with other proteins.[2]

Core histones undergo a variety of reversible post-translational modifications (PTMs), including acetylation, methylation, and phosphorylation (Fig. 23.1). Core histone PTMs are found within the N-terminal tails, the histone fold, and C-terminal tails, with most PTMs being located in the N-terminal tails (Table 23.1). Histone PTMs function to alter chromatin structure and/or provide a "code" for recruitment or occlusion of nonhistone chromosomal proteins to chromatin. These recruited proteins are referred to as "readers."[3] Some PTMs (active marks) are associated with transcriptionally active chromatin regions, while others (repressive marks) are present in silent regions. Histone acetylation usually marks active genes as does trimethylation of H3 at K4 (H3K4me3), whereas dimethylation of H3

Human Stem Cell Technology and Biology, edited by Stein, Borowski, Luong, Shi, Smith, and Vazquez
Copyright © 2011 Wiley-Blackwell.

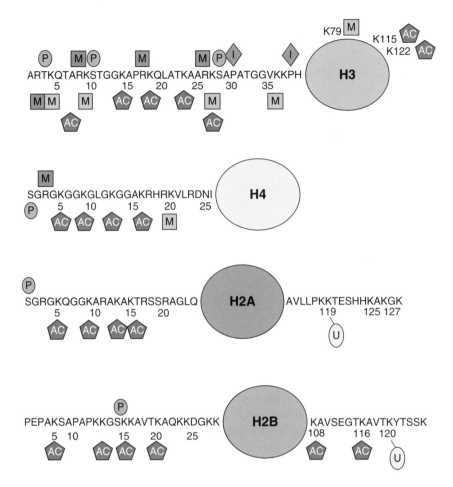

FIGURE 23.1. *Site-specific histone modifications. Sequence specificity and type of post-translational modifications by modifying enzymes are depicted on core histone amino acid sequences for H2A, H2B, H3, and H4. Ac, acetylation; P, phosphorylation; M, methylation; I, isomerization; U, ubiquitination.*

TABLE 23.1 DIFFERENT HISTONE MODIFICATIONS AND CORRESPONDING CHROMATIN MODULATORS[a]

Modification	Function
Acetyltion (**K**-ac)	Transcription, repair, replication, condensation
Methylation (Lysines, **K**-me1,2,3)	Transcription, repair
Methylation (Arginines, **R**-me1,2,3)	Transcription
Phosphorylation (**S**-ph, **T**-ph)	Transcription, repair, condensation
Ubiquitylation (**K**-ub)	Transcription, repair
Sumoylation (**K**-su)	Transcription
ADP ribosylation (**E**-ar)	Transcription
Proline isomerization (**P**-cis > **P**-trans)	Transcription

[a]The altered residues are shown in bold.
Source: Kouzarides.[46]

at K9 (H3K9me2) constitutes a repressive mark.[4,5] Added to the complexity is the dynamics of histone PTMs and histone variants.[6,7] H2A, H2B, and H3 have variants that are expressed at the time of DNA synthesis (e.g., H3.1, H3.2, H2A.1; replication dependent) and those that are expressed throughout the cell cycle (H3.3, H2A.Z; replication independent). H3.3 is located in regulatory regions of genes,[8,9] and H3.3 is enriched in active marks (K4me3 and acetylated K9, 14, 18 and 23).[10–12] Genome mapping of histone PTMs has revealed that some histone

PTMs such as H3K4me3 are located at the promoter, while H3K36me3 is located throughout the coding region of transcribed genes.

The activation or repression of mammalian genes involves chromatin remodeling by histone modifying enzymes (Table 23.2) and ATP-dependent chromatin remodeling complexes (e.g., SWI/SNF). Lysine acetyltransferases (KATs) and histone deacetylases (HDACs) catalyze reversible histone acetylation. Transcription factors recruit coactivators with KAT activity (e.g., p300/CBP; PCAF) to regulatory DNA sites, while transcriptional repressors recruit corepressors with HDAC activity.[13] In transcriptionally poised and active chromatin regions, histone acetylation is a dynamic process, with the steady state of acetylated histones being decided by the relative activities of the recruited KATs and HDACs.[14,15] Similarly, the steady state of histone phosphorylation is decided by histone kinases and protein phosphatases. The ATP-dependent chromatin remodeling complexes (e.g., SWI/SNF) remodel nucleosomes, allowing transcription factors and the transcription initiation factors access to regulatory DNA sequences.[16]

DNA METHYLATION

DNA methylation is associated with gene silencing and occurs on cytosine at CpG dinucleotides across the human genome. Associated with DNA methylation are methyl-CpG-binding-domain proteins (MBDs), which belong to repressor complexes with histone deacetylase (HDAC) activity.[17,18] Only 1–2% of the genome comprising CpG islands, short CpG-rich regions surrounding promoters, may escape methylation.[19] DNA methylation is involved in X-chromosome inactivation in females and DNA imprinting, events that result in monoallelic gene expression. It also contributes to genome stability by preventing translocations of repetitive and transposable sequences, and most likely plays a dynamic role in development.[20,21] About half of the CpG islands are not associated with annotated promoters, but are intra- or intergenic. It has been suggested that these CpG islands mark the transcription start site of noncoding RNAs.[21] This hypothesis is supported by the detection in various tumors of abnormal expression of numerous miRNAs linked to abnormal DNA methylation.[22] Overall methylation might also affect expression potential as it was reported that, compared to inactive X chromosome, active X chromosome is hypomethylated at promoter regions but hypermethylated at gene bodies.[23] Mechanisms ruling the targeting of de novo DNA methyltransferases (DNMT3a and DNMT3b) to specific CpG sequences in early development are mostly unknown. In somatic cells, DNA methylation patterns are copied by the maintenance DNA methyltransferase I (DNMTI) positioned at the replication forks, with some cooperation of DNMT3a and DNMT3b.[20] Thus DNA methylation, maintained through mitosis, is considered a stable epigenetic mark. However, this conventional standpoint is challenged by the existence of active DNA demethylation,[24–26] as well as the role of rapid and dynamic DNA methylation/demethylation in long-term memory formation in adult neurons.[27] It has been proposed that DNA methylation patterns in many or all cell types result from the balance of methylating and demethylating activities, with both methyltransferases and demethylases being targeted to specific genes by transcription factors acting downstream of signaling pathways yet to be unraveled.[24,25]

ANALYSES OF EPIGENETIC EVENTS

The task to decipher pluripotent and differentiated epigenomes and dysfunctional epigenomes leading to the vast array of diseases and cancers is colossal. However, thanks to molecular biology techniques such as chromatin immunoprecipitation (including large-scale variants ChIP-on-chip and ChIP-seq), fluorescent

TABLE 23.2 CHROMATIN REMODELING FACTORS[a] AND CORRESPONDING HISTONE MARKS[110,111]

Lysine Acetyltransferases	Substrates
HAT1 (**KAT1**)	H4 (K5, K12)
GCN5 (**KAT2A**), PCAF (**KAT2B**)	H3 (K9, K14, K18)
CBP (**KAT3A**), P300 (**KAT3B**)	H3 (K14, K18), H4 (K5, K8), H2AK5, H2B (K12,K15)
TIP60/PLIP (**KAT5**)	H4 (K5, K8, K12, K16), H3K14
HBO1 (**KAT7**)	H4 (K5, K8, K12)

Lysine Methyltransferases	Substrates
SUV39H1-2 (**KMT1A-B**)	H3K9
G9a (**KMT1C**)	H3K9
EuHMTase/GLP (**KMT1D**)	H3K9
ESET/SETDB1 (**KMT1E**)	H3K9
CLL8 (**KMT1F**)	H3K9
MLL1-5 (**KMT2A-E**)	H3K4
SET1A-B (**KMT2F-G**)	H3K4
ASH1 (**KMT2H**)	H3K4
SET2 (**KMT3A**)	H3K36
NSD1 (**KMT3B**)	H3K36
SYMD2 (**KMT3C**)	H3K36
DOT1 (**KMT4**)	H3K79
Pr-SET7/8 (**KMT5A**)	H4K20
SUV4 20H1-2 (**KMT5B-C**)	H4K20
EZH2 (**KMT6**)	H3K27
SET7/9 (**KMT7**)	H3K4
RIZ1 (**KMT8**)	H3K9
SirT2-3	H4K16

Lysine Demethylase	Substrates
LSD1/BHC110 (**KDM1**)	H3(K4, K9)
JHDM1a-b (**KDM2A-B**)	H3K36
JHDM2a-b (**KDM3A**)	H3K9
JMJD2A/JHDM3A (**KDM4A**)	H3 (K9, K36)
JMJD2B-C (**KDM4B-C**)	H3 (K9, K36)
JMJD2D (**KDM4D**)	H3K9
JARID1A-D (**KDM5A-D**)	H3K4
UTX (**KDM6A**)	H3K27
JMJD3 (**KDM6B**)	H3K27

TABLE 23.2 (Continued)

Histone Serine/ Threonine Kinases	Substrates	Arginine Methyltransferases	Substrates
Haspin	H3T3	CARM1	H3 (R2, R17, R26)
MSK1-2	H3S28	PRMT4	H4R3
CKII	H4S1	PRMT5	H3R8, H4R3
Mst1	H2BS14		
Rsk2	H3S10		

Histone Ubiquitilases		Substrates	
Bmi/Ring1a		H2AK119	
RNF20/RNF40		H2BK120	

[a]The new nomenclatures are in bold.[112]

in situ hybridization, DamID,[28] or bisulfite sequencing PCR, associations of epigenetic marks with specific DNA sequences can be directly elucidated. To identify methylated cytosines along a DNA sequence, the bisulfite sequencing PCR and pyrosequencing methods are applied. Bisulfite treatment of DNA converts unmethylated cytosines to uracil. Following treatment of the DNA, the region of interest is amplified by PCR amplification followed by sequencing by conventional methods or by pyrosequencing of the amplified product.[29] Mapping of transcription factors, histone PTMs, histone variants, nonhistone proteins associated with histone PTMs, chromatin modifying enzymes, and chromatin remodelers along DNA sequences is accomplished by the chromatin immunoprecipitation (ChIP) assay. Regarding the ChIP technique, a note of caution is warranted. Indeed, Solomon and Varshavsky had suggested that the crosslinking efficiency of formaldehyde was unpredictable.[30] Our results demonstrated that a standard ChIP assay using formaldehyde crosslinking (sometimes referred to as X-ChIP) and anti-HDAC2 antibodies would only detect DNA fragments associated with complexes like Sin3 or NuRD containing phosphorylated HDAC2. However, unmodified HDAC2 and regions associated with this more abundant form of HDAC2 would not be detected.[31] To monitor the distribution of unmodified HDAC2 in chromatin, the native ChIP assay (also called N-ChIP) or dual crosslink assay (e.g., DSP-X-ChIP) should be applied.[31] Thus it must be kept in mind that the lack of detection of an association between a protein and a DNA sequence does not necessarily mean that this association does not occur.

EPIGENETIC CONTROL OF STEM CELL PLURIPOTENCY

Mammalian development occurs through a succession of directed cellular divisions beginning from a single cell. Decisions leading to the commitment of cellular progeny toward a specific fate are precisely programmed at the genomic, proteomic, and epigenomic levels during each cellular division. Considering the number of cell types and tissues contributing to the development and maintenance of a living organism, it is easy to appreciate how exquisitely complex the process of early embryogenesis truly is. Understandably, much attention has been focused on elucidating the underlying mechanisms regulating cellular commitment and differentiation during mammalian development. In 1981, Gail Martin successfully isolated embryonic stem (ES) cells from the inner cell mass (ICM) of preimplantaion mouse blastocysts[32] and addressed the longstanding search for a proper system to model early development.

However, human ES cells were not isolated until 1998 when James Thomson reported the first isolation and culture of human ES cells.[33] Both human and mouse ES cells show pluripotency, the unique property of infinite self-renewal and differentiation into all three germ layers: the ectoderm, the mesoderm, and the endoderm.[32,34] The ability of ES cells to generate every cell type within the human body has brought ES cells to the forefront of biomedical research for studying cellular differentiation and early development, while providing an excellent system for human disease modeling. Moreover, ES cells present an exciting potential as an endless resource for cell-based therapies in human degenerative diseases. A network of limited transcription factors, namely, OCT4, NANOG, SOX2, and SALL4,[35–37] maintain the pluripotency in self-renewing ES cells. However, upon differentiation specific sets of lineage-restricted factors promote directing cellular fates toward specific lineages. Most noteworthy perhaps is the suggestion that ES cells possess a phenotypic heterogeneity that is controlled in part by these very same pluripotency factors. ES cell heterogeneity could be the result of fluctuation between dynamically higher or lower expression levels of these factors, as has been demonstrated to be true at least for NANOG.[38] In partnership with pluripotent transcription factors, epigenetic mechanisms balance the cellular identity between pluripotency and differentiation and determine the cellular fate during proliferation. Current epigenome mapping of pluripotent ES cells versus lineage-specific stem cells and their differentiated progeny indicate a central role for epigenetics in both self-renewal and differentiation.[39,40] Epigenetic mechanisms that contribute to gene regulation in ES cells include nucleosome positioning, exchange of histones and histone variants, histone modifications, DNA methylation, chromatin remodeling, and small RNA molecules.[41] These mechanisms work in harmony in multiple layers[42] to affect the expression profiles of stem cells during self-renewal and differentiation, thereby influencing the development of specific cell lineages.

CHROMATIN STRUCTURE IN ES CELLS

In undifferentiated ES cells, the interplay between transcription factors and other classes of protein or nonprotein molecules contributes in maintaining the pluripotency. Histone modifications regulate gene expression via controlling chromatin structure that affects higher order interactions between neighboring nucleosomes and affecting either the spacing between the nucleosomes or the contact of histones with DNA.[43] Chromatin organization in ES cells is transcriptionally permissive, highly dynamic, and enriched in euchromatin with no sign of condensed heterochromatic foci.[39] However, upon differentiation it becomes more compact in a subset of specific repressed loci. In support of this, studies using fluorescence recovery after photobleaching (FRAP) and biochemical experiments indicate an elevated fraction of soluble and loosely bound architectural chromatic proteins in ES cells compared to differentiated cells.[39] The existence of such a hyperdynamic chromatin structure is vital for ES cell pluripotency, and limitation of histone H1 exchange prevents cellular differentiation.[39] Usually, chromatin structure plays a preventive role in any process that requires accessibility to the naked DNA, which is controlled in a genome-wide and site-specific manner by the action of chromatin remodeling factors[44] (Table 23.2). There are two major classes of chromatin remodeling enzymes: (1) protein complexes with enzymatic activity specific for modifying the amino terminus of the histone tails and (2) enzymes that utilize ATP hydrolysis to interrupt the contact of histone proteins and DNA, thereby altering the nucleosomal and chromatin structure.[45] In general, histone modifications such as acetylation, methylation, phosphorylation, and ubiquitination are associated with transcriptional activation.[46] Conversely, modifications caused by methylation, ubiquitination, sumolization, deamination, and proline isomerization are implicated in repression[46] (Table 23.1). However, depending on the context of the underlying DNA sequence, association with a specific histone modification mark can serve as either a positive or a negative effector.[46] In differentiated cells, histone modification patterns play a defining role in the active or inactive state of individual genes and this epigenetic memory will be heritably passed on during cellular proliferation. Overall, chromatin structure and in turn the histone marks are more permissive in ES cells in comparison to differentiated cells. Furthermore, lineage-specific differentiation of ES cells is accompanied by a global reduction of histone modifications in active chromatin (such as histone acetylation of H3 and H4), and a significant increase in repressive marks (such as H3K9m).[39,40] In pluripotent ES cells a group of developmentally important genes carry both repressive histone marks (such as H3K9me3 or H3K27me3) and activating (such as H3K9ac3 and H3K4me3) modifications. These genes make up the category of genes with "bivalent marks" and are primed for transcription.[47] Importantly, genes that carry these bivalent marks replicate earlier in ES cells compared to more differentiated cells, which is in agreement with the chromatin being more relaxed in ES cells.[41] During differentiation, the bivalent marks at the poised genes are resolved and active histone marks are erased from silenced genes, while inactive histone marks are lost from activated genes. As the H3K27me3 marks are overrepresented compared to the H3K4me3, these genes stay repressed in ES cells, but in a poised transcriptional state bound by RNA Pol II.[47] Although most developmentally important genes such as *Sox, Hox, Pax,* and *Pou* gene family members carry the bivalent marks,[47] the existence of such chromatin domains has not been found to be associated with certain lineage control genes such as *Myf5* and *Mash1*.[47–49] Moreover, examples of developmentally important genes such as the *Foxp1* gene are not found to carry either H3K4me or H3K9me histone marks.[47] As it is unlikely that there is another epigenetic permissive or repressive mechanism of control for these genes, an emerging new concept of "histone modification pulsing" explains that perhaps the association of these marks at these specific regions happens in a highly dynamic manner, which needs much more precise and time-sensitive detection techniques than are currently in use[50] (Fig. 23.2). To address the cellular heterogeneity during early embryogenesis, this theory suggests that individual ICM derived ES cells experience asynchronous histone modifications at the important developmental genes and respond differently to the environmental signals and morphogenic gradients that are vital for cell lineage determination.[50] Such an hypothesis would predict that a rapid and highly dynamic histone modification exchange becomes more stable as differentiation proceeds and specialization toward specific lineages progresses.[50] Recent studies in search for the co-occupancy of the active and inactive marks during vertebrate embryonic development in *Xenopus tropicalis* detected very limited genes carrying the bivalent marks with the majority of genes being expressed with no sign of RNA Pol II pausing.[51,52] This indicates the bivalency concept does not exist in *Xenopus* embryos,[51,52] similar to what is found in *Drosophila*.[53] As both *Xenopus* and *Drosophila* contain closely related Trithorax (TrxG) and Polycomb (PcG) proteins, the discovery of early embryogenesis without bivalency demands further research in other model organisms, as chromatin bivalency with no doubt stands true in mammalian ES cells.[47]

POLYCOMB AND TRITHORAX PROTEINS

Polycomb and Trithorax proteins belong to the histone methyltransferase (HMT) family of chromatin remodeling factors, which consist of histone–lysine methyltransferases and histone–arginine methyltransferases (Table 23.2). PcG proteins are transcriptional silencers and epigenetic modifiers involved in ES cell pluripotency.[54] They were originally identified in *Drosophila* as repressors of homeotic genes, which

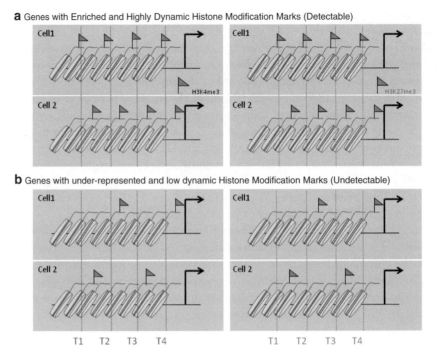

a Genes with Enriched and Highly Dynamic Histone Modification Marks (Detectable)

b Genes with under-represented and low dynamic Histone Modification Marks (Undetectable)

FIGURE 23.2. *Detection of histone modifications. Lineage-specific controlled genes in ES cells may undergo dynamically different histone modification kinetics. At specific time points (T1, T2, T3, T4) both H3K4me3 and H3K27me3 (green and red flags, respectively) are detectable in bivalent genes (a), as histone modifications are enriched in this category of genes. It is still possible to miss detection of histone marks through random sampling times (comparison of cell 1 and cell 2). Furthermore, if the theory of histone pulsing is correct, it is still less likely to detect any histone modification mark where lower dynamics of such modifications exist (b). (Adapted from Gan et al.[50]).*

act in an opposite direction to the TrxG proteins. PcG repression is mediated through two major protein complexes, PRC (polycomb-group repressive complex) 1 and PRC2 that in ES cells silence a wide range of developmentally important genes.[55] These complexes exert their repressor activity through maintenance of the silenced state or inhibiting transcriptional initiation, respectively.[55,56] Core components of the PRC2 complex are Ezh2, Suz12, and EeD, which catalyze di- and trimethylation of H3K27. Almost 10 subunits make up the PRC1; these include Ring 1A, Ring 1B, Bmi1, and Cbx8, which is suggested to mediate the mono-ubiquitination of the histone H2A at K119, increasing chromatin condensation and maintaining gene silencing.[57] Recent data suggest that PRC1 and PRC2 have critical roles in ES cell pluripotency. Indeed, PRC1 and PRC2 co-occupy the genomic regions that are highly enriched for H3K27me3 at the regulatory regions of developmental genes and signaling factors.[55,58] During ES cell differentiation, the majority of regions with bivalent marks tend to change their status toward univalent marks (Fig. 23.3). Active genes become enriched in H3K4me3 through TrxG proteins in parallel with the activity of UTX and JMJD3, which demethylate H3K27me3.[59] PRC2 complex recruits RBP2 (JARID1a), which is a H3K4me3 demethylase, and this association is sequentially followed up by PRC2-mediated repression during ES cell differentiation.[1,60,61] The genes with bivalent domains could be classified into two classes based on the association with PcG proteins: one class is associated with both PRC1 and PRC2, while the second class only associates with PRC2.[1,62] The first class covers the genes encoding transcription factors, morphogenes and

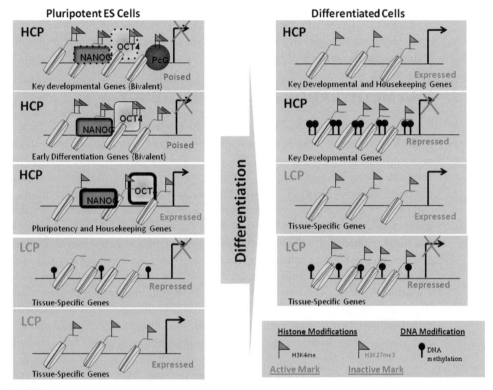

FIGURE 23.3. *A summary of epigenetic control and transcriptional regulation in mouse pluripotent embryonic stem cells and after differentiation. HCP, high-CpG-density promoter; LCP, low-CpG-density promoter.*[1,83,86]

cytokines, while the second group does not include transcription factors in general. In ES cells, PRC2 association with its target sites is in parallel with the binding of the pluripotent factors SOX2, OCT4, and NANOG, while this is not the case in differentiated cells, which indicates that PRC2 leaves these regulatory sequences[58] upon differentiation. Both PcG and TrxG transfer 1 to 3 methyl groups to the lysine or arginine residues of histones and often contain a SET [Su(var)3–9, Enhancer of Zeste, Trithorax] domain.[63] Methylated histones bind DNA more firmly, which is inhibitory to transcription. Among histone modifications, histone H3K4me3 is deposited by Set1/Trithorax-related COMPASS-like complexes that are associated with active genes[64]; however, H3K27me3 is executed by the polycomb group EZH2 (PRC2) and is associated with repressed genes.[64] Several epigenetic factors including EdD, Suz12, Ezh2, and G9a, which contribute to gene repression, are bound by OCT4, SOX2, NANOG, and SALL4.[37] However, it seems that SALL4 has a particular part in the crosstalk between pluripotent factors and epigenetic modulators, as it binds with both PRC1 and PRC2 to H3K27me3 at specific sites, but can also bind to other loci independently of these PcG complexes.[65] SALL4 forms heteromeric complexes with OCT4 and NANOG, suggested to contribute in ES cell pluripotency, and is found to be bound to many genes that are controlled by chromatin structure and epigenetic mechanisms through PcG complexes and bivalent marks.[66] Notably, OCT4, NANOG, SOX2, and SALL4 control their own expression and regulate a set of target genes through complex formation or promoter co-occupation resulting in an autoregulatory loop of pluripotency.[37] During ES cell differentiation, the majority of PcG targets are upregulated, and differentiation toward neural precursor cells results in significant increase in the expression of neuronal-specific genes.[55,56] Decrease and loss of H3K27me3 was found to be associated with H3K4me3 deposition through TrxG proteins in parallel with RNA

Pol II recruitment.[54] In ES cells, about 30% of genes that do not carry either repressive or active marks are repressed and actively marked by DNA methylation, a complementary epigenetic modification to the histone codes that precisely control heritable gene expression[37] (Fig. 23.3).

DNA METHYLATION AND PLURIPOTENCY

In mammals, DNA methylation is the only known epigenetic modification of DNA that occurs at 70% of cytosine residues, almost exclusively in the sequence of 5′-CpG-3′.[67,68] This is critical for gene regulation and control of chromatin structure during development and differentiation.[69] DNA methylation is required for transcriptional silencing and developmental processes such as imprinting, X-chromosome inactivation, genomic stability, repression of endogenous retroviral and transposon elements, and the control of chromatin structure during development and differentiation.[67,69,70] The global DNA methylation is dynamic and controlled precisely during mammalian development with genome-wide reprogramming in two major steps,[71] suggested to be necessary for directing the embryonic genome toward a pluripotent state. The first genome-wide reprogramming of DNA methylation happens shortly after fertilization during early embryogenesis.[72] This occurs through demethylation of the male pronucleus, while the maternal pronucleus does not undergo the demethylation process.[72] This might be the result of an accessibility of the paternal DNA to the demethylase enzymes during active exchange of protamines to histones. From the first zygote division, up to the blastocyst stage, DNA methyltransferases are underrepresented in the nucleus and therefore both the maternal and the paternal DNA experience a passive demethylation. This results in a global demethylation of the embryo at the implantation stage.[72] The second major DNA demethylation happens in germ cells of both males and females at early stages of primordial germ cell (PGC) development (about E8.0 in mice) simultaneous with the loss of H3K9 dimethylation,[71] followed by demethylation of the imprinted genes and repetitive elements (between E10.5 and E12.5).[73–76] Two classes of DNA methyltransferses exist, the de novo (DNMT3A and DNMT3B, which take care of remethylation at the postimplantation embryos and germ cells) and the maintenance DNA methyltransferases. The activity of the de novo DNA methyltransferases DNMT3A and DNMT3B is further increased by DNMT3L, which does not have any DNA methyltransferase activity on its own.[77] The maintenance DNA methyltransferase is the DNMT1, which hemimethylates the templates and is actively present at the DNA replication site and methylates the nascent DNA strand at the replication fork. Absence of DNA methylation is lethal in mice[78,79] and abnormal DNA methylation is associated with human mental retardation and neurological disorders, including fragile X syndrome,[80] Rett syndrome,[1] and ICF (immunodeficiency, centromeric instability, and facial abnormalities) syndrome.[81] Indeed, genome-wide data of chromatin structure[82] and DNA methylation[83] in ES cells versus differentiated cells suggest that cellular identity is a reflection of cellular epigenetic status. Comparison of the global DNA methylation in mouse ES cells, embryonic germ (EG) cells, trophoblast stem (TS) cells, and primary mouse embryonic fibroblasts (MEFs) shows a very similar pattern of DNA methylation in ES, EG, and TS cells that is distinct from MEFs.[84] Also, human ES cells show very minor alteration in their DNA methylation pattern between different lines and during long-term cultivation, while having a distinct genome-wide pattern compared to differentiated cells.[85] Global mapping of DNA methylation at the proximal promoter of mouse ES cells indicates that most genes associated with DNA methylation marks are induced upon differentiation, but exist in a repressed state in ES cells; however, the unmethylated genes encode for housekeeping genes and pluripotency factors.[86] Recent genome-wide studies of DNA methylation[83] and chromatin structure,[82] in self-renewing ES cells and after differentiation, suggest that cellular identity is indeed a close reflection of an epigenetic cellular status. In ES cells, distribution of methylated CpG is found to be related to distinct histone methylation patterns more than the CpG density.[83] Meissner et al.[83] classify the promoters as HCPs (high-CpG-density promoters) and LCPs (low-CpG-density promoters). They demonstrate that the HCPs encode for two groups of genes: housekeeping genes and tightly regulated key developmental genes, respectively, enriched with H3K4me3 alone (univalent) or H3K4me3 in combination with H3K27me3 (bivalent)[83] (Fig. 23.3). The first HCP class defines genes that are highly expressed in ES cells, while genes within the later group are generally silent.[47,82,83] Moreover, the LCPs are associated mainly with tissue-specific genes[3] with a small group enriched for H3K4me3 or H3K4me2.[83] Moreover, most CpGs of LCPs are methylated, but not that small group enriched with H3K4me marks.[83] In general, the conclusion is that, in ES cells, the presence of H3K4me and absence of H3K9me is a more reliable predictor of DNA methylation than the underlying DNA sequence.[83] During NPC (neural precursor cell) differentiation, the majority of CpGs in HCPs maintained their unmethylated status; however, loss of H3K4me3 and memory of H3K4me coincides with an increase in DNA methylation and total loss of H3K4me and H3K27me resulting in DNA hypermethylation.[83] Most LCPs that carry H3K4me lose this mark during differentiation, while LCPs of genes expressed after differentiation gain H3K4me mark.[83] Overall, the enrichment of CpGs within a gene regulatory sequence could determine what category of epigenetic regulation the gene would follow. It seems that the 95% of HCPs, which are ubiquitously unmethylated, could be subject to regulation by members of polycomb and trithorax group.[83]

EPIGENETICS AND CELLULAR REPROGRAMMING

Biomedical research has been advanced enormously by the use of ES cells to address key questions about early development, human disease, regenerative medicine, and drug screens for therapeutic studies. However, because of the social and ethical issues associated with the generation and application of human ES cells, a new concept of cellular reprogramming from somatic cells has emerged. The first mammalian reprogramming event was achieved by somatic cell nuclear transfer[87,88] that was soon followed by somatic cell fusion with ES cells,[89,90] and more recently by introducing defined transcription factors into somatic cells to generate induced pluripotent stem (iPS) cells.[91] These studies were the proof of principle that the epigenetic marks of cellular identity can be reestablished toward pluripotency[92] and demonstrated the experimental possibility of reversing the developmental clock, once believed to be unidirectional. In 2006, Yamanaka reported his outstanding discovery of generating iPS cells through retroviral gene transfer of only four pluripotency factors (Oct4, Sox2, Klf4, and c-Myc),[91] thereby transforming the future promise of regenerative medicine forever. IPS cells are shown to be pluripotent with the ability of generating teratomas[91] and live chimeras.[93] While OCT4 and SOX2 are transcription factors that are essential for pluripotency,[92,94] the role of KLF4 and C-MYC is not yet fully understood and reprogramming by omitting C-MYC is shown to be possible at the expense of reprogramming efficiency.[95–97] Generation of iPS cells was originally achieved through transduction of Moloney-based retroviral vectors[91]; however, other strategies using HIV-based lentiviruses,[98,99] plasmid transfection,[100] adenoviral infection,[101] and PiggyBac transposon vectors[102,103] were soon developed. Following iPS cell generation, the endogenous expression of other pluripotent factors such as Nanog occurs and may likely be the result of a pluripotent autoregulatory loop described earlier. Microarray data indicate that while the genomic profiles of iPS cells are similar, they are not exactly identical to

ES cells.[94] Regardless, the process of cellular reprogramming gives an exceptional opportunity to study genomic and proteomic profiles in parallel with the epigenetic modifications associated with pluripotency. A change in pluripotency requires drastic rearrangements of cellular epigenome and transcriptional profiles. Such alterations are induced by defined transcription factors, involving the interaction of transcription factors with the chromatin remodeling proteins and epigenetic modulators. During the process of cellular reprogramming not all iPS cell lines are fully reprogrammed, an observation suggested to be the result of a failure in complete epigenomic reprogramming.[104,105] Importantly, treatment with epigenetic modifiers to alter chromatin modifications including DNA and histone methylation inhibitors drastically increases the efficiency of reprogramming, while promoting the reprogramming completion by directing the partially reprogrammed iPS cell clones toward a pluripotent state.[104–108]

Recent studies have shown that key epigenetic marks for histone modifications in iPS cells are very similar to ES cells.[109] However, DNA hypomethylation at the regulatory regions of pluripotent genes such as Oct4 and Nanog is not always fully restored and intermediate DNA hypomethylation at these specific loci could be an indication of restricted developmental potential as has been reported for the Fbx15-selected iPS cells.[91] Consistent with this is the fact that establishment of the DNA hypomethylation at these specific sites is a late event during reprogramming.[104] The epigenetic landscape of a differentiated cell is quite distinct from a pluripotent cell and achieving in vitro reprogramming requires a genome-wide epigenetic reset. To address whether or not somatic cell reprogramming toward pluripotency and iPS cell requires restoration of the epigenomic signature of the reprogrammed cells, Hochedlinger and colleagues performed a genome-wide study of H3K4me3 and H3K27me3 in Nanog-selected iPS cell lines from both male and female MEFs.[109] They showed that not only are iPS cells similar to ES cells at the level of transcriptomes, but the two cell types are highly similar in the epigenetic signatures. These results indicated that indeed iPS cells and ES cells are quite similar in the H3K27me3 pattern, while both remaining very distinct from MEFs. Moreover, a greater degree of similarity in the H3K4me3 distribution pattern was observed between MEFs, ES cells, and iPS cells, indicating that reprogramming is more associated with alterations in H3K27me3 rather than H3K4me3.[109] Examination of approximately 16,500 genes from a single iPS cell line revealed that 94.4% of the "signature" genes (957 genes in total, referred to as genes with a distinct epigenetic state between MEFs and ES cells) exhibited a pluripotent state similar to ES cells in that particular iPS cell line.[109] The final conclusion would be that the in vitro reprogramming of the somatic cells could indeed reset the epigenetic memory of a somatic cell into a pluripotent state. In agreement with this, female mouse iPS cells exhibited correct demethylation of regulatory elements of pluriptency factors, resulting in reactivation of the repressed X chromosome previously silenced through random X inactivation in the original female MEFs.[109] Despite these results, it remains to be determined whether or not somatic cell reprogramming into iPS cells is accompanied by a complete or partial "epigenetic memory" loss. Overall, the reprogramming phenomenon is a step-by-step process and the complete establishment of the epigenomic signature is required for full reprogramming toward pluripotency. In support of this, treatment of cells with chemicals that influence chromatin modifications leading to inhibition of DNA and histone methylation increases the reprogramming efficiency and facilitates a faithful commitment toward pluripotency.[104,106–108] However, whether or not these chemical compounds specifically influence the epigenetic marks found in pluripotency genes or affect a broader range of target genes causing iPS cells to reach a fully reprogrammed state remains to be determined. Full reprogramming of iPS cells requires DNA demethylation at the pluripotency genes[105]; however, how the process of demethylation happens and whether or not it is an active or passive process requires further study.

CONCLUSION

Despite the huge effort toward understanding the precise epigenetic signature of iPS cells and perfecting the experimental process, because of the unique platform of genomic and epigenomic profiles of each iPS clone, the clinical application of iPS cells will still have to wait. Overall, epigenetic mechanisms are profoundly involved in mammalian development, stem cell pluripotency, and cellular differentiation. The life of an individual is not only defined by his/her genome, but also by his/her numerous epigenomes, with different epigenomes being generated through development, not only during fetal development but also during the plastic phase of early childhood, and existing in different cell types. Moreover, epigenomes react to environmental influence including maternal care, diet, and exposure to toxins and xenobiotics, and epigenetic responses to environmental stimuli may have long-term consequences, even affecting future generations.[25] In today's scientific community, an intensive interest is directed to both the epigenetic mechanisms of directed stem cell differentiation as well as somatic cell reprogramming toward pluripotency. A comprehensive knowledge of cell type-specific transcriptional and proteomic profiles and their interplay with epigenomics might eventually enable us to reprogram accessible adult tissues directly toward pluripotency or to perform cell fate switch into any differentiated cell types for therapeutic applications in human degenerative disease.

Acknowledgments

The authors acknowledge the following grant sponsors:

Canadian Institute of Health Research (MOP-9186); CancerCare Manitoba Foundation, Inc.; Canadian Breast Cancer Foundation; Canadian Cancer Society Research Institute; Canada Research Chair to JRD; Natural Sciences and Engineering Research Council of Canada (372405–2009); Manitoba Health Research Council; Manitoba Institute of Child Health; and University of Manitoba (funds from the Dr. Paul HT Thorlakson and the University Research Grants Program (URGP)).

REFERENCES

1. Delcuve GP, Rastegar M, Davie JR. Epigenetic control. *J Cell Physiol*. 2009;219(2):243–250.

2. Luger K, Richmond TJ. The histone tails of the nucleosome. *Curr Opin Genet Dev*. 1998;8:140–146.

3. Taverna SD, Li H, Ruthenburg AJ, Allis CD, Patel DJ. How chromatin-binding modules interpret histone modifications: lessons from professional pocket pickers. *Nat Struct Mol Biol*. 2007;14(11):1025–1040.

4. Peterson CL, Laniel MA. Histones and histone modifications. *Curr Biol*. 2004;14(14):R546–R551.

5. Sims RJ III, Reinberg D. Histone H3 Lys 4 methylation: caught in a bind? *Genes Dev*. 2006;20(20):2779–2786.

6. Clayton AL, Hazzalin CA, Mahadevan LC. Enhanced histone acetylation and transcription: a dynamic perspective. *Mol Cell*. 2006;23(3):289–296.

7. Ausio J. Histone variants—the structure behind the function. *Brief Funct Genomic Proteomic*. 2006;5(3):228–243.

8. Jin C, Felsenfeld G. Distribution of histone H3.3 in hematopoietic cell lineages. *Proc Natl Acad Sci USA*. 2006;103(3):574–579.

9. Mito Y, Henikoff JG, Henikoff S. Histone replacement marks the boundaries of cis-regulatory domains. *Science*. 2007;315(5817):1408–1411.

10. McKittrick E, Gafken PR, Ahmad K, Henikoff S. Histone H3.3 is enriched in covalent modifications associated with active chromatin. *Proc Natl Acad Sci USA*. 2004;101(6):1525–1530.

11. Loyola A, Bonaldi T, Roche D, Imhof A, Almouzni G. PTMs on H3 variants before chromatin assembly potentiate their final epigenetic state. *Mol Cell*. 2006;24(2):309–316.

12. Loyola A, Almouzni G. Marking histone H3 variants: how, when and why ? *Trends Biochem Sci.* 2007;32(9):425–433.

13. Davie JR, Moniwa M. Control of chromatin remodeling. *Crit Rev Eukaryot Gene Expr.* 2000;10(3–4):303–325.

14. Davie Jr. Inhibition of histone deacetylase activity by butyrate. *J Nutr.* 2003;133:2485S–2493S.

15. Katan-Khaykovich Y, Struhl K. Dynamics of global histone acetylation and deacetylation in vivo: rapid restoration of normal histone acetylation status upon removal of activators and repressors. *Genes Dev.* 2002;16(6):743–752.

16. Langst G, Becker PB. Nucleosome remodeling: one mechanism, many phenomena? *Biochim Biophys Acta.* 2004;1677(1–3):58–63.

17. Nan X, Ng HH, Johnson CA, et al. Transcriptional repression by the methyl-CpG-binding protein MeCP2 involves a histone deacetylase complex. *Nature.* 1998;393:386–389.

18. Jones PL, Veenstra GJ, Wade PA, et al. Methylated DNA and MeCP2 recruit histone deacetylase to repress transcription. *Nat Genet.* 1998;19:187–191.

19. Suzuki MM, Bird A. DNA methylation landscapes: provocative insights from epigenomics. *Nat Rev Genet.* 2008;9(6):465–476.

20. Miranda TB, Jones PA. DNA methylation: the nuts and bolts of repression. *J Cell Physiol.* 2007;213(2):384–390.

21. Illingworth R, Kerr A, Desousa D, et al. A novel CpG island set identifies tissue-specific methylation at developmental gene loci. *PLoS Biol.* 2008;6(1):e22.

22. Guil S, Esteller M. DNA methylomes, histone codes and miRNAs: tying it all together. *Int J Biochem Cell Biol.* 2008;41:87–95.

23. Hellman A, Chess A. Gene body-specific methylation on the active X chromosome. *Science.* 2007;315(5815):1141–1143.

24. Szyf M. The dynamic epigenome and its implications in toxicology. *Toxicol Sci.* 2007;100(1):7–23.

25. Szyf M, McGowan P, Meaney MJ. The social environment and the epigenome. *Environ Mol Mutagen.* 2008;49(1):46–60.

26. Ooi SK, Bestor TH. The colorful history of active DNA demethylation. *Cell.* 2008;133(7):1145–1148.

27. Miller CA, Sweatt JD. Covalent modification of DNA regulates memory formation. *Neuron.* 2007;53(6):857–869.

28. van Steensel B, Henikoff S. Epigenomic profiling using microarrays. *Biotechniques.* 2003;35(2):346–356.

29. Reed K, Poulin ML, Yan L, Parissenti AM. Comparison of bisulfite sequencing PCR with pyrosequencing for measuring differences in DNA methylation. *Anal Biochem.* 2009;397:96–106.

30. Solomon MJ, Varshavsky A. Formaldehyde-mediated DNA–protein crosslinking: a probe for in vivo chromatin structures. *Proc Natl Acad Sci USA.* 1985;82:6470–6474.

31. Sun JM, Chen HY, Davie JR. Differential distribution of unmodified and phosphorylated histone deacetylase 2 in chromatin. *J Biol Chem.* 2007;282(45):33227–33236.

32. Martin GR. Isolation of a pluripotent cell line from early mouse embryos cultured in medium conditioned by teratocarcinoma stem cells. *Proc Natl Acad Sci USA.* 1981;78(12):7634–7638.

33. Thomson JA, Itskovitz-Eldor J, Shapiro SS, et al. Embryonic stem cell lines derived from human blastocysts. *Science.* 1998;282(5391):1145–1147.

34. Pan G, Thomson JA. Nanog and transcriptional networks in embryonic stem cell pluripotency. *Cell Res.* 2007;17(1):42–49.

35. Boyer LA, Lee TI, Cole MF, et al. Core transcriptional regulatory circuitry in human embryonic stem cells. *Cell.* 2005;122(6):947–956.

36. Loh YH, Wu Q, Chew JL, et al. The Oct4 and Nanog transcription network regulates pluripotency in mouse embryonic stem cells. *Nat Genet.* 2006;38(4):431–440.

37. Hemberger M, Dean W, Reik W. Epigenetic dynamics of stem cells and cell lineage commitment: digging Waddington's canal. *Nat Rev Mol Cell Biol.* 2009;10(8):526–537.

38. Kalmar T, Lim C, Hayward P, et al. Regulated fluctuations in nanog expression mediate cell fate decisions in embryonic stem cells. *PLoS Biol.* 2009;7(7):e1000149.

39. Meshorer E, Yellajoshula D, George E, Scambler PJ, Brown DT, Misteli T. Hyperdynamic plasticity of chromatin proteins in pluripotent embryonic stem cells. *Dev Cell.* 2006;10(1):105–116.

40. Lee JH, Hart SR, Skalnik DG. Histone deacetylase activity is required for embryonic stem cell differentiation. *Genesis.* 2004;38(1):32–38.

41. Keenen B, de la Serna, I. Chromatin remodeling in embryonic stem cells: regulating the balance between pluripotency and differentiation. *J Cell Physiol.* 2009;219(1):1–7.

42. Yeo GW, Coufal N, Aigner S, et al. Multiple layers of molecular controls modulate self-renewal and neuronal lineage specification of embryonic stem cells. *Hum Mol Genet.* 2008;17(R1):R67–R75.

43. Narlikar GJ, Fan HY, Kingston RE. Cooperation between complexes that regulate chromatin structure and transcription. *Cell.* 2002;108(4):475–487.

44. de Napoles M, Mermoud JE, Wakao R, et al. Polycomb group proteins Ring1A/B link ubiquitylation of histone H2A to heritable gene silencing and X inactivation. *Dev Cell.* 2004;7(5):663–676.

45. Li B, Carey M, Workman JL. The role of chromatin during transcription. *Cell.* 2007;128(4):707–719.

46. Kouzarides T. Chromatin modifications and their function. *Cell.* 2007;128(4):693–705.

47. Bernstein BE, Mikkelsen TS, Xie X, et al. A bivalent chromatin structure marks key developmental genes in embryonic stem cells. *Cell.* 2006;125(2):315–326.

48. Williams RR, Azuara V, Perry P, et al. Neural induction promotes large-scale chromatin reorganisation of the Mash1 locus. *J Cell Sci.* 2006;119(Pt 1):132–140.

49. Azuara V, Perry P, Sauer S, et al. Chromatin signatures of pluripotent cell lines. *Nat Cell Biol.* 2006;8(5):532–538.

50. Gan Q, Yoshida T, McDonald OG, Owens GK. Concise review: epigenetic mechanisms contribute to pluripotency and cell lineage determination of embryonic stem cells. *Stem Cells*. 2007;25(1):2–9.

51. Herz HM, Nakanishi S, Shilatifard A. The curious case of bivalent marks. *Dev Cell*. 2009;17(3):301–303.

52. Akkers RC, van Heeringen SJ, Jacobi UG, et al. A hierarchy of H3K4me3 and H3K27me3 acquisition in spatial gene regulation in *Xenopus* embryos. *Dev Cell*. 2009;17(3):425–434.

53. Rudolph T, Yonezawa M, Lein S, et al. Heterochromatin formation in *Drosophila* is initiated through active removal of H3K4 methylation by the LSD1 homolog SU(VAR)3–3. *Mol Cell*. 2007;26(1):103–115.

54. Ringrose L, Paro R. Epigenetic regulation of cellular memory by the Polycomb and Trithorax group proteins. *Annu Rev Genet*. 2004;38:413–443.

55. Boyer LA, Plath K, Zeitlinger J, et al. Polycomb complexes repress developmental regulators in murine embryonic stem cells. *Nature*. 2006;441(7091):349–353.

56. Cao R, Wang L, Wang H, et al. Role of histone H3 lysine 27 methylation in Polycomb-group silencing. *Science*. 2002;298(5595):1039–1043.

57. Wang H, Wang L, Erdjument-Bromage H, et al. Role of histone H2A ubiquitination in Polycomb silencing. *Nature*. 2004;431(7010):873–878.

58. Lee TI, Jenner RG, Boyer LA, et al. Control of developmental regulators by Polycomb in human embryonic stem cells. *Cell*. 2006;125(2):301–313.

59. Soshnikova N, Duboule D. Epigenetic regulation of Hox gene activation: the waltz of methyls. *BioEssays*. 2008;30(3):199–202.

60. Cloos PA, Christensen J, Agger K, Helin K. Erasing the methyl mark: histone demethylases at the center of cellular differentiation and disease. *Genes Dev*. 2008;22(9):1115–1140.

61. Pasini D, Hansen KH, Christensen J, Agger K, Cloos PA, Helin K. Coordinated regulation of transcriptional repression by the RBP2 H3K4 demethylase and Polycomb-Repressive Complex 2. *Genes Dev*. 2008;22(10):1345–1355.

62. Ku M, Koche RP, Rheinbay E, et al. Genomewide analysis of PRC1 and PRC2 occupancy identifies two classes of bivalent domains. *PLoS Genet*. 2008;4(10):e1000242.

63. Niessen HE, Demmers JA, Voncken JW. Talking to chromatin: post-translational modulation of polycomb group function. *Epigenetics Chromatin*. 2009;2(1):10.

64. Shilatifard A. Chromatin modifications by methylation and ubiquitination: implications in the regulation of gene expression. *Annu Rev Biochem*. 2006;75:243–269.

65. Loh YH, Zhang W, Chen X, George J, Ng HH. Jmjd1a and Jmjd2c histone H3 Lys 9 demethylases regulate self-renewal in embryonic stem cells. *Genes Dev*. 2007;21(20):2545–2557.

66. Yang J, Chai L, Fowles TC, et al. Genome-wide analysis reveals Sall4 to be a major regulator of pluripotency in murine-embryonic stem cells. *Proc Natl Acad Sci USA*. 2008;105(50):19756–19761.

67. Bird AP. DNA methylation and the frequency of CpG in animal DNA. *Nucleic Acids Res*. 1980;8(7):1499–1504.

68. Bird A. The methyl-CpG-binding protein MeCP2 and neurological disease. *Biochem Soc Trans*. 2008;36(Pt 4):575–583.

69. Li E. Chromatin modification and epigenetic reprogramming in mammalian development. *Nat Rev Genet*. 2002;3(9):662–673.

70. Jaenisch R, Bird A. Epigenetic regulation of gene expression: how the genome integrates intrinsic and environmental signals. *Nat Genet*. 2003;33(Suppl):245–254.

71. Reik W, Dean W, Walter J. Epigenetic reprogramming in mammalian development. *Science*. 2001;293(5532):1089–1093.

72. Zernicka-Goetz M, Morris SA, Bruce AW. Making a firm decision: multifaceted regulation of cell fate in the early mouse embryo. *Nat Rev Genet*. 2009;10(7):467–477.

73. Hajkova P, Erhardt S, Lane N, et al. Epigenetic reprogramming in mouse primordial germ cells. *Mech Dev*. 2002;117(1–2):15–23.

74. Lees-Murdock DJ, De FM, Walsh CP. Methylation dynamics of repetitive DNA elements in the mouse germ cell lineage. *Genomics*. 2003;82(2):230–237.

75. Lane N, Dean W, Erhardt S, et al. Resistance of IAPs to methylation reprogramming may provide a mechanism for epigenetic inheritance in the mouse. *Genesis*. 2003;35(2):88–93.

76. Lee J, Inoue K, Ono R, et al. Erasing genomic imprinting memory in mouse clone embryos produced from day 11.5 primordial germ cells. *Development*. 2002;129(8):1807–1817.

77. Ooi SK, Qiu C, Bernstein E, et al. DNMT3L connects unmethylated lysine 4 of histone H3 to de novo methylation of DNA. *Nature*. 2007;448(7154):714–717.

78. Chen T, Li E. Structure and function of eukaryotic DNA methyltransferases. *Curr Top Dev Biol*. 2004;60:55–89.

79. Goll MG, Bestor TH. Eukaryotic cytosine methyltransferases. *Annu Rev Biochem*. 2005;74:481–514.

80. Warren ST. The epigenetics of fragile X syndrome. *Cell Stem Cell*. 2007;1(5):488–489.

81. Jin B, Tao Q, Peng J, et al. DNA methyltransferase 3B (DNMT3B) mutations in ICF syndrome lead to altered epigenetic modifications and aberrant expression of genes regulating development, neurogenesis and immune function. *Hum Mol Genet*. 2008;17(5):690–709.

82. Mikkelsen TS, Ku M, Jaffe DB, et al. Genome-wide maps of chromatin state in pluripotent and lineage-committed cells. *Nature*. 2007;448(7153):553–560.

83. Meissner A, Mikkelsen TS, Gu H, et al. Genome-scale DNA methylation maps of pluripotent and differentiated cells. *Nature*. 2008;454(7205):766–770.

84. Farthing CR, Ficz G, Ng RK, et al. Global mapping of DNA methylation in mouse promoters reveals epigenetic reprogramming of pluripotency genes. *PLoS Genet*. 2008;4(6):e1000116.

85. Bibikova M, Chudin E, Wu B, et al. Human embryonic stem cells have a unique epigenetic signature. *Genome Res.* 2006;16(9):1075–1083.

86. Fouse SD, Shen Y, Pellegrini M, et al. Promoter CpG methylation contributes to ES cell gene regulation in parallel with Oct4/Nanog, PcG complex, and histone H3 K4/K27 trimethylation. *Cell Stem Cell.* 2008;2(2):160–169.

87. Campbell KH, McWhir J, Ritchie WA, Wilmut I. Sheep cloned by nuclear transfer from a cultured cell line. *Nature.* 1996;380(6569):64–66.

88. Wilmut I, Schnieke AE, McWhir J, Kind AJ, Campbell KH. Viable offspring derived from fetal and adult mammalian cells. *Nature.* 1997;385(6619):810–813.

89. Ambrosi DJ, Rasmussen TP. Reprogramming mediated by stem cell fusion. *J Cell Mol Med.* 2005;9(2):320–330.

90. Hochedlinger K, Jaenisch R. Nuclear reprogramming and pluripotency. *Nature.* 2006;441(7097):1061–1067.

91. Takahashi K, Yamanaka S. Induction of pluripotent stem cells from mouse embryonic and adult fibroblast cultures by defined factors. *Cell.* 2006;126(4):663–676.

92. Jaenisch R. Stem cells, pluripotency and nuclear reprogramming. *J Thromb Haemost.* 2009;7(Suppl 1):21–23.

93. Okita K, Ichisaka T, Yamanaka S. Generation of germline-competent induced pluripotent stem cells. *Nature.* 2007;448(7151):313–317.

94. Jaenisch R, Hochedlinger K, Blelloch R, Yamada Y, Baldwin K, Eggan K. Nuclear cloning, epigenetic reprogramming, and cellular differentiation. *Cold Spring Harb Symp Quant Biol.* 2004;69:19–27.

95. Wernig M, Meissner A, Cassady JP, Jaenisch R. c-Myc is dispensable for direct reprogramming of mouse fibroblasts. *Cell Stem Cell.* 2008;2(1):10–12.

96. Yu J, Vodyanik MA, Smuga-Otto K, et al. Induced pluripotent stem cell lines derived from human somatic cells. *Science.* 2007;318(5858):1917–1920.

97. Nakagawa M, Koyanagi M, Tanabe K, et al. Generation of induced pluripotent stem cells without Myc from mouse and human fibroblasts. *Nat Biotechnol.* 2008;26(1):101–106.

98. Stadtfeld M, Maherali N, Breault DT, Hochedlinger K. Defining molecular cornerstones during fibroblast to iPS cell reprogramming in mouse. *Cell Stem Cell.* 2008;2(3):230–240.

99. Brambrink T, Foreman R, Welstead GG, et al. Sequential expression of pluripotency markers during direct reprogramming of mouse somatic cells. *Cell Stem Cell.* 2008;2(2):151–159.

100. Okita K, Nakagawa M, Hyenjong H, Ichisaka T, Yamanaka S. Generation of mouse induced pluripotent stem cells without viral vectors. *Science.* 2008;322(5903):949–953.

101. Stadtfeld M, Nagaya M, Utikal J, Weir G, Hochedlinger K. Induced pluripotent stem cells generated without viral integration. *Science.* 2008;322(5903):945–949.

102. Woltjen K, Michael IP, Mohseni P, et al. piggyBac transposition reprograms fibroblasts to induced pluripotent stem cells. *Nature.* 2009;458(7239):766–770.

103. Kaji K, Norrby K, Paca A, Mileikovsky M, Mohseni P, Woltjen K. Virus-free induction of pluripotency and subsequent excision of reprogramming factors. *Nature.* 2009;458(7239):771–775.

104. Mikkelsen TS, Hanna J, Zhang X, et al. Dissecting direct reprogramming through integrative genomic analysis. *Nature.* 2008;454(7200):49–55.

105. Hochedlinger K, Plath K. Epigenetic reprogramming and induced pluripotency. *Development.* 2009;136(4):509–523.

106. Huangfu D, Osafune K, Maehr R, et al. Induction of pluripotent stem cells from primary human fibroblasts with only Oct4 and Sox2. *Nat Biotechnol.* 2008;26(11):1269–1275.

107. Huangfu D, Maehr R, Guo W, et al. Induction of pluripotent stem cells by defined factors is greatly improved by small-molecule compounds. *Nat Biotechnol.* 2008;26(7):795–797.

108. Shi Y, Do JT, Desponts C, Hahm HS, Scholer HR, Ding S. A combined chemical and genetic approach for the generation of induced pluripotent stem cells. *Cell Stem Cell.* 2008;2(6):525–528.

109. Maherali N, Sridharan R, Xie W, et al. Directly reprogrammed fibroblasts show global epigenetic remodeling and widespread tissue contribution. *Cell Stem Cell.* 2007;1(1):55–70.

110. Ohm JE, Baylin SB. Stem cell chromatin patterns: an instructive mechanism for DNA hypermethylation? *Cell Cycle.* 2007;6(9):1040–1043.

111. Kim JK, Samaranayake M, Pradhan S. Epigenetic mechanisms in mammals. *Cell Mol Life Sci.* 2009;66(4):596–612.

112. Allis CD, Berger SL, Cote J, et al. New nomenclature for chromatin-modifying enzymes. *Cell.* 2007;131(4):633–636.

DIFFERENTIATION OF PLURIPOTENT STEM CELLS: AN OVERVIEW

24

Jeremy Micah Crook

INTRODUCTION

Human pluripotent stem cells (hPSCs) hold promise for basic research and medicine. They may be used to model development and disease and offer new biologically relevant strategies for drug discovery and cell therapy. Fulfilling this potential will depend on differentiating cells by efficient, safe, reproducible, and cost-effective methods, ideally with selection techniques to generate populations of specialized and genetically stable cells that are free of unwanted cells.

To date, hPSC differentiation has predominantly focused on human embryonic stem cells (hESCs). While it is anticipated that methods established for hESCs will be adaptable to other cell types such as induced pluripotent stem cells (iPSCs), this may not always be the case and may require alternative induction strategies. For example, differences may arise from mechanisms related to genetic or epigenetic memory of reprogrammed cells.[1] Interestingly, initial concerns of persistent expression of reprogrammed genes interfering with iPSC differentiation have ostensibly been dispelled, with methylation of vector promoters silencing viral transgenes and demethylation of endogenous Oct4 and Nanog promoters reactivating normal pluripotency gene transcription.[2] Importantly, the advent of iPSCs has expanded the toolbox for stem cell biologists and so increased interest in hPSCs by both the scientific and lay communities. This will undoubtedly bolster efforts to improve hPSC differentiation, for research and translational application.

In this chapter, a brief overview of the concepts and current approaches to differentiating hPSCs is provided. For reasons stated above, seminal methods are highlighted, many of which were first and foremost devised for hESC differentiation, but are or potentially adaptable to other hPSC types such as iPSCs. While methods can be broadly categorized as enabling directed or spontaneous differentiation, to date few if any can claim controlled induction to pure populations of usable, safe, and specific cell types. Challenges aside, progress is being made with greater insight into the biology of these complex and intriguing cells.

DIRECTED DIFFERENTIATION

Directed and controlled differentiation of hPSCs remains the major challenge for the stem cell field. Although there has been some success with manipulating culture conditions to promote differentiation, most specialized cells are derived from

Human Stem Cell Technology and Biology, edited by Stein, Borowski, Luong, Shi, Smith, and Vazquez
Copyright © 2011 Wiley-Blackwell.

spontaneously differentiated hESCs.[3] At best, chosen conditions select and enrich for a specific lineage from a mixed cell population. Nevertheless, there are protocols that provide the foundation for generating specific cells in varying degrees of purity for preclinical and clinical application. Certainly, the latter will require much more refined strategies, likely involving chemical agents, growth factors, and/or genetic manipulation.

Despite the different requirements for research and therapeutic applications of hPSC derivatives, there is merit in devising standardized differentiation platforms used for preclinical research and development (R&D) that require minimal adaptation for clinical compliance, where the work has translational potential. Ideally, this would include defined conditions where individual components of culture including chemicals and biologicals used in media, constituents of growth substrates, and other materials that support or contact cells during differentiation are known. Although the regulatory requirements for clinical use of hPSCs remain uncertain, the International Society of Stem Cell Research (ISSCR), the US Food and Drug Administration, and European regulators provide relevant guidelines.[4-6]

More or less defined, cheaper research-grade variants of materials and reagents used to differentiate stem cells can be incorporated early in the development pipeline, with the option to employ more expensive clinical-grade versions for late-phase preclinical and clinical use. This will ensure a most cost- and time-effective approach to developing and optimizing protocols and avoid the need to reoptimize and revalidate for regulatory approval and clinical application. Where possible, it is recommended that a similar approach to stem cell line selection also be taken to sidestep the inherent variation between the ability to differentiate different lines. Thus optimization of differentiation using research-grade cells of a clinical-grade cell line will enable consistency between research and development (R&D) and clinical work by avoiding the need to adapt protocols from one line to another. Importantly, in the case of cell therapy, a specific-cell-line approach is possible for allogeneic transplantation but less amenable for autologous transplantation, with the latter likely requiring tailoring of a differentiation method to a recipient's own cell line.

Regardless of the preferred schema for R&D, hPSC differentiation must enable the production of large numbers of specialized cells with a desired phenotype and ability to integrate with surrounding cells or tissue in vitro or in vivo, depending on the desired application. For novel cell therapies, defined and safe populations of functional cells must repair damaged tissues, augment tissue function, and/or correct genetic aberrances, while surviving grafting or implantation. While it is not within the scope of this chapter to consider in detail all the requirements of cell transplantation, they should not be overlooked wherever practicable when devising a differentiation strategy intended for clinical translation. To reiterate, proof-of-concept studies will benefit from incorporating principles and practices that are regulatable.

Cardiomyocytes

Differentiation of hPSCs toward cardiac lineage has attracted much interest since the first derivation of bona fide hESC lines.[7] While no doubt driven by therapeutic potential and necessity, early interest was indubitably fueled by the propensity for hESC colonies to spontaneously differentiate into conspicuously contracting cardiac foci. Regrettably, the apparent ease of deriving functional cardiomyocytes was deceptive, since like many other specialized cell types, generating homogeneous populations of functional cells to repair hearts has proved challenging. Nonetheless, the state-of-the-art has evolved considerably since 1998. Refinement of methods for culture and differentiation, complemented by strategies for cell enrichment,[8] is anticipated to enable short-term translational application of hPSC-cardiomyocytes by the pharmaceutical industry for drug discovery and toxicity testing.

Many protocols for directed differentiation of hPSCs rely on initially forming embryoid bodies (EBs), where aggregates of cells undergo partial embryogenesis.[9] EBs are in effect spontaneously generated (see below) and contain derivatives of the three primary germ layers. Induction of cardiomyoctye mesodermal progenitors from EBs can be enhanced by co-culture of hESCs with the endodermal cell line END-2.[10,11] Induction with END-2-conditioned medium can be enhanced by insulin depletion,[12] inhibiting p38 MAPK,[13] and supplementation with prostaglandin E inhibition.[14]

A second approach to cardiac differentiation depends on using growth factors to recapitulate the core signaling pathways underlying embryonic cardiac mesoderm development. For example, treatment with basic fibroblast growth factor (bFGF), bone morphogenic protein 4 (BMP4), and activin A followed by vascular endothelial growth factor (VEGF) and WNT inhibition supports induction to cardiac lineage.[15] Notably, despite the importance of WNT signaling for cardiac specification, its mechanism of control is undoubtedly complex, with evidence for switching between activation and inhibition during the course of renewal and differentiation.[16]

Despite improvements to differentiation, cardiac cell enrichment is necessary for the isolation of more homogeneous populations of cardiomyocytes. Strategies include manual dissection of beating regions of cultures,[17,18] fluorescence activated cell sorting (FACS),[17,18] Percoll® density gradient sedimentation,[19,20] and/or drug resistance selection of functional cardiomyoctyes.[8]

Hepatocytes

Human hepatocytes are valuable for drug and toxicity testing as primary cells of the liver. Moreover, hepatocyte transplantation or extracorporeal application could be employed to treat liver failure as an alternative to whole organ transplantation.[21] Several studies have described hESC-derived hepatocyte-like cells, with recent methods guided by embryogenesis and applying serum-free conditions.[22] Essentially, activin A initiates endodermal induction of hESCs.[23,24] Fibroblast growth factor 4 (FGF4) and BMP2 induce hepatic differentiation from definitive endoderm cells. Further maturation of early hepatic cells with hepatocyte growth factor (HGF) causes expression of the adult liver cell markers tyrosine aminotransferase, tryptophan oxygenase 2, phospho*enol*pyruvate carboxykinase (PEPCK), Cyp7A1, Cyp3A4, and Cyp2B6. Cells should exhibit functions associated with mature hepatocytes including albumin secretion, glycogen storage, indocyanine green, low-density lipoprotein uptake, and inducible cytochrome P450 activity.[24]

Neural Cells

Despite the myriad of protocols published over the past decade for differentiating neural cells from hPSCs, there has been an escalation in recent years of methods skewed toward generating more specific neural subtypes, including oligodendrocytes[25] and motor,[26–30] dopaminergic,[31–34] glutamatergic,[35,36] gamma aminobutyric acid (GABA)ergic,[35,36] and cholinergic[37] neurons. Significant effort is being directed to overcome cell heterogeneity by improving induction efficiency combined with enrichment strategies.[29] Furthermore, xeno-free, feeder-free, and/or defined conditions for improved clinical compliance and directed differentiation have been reported.[28,34,38]

Retinal Cells

Production of retinal progenitor cells, retinal pigment epithelium (RPE) cells, and photoreceptors from hPSCs (including both hESCs and iPSCs) has been described using the defined culture method.[39] Retinal progenitors were derived from "EB-like aggregates" using inhibitors of activin type I receptor-like kinase 4 (ALK4), casein kinase I, and Rho-associated kinase under serum- and feeder-free conditions. Extended treatment should induce pigmented cells that exhibit typical hexagonal morphology of RPE cells. Further treatment with retinoic acid and taurine gives rise to photoreceptors.

A second more recent report describes defined directed differentiation of hESCs to an RPE fate.[40] The authors recommend using nicotinamide to induce hESCs to neural and subsequently RPE fate. RPE induction is further directed with concomitant use of factors from the transforming growth factor (TGF)-β superfamily, thought to pattern RPE development during embryogenesis. The hESC-derived pigmented cells were shown to rescue retinal structure and function after transplantation to an animal model of retinal degeneration.

Other methods of retinal cell specification have tended to focus on production of retinal progenitors,[41,40] RPE,[42,43] or photoreceptors[44] using exogenous factors for enrichment. A recently described protocol by Meyer et al.[45] models retinogenesis by judiciously directing hESCs and iPSCs through each of the key stages of human retinal development. According to the authors, the method provides lineage-specific cells representative of all major stages of retinal development and, useful to understanding retinal specification from hPSCs, modeling retinal degeneration diseases, pharmaceutical R&D, and future transplantation of photoreceptor or RPE cells.

Islet Cells

Several groups have reported stepwise protocols to generate insulin-producing cells from hPSCs by mimicking in vivo pancreas development.[46–51] However, in most instances the differentiation efficiency has been low ($\sim$4–7%) and cells exhibited immature islet characteristics with low levels of insulin/C-peptide. Contemporary methods are tackling both hESC and iPSC induction using chemically defined systems with promising results.[51] First, definitive endoderm was derived from hPSC aggregates on Matrigel™ using activin A and wortmannin, followed by a further 4 days of treatment with retinoic acid FGF7 and noggin. Progenitors were subsequently expanded by culture with EGF followed by maturation in bFGF, nicotinamide, exendin-4, and BMP4. Insulin-positive cells were observed after approximately 20 days of induction, with a relatively high efficiency of $\sim$25%. The authors assert the utility of their approach for efficient induction of mature β-cells able to respond to glucose stimuli similar to adult human islets. If confirmed, the strategy represents a significant leap forward to generating functional and transplantable cells for diabetes research and therapy.

SPONTANEOUS DIFFERENTIATION

Notwithstanding the need for standardized methods of differentiation for directed hPSC induction, there is a requirement for less controlled approaches reliant on spontaneous in vitro or in vivo differentiation to mixed cell tissues. As mentioned above, spontaneous differentiation to EBs is commonly used as a starting point for subsequent and more directed differentiation to desired lineage-specific cells.[9] Moreover, EB formation is a useful in vitro assay of pluripotency, where derivatives of the three embryonic germ layers are the order of the day.

A second more onerous approach to testing pluripotency is the ability for hPSCs to form teratomas in immunodeficient mice. Other than producing chimeras, in vivo teratoma formation is the "gold standard" for determining pluripotency.

Embryoid Bodies

While a detailed protocol for forming EBs is provided elsewhere in this book, considerable effort has gone into devising methods to achieve homogeneous and efficient differentiation.[52] While conventional methods include dissociated suspension[9] and hanging drop,[53] 96-well plate,[54] and conical tube[55] culture, more contemporary approaches include spinner flask[56] and bioreactor[57,58] based differentiation for large-scale EB production. Unlike traditional suspension culture, which provides heterogeneous EBs, hanging drop, 96-well plate, and conical tube methods enable more uniform EB size and shape. The use of spinner flasks and bioreactors may be technically challenging to set up although efficient once mastered.

Teratomas

Teratomas arise from introducing hPSCs into, for example, SCID mice by subdermal, intramuscular, peritoneal cavity, liver, epididymal fat pad, intratesticular, or kidney-capsule injection. While some recommend intramuscular inoculation,[59] others prefer subcutaneous delivery.[60] The time taken for a teratoma to arise can vary from cell line to cell line and mouse to mouse and is site dependent. Regardless, teratoma formation will range from 5 to 12 weeks and requires histological evaluation by an expert to identify the cell lineages of all three germ layers, ideally combined with supporting immunohistochemistry of relevant marker proteins.[61] Recently, teratoma formation has been correlated with fibroblast cell reprogramming where bona fide iPSCs readily formed teratomas compared to partially reprogrammed cells that formed poorly differentiated teratomas if at all.[62]

CONCLUSION

The complexity of processes regulating hPSC fate make directed and controlled differentiation a major challenge for the stem cell field. However, with improved understanding of embryogenesis, the in vivo and in vitro microenvironment, and molecular mechanisms of lineage specification, better differentiation strategies are being devised. In the mean time, there are protocols that enable the generation of many types of organ specific cells for further research and in some instances limited short-term translational application.

REFERENCES

1. Marchetto MC, Yeo GW, Kainohana O, et al. Transcriptional signature and memory retention of human-induced pluripotent stem cells. *PLoS One*. 2009;4:e7076.

2. Muller LU, Daley GQ, Williams DA. Upping the ante: recent advances in direct reprogramming. *Mol Ther*. 2009;17:947–953.

3. Gokhale PJ, Andrews PW. Human embryonic stem cells: 10 years on. *Lab Invest*. 2009;89:259–262.

4. Research ISoSC. Guidelines for the Clinical Translation of Stem Cells.

5. Title 21. In: Code of Federal Regulations FaD, ed; 2006.

6. In: Directive EU, ed; 2006:17/EC, 86/EC.

7. Thomson JA, Itskovitz-Eldor J, Shapiro SS, et al. Embryonic stem cell lines derived from human blastocysts. *Science*. 1998;282:1145–1147.

8. Kita-Matsuo H, Barcova M, Prigozhina N, et al. Lentiviral vectors and protocols for creation of stable hESC lines for fluorescent tracking and drug resistance selection of cardiomyocytes. *PLoS One*. 2009;4:e5046.

9. Itskovitz-Eldor J, Schuldiner M, Karsenti D, et al. Differentiation of human embryonic stem cells into embryoid bodies compromising the three embryonic germ layers. *Mol Med*. 2000;6:88–95.

10. Mummery C, Ward-van Oostwaard D, Doevendans P, et al. Differentiation of human embryonic stem cells to cardiomyocytes: role of coculture with visceral endoderm-like cells. *Circulation*. 2003;107:2733–2740.

11. Passier R, Oostwaard DW, Snapper J, et al. Increased cardiomyocyte differentiation from human embryonic stem cells in serum-free cultures. *Stem Cells*. 2005;23:772–780.

12. Freund C, Ward-van Oostwaard D, Monshouwer-Kloots J, et al. Insulin redirects differentiation from cardiogenic mesoderm and endoderm to neuroectoderm in differentiating human embryonic stem cells. *Stem Cells*. 2007;26:724–733.

13. Graichen R, Xu X, Braam SR, et al. Enhanced cardiomyogenesis of human embryonic stem cells by a small molecular inhibitor of p38 MAPK. *Differentiation*. 2008;76:357–370.

14. Xu XQ, Graichen R, Soo SY, et al. Chemically defined medium supporting cardiomyocyte differentiation of human embryonic stem cells. *Differentiation*. 2008;76:958–970.

15. Yang L, Soonpaa MH, Adler ED, et al. Human cardiovascular progenitor cells develop from a KDR+ embryonic-stem-cell-derived population. *Nature*. 2008;453:524–528.

16. Qyang Y, Martin-Puig S, Chiravuri M, et al. The renewal and differentiation of Isl1+ cardiovascular progenitors are controlled by a Wnt/beta-catenin pathway. *Cell Stem Cell*. 2007;1:165–179.

17. Mummery CL, Ward D, Passier R. Differentiation of human embryonic stem cells to cardiomyocytes by coculture with endoderm in serum-free medium. *Curr Protoc Stem Cell Biol*. 2007;Chapter 1:Unit 1F 2.

18. Muller M, Fleischmann BK, Selbert S, et al. Selection of ventricular-like cardiomyocytes from ES cells in vitro. *FASEB J*. 2000;14:2540–2548.

19. Laflamme MA, Chen KY, Naumova AV, et al. Cardiomyocytes derived from human embryonic stem cells in pro-survival factors enhance function of infarcted rat hearts. *Nat Biotechnol*. 2007;25:1015–1024.

20. Xu C, Police S, Hassanipour M, et al. Cardiac bodies: a novel culture method for enrichment of cardiomyocytes derived from human embryonic stem cells. *Stem Cells Dev*. 2006;15:631–639.

21. Soto-Gutierrez A, Kobayashi N, Rivas-Carrillo JD, et al. Reversal of mouse hepatic failure using an implanted liver-assist device containing ES cell-derived hepatocytes. *Nat Biotechnol*. 2006;24:1412–1419.

22. Hay DC, Zhao D, Fletcher J, et al. Efficient differentiation of hepatocytes from human embryonic stem cells exhibiting markers recapitulating liver development in vivo. *Stem Cells*. 2008;26:894–902.

23. Chen Y, Soto-Gutierrez A, Navarro-Alvarez N, et al. Instant hepatic differentiation of human embryonic stem cells using activin A and a deleted variant of HGF. *Cell Transplant*. 2006;15:865–871.

24. Cai J, Zhao Y, Liu Y, et al. Directed differentiation of human embryonic stem cells into functional hepatic cells. *Hepatology*. 2007;45:1229–1239.

25. Hu BY, Du ZW, Li XJ, et al. Human oligodendrocytes from embryonic stem cells: conserved SHH signaling networks and divergent FGF effects. *Development*. 2009;136:1443–1452.

26. Li XJ, Du ZW, Zarnowska ED, et al. Specification of motoneurons from human embryonic stem cells. *Nat Biotechnol*. 2005;23:215–221.

27. Li XJ, Hu BY, Jones SA, et al. Directed differentiation of ventral spinal progenitors and motor neurons from human embryonic stem cells by small molecules. *Stem Cells*. 2008;26:886–893.

28. Hu BY, Zhang SC. Differentiation of spinal motor neurons from pluripotent human stem cells. *Nat Protoc*. 2009;4:1295–1304.

29. Wada T, Honda M, Minami I, et al. Highly efficient differentiation and enrichment of spinal motor neurons derived from human and monkey embryonic stem cells. *PLoS One*. 2009;4:e6722.

30. Karumbayaram S, Novitch BG, Patterson M, et al. Directed differentiation of human-induced pluripotent stem cells generates active motor neurons. *Stem Cells*. 2009;27:806–811.

31. Perrier AL, Tabar V, Barberi T, et al. Derivation of midbrain dopamine neurons from human embryonic stem cells. *Proc Natl Acad Sci USA*. 2004;101:12543–12548.

32. Aubry L, Bugi A, Lefort N, et al. Striatal progenitors derived from human ES cells mature into DARPP32 neurons in vitro and in quinolinic acid-lesioned rats. *Proc Natl Acad Sci USA*. 2008;105:16707–16712.

33. Yan Y, Yang D, Zarnowska ED, et al. Directed differentiation of dopaminergic neuronal subtypes from human embryonic stem cells. *Stem Cells*. 2005;23:781–790.

34. Swistowski A, Peng J, Han Y, et al. Xeno-free defined conditions for culture of human embryonic stem cells, neural stem cells and dopaminergic neurons derived from them. *PLoS One*. 2009;4:e6233.

35. Heikkila TJ, Yla-Outinen L, Tanskanen JM, et al. Human embryonic stem cell-derived neuronal cells form spontaneously active neuronal networks in vitro. *Exp Neurol*. 2009;218:109–116.

36. Li XJ, Zhang X, Johnson MA, et al. Coordination of sonic hedgehog and Wnt signaling determines ventral and dorsal telencephalic neuron types from human embryonic stem cells. *Development*. 2009;136:4055–4063.

37. Nilbratt M, Porras O, Marutle A, et al. Neurotrophic factors promote cholinergic differentiation in human embryonic stem cell-derived neurons. *J Cell Mol Med*. 2009; Epub.

38. Axell MZ, Zlateva S, Curtis M. A method for rapid derivation and propagation of neural progenitors from human embryonic stem cells. *J Neurosci Methods*. 2009;184:275–284.

39. Osakada F, Jin ZB, Hirami Y, et al. In vitro differentiation of retinal cells from human pluripotent stem cells by small-molecule induction. *J Cell Sci*. 2009;122:3169–3179.

40. Idelson M, Alper R, Obolensky A, et al. Directed differentiation of human embryonic stem cells into functional retinal pigment epithelium cells. *Cell Stem Cell*. 2009;5:396–408.

41. Lamba DA, Karl MO, Ware CB, et al. Efficient generation of retinal progenitor cells from human embryonic stem cells. *Proc Natl Acad Sci USA*. 2006;103:12769–12774.

42. Klimanskaya I, Hipp J, Rezai KA, et al. Derivation and comparative assessment of retinal pigment epithelium from human embryonic stem cells using transcriptomics. *Cloning Stem Cells*. 2004;6:217–245.

43. Vugler A, Carr AJ, Lawrence J, et al. Elucidating the phenomenon of HESC-derived RPE: anatomy of cell genesis, expansion and retinal transplantation. *Exp Neurol*. 2008;214:347–361.

44. Osakada F, Ikeda H, Mandai M, et al. Toward the generation of rod and cone photoreceptors from mouse, monkey and human embryonic stem cells. *Nat Biotechnol*. 2008;26:215–224.

45. Meyer JS, Shearer RL, Capowski EE, et al. Modeling early retinal development with human embryonic and

induced pluripotent stem cells. *Proc Natl Acad Sci USA.* 2009;106:16698–16703.

46. D'Amour KA, Bang AG, Eliazer S, et al. Production of pancreatic hormone-expressing endocrine cells from human embryonic stem cells. *Nat Biotechnol.* 2006;24:1392–1401.

47. Jiang J, Au M, Lu K, et al. Generation of insulin-producing islet-like clusters from human embryonic stem cells. *Stem Cells.* 2007;25:1940–1953.

48. Jiang W, Shi Y, Zhao D, et al. In vitro derivation of functional insulin-producing cells from human embryonic stem cells. *Cell Res.* 2007;17:333–344.

49. Phillips BW, Hentze H, Rust WL, et al. Directed differentiation of human embryonic stem cells into the pancreatic endocrine lineage. *Stem Cells Dev.* 2007;16:561–578.

50. Kroon E, Martinson LA, Kadoya K, et al. Pancreatic endoderm derived from human embryonic stem cells generates glucose-responsive insulin-secreting cells in vivo. *Nat Biotechnol.* 2008;26:443–452.

51. Zhang D, Jiang W, Liu M, et al. Highly efficient differentiation of human ES cells and iPS cells into mature pancreatic insulin-producing cells. *Cell Res.* 2009;19:429–438.

52. Kurosawa H. Methods for inducing embryoid body formation: in vitro differentiation system of embryonic stem cells. *J Biosci Bioeng.* 2007;103:389–398.

53. Hopfl G, Gassmann M, Desbaillets I. Differentiating embryonic stem cells into embryoid bodies. *Methods Mol Biol.* 2004;254:79–98.

54. Koike M, Kurosawa H, Amano Y. A Round-bottom 96-well polystyrene plate coated with 2-methacryloyloxyethyl phosphorylcholine as an effective tool for embryoid body formation. *Cytotechnology.* 2005;47:3–10.

55. Kurosawa H, Imamura T, Koike M, et al. A simple method for forming embryoid body from mouse embryonic stem cells. *J Biosci Bioeng.* 2003;96:409–411.

56. Carpenedo RL, Sargent CY, McDevitt TC. Rotary suspension culture enhances the efficiency, yield, and homogeneity of embryoid body differentiation. *Stem Cells.* 2007;25:2224–2234.

57. Come J, Nissan X, Aubry L, et al. Improvement of culture conditions of human embryoid bodies using a controlled perfused and dialyzed bioreactor system. *Tissue Eng Part C Methods.* 2008;14:289–298.

58. Yirme G, Amit M, Laevsky I, et al. Establishing a dynamic process for the formation, propagation, and differentiation of human embryoid bodies. *Stem Cells Dev.* 2008;17:1227–1241.

59. Hentze H, Soong PL, Wang ST, et al. Teratoma formation by human embryonic stem cells: evaluation of essential parameters for future safety studies. *Stem Cell Res.* 2009; Epub.

60. Prokhorova TA, Harkness LM, Frandsen U, et al. Teratoma formation by human embryonic stem cells is site-dependent and enhanced by the presence of Matrigel. *Stem Cells Dev.* 2009;18(1):47–54.

61. Crook JM, Peura TT, Kravets L, et al. The generation of six clinical-grade human embryonic stem cell lines. *Cell Stem Cell.* 2007;1:490–494.

62. Chan EM, Ratanasirintrawoot S, Park IH, et al. Live cell imaging distinguishes bona fide human iPS cells from partially reprogrammed cells. *Nat Biotechnol.* 2009;27:1033–1037.

CELLULAR REPROGRAMMING: CURRENT TECHNOLOGY, PERSPECTIVES, AND GENERATION OF INDUCED PLURIPOTENT CELLS

25

Tanja Dominko

INTRODUCTION

During early mammalian embryogenesis, a developmental program is triggered that assures all somatic tissues of an organism will develop in an appropriate spatial and temporal pattern. Sequential progression through embryonic cell divisions, cell positioning, and cell-to-cell communications leads to specifications of endoderm, mesoderm, and ectoderm and further to differentiation of cell types with distinct phenotypes and functions. These orchestrated changes assure progressively more restrictive, unidirectional developmental pathways ultimately producing terminally differentiated cells. While differentiated cell types clearly appear to be fully committed to their fate in vivo, at least most of the time, it is becoming increasingly obvious that their "terminal commitment" is reversible.

Developmental unidirectionality that is defined by progression from a less differentiated to a more differentiated state has been challenged over the past decades. Exposure of a differentiated cell nucleus to the cytoplasm of a less differentiated cell leads to erasure of the stable epigenetic code that maintains the differentiated cell's phenotype. Gradually, the nucleus acquires a new epigenetic code that is characteristic of the dedifferentiated cell donating the cytoplasm, a process termed cellular reprogramming.

Cellular reprogramming that occurs after fertilization and somatic cell nuclear transfer (SCNT) within egg cytoplasm results in development of embryos containing fully reprogrammed stem cells restricted to the inner cell mass (ICM). Embryonic stem cells (ESCs) isolated from ICMs have the capacity to proliferate indefinitely in vitro by self-renewal and retain the ability to differentiate into all cells of the three germ layers. Due to their unprecedented ability to generate all adult cell types in vitro and in vivo, the use of ESCs is being pursued in development of cell replacement strategies for various human pathologies. The utility of these cells for derivation of useful targets, however, remains largely untested in human patients. To date, only one human clinical trial using ESC-derived cells (for

Human Stem Cell Technology and Biology, edited by Stein, Borowski, Luong, Shi, Smith, and Vazquez
Copyright © 2011 Wiley-Blackwell.

spinal cord injury) has been approved by the FDA (Geron Corporation, 2009). The main concerns that have contributed to the slow transition of ESC-derived therapeutic cells into clinical trials are their controversial embryonic origin, the need for patient immunosuppression (allogenicity), and their potential for unregulated developmental escape and tumor formation. Efforts have since been underway to recapitulate the pluripotent stem cell phenotype in a way that would alleviate the above mentioned concerns. Recently developed induced pluripotent stem (iPS) cells, cells derived from patients' own differentiated somatic cells (autologous) without the need for embryonic components, may begin to provide a viable alternative.

This chapter will summarize what we have learned about cellular reprogramming and will discuss the experimental reprogramming systems. While exploration of several of these systems has led to derivation of reprogrammed pluripotent cells, the molecular mechanisms that are responsible for the reprogramming remain largely unknown. The last part of this chapter will summarize the accomplishments during the short three years since Takahashi and Yamanaka[1,2] described for the first time that introduction of a few genes into a differentiated cell can cause its conversion into a fully dedifferentiated pluripotent stem cell.

REPROGRAMMING DURING FERTILIZATION

The most efficient cellular reprogramming event in mammals occurs after fertilization. Reprogramming of mammalian sperm and egg genomes to totipotency is completed in less than 30 hours in mouse (two-cell stage[3]), 72 hours in human (4–8 cell stage[4]), and 96 hours in cow embryos (8–16 cell stage[5]). Separation of blastomeres at these early stages leads to development of totipotent embryos as they develop to term after transfer into recipient animals.[6–8]

This fast but complex process depends on DNA replication (aphidicolin sensitive[9,10]) but is executed without the need for de novo transcription from the embryonic genome[11] and hence relies solely on maternal components present in an egg. Insensitivity of early embryonic transcripts to a transcriptional inhibitor α-amanitin prior to the maternal embryonic transition (MET),[10] however, does not imply that these early embryonic stages are transcriptionally incompetent. Injection of murine Oct4 upstream region-driven EGFP constructs into mouse, pig, and cow zygotes shows that the transgene is transcribed in all species at all embryonic stages.[12] This indicates that the zygote cytoplasm contains all the components of transcriptional machinery and that, at least for mouse Oct4, transcription could be conferred by interspecies cytoplasm.

It is likely that regulatory regions of genes required for successful MET and continuation of embryonic development is transcriptionally repressed during this time. In fact, it has been shown that changes in epigenetic chromatin structure are essential for and accompany pre-MET embryonic development.[13] Heritable nongenetic changes that lead to embryonic gene expression are a consequence of two types of tightly regulated modifications: DNA methylation and post-translational modification of DNA-associated histones. Indeed, maternally encoded transcripts for chromatin modifiers have been described in oocytes and eggs of several species and include histone acetyl transferases (CBP, p300), deacetylases (HDAC2, HDAC6), SWI/SNF related transcriptional regulators (SMARCA2, ARID1A, BR140), regulators of chromatin accessibility (HMGs, ACF1, SMARCA5), and other transcription modifiers YY1, YAF2, RY-1 CREB, and YAP65.[9,14–16] The general pattern of gradual decrease of mRNAs during first cleavage divisions is likely due to their translational depletion and several chromatin modifiers begin to be expressed again in α-amanitin-dependent manner after the activation of the embryonic genome.[14,15] Orchestrated action of these and other maternal factors establishes transcriptionally permissive chromatin that is generally characterized by hyperacetylation of histones H3 and H4 and by hypermethylation of specific lysine residues in H3.

Continuation of development beyond the maternally driven embryonic program is strictly dependent on the activation of the embryonic genome during the maternal-to-embryonic transition. Among the first genes that become activated at the time or shortly after the MET is a set of transcription factors, including Oct4, Nanog, and Sox2.[17,18] Presence of these transcription factors is critical for MET and continuation of embryonic development[19] and their inefficient activation leads to loss of pluripotency and embryonic mortality.[20] Transcriptional profiling studies of female gametes during oogenesis, maturation, and early embryonic development in mice[14–16,21,22] have examined temporal correlation between the activation of maternal messages versus embryonic transcriptional activation. RT-PCR and examination of expression libraries during oocyte and embryonic stages revealed that lin28 and Oct4 were expressed before, during, and after MET, while nanog, Sox2, Klf4, and c-myc began to be activated sequentially well after the MET.[20,23] Oct4 mRNA and protein could be detected in mouse and human oocytes and preimplantation embryos with varying degree of mosaicism.[24–27]

Examination of Oct4 protein expression in human oocytes is complicated by alternative splicing of the Oct4 transcript into messages coding for the nuclear transcription factor variant Oct4A and cytoplasmic variant Oct4B,[28] indicating their potential functional difference. By using antibodies that can distinguish Oct4A from Oct4B variant,[26] Oct4B can be observed in the oocyte cytoplasm and persists throughout the preimplantation stages, while Oct4A protein becomes detectable in the nuclei only after the MET at the morula and blastocysts stages.[26] Expression of Oct4 is restricted to the pluripotent inner cell mass (ICM) in mouse preimplantation embryos,[12,29] but its expression in both ICM and trophectoderm in blastocysts of large mammals indicates that sole presence of Oct4 in these species is not by itself sufficient for identification or selection of pluripotent cells.[12,26] Together, Oct4, nanog, and Sox2 regulate a vast number of downstream genes by occupying promoters either individually or in combination[30] to cooperatively establish a master regulatory circuit that maintains pluripotency.

All of the currently available human pluripotent ESC lines have been derived from ICMs of fertilized embryos, first described by Thomson et al.[31] Their molecular, morphological and developmental characteristics remain the gold standard against which all other pluripotent stem cells are being compared and provide an irreplaceable model system for studying reprogramming.

REPROGRAMMING DURING SOMATIC CELL NUCLEAR TRANSFER (SCNT)

In 1958, John Gurdon reported successful SCNT in *Xenopus laevis* frogs by demonstrating that egg cytoplasm can impose fertilization-like reprogramming changes not only onto sperm and egg genomes, but also onto a genome of differentiated somatic cells as well. It took another forty years before similar success was reported in mammals with births of live animals[32–34] and derivation of SCNT-derived pluripotent ESCs in mice.[35,36]

Two sequential series of developmental events, both of which have to be faithfully recapitulated, take place during SCNT. The first defines dedifferentiation and reprogramming changes in the somatic nucleus allowing preimplantation embryonic development to the blastocyst stage; the second defines postimplantation changes that assure proper spatial and temporal activation of differentiation events leading to normal fetal tissue development. The use of SCNT for production of viable animal offspring needs to successfully recapitulate both series of events and has been successful in a range of animal species.

This paradigm-shifting demonstration of SCNT-induced nuclear plasticity was quickly recognized as an alternative for derivation of embryos that could be used for isolation of ESCs. ESCs derived from SCNT embryos would offer a unique advantage as they could be genetically customized for a specific patient. While pluripotent stem cells from early primate SCNT preimplantation embryos have been isolated,[37,38] such success in humans, however, remains unrealized.

Understanding the events taking place shortly after SCNT in other species may lead to development of successful strategies for derivation of human SCNT-derived ESCs. Embryonic development after nuclear transplantation requires faithful recapitulation of events taking place during and after fertilization. Synchronization of the cell cycle stage between the oocyte (G2) or egg (MII) and a cycling somatic cell is critical to avoid chromosomal damage, aneuploidy, or partial reduplication of DNA (reviewed by Campbell et al.[32]). Other factors that need to be considered include the timing of the activation stimulus (replacing sperm-mediated activation during fertilization) after nuclear transplantation, and the choice of the activation stimulus (reviewed by Campbell[39]). Finally, the choice of a somatic cell as a nuclear donor appears to be equally important, and less differentiated cells (i.e., blastomeres or embryonic fibroblasts) appear to be easier to reprogram than more differentiated cells (lymphocytes or neurons); a property likely defined by the extent of epigenetic chromatin modifications of the cells.[40] When combined, these factors play a significant role in determining the outcome of SCNT in terms of development of preimplantation embryos suitable for isolation of pluripotent ESCs.

Low efficiency of SCNT-derived embryo development can be, in part, attributed to incomplete reprogramming of the somatic cell genome and consequently incomplete embryonic gene activation. Among the changes following SCNT, DNA methylation and histone modifications seem to play a critical role in establishing transcriptionally permissive embryonic epigenome[41,42] allowing for expression of early embryonic genes. Failure to establish expression[43] is often correlated to the methylation status of the Oct4 promoter. Oct4 promoter is unmethylated in early embryos and hypermethylated in differentiated cells,[44] requiring extensive promoter demethylation after SCNT.[45] Oct4 and Nanog genomic sequences in cells carrying a hypomorphic allele of the DNA methyltransferase Dnmt1 are reprogrammed more efficiently than their wild-type counterparts,[46] suggesting that the ability of the egg cytoplasm to demethylate DNA may be limiting and responsible, at least in part, for low efficiency of SCNT generated ESCs. Similarly, increasing global histone acetylation by inhibition of HDACs with trichostatin A (TCA) could lead to higher efficiency of reprogramming[47,48] in SCNT embryos by altering the expression of both chromatin modifiers and embryonic stem cell transcription factor Sox2.[48] Trichostatin A not only inhibits histone deacetylases but also decreases their expression. Collaboration between histone acetylation and DNA methylation has been observed in reactivation of X chromosome in female SCNT embryos, even in the absence of a demethylating agent 5-aza-2′-deoxyctidine.[49] Reduced levels of de novo methyltransferase Dnmt3b and "maintenance" methyltransferase Dnmt1 in TSA treated SCNT embryos were comparable to levels in in vivo derived embryos.[48] Similarities between ESCs derived from fertilized and SCNT-derived embryos (in terms of imprinted gene expression, X chromosome reactivation, expression of Oct4, Nanog, Sox2, and other embryonic genes, and developmental potency) suggest that successful, complete, and faithful nuclear reprogramming can be achieved using SCNT.[35,38,50]

Some mechanisms that establish transcriptionally permissive chromatin in early mammalian embryos appear to be conserved among species, as transcription of human embryonic genes in primate cell nuclei can be induced by *Xenopus laevis*,[51] and bovine oocyte cytoplasm[52,53] and interspecies embryos develop past the MET.[52,54,55] Efforts are underway in the United Kingdom to explore this strategy for derivation of therapeutically useful human pluripotent cells.[56]

REPROGRAMMING WITH CELL EXTRACTS

It has been shown decades ago that gene expression of mammalian cells can be influenced by exposure to different intracellular or extracellular molecular environments, demonstrating inherent plasticity of differentiated cells. These hypotheses have been examined and accepted for driving cellular changes from one differentiated cell type into another (transdifferentiation),[57,58] from a less differentiated cell to a more differentiated cell (differentiation[59]), and from a more to a less differentiated state (dedifferentiation). The latter application, leading to dedifferentiation, is based on introduction of differentiated nuclei into the cytoplasm of less differentiated cells,[51] or introduction of nuclear and cytoplasmic extracts derived from Xenopus oocytes,[60–62] embryonal carcinoma cells,[63–65] embryonic germ cells,[66] and embryonic stem cells[64,67–69] into permeabilized somatic cells.

Active remodeling of Xenopus somatic cells could be achieved by the nucleosomal ATPase imitation switch (ISWI) present in egg extracts.[60] Nucleoplasmin, a protein present in Xenopus eggs, plays an important role in the chromatin decondensation through centromeric DNA decondensation, reduction of chromatin-associated histone H1 protein, loss of histone H3K9 trimethylation, and reactivation of the oocyte-specific genes c-mos, Msy-2, and H1foo.[70] Reactivation of Oct4 gene and inhibition of somatic Thy1 gene in human lymphocytes injected into Xenopus oocytes[51] demonstrates conservation of reprogramming machinery across species. The activation of Oct4 requires demethylation of its promoter and the speed of Oct4 reactivation is dependent on the degree of chromatin organization.[44] How the Oct4 promoter becomes demethylated in the absence of DNA synthesis is not understood and there is no clear evidence of a DNA demethylase enzyme in frog oocytes. It is possible that DNA demethylation may be the result of a remodeling activity mediated by methyl-binding domain or MBD-proteins and ATP-dependent chromatin remodeling factors.

Human leukocytes exposed to Xenopus egg extracts reactivate Oct4 and germ cell-specific alkaline phosphatase (GCAP) after 1 week in culture.[61] The remodeling taking place under these conditions may depend on the chromatin remodeling factor, Brahma-related group 1 (BRG1), which is needed for this gene activation.[61] BRG1 has been described in primate oocytes and preimplantation embryos where its transcriptional regulation follows a predetermined maternal to embryonic pattern.[14]

Cellular activities induced in bovine fetal fibroblasts by either Xenopus oocyte or egg extracts were recently described.[62] An important difference between these two extracts is their different capacities to sustain transcription and DNA replication. Oocyte extracts promote transcription but not DNA replication in fibroblasts, whereas egg extracts promote replication but not transcription. Reprogramming has been achieved recently with primary adult human fibroblasts after exposure to Xenopus egg extracts. Select extract fractions obtained after gel filtration chromatography displayed activities inducing rapid histone H1 removal, reactivation of Rex1, lin28, and Sox2 gene expression, translation and nuclear localization of Oct4, Sox2, and Nanog transcription factors, and acquisition of multilineage differentiation potential (D. Kole, personal communication).

Similarly, extracts from various mammalian cell types can drive new gene-expression profiles in somatic cells. This approach has recently been used for dedifferentiation of 293T cells and NIH3T3 fibroblasts after incubation in extracts derived from pluripotent embryonic carcinoma (EC) cells and ESCs.[64] The brief treatment of cells with these extracts triggers the formation of colonies with morphology similar to ESCs. Upregulation of a number of pluripotency genes and downregulation of somatic genes, such as Lamin A, are detected and maintained for up to 4 weeks after treatment. In addition, these cells are able to differentiate to mesoderm and ectoderm lineages. The phenotypic changes in these cells are the result of epigenetic modifications of the chromatin mediated by chromatin remodeling factors, histone acetylation, expression of specific genes, and protein synthesis.

Immunocytochemical and ChIP (chromatin immunoprecipitation) studies have shown that global hyperacetylation and hypermethylation of histones H3 and H4 takes place in mES cell/thymocyte hybrids, a pattern identified in mouse ESCs but not in thymocytes.[68] The hyperacetylated and hypermethylated histone H3 and H4 occupied regions included the promoter of the stem cell gene Oct4 and corresponded to DNA demethylation and Oct4 transcriptional activation. In addition, the transcriptionally active Oct4 was characterized by hyperacetylation of H3 and hypomethylation of H3K9—changes marking active Oct4 gene.[68] H3K4me2 and H3K4me3 are tightly linked to the activity of Isw1p ATP-ase activity in regulation of transcription[71] and H3K4 methylation may function as an epigenetic mark to induce recruitment of the Brm-associated SWI/SNF chromatin remodeling factors.[72] A more extensive analysis of Oct4 regulatory regions, including proximal enhancer, distal enhancer, and proximal promoter revealed a number of histone H3 changes marking transcriptionally active Oct4.[65] Transient exposure of human 293T epithelial cells to EC extract resulted in changes in H3K4, K9, and K27. Prominent acetylation of H3K9, demethylation of H3K9me2 and H3K9me3, and demethylation of H3K27me3, but no significant changes in H3K4me2 and H3K4me3 methylation may indicate that the threshold that needs to be crossed in somatic cells for Oct4 expression involves changes from transcriptionally permissive to transcriptionally active chromatin. If this is indeed the case, it may signify a different threshold for Oct4 activation between mouse and human cells, where the gene is transcriptionally silent[68] in mouse and low levels of Oct4 have been described in human cells.[65,73]

In summary, the exposure of cells' nuclei to a less differentiated molecular environment can initiate dedifferentiation of the cells' phenotype through chromatin remodeling[65,68,74] and leads to reprogramming of gene expression,[51,69] X-chromosome reactivation,[75,76] and acquisition of pluripotency, including contribution to all three germ layers in teratomas.[67,77]

REPROGRAMMING WITH TRANSCRIPTION FACTORS: INDUCED PLURIPOTENT STEM (iPS) CELLS

Direct reprogramming of somatic cells to pluripotent stem cells appears to be an attractive alternative to circumvent some of the problems associated with ESCs. Maintenance of ESC phenotype depends on Oct4, Sox2, and Nanog transcription factors. These transcription factors are part of the molecular signature of ES cells, known as "stemness,"[78,79] and form a master regulatory circuitry for the maintenance of pluripotency.[30] Reactivation of this circuitry is an important step for achieving stably reprogrammed cells.

The observations that extracts from ES cells contain factors that can confer pluripotency to somatic cells[64,77] led Yamanaka's laboratory to explore the hypothesis that factors which play important roles in the maintenance of ES cell identity may play a pivotal role in induction of pluripotency as well. They selected a group of 24 candidate genes identified in ES cells and inserted each gene into Fbx15$^{bgeo/bgeo}$ mouse embryonic fibroblasts (MEFs) by retroviral transduction.[1] Induction of the Fbx15 locus, expressed exclusively in mouse embryonic stem cells and embryos,[80] was monitored by colony resistance to G418. While cells transduced with any single candidate gene failed to confer G418 resistance, cells transduced with all 24 genes acquired stem cell phenotype, albeit at very low efficiency (0.01%). By systematically removing one gene at a time and introducing the remaining 23 genes, and monitoring the efficiency and timeline of ES-like colony formation, they were able to identify genes whose absence in the transduction mix either delayed or obliterated the ability of the cells to form colonies altogether. The four genes whose removal slowed down or prevented G418-resistant colony formation were identified as c-myc and Oct4, Sox2, and Klf4, respectively. A systematic examination of combinations of these genes in groups of two and three in transduction protocols revealed that all four transgenes were needed for G418-resistant colony formation, they could be differentiated in vitro and formed teratomas containing tissues from all three germ layers. The name for the cells derived by four candidate genes was coined induced pluripotent stem (iPS) cells. iPS cells with similar properties were also derived from adult FBXbneo/bneo tail tip mouse fibroblasts demonstrating that adult cells, too, were permissive to reprogramming into iPS cell phenotype. Global gene expression profile pointed to an efficient induction of stem cell genes as well as cell morphology only in cells derived with all four transgenes. While very similar to ESCs, these cells remained clearly distinct from true ESCs based on global gene expression profiling. Blastocyst injections did not result in production of chimeric mice, suggesting that the potency of iPS cells was not that of ESCs.

Nevertheless, the observations from this and a closely related study employing the same approach for derivation of iPS cells from human embryonic and adult fibroblasts[81,82] are groundbreaking. Demonstration that introduction of a limited number of known genes into a differentiated cell is sufficient for its reprogramming and dedifferentiation to a pluripotent state marked the beginning of a new era in stem cell biology.

Unlike sperm nucleus reprogramming after fertilization and reprogramming of somatic nucleus in oocyte or ESC cytoplasm, reprogramming of a somatic cell with forced gene expression requires several weeks.[81] Regardless of the protocol employed, continuous forced expression of reprogramming factors is generally required for at least 10–12 days. Different endogenous stem cell genes become activated gradually and at different times. In mouse, the appearance of SSEA1 is observed on day 3, Oct4 on day 12, followed by Klf4 and Sox2 much later.[83] Increased histone H3 acetylation and decreased H3K9me2 were associated with transcriptional activation of the Oct4 gene, while its promoter remained partially methylated,[1] similar to previously reported partial demethylation of Oct4 exposed to EC extracts.[65] A much more efficient demethylation profile of the Oct region was achieved using the same four transgenes in human cells.[81] In parallel with endogenous stem cell gene activation, small ESC-like colonies begin to form, but it is not until 6 weeks that these colonies become undistinguishable from true pluripotent ESCs. Continuous forced expression has been attained by constitutively expressed retroviral vectors or constitutive or inducible lentiviral vectors. Genomic integration of virus-mediated transgenes delivery can, however, lead to insertional mutagenesis and/or reactivation of transgene expression leading to tumor formation. None of these consequences is acceptable if iPS cells are to be used for human therapeutic applications.

To avoid these concerns, nonintegrating adenoviruses,[84] polycistronic plasmids,[85–89] and episomal vectors[90] have been explored for delivery of the reprogramming factors with some success. However, the efficiency of iPS cell derivation, which is already low with integrating viruses, progressively declines further when nonintegrating vectors are used or when c-myc is omitted. Furthermore, the number of iPS cell lines derived with these vectors that are free of detectable genomic integrations is low, necessitating their rigorous screening. Recent successful demonstration of iPS phenotype with piggyBac transposition of the four transcription factors enabled derivation of transgene-free mouse iPS cells and chimeric mice, as well as human iPS cells.[91,92]

Low efficiency of iPS cell induction, regardless of the protocol, raises the possibility that differentiated somatic cell culture contains distinct subpopulations of cells that may be more susceptible to phenotypic conversion. Identification of subpopulations of human dermal fibroblasts that express stem cell related genes SSEA3[93] and Oct4 and Nanog[73] indicates that selection of specific subpopulations of somatic cells may result in higher efficiency of iPS cell generation. SSEA3 positive human fibroblasts derived from an adult male could indeed be converted to iPS cells with eightfold higher efficiency than unsorted cells.[93]

Due to the low efficiency of iPS cell phenotype induction, strategies used for identification and selection of "true" iPS cells may have distinct consequences. Choice of a reporter that does not uniquely select for the true pluripotent stem cells may skew the interpretation of the efficiency of the induction protocol employed. Replacing the FBX15neo selection with GFP-IRES-puro cassette introduced into the endogenous nanog locus in MEFs increased the proportion of iPS cells that resembled ESCs in their global histone acetylation profile and displayed demethylation of endogenous Oct4 and Nanog regulatory regions that were almost identical to the methylation status of these regions in ESCs. These iPS cells formed colonies indistinguishable from ESC colonies, reactivated X chromosome, differentiated in vitro, formed teratomas, and gave rise to viable chimeras with contribution to the germline.[83,94] The Nanog-GFPpuro derived iPS cells further demonstrated the potency to reprogram endogenous Oct4 locus in Oct4-hygromycin nuclei upon creation of cell hybrids and puromycin/G418 colonies were obtained. FBX15 selected iPS cells expressed lower levels of endogenous Oct4, Sox2, and Nanog than ESCs, and likely required continuous expression of the transgenes to maintain their iPS phenotype.[1] iPS cells derived from doxycycline inducible Oct4 fibroblasts, transduced with Sox2, Klf4, and c-myc on the other hand, could be maintained from endogenous expression of Oct4. Removal of doxycycline 3 weeks after transfection did not affect expression of the Oct-neo allele and G418 colonies could be maintained without changes in their growth or morphology.[83] Using inducible transgene constructs, it became evident that the process of endogenous stem cell gene activation is gradual and occurs over several weeks.[95] Doxycycline inducible Oct4, Sox2, Klf4, and c-myc transgenes carried by lentivirus were introduced into Oct4-GFP mouse fibroblasts[96] and the first weakly GFP positive colonies, indicating activation of the endogenous Oct4 gene, were observed 9 days after transduction. At this time the iPS colonies could be maintained in culture without doxycyclin.[84] FACS sorting of cells based on presentation of Thy1 (expressed in differentiated cells, including fibroblasts) and SSEA1 (expressed in mESCs) revealed that changes from an exclusive Thy+/SSEA− phenotype in fibroblasts started shifting gradually following 3 days of doxycyclin administration. Loss of Thy+ cells preceded the appearance of SSEA+ populations and at 12 days of culture resulted in the phenotype reversal to Thy1−/SSEA+. While expression of some of the endogenous stem cell genes (Oct4) could be detected on day 8, detection of mTert-GFP, Sox2-GFP expression, and X-GFP detectable chromosome reactivation required an additional week.[95] Sorting of transduced cells at various times after transduction identified distinct populations of intermediates and late intermediates characterized by Thy1−/SSEA+ and SSEA+/Oct+/Sox+, respectively. Retrovirus inactivation in embryonic cells,[97] observed in established iPS cultures,[83,85,98] occurs in parallel with the acquisition of the late stem cell gene expression in iPS cells.[95] Combined, these studies laid out a pattern of select molecular events that take place during induction of pluripotency in mouse cells using integrating viral vectors. As the kinetics of the reprogramming events became described in some detail,[99] selection and identification of iPS colonies started to rely on cell morphology, gene expression, and cells' developmental potential instead of expression of transgenic reporters. Consequently, primary, genetically unmodified human fibroblasts could be used for derivation of iPS cells with relevant therapeutic potential.[90,93,100,101]

To reduce the number of genomic integrations, which ranged anywhere from 1–4 integrations per retroviral transgene in iPS cells derived from mouse liver and stomach cells to 1–9 retroviral integrations per transgene in MEFs,[102] a strategy incorporating a combination of delivery vectors was developed. iPS cells could be developed from a nanog-IRES-puro hepatic cell line only when adenovirus delivered either Sox2 or Klf4 and the remaining two genes were delivered using retrovirus.[85] Oct4 had to be introduced with a retrovirus in order to derive functional iPS cells. The number of integration sites was reduced as none of the adenoviral sequences were detected in iPS cell genome, neither by qPCR or Southern blotting.[85] As an alternative, a single retroviral polycistronic expression vector carrying Oct4, Klf4, and Sox2 and the 2A self-cleaving peptide[103] allowing for polycistronic expression under the constitutive CAG promoter (pMSX-OKS-2A) was introduced into the nanog-GFP-puro reporter mouse hepatocyte cell line[85] and iPS cells were derived. However, all retroviral protocols induced tumor formation in a significant proportion of chimeric animals.[1,94,102]

These observations called for development of new strategies that would not cause insertional mutagenesis and strategies that would not depend on the presence of c-myc. It was soon demonstrated that insertional mutagensis is not required for reprogramming. Introduction of the same sequence of transgenes in a plasmid vector (pCX-OKS-2A) together with a separate pCX-Myc expression plasmid into hepatocytes by repeated transfection on days 1, 3, 5, and 7 produced iPS cells without any plasmid integration. Similarly, Sox2, Klf4, and c-myc delivered with replication deficient adenoviruses under the control of the human CMV immediate early promoter into puromycine inducible Oct4 fetal mouse liver cells[84] or all four factors delivered using adenovirus into Oct4-GFP hepatocytes resulted in derivation of iPS without the transgene integration, cells displayed molecular and developmental characteristics described for mESCs, including contribution to the germline in chimeric animals. The efficiency of nonintegrating transduction protocols, however, was significantly reduced to 0.0001–0.001%[84] and appeared to be applicable to select cell types, such as liver and stomach cells.[85,102]

The bigger problem with induction protocols, however, was the requirement for the oncogene c-myc in the combination of the four genes required for reprogramming to a pluripotent state. To examine whether transduction with c-myc transgene and its forced expression could be omitted, the remaining three transgenes (Oct4, Sox2, and Klf4) were introduced into mouse and human fibroblasts. Unlike in the original studies,[1,81] transduced cells were cultured without selection for at least 2 or 3 weeks. nanog-GFP expressing colonies started appearing on day 30 and properties of iPS cells more closely resembled those of true ESCs.[104] Omitting c-myc significantly reduces the efficiency, but also reduces non-iPS background cells and selects for iPS cells of better quality. Chimeras do not develop tumors, thus alleviating a major concern from previous studies. Human iPS cells derived without c-myc display stem cell gene expression characteristics of hESCs, express hESC surface markers (SSEA3, SSEA4, TRA-1-60, TRA-181), and differentiate into cell types from the three germ layers.[104]

A similar, but distinct combination of genes that could accomplish conversion from a differentiated to pluripotent human cell was described by Thomson's group.[105] From 14 initially identified genes, Oct4, Sox2, nanog, and lin28 were established as the minimally required set for conversion of ES-derived mesenchymal cells and IMR90 fibroblasts to iPS phenotype. Substituting Klf4 and c-myc with lin28 and nanog without compromising iPS efficiency suggests that Klf4 and c-myc are not absolutely required for the reprogramming process in human cells.

Common targets for chemically defined interventions are emerging and focus on inhibition of repressive chromatin modifying enzymes, including G9a histone methyltransferase,[106] histone deacetylases,[107,108] and DNA methyltransferase.[100]

Inhibition of any of these enzymes generally results in improved efficiency but does not alleviate the need for transgenes nor does it shorten the time required for acquisition of iPS phenotype. These studies indicate that genome-wide approaches to modifying the chromatin structure alone are not sufficient for activation of endogenous stem cell genes, but may be beneficial in relaxing the chromatin structure of their downstream targets (reviewed by Feng et al.[109]).

Membrane permeable recombinant Oct4 and Sox2 transcription factor proteins, fused with polyarginine sequences, have been shown to functionally replace endogenous Oct4 and Sox2 in mouse ESCs.[110] It was recently demonstrated that introducing Oct4, Sox2, Klf4, and c-myc transcription factor proteins instead of forcing their expression through transgenes results in generation of stable iPS cells from human fibroblasts.[108,111] These latest protocols for derivation of iPS cells alleviate the many concerns of transgenic approaches, including insertional mutagenesis and unregulated transgene reactivation (reviewed by O'Malley et al.[112]).

Somatic cell types that have to date been successfully reprogrammed into iPS cells include fibroblasts, keratinocytes,[87] liver and stomach cells,[102] lymphocytes,[113] pancreatic beta-cells,[114] adipose stem cells,[115] CD34$^+$ cells,[116] and adult neural stem cells,[117,118] generally at very low efficiencies. Augmenting iPS induction protocols with selection of reprogramming permissive cells,[93] and alteration of in vitro culture conditions,[73,119,120] promises to increase the efficiency of iPS methods.

CONCLUSION

Since the initial reports outlining the generation of induced pluripotent stem cells from human and mouse somatic cells, many groups are further refining the process. The inherent and significant differences between mouse and human pluripotent stem cell biology caution that "what works in mouse" may not be successfully applied to humans and vice versa. Unique and distinct signal transduction pathways, gene expression properties, in vitro cell culture demands, and dependency on specific cytokines to establish and maintain the self-renewing identity of pluripotent cells have to be carefully considered when attempting new protocols. At the same time, while some of the approaches to induce pluripotency may be successful in both species, it is unlikely that they exert their effect through identical mechanisms.

While a number of challenges remain, reprogramming of human differentiated cells toward functional pluripotency will have long-term implications in regenerative medicine. The first encouraging developments have already been reported. iPS cells have been differentiated into neurons and glial cells and could improve behavior in rat models of Parkinson disease.[121] iPS cell-derived motor neurons from amyotrophic lateral sclerosis (ALS) patients,[122] cardiac myocytes,[123–125] and hepatocytes from mouse iPS cells[126] are likely the first in development of many iPS-derived therapeutically promising cells. Development of customized, patient-specific cells for therapeutic interventions in degenerative diseases, interventions in loss of tissue functionality, and eventually for functional replacement of lost complex tissues in humans are becoming rational targets for the research efforts described in this chapter. Development of customized cell therapeutics is becoming a reality, one step at a time.

REFERENCES

1. Takahashi K, Yamanaka S. Induction of pluripotent stem cells from mouse embryonic and adult fibroblast cultures by defined factors. *Cell*. 2006;126:663–676.

2. Sakaguchi M, Dominko T, Leibfried-Rutledge ML, et al. A combination of EGF and IGF-I accelerates the progression of meiosis in bovine follicular oocytes in vitro and fetal

calf serum neutralizes the acceleration effect. *Theriogenology*. 2000;54:1327–1342.

3. Schultz RM. Regulation of zygotic gene activation in the mouse. *Bioessays*. 1993;15:531–538.

4. Telford NA, Watson AJ, Schultz GA. Transition from maternal to embryonic control in early mammalian development: a comparison of several species. *Mol Reprod Dev*. 1990;26:90–100.

5. Frei RE, Schultz GA, Church RB. Qualitative and quantitative changes in protein synthesis occur at the 8–16-cell stage of embryogenesis in the cow. *J Reprod Fertil*. 1989;86:637–641.

6. Tagawa M, Matoba S, Narita M, et al. Production of monozygotic twin calves using the blastomere separation technique and Well of the Well culture system. *Theriogenology*. 2008;69:574–582.

7. Johnson WH, Loskutoff NM, Plante Y, et al. Production of four identical calves by the separation of blastomeres from an in vitro derived four-cell embryo. *Vet Rec*. 1995;137:15–16.

8. Chan AW, Dominko T, Luetjens CM, et al. Clonal propagation of primate offspring by embryo splitting. *Science*. 2000;287:317–319.

9. Stein P, Worrad DM, Belyaev ND, et al. Stage-dependent redistributions of acetylated histones in nuclei of the early preimplantation mouse embryo. *Mol Reprod Dev*. 1997;47:421–429.

10. Memili E, First NL. Control of gene expression at the onset of bovine embryonic development. *Biol Reprod*. 1999;61:1198–1207.

11. Bilodeau-Goeseels S, Schultz GA. Changes in the relative abundance of various housekeeping gene transcripts in in vitro-produced early bovine embryos. *Mol Reprod Dev*. 1997;47:413–420.

12. Kirchhof N, Carnwath JW, Lemme E, et al. Expression pattern of Oct-4 in preimplantation embryos of different species. *Biol Reprod*. 2000;63:1698–1705.

13. Wolffe AP. Chromatin and gene regulation at the onset of embryonic development. *Reprod Nutr Dev*. 1996;36:581–606.

14. Zheng P, Patel B, McMenamin M, et al. Expression of genes encoding chromatin regulatory factors in developing rhesus monkey oocytes and preimplantation stage embryos: possible roles in genome activation. *Biol Reprod*. 2004;70:1419–1427.

15. Vigneault C, McGraw S, Massicotte L, et al. Transcription factor expression patterns in bovine in vitro-derived embryos prior to maternal-zygotic transition. *Biol Reprod*. 2004;70:1701–1709.

16. Misirlioglu M, Page GP, Sagirkaya H, et al. Dynamics of global transcriptome in bovine matured oocytes and preimplantation embryos. *Proc Natl Acad Sci USA*. 2006;103:18905–18910.

17. Pesce M, Scholer HR. Oct-4: gatekeeper in the beginnings of mammalian development. *Stem Cells*. 2001;19:271–278.

18. Chambers I, Colby D, Robertson M, et al. Functional expression cloning of Nanog, a pluripotency sustaining factor in embryonic stem cells. *Cell*. 2003;113:643–655.

19. Nichols J, Zevnik B, Anastassiadis K, et al. Formation of pluripotent stem cells in the mammalian embryo depends on the POU transcription factor Oct4. *Cell*. 1998;95:379–391.

20. Silva J, Nichols J, Theunissen TW, et al. Nanog is the gateway to the pluripotent ground state. *Cell*. 2009;138:722–737.

21. Adenot PG, Mercier Y, Renard JP, et al. Differential H4 acetylation of paternal and maternal chromatin precedes DNA replication and differential transcriptional activity in pronuclei of 1-cell mouse embryos. *Development*. 1997;124:4615–4625.

22. Zheng P, Patel B, McMenamin M, et al. The primate embryo gene expression resource: a novel resource to facilitate rapid analysis of gene expression patterns in non-human primate oocytes and preimplantation stage embryos. *Biol Reprod*. 2004;70:1411–1418.

23. de Vries WN, Evsikov AV, Brogan LJ, et al. Reprogramming and differentiation in mammals: motifs and mechanisms. *Cold Spring Harb Symp Quant Biol*. 2008;73:33–38.

24. Scholer HR, Balling R, Hatzopoulos AK, et al. Octamer binding proteins confer transcriptional activity in early mouse embryogenesis. *EMBO J*. 1989;8:2551–2557.

25. Rosner MH, Vigano MA, Ozato K, et al. A POU-domain transcription factor in early stem cells and germ cells of the mammalian embryo. *Nature*. 1990;345:686–692.

26. Cauffman G, Liebaers I, Van Steirteghem A, et al. POU5F1 isoforms show different expression patterns in human embryonic stem cells and preimplantation embryos. *Stem Cells*. 2006;24:2685–2691.

27. Cauffman G, Van de Velde H, Liebaers I, et al. Oct-4 mRNA and protein expression during human preimplantation development. *Mol Hum Reprod*. 2005;11:173–181.

28. Takeda J, Seino S, Bell GI. Human Oct3 gene family: cDNA sequences, alternative splicing, gene organization, chromosomal location, and expression at low levels in adult tissues. *Nucleic Acids Res*. 1992;20:4613–4620.

29. Pesce M, Wang X, Wolgemuth DJ, et al. Differential expression of the Oct-4 transcription factor during mouse germ cell differentiation. *Mech Dev*. 1998;71:89–98.

30. Boyer LA, Lee TI, Cole MF, et al. Core transcriptional regulatory circuitry in human embryonic stem cells. *Cell*. 2005;122:947–956.

31. Thomson JA, Itskovitz-Eldor J, Shapiro SS, et al. Embryonic stem cell lines derived from human blastocysts. *Science*. 1998;282:1145–1147.

32. Campbell KH, Loi P, Otaegui PJ, et al. Cell cycle coordination in embryo cloning by nuclear transfer. *Rev Reprod*. 1996;1:40–46.

33. Wilmut I, Schnieke AE, McWhir J, et al. Viable offspring derived from fetal and adult mammalian cells. *Nature*. 1997;385:810–813.

34. Wakayama T, Perry AC, Zuccotti M, et al. Full-term development of mice from enucleated oocytes injected with cumulus cell nuclei. *Nature*. 1998;394:369–374.

35. Wakayama S, Mizutani E, Kishigami S, et al. Mice cloned by nuclear transfer from somatic and ntES cells derived from the same individuals. *J Reprod Dev*. 2005;51:765–772.

36. Wakayama T, Tabar V, Rodriguez I, et al. Differentiation of embryonic stem cell lines generated from adult somatic cells by nuclear transfer. *Science*. 2001;292:740–743.

37. Byrne JA, Pedersen DA, Clepper LL, et al. Producing primate embryonic stem cells by somatic cell nuclear transfer. *Nature*. 2007;450:497–502.

38. Sparman M, Dighe V, Sritanaudomchai H, et al. Epigenetic reprogramming by somatic cell nuclear transfer in primates. *Stem Cells*. 2009;27:1255–1264.

39. Campbell KH. Nuclear equivalence, nuclear transfer, and the cell cycle. *Cloning*. 1999;1:3–15.

40. McLean CA, Wang Z, Babu K, et al. Normal development following chromatin transfer correlates with donor cell initial epigenetic state. *Anim Reprod Sci*. 2010;118(2):388–393.

41. Dean W, Santos F, Stojkovic M, et al. Conservation of methylation reprogramming in mammalian development: aberrant reprogramming in cloned embryos. *Proc Natl Acad Sci USA*. 2001;98:13734–13738.

42. Santos F, Zakhartchenko V, Stojkovic M, et al. Epigenetic marking correlates with developmental potential in cloned bovine preimplantation embryos. *Curr Biol*. 2003;13:1116–1121.

43. Bortvin A, Eggan K, Skaletsky H, et al. Incomplete reactivation of Oct4-related genes in mouse embryos cloned from somatic nuclei. *Development*. 2003;130:1673–1680.

44. Simonsson S, Gurdon J. DNA demethylation is necessary for the epigenetic reprogramming of somatic cell nuclei. *Nat Cell Biol*. 2004;6:984–990.

45. Yamazaki Y, Fujita TC, Low EW, et al. Gradual DNA demethylation of the Oct4 promoter in cloned mouse embryos. *Mol Reprod Dev*. 2006;73:180–188.

46. Blelloch R, Wang Z, Meissner A, et al. Reprogramming efficiency following somatic cell nuclear transfer is influenced by the differentiation and methylation state of the donor nucleus. *Stem Cells*. 2006;24:2007–2013.

47. Kishigami S, Mizutani E, Ohta H, et al. Significant improvement of mouse cloning technique by treatment with trichostatin A after somatic nuclear transfer. *Biochem Biophys Res Commun*. 2006;340:183–189.

48. Li X, Kato Y, Tsuji Y, et al. The effects of trichostatin A on mRNA expression of chromatin structure-, DNA methylation-, and development-related genes in cloned mouse blastocysts. *Cloning Stem Cells*. 2008;10:133–142.

49. Xiong Y, Dowdy SC, Podratz KC, et al. Histone deacetylase inhibitors decrease DNA methyltransferase-3B messenger RNA stability and down-regulate de novo DNA methyltransferase activity in human endometrial cells. *Cancer Res*. 2005;65:2684–2689.

50. Brambrink T, Hochedlinger K, Bell G, et al. ES cells derived from cloned and fertilized blastocysts are transcriptionally and functionally indistinguishable. *Proc Natl Acad Sci USA*. 2006;103:933–938.

51. Byrne JA, Simonsson S, Western PS, et al. Nuclei of adult mammalian somatic cells are directly reprogrammed to oct-4 stem cell gene expression by amphibian oocytes. *Curr Biol*. 2003;13:1206–1213.

52. Li F, Cao H, Zhang Q, et al. Activation of human embryonic gene expression in cytoplasmic hybrid embryos constructed between bovine oocytes and human fibroblasts. *Cloning Stem Cells*. 2008;10:297–305.

53. Wang K, Beyhan Z, Rodriguez RM, et al. Bovine ooplasm partially remodels primate somatic nuclei following somatic cell nuclear transfer. *Cloning Stem Cells*. 2009;11:187–202.

54. Dominko T, Mitalipova M, Haley B, et al. Bovine oocyte cytoplasm supports development of embryos produced by nuclear transfer of somatic cell nuclei from various mammalian species. *Biol Reprod*. 1999;60:1496–1502.

55. Sugimura S, Narita K, Yamashiro H, et al. Interspecies somatic cell nucleus transfer with porcine oocytes as recipients: a novel bioassay system for assessing the competence of canine somatic cells to develop into embryos. *Theriogenology*. 2009;72:549–559.

56. Vogel G. Bioethics. U.K. approves new embryo law. *Science*. 2008;322:663.

57. Hakelien AM, Landsverk HB, Robl JM, et al. Reprogramming fibroblasts to express T-cell functions using cell extracts. *Nat Biotechnol*. 2002;20:460–466.

58. Condorelli G, Borello U, De Angelis L, et al. Cardiomyocytes induce endothelial cells to trans-differentiate into cardiac muscle: implications for myocardium regeneration. *Proc Natl Acad Sci USA*. 2001;98:10733–10738.

59. Qin M, Tai G, Collas P, et al. Cell extract-derived differentiation of embryonic stem cells. *Stem Cells*. 2005;23:712–718.

60. Kikyo N, Wade PA, Guschin D, et al. Active remodeling of somatic nuclei in egg cytoplasm by the nucleosomal ATPase ISWI. *Science*. 2000;289:2360–2362.

61. Hansis C, Barreto G, Maltry N, et al. Nuclear reprogramming of human somatic cells by *Xenopus* egg extract requires BRG1. *Curr Biol*. 2004;14:1475–1480.

62. Alberio R, Johnson AD, Stick R, et al. Differential nuclear remodeling of mammalian somatic cells by *Xenopus laevis* oocyte and egg cytoplasm. *Exp Cell Res*. 2005;307:131–141.

63. Shimazaki T, Okazawa H, Fujii H, et al. Hybrid cell extinction and re-expression of Oct-3 function correlates with differentiation potential. *EMBO J*. 1993;12:4489–4498.

64. Taranger CK, Noer A, Sorensen AL, et al. Induction of dedifferentiation, genomewide transcriptional programming, and epigenetic reprogramming by extracts of carcinoma and embryonic stem cells. *Mol Biol Cell*. 2005;16:5719–5735.

65. Freberg CT, Dahl JA, Timoskainen S, et al. Epigenetic reprogramming of OCT4 and NANOG regulatory regions by embryonal carcinoma cell extract. *Mol Biol Cell*. 2007;18:1543–1553.

66. Tada M, Tada T, Lefebvre L, et al. Embryonic germ cells induce epigenetic reprogramming of somatic nucleus in hybrid cells. *EMBO J*. 1997;16:6510–6520.

67. Tada M, Takahama Y, Abe K, et al. Nuclear reprogramming of somatic cells by in vitro hybridization with ES cells. *Curr Biol*. 2001;11:1553–1558.

68. Kimura H, Tada M, Nakatsuji N, et al. Histone code modifications on pluripotential nuclei of reprogrammed somatic cells. *Mol Cell Biol*. 2004;24:5710–5720.

69. Bru T, Clarke C, McGrew MJ, et al. Rapid induction of pluripotency genes after exposure of human somatic cells to mouse ES cell extracts. *Exp Cell Res*. 2008;314:2634–2642.

70. Tamada H, Van Thuan N, Reed P, et al. Chromatin decondensation and nuclear reprogramming by nucleoplasmin. *Mol Cell Biol*. 2006;26:1259–1271.

71. Santos-Rosa H, Schneider R, Bernstein BE, et al. Methylation of histone H3 K4 mediates association of the Isw1p ATPase with chromatin. *Mol Cell*. 2003;12:1325–1332.

72. Kingston RE, Narlikar GJ. ATP-dependent remodeling and acetylation as regulators of chromatin fluidity. *Genes Dev*. 1999;13:2339–2352.

73. Page RL, Ambady S, Holmes WF, et al. Induction of stem cell gene expression in adult human fibroblasts without transgenes. *Cloning Stem Cells*. 2009;11:417–426.

74. Kikyo N, Wolffe AP. Reprogramming nuclei: insights from cloning, nuclear transfer and heterokaryons. *J Cell Sci*. 2000;113(Pt 1):11–20.

75. Kimura H, Tada M, Hatano S, et al. Chromatin reprogramming of male somatic cell-derived XIST and TSIX in ES hybrid cells. *Cytogenet Genome Res*. 2002;99:106–114.

76. Tada T, Tada M. Toti-/pluripotential stem cells and epigenetic modifications. *Cell Struct Funct*. 2001;26:149–160.

77. Cowan CA, Atienza J, Melton DA, et al. Nuclear reprogramming of somatic cells after fusion with human embryonic stem cells. *Science*. 2005;309:1369–1373.

78. Ivanova NB, Dimos JT, Schaniel C, et al. A stem cell molecular signature. *Science*. 2002;298:601–604.

79. Ramalho-Santos M, Yoon S, Matsuzaki Y, et al. "Stemness": transcriptional profiling of embryonic and adult stem cells. *Science*. 2002;298:597–600.

80. Tokuzawa Y, Kaiho E, Maruyama M, et al. Fbx15 is a novel target of Oct3/4 but is dispensable for embryonic stem cell self-renewal and mouse development. *Mol Cell Biol*. 2003;23:2699–2708.

81. Takahashi K, Tanabe K, Ohnuki M, et al. Induction of pluripotent stem cells from adult human fibroblasts by defined factors. *Cell*. 2007;131:861–872.

82. Takahashi K, Okita K, Nakagawa M, et al. Induction of pluripotent stem cells from fibroblast cultures. *Nat Protoc*. 2007;2:3081–3089.

83. Maherali N, Sridharan R, Xie W, et al. Directly reprogrammed fibroblasts show global epigenetic remodeling and widespread tissue contribution. *Cell Stem Cell*. 2007;1:55–70.

84. Stadtfeld M, Nagaya M, Utikal J, et al. Induced pluripotent stem cells generated without viral integration. *Science*. 2008;322:945–949.

85. Okita K, Nakagawa M, Hyenjong H, et al. Generation of mouse induced pluripotent stem cells without viral vectors. *Science*. 2008;322:949–953.

86. Gonzales C, Pedrazzini T. Progenitor cell therapy for heart disease. *Exp Cell Res*. 2009;315:3077–3085.

87. Carey BW, Markoulaki S, Hanna J, et al. Reprogramming of murine and human somatic cells using a single polycistronic vector. *Proc Natl Acad Sci USA*. 2009;106:157–162.

88. Sommer CA, Stadtfeld M, Murphy GJ, et al. Induced pluripotent stem cell generation using a single lentiviral stem cell cassette. *Stem Cells*. 2009;27:543–549.

89. Chang CW, Lai YS, Pawlik KM, et al. Polycistronic lentiviral vector for "hit and run" reprogramming of adult skin fibroblasts to induced pluripotent stem cells. *Stem Cells*. 2009;27:1042–1049.

90. Yu J, Hu K, Smuga-Otto K, et al. Human induced pluripotent stem cells free of vector and transgene sequences. *Science*. 2009;324:797–801.

91. Kaji K, Norrby K, Paca A, et al. Virus-free induction of pluripotency and subsequent excision of reprogramming factors. *Nature*. 2009;458:771–775.

92. Woltjen K, Michael IP, Mohseni P, et al. piggyBac transposition reprograms fibroblasts to induced pluripotent stem cells. *Nature*. 2009;458:766–770.

93. Byrne JA, Nguyen HN, Reijo Pera RA. Enhanced generation of induced pluripotent stem cells from a subpopulation of human fibroblasts. *PLoS One*. 2009;4:e7118.

94. Okita K, Ichisaka T, Yamanaka S. Generation of germline-competent induced pluripotent stem cells. *Nature*. 2007;448:313–317.

95. Stadtfeld M, Maherali N, Breault DT, et al. Defining molecular cornerstones during fibroblast to iPS cell reprogramming in mouse. *Cell Stem Cell*. 2008;2:230–240.

96. Lengner CJ, Camargo FD, Hochedlinger K, et al. Oct4 expression is not required for mouse somatic stem cell self-renewal. *Cell Stem Cell*. 2007;1:403–415.

97. Wolf D, Goff SP. TRIM28 mediates primer binding site-targeted silencing of murine leukemia virus in embryonic cells. *Cell*. 2007;131:46–57.

98. Wernig M, Meissner A, Foreman R, et al. In vitro reprogramming of fibroblasts into a pluripotent ES-cell-like state. *Nature*. 2007;448:318–324.

99. Maherali N, Ahfeldt T, Rigamonti A, et al. A high-efficiency system for the generation and study of human induced pluripotent stem cells. *Cell Stem Cell*. 2008;3:340–345.

100. Meissner A, Wernig M, Jaenisch R. Direct reprogramming of genetically unmodified fibroblasts into pluripotent stem cells. *Nat Biotechnol*. 2007;25:1177–1181.

101. Lowry WE, Richter L, Yachechko R, et al. Generation of human induced pluripotent stem cells from dermal fibroblasts. *Proc Natl Acad Sci USA*. 2008;105:2883–2888.

102. Aoi T, Yae K, Nakagawa M, et al. Generation of pluripotent stem cells from adult mouse liver and stomach cells. *Science*. 2008;321:699–702.

103. Hasegawa K, Cowan AB, Nakatsuji N, et al. Efficient multicistronic expression of a transgene in human embryonic stem cells. *Stem Cells*. 2007;25:1707–1712.

104. Nakagawa M, Koyanagi M, Tanabe K, et al. Generation of induced pluripotent stem cells without Myc from mouse and human fibroblasts. *Nat Biotechnol*. 2008;26:101–106.

105. Yu J, Vodyanik MA, Smuga-Otto K, et al. Induced pluripotent stem cell lines derived from human somatic cells. *Science*. 2007;318:1917–1920.

106. Shi Y, Do JT, Desponts C, et al. A combined chemical and genetic approach for the generation of induced pluripotent stem cells. *Cell Stem Cell*. 2008;2:525–528.

107. Huangfu D, Maehr R, Guo W, et al. Induction of pluripotent stem cells by defined factors is greatly improved by small-molecule compounds. *Nat Biotechnol*. 2008;26:795–797.

108. Zhou H, Wu S, Joo JY, et al. Generation of induced pluripotent stem cells using recombinant proteins. *Cell Stem Cell*. 2009;4:381–384.

109. Feng B, Ng JH, Heng JC, et al. Molecules that promote or enhance reprogramming of somatic cells to induced pluripotent stem cells. *Cell Stem Cell*. 2009;4:301–312.

110. Bosnali M, Edenhofer F. Generation of transducible versions of transcription factors Oct4 and Sox2. *Biol Chem*. 2008;389:851–861.

111. Kim D, Kim CH, Moon JI, et al. Generation of human induced pluripotent stem cells by direct delivery of reprogramming proteins. *Cell Stem Cell*. 2009;4:472–476.

112. O'Malley J, Woltjen K, Kaji K. New strategies to generate induced pluripotent stem cells. *Curr Opin Biotechnol*. 2009;20(5):516–521.

113. Hanna J, Carey BW, Jaenisch R. Reprogramming of somatic cell identity. *Cold Spring Harb Symp Quant Biol*. 2008;73:147–155.

114. Stadtfeld M, Brennand K, Hochedlinger K. Reprogramming of pancreatic beta cells into induced pluripotent stem cells. *Curr Biol*. 2008;18:890–894.

115. Sun N, Panetta NJ, Gupta DM, et al. Feeder-free derivation of induced pluripotent stem cells from adult human adipose stem cells. *Proc Natl Acad Sci USA*. 2009;106:15720–15725.

116. Ye Z, Zhan H, Mali P, et al. Human induced pluripotent stem cells from blood cells of healthy donors and patients with acquired blood disorders. *Blood*. 2009;114(27):5473–5480.

117. Kim JB, Zaehres H, Wu G, et al. Pluripotent stem cells induced from adult neural stem cells by reprogramming with two factors. *Nature*. 2008;454:646–650.

118. Kim JB, Zaehres H, Arauzo-Bravo MJ, et al. Generation of induced pluripotent stem cells from neural stem cells. *Nat Protoc*. 2009;4:1464–1470.

119. Yoshida Y, Takahashi K, Okita K, et al. Hypoxia enhances the generation of induced pluripotent stem cells. *Cell Stem Cell*. 2009;5:237–241.

120. Silvan U, Diez-Torre A, Arluzea J, et al. Hypoxia and pluripotency in embryonic and embryonal carcinoma stem cell biology. *Differentiation*. 2009;78:159–168.

121. Wernig M, Zhao JP, Pruszak J, et al. Neurons derived from reprogrammed fibroblasts functionally integrate into the fetal brain and improve symptoms of rats with Parkinson's disease. *Proc Natl Acad Sci USA*. 2008;105:5856–5861.

122. Dimos JT, Rodolfa KT, Niakan KK, et al. Induced pluripotent stem cells generated from patients with ALS can be differentiated into motor neurons. *Science*. 2008;321:1218–1221.

123. Mauritz C, Schwanke K, Reppel M, et al. Generation of functional murine cardiac myocytes from induced pluripotent stem cells. *Circulation*. 2008;118:507–517.

124. Narazaki G, Uosaki H, Teranishi M, et al. Directed and systematic differentiation of cardiovascular cells from mouse induced pluripotent stem cells. *Circulation*. 2008;118:498–506.

125. Gai H, Leung EL, Costantino PD, et al. Generation and characterization of functional cardiomyocytes using induced pluripotent stem cells derived from human fibroblasts. *Cell Biol Int*. 2009;33(11):1184–1193.

126. Song Z, Cai J, Liu Y, et al. Efficient generation of hepatocyte-like cells from human induced pluripotent stem cells. *Cell Res*. 2009;19:1233–1242.

Applications of Human Embryonic Stem Cells

SECTION V

Human Stem Cell Technology and Biology, edited by Stein, Borowski, Luong, Shi, Smith, and Vazquez
Copyright © 2011 Wiley-Blackwell.

HUMAN PLURIPOTENT CELLS: THE BIOLOGY OF PLURIPOTENCY

26

Li-Fang Chu and Thomas P. Zwaka

INTRODUCTION

The ability of pluripotent stem cells to differentiate into all cell types of the adult organism holds great promise for future applications in regenerative medicine. Crucial knowledge gained from the mouse and other model organisms eventually led to derivation of human pluripotent stem cells from embryos and the discovery of induced pluripotent stem (iPS) cells derived from somatic cells. These breakthroughs create the possibility of using pluripotent stem cells in cell-based transplantation therapies.

Derivation of patient-specific iPS cells is currently underway and this work is likely to yield new insights into the molecular mechanisms of specific diseases as well as the discovery of novel drugs. Here, we describe some of the most fundamental findings in pluripotent stem cell biology and highlight recent studies that bring us closer to the goal of future regenerative medicine.

PLURIPOTENT STEM CELLS FROM MOUSE TO HUMAN: THE DISCOVERY OF EMBRYONAL CARCINOMA CELLS AND THE DERIVATION OF EMBRYONIC STEM CELLS

Pluripotent stem cells (Table 26.1) are characterized by two defining features: (1) self-renewal, namely, the capacity to divide and renew in the undifferentiated state for a long period in vitro; and (2) pluripotency, the ability to give rise to all specialized cells found in the adult body.

Detailed studies conducted more than half a century ago on a specialized type of gonadal tumor, the teratocarcinoma, unexpectedly led to the discovery of pluripotent stem cells. The prefix "terato-" (from the Greek word for "monster") was used to describe the bizarre biology of this tumor type. Teratocarcinoma is a puzzling form of rare testicular tumor, and histological studies have revealed a mixture of differentiated adult tissues in the tumor mass. Tissues derived from the three germ layers of the early developing embryo (including skin, bone, muscles, and gut-like structures) were found within the tumor. The rare incidence of teratocarcinoma made the tumor difficult to study until Leroy Stevens described an inbred mouse line, known as the 129 strain, that develops spontaneous teratocarcinomas at a rate of about 1%.[1] It was soon realized that a small population of cells in the tumor

Human Stem Cell Technology and Biology, edited by Stein, Borowski, Luong, Shi, Smith, and Vazquez
Copyright © 2011 Wiley-Blackwell.

TABLE 26.1 SUMMARY OF DIFFERENT TYPES OF PLURIPOTENT STEM CELL LINES

Pluripotent Cell Line	Source of Tissue for Derivation	Differentiation Potential/Potential/Potency	Derivation Strategy	Growth Factor(s) Requirement	Reference
ES	ICM/epiblast	Tetraploid complementation	Direct explantation of preimplantation embryos	LIF, BMP	9,10
EG	Primordial germ cells	Germline transmission	Prolonged culture with addition of Fgf2, Lif, and SCF	LIF, SCF, FGF	39,40
EpiSC	Epiblast	Teratoma	Direct explantation of postimplantation embryos	FGF, Activin	31, 32
Mouse iPS	Mouse embryonic/adult fibroblast	Tetraploid complementation	Forced expression of Oct4, Sox2, c-Myc, and Klf4 combined with selection of Oct4 or Nanog	LIF	103, 104, 106, 107
hES	Human blastocysts	Teratoma	Direct of preimplantation explantation embryos	FGF, Activin	65
hEG	Human primordial germ cells	Teratoma	Prolonged culture of primordial germ Cells	FGF, LIF, Forskolin	49
hiPS	Human foreskin/adult fibroblast	Teratoma	Forced expression of OCT4, SOX2, c-MYC, KLF4 and/or NANOG, and LIN28	FGF	108, 109

mass, termed embryonal carcinoma (EC) cells, were capable of differentiating into a variety of tissues, even after serial transplantation.[2] Notably, Barry Pierce and colleagues demonstrated that, upon transplantation into adult animals, a single EC cell could form teratocarcinomas containing the same variety of differentiated tissues originating from all three germ layers.[3] These EC cells were the first self-renewing pluripotent stem cells reported. When EC cells are transferred to an in vitro environment, they quickly adapt and are maintained as undifferentiated self-renewing cell lines or can be induced to differentiate.[4,5] In all experiments performed to characterize EC cells, the most important finding is that when EC cells are introduced into blastocyst-stage embryos, the cells can participate in normal embryonic development and contribute to organ and tissue formation in the host.[6-8] The results of these experiments inspired a search for the embryonic counterpart of EC cells in the preimplantation mouse embryo. Indeed, the knowledge gained from manipulating EC cells paved the way for successful derivation of embryonic stem (ES) cells from the inner cell mass (ICM) of the blastocyst-stage embryo.[9,10] ES cell lines were established using EC-conditioned medium and using mouse embryonic fibroblasts (MEFs) as a feeder layer, a common culture approach for propagation of EC cells. ES and EC cells share similar characteristics including morphology and expression of specific markers, and both cell types form teratocarcinomas when introduced into immunodeficient host animals.

Perhaps the most unique feature of ES cells, compared with EC cells, is the ability of ES cells to colonize the chimera at a much higher frequency.[11] Moreover, when ES cells are introduced into a host embryo, the development of which is compromised by tetraploidy, ES cells can autonomously give rise to an entire fetus.[12] One of the key factors in successful derivation of ES cells is the supply of MEF feeder layers for initial embryo outgrowth. It was later realized that leukemia inhibitory factor (Lif) was the key factor secreted by MEFs sustaining ES cell self-renewal.[13,14] The Lif receptor gp130 activates Jak/Stat3 signaling pathways that are likely to mediate the expression of a unique set of target genes that maintain ES cells in the undifferentiated state.[15] In addition, activation of Bmp4 signaling pathways in combination with Lif are sufficient to substitute for exogenous stimuli from serum to maintain the undifferentiated state.[16] Recently, Lif, combined with specific inhibitors for Erk and GSK3 signaling pathways, has been shown to efficiently block differentiation and to enhance the biosynthetic growth of ES cells; this technique thus maintains the self-renewing pluripotent state.[17] In addition to extrinsic factors, genes encoding key pluripotent proteins, namely, *Oct4, Nanog,* and *Sox2,* are crucial for ES cell self-renewal. Reduction in Oct4 levels causes ES cells to differentiate into trophectoderm, whereas overexpression of Oct4 results in formation of primitive endoderm/mesoderm.[18] Genetic ablation of *Nanog* causes embryonic lethality at the blastocyst stage and ES cells can never be established without *Nanog.* Interestingly, overexpression of *Nanog* can sustain ES cell self-renewal in a Lif-independent manner.[19,20] The pluripotent state exhibited by ES cells is characterized by a unique transcriptional and epigenetic signature unique to somatic cells.[21,22] The continuous search for additional key pluripotent factors will likely provide a better picture of how the pluripotent state is maintained at the molecular level.[23–25]

Mouse ES cells are now used as a vehicle to introduce desired mutations and/or modifications by homologous recombination into chosen genomic loci.[26,27] This "knockout" technology has a substantial impact on the advancement of biomedical science. Knockout mice provide models for human disorders[28,29] and have led to several major discoveries and insights into cancer, diabetes, obesity, and neurodegenerative diseases. To date, laboratories worldwide have generated nearly 4000 targeted knockouts of mouse genes[30] and both the scientific community and the general public will likely benefit from this extremely useful technology.

EPIBLAST STEM CELLS

In mice, the inner cell mass (ICM) of the preimplantation embryo matures into the epiblast of the pre- and postimplantation embryo. The epiblast next gives rise to the embryo proper and all embryonic lineages. Pluripotent ES cells derived from the ICM have been extensively studied over the past few decades, and pluripotent stem cells directly isolated from the postimplantation epiblast, termed epiblast stem cells (EpiSCs), have been discovered recently.[31,32] Interestingly, EpiSC colonies are flat, thus markedly different from the three-dimensional, tightly packed colonies of ES cells. EpiSCs are pluripotent, as shown by formation of teratomas containing derivatives of all three germ layers after injection into immunodeficient mice. However, when EpiSCs are injected into blastocyst-stage embryos, they do not participate in normal embryonic development. Moreover, female EpiSCs carry inactivated X chromosomes, and thus differ from female ES cells in which both X chromosomes are reactivated.[33] Although EpiSCs express key pluripotent factors such as those encoded by *Oct4, Nanog,* and *Sox2,* they also synthesize higher levels of markers characteristic of mature epiblasts, namely, proteins encoded by the *Fgf5, Lefty1,* and *Cer1* genes. Importantly, stable propagation of EpiSCs is supported by the Fgf2 and Tgf β/Activin/Nodal signaling pathways; this creates markedly similar conditions as required for maintenance of human ES cell self-renewal, but the cells are refractory to Lif and Jak/Stat3 signaling. These observations support the hypothesis that EpiSCs may be the true mouse counterpart of human ES cells.[34] Recently, it has been demonstrated in vitro that EpiSCs are capable of continuously transitioning to become the precursors of the germ cell lineage, the primordial germ cells (PGCs).[35] Thus EpiSCs functionally resemble their in vivo counterpart, the epiblast, which can be induced to form PGCs in a BMP signaling pathway-dependent manner.[35] Furthermore, if EpiSCs are maintained under culture conditions that favor the propagation of mouse ES cells (Lif and bovine serum), they can be epigenetically reprogrammed into a pluripotent state identical to that of mouse ES cells.[33] It has been suggested that mouse ES cells and EpiSCs represent two distinct but metastable pluripotent states captured from the ICM and the epiblast, respectively.[36,37] Future studies on the epigenetic mechanisms controlling these two distinct and stable pluripotent states will likely provide profound insight into the pluripotent nature of human ES cells, as the behavior of such cells is extremely similar to that of EpiSCs.

EMBRYONIC GERM CELLS AND PLURIPOTENT CELLS FROM THE GERMLINE

The nature of the testicular teratocarcinoma suggested an intimate relationship between the germline and pluripotent stem cells. PGCs, the precursors of eggs and sperm, when cultured with Lif, Fgf2, and stem cell factor (also known as the c-kit ligand), along with an MEF feeder layer, are able to develop into pluripotent stem cells termed embryonic germ (EG) cells.[38–40] This conversion is referred to as "dedifferentiation" or "epigenetic reprogramming." Remarkably, PGCs behave as unipotent germ cell precursors in vivo. Freshly isolated PGCs do not effectively contribute to development if aggregated with morulae or injected into blastocyst-stage embryos.[41] However, once PGCs are converted into EG cells, they promptly reenter the germline of the chimera when injected into host blastocysts.[42,43] ES cells and EG cells are indistinguishable in many respects, including morphological and biochemical features as well as signature gene expression; hence it is of great interest to understand the reprogramming event(s) that allow(s) PGCs to dedifferentiate into EG cells. It has been shown that Fgf2 is required for the initial phase of this conversion, followed by decreased expression of *Blimp1* (also known as *Prdm1*), a key PGC fate determinant. This is a critical step in establishment of EG cells.[44]

Several distinct pluripotent cell lines have been established from the neonate and adult testis, specifically from the spermatogonia. Based on morphological features, multipotent germline stem cells (mGSCs) can be derived under tissue culture conditions that normally support ES cell propagation.[45] Multipotent adult germline stem cells (maGSCs) can be isolated from adult testis using a *Stra8-GFP* reporter, a spermatogonia-specific marker.[46] Alternatively, when the spermatogonia marker *GPR125* is used, multipotent adult spermatogonia-derived stem cells (MASCs) can become established.[47] Finally, using an *Oct4-GFP* reporter line, germline-derived pluripotent stem (gPS) cells have been shown to originate from germline stem cells.[48] Although all cells were derived from the testis, the various cell lines differed in global gene expression pattern, and morphological differences were also evident. The differences among mGSCs, maGSCs, MASCs, and gPS cells may result from the use of different reporters, mouse strains, and tissue culture conditions. Nevertheless, the ability to acquire the pluripotent state in the absence of any exogenous reprogramming factor seems to be unique to the germline. Subsequently, human EG cells were derived from cultured human PGCs.[49] Human EG cells were shown to form EBs and differentiate into derivatives of all three germ layers in vitro. Interestingly, undifferentiated human EG cells are SSEA1-positive, which is a characteristic of mouse ES cells and human PGCs, and not of human ES or human EC cells. Recently, several groups have reported the successful derivation of human ES-like pluripotent cell lines from adult human testis.[50–53] These protocols may provide an alternative source for generating patient-specific pluripotent cells for cell-based transplantation therapy.

PLURIPOTENT STEM CELLS DERIVED FROM HUMAN EMBRYOS

Although the generation of mouse ES cells provides a model for the derivation of similar pluripotent cell lines from other species, it has been difficult to establish such cell lines from nonmouse mammals.[54,55] In fact, germline competent ES cells derived from rat embryos were only recently described, almost three decades after the first derivation of the mouse counterpart.[56,57] In parallel with the discovery of mouse EC cells described above, human EC cells have been derived and characterized.[58] It has been shown that clonal EC cell lines isolated from human teratocarcinomas can form tumors displaying a variety of adult tissue properties when introduced into adult mice, constituting a demonstration of the pluripotent nature of such cells.[59,60] However, distinct antigenic features are presented by human EC cells compared with mouse EC cells. For example, SSEA1 is highly expressed in undifferentiated mouse EC cells, but not in human EC cells.[60,61] Undifferentiated human EC cells express SSEA3, SSEA4, TRA-1-60, and TRA-1-81, all of which are absent from mouse EC cells.[62] Species-specific factors are thought to contribute to the morphological and biochemical differences between mouse and human EC cells, and studies on the latter cells provide essential knowledge and tools for the derivation and identification of human ES cells. The possibility of derivation of human pluripotent ES cells was first suggested following the successful description of primate ES cells.[63,64] As expected, self-renewing rhesus monkey ES cells, like human EC cells, express markers such as SSEA3, SSEA4, TRA-1-60, and TRA-1-80, at similar levels to those seen in human EC cells. Clonal primate ES cells are able to differentiate into derivatives of all three germ layers. Interestingly, primate ES cells can express trophoblast markers upon differentiation, suggesting that the cells might have been isolated from an earlier developmental phase of primate embryos, compared with mouse ES cells.[64] Pluripotent ES cells from the marmoset and the rhesus monkey provide an important new in vitro model for the understanding of early primate embryonic development, which differs markedly from rodent development.[64] Knowledge acquired from work with primate ES cells allowed the successful derivation of human ES cells.[65] These cells are maintained in the undifferentiated state, without compromising the normal karyotype, and are capable of differentiation into derivatives of the three germ layers in vitro.[66] For ethical reasons, pluripotent properties such as chimera formation and germline transmission cannot be tested in humans or primates. However, clonal human ES cell lines can be maintained for long periods and have displayed no obvious chromosomal abnormality, although pluripotency is retained.[66] In contrast to mouse ES cells, human ES cell self-renewal appears to be Lif-independent. Indeed, the components of the LIF-JAK-STAT3 axis are underrepresented in human ES cells.[67] It was later demonstrated that high levels of FGF2 in the culture medium could sustain the undifferentiated state of human ES cells even in the absence of the MEF feeder layer.[68–71] FGF2 and TGF β/Activin/Nodal signaling pathways play central roles in the self-renewal of human ES cells.[72,73] Unlike mouse ES cells, in which Bmp4 signaling pathways seem to play beneficial roles in self-renewal, differentiation processes can be initiated by BMP4 signaling in human ES cells.[74] Therefore fine-tuning of the balance between TGF β/Activin/Nodal and BMP signaling levels can enhance maintenance of the undifferentiated state, even without the MEF feeder layer.[75]

Additional growth factors and pathways are likely to participate in growth enhancement with maintenance of the undifferentiated state of human ES cells. For example, sphingosine-1-phosphate and platelet-derived growth factor,[76] insulin-like growth factor,[77] Rho-associated kinase inhibitor,[78] and inhibition of glycogen synthase kinase—3[79] have all been shown to positively influence human ES cell self-renewal. Thus the complex molecular network that controls the human ES cell pluripotent state is still far from understood. The marked variations between mouse and human ES cells can be attributed to dramatic differences between rodent and primate early embryonic development. For example, the mouse early embryo develops into a cup-shaped egg-cylinder form, but the corresponding stage in the human embryo is the embryonic disk. Nevertheless, human ES cells can serve as a powerful in vitro model for investigating early human embryonic development, which is normally difficult to study in vivo. Furthermore, human ES cells offer an attractive model for study of human diseases, allowing drug screens and guided differentiation to be performed under in vitro culture conditions, with a near-unlimited supply of cells.

PLURIPOTENCY OF HUMAN ES CELLS: IN VITRO DIFFERENTIATION

Strategies for differentiation of human ES cells have been derived from experience with mouse ES cell differentiation protocols. Two major methods have been used to promote human ES cell differentiation. The first is the "embryoid body" (EB) protocol. Human ES cells can be aggregated in suspension and small aggregates will further develop into complex three-dimensional EBs, which (it is often assumed) somewhat resemble a stage in early embryonic development but show less organization. One observation supporting this assumption is that derivatives of all three primary germ layers can be found in mature EBs.[80] The EBs can be dissociated and replated onto gelatinized dishes and, when specific protocols are used, precursors of neurons[81] and cardiomyocytes[82] can be enriched in vitro. In addition, blood vessel-like structures as well as PECAM1-positive endothelial cells are found in human EBs.[83] The second method is simply to induce spontaneous differentiation of human ES cells, as adherent monolayers, using modified culture conditions. In prolonged culture systems, progenitors of neurons expressing the neuronal marker *PAX6* can be identified.[84] If human ES cells are co-cultured with mouse bone marrow-derived cell lines or stromal cell lines, derivatives of hematopoietic systems can be found.[85] Indeed, even the progenitors of the hematopoietic system, CD34-positive cells, can be generated at moderate efficiency.[86] When human ES cells are co-cultured with END-2, a visceral endodermal cell line, they can be stimulated to differentiate into contracting cardiomyocytes.[87] One of the caveats of differentiation procedures is the heterogeneous nature of resulting cell populations, although the desired cell type may indeed be present. From a regenerative medicine perspective, purification of the cell population of interest from cells differentiated from pluripotent human ES cells is essential before such cells can be transplanted into a patient. It is critical to implement procedures that prevent contamination by residual undifferentiated ES cells, which are likely to be tumorigenic.[88] One popular method that has been experimentally successful is to express a reporter or drug-resistance gene under the control of a lineage-specific promoter. By fluorescence-activated cell sorting (FACS) or use of the appropriate drug, the population of interest can be purified. For example, differentiation of human ES cells carrying a *SOX1* promoter-driven *GFP* reporter transgene results in GFP expression in neuronal precursors.[89] Using FACS followed by grafting of purified cells into mouse brain, the majority of SOX1 (GFP-positive) donor cells became incorporated into the recipient brain and differentiated into neurons.[89]

Recently, purification of neuron precursors differentiated from pluripotent stem cells using SSEA1-negative cell sorting reduced the incidence of teratoma formation in recipient animals.[88] Small molecule inhibitors have been shown to enhance lineage specification. For example, human PAX6-positive neuronal precursors were efficiently produced in the presence of a combination of inhibitors blocking the TGF β/ACTIVIN/NODAL signaling pathways.[90]

SOMATIC REPROGRAMMING AND INDUCED PLURIPOTENT STEM CELLS: OVERCOMING EPIGENETIC BARRIERS

The development of nuclear transfer (NT) represents an elegant approach to address the developmental potency of a given cell genome.[91] Pioneering experiments used the frog as a model system and showed that the nucleus from an adult frog cell, injected into an enucleated egg, enabled the cloned frog to grow to full term, and such animals were fertile.[92,93] This method exploited the possibility that "nuclear reprogramming" would occur in the presence of trans-acting factors in the egg, and unequivocally demonstrated that the genome of any given cell of an organism is essentially identical, regardless of source or differentiation state.[94–97] Importantly, reprogrammable genomes were obtained even from terminally differentiated cell types, indicating that the differentiation process is a series of reversible epigenetic modifications and not a permanent change. The birth of Dolly the sheep showed that the same principle could be applied to mammals.[98] Given that laboratory mice have a relative short lifespan and are amenable to embryological and genetic manipulation, cloned mice were next obtained.[99] Subsequently, mouse ES cell lines derived from cloned embryos were generated.[100] Following the successful derivation of human ES cells,[65] it was soon suggested that therapeutic cloning, generating patient-specific pluripotent stem cell lines, might be possible; these cells could be differentiated into cell types of interest for cell-based transplantation. Indeed, primate ES cells can be generated from nuclear-transferred embryos, but at relatively low efficiency.[101] Although NT is an elegant method of nuclear reprogramming, this highly specialized method remains technically challenging, with a low success rate even when performed by experienced scientists. The major drawback is the requirement for many healthy donor eggs to ensure successful NT, which makes the technique unsuitable for most clinical purposes. Nevertheless, the proof-of-concept gained from NT experiments led to attempts to identify the trans-acting factors responsible for nuclear reprogramming. Concepts derived from earlier knowledge paved the way for the next seminal studies. Perhaps the best illustration is the discovery of iPS cells.[102] Shinya Yamanaka and his team devised a method to reprogram mouse fibroblasts by introducing a pool of 24 transcription factors that were known to be important in the pluripotent state. By gradually omitting factors one-by-one, the group narrowed the requisite proteins to only four transcription factors that were sufficient for reprogramming. These were *Oct4, Sox2, Klf4*, and *c-Myc*. In this set of experiments, a crucial strategy was the use of the *fbx15* promoter, an important but dispensable gene in the pluripotent state, as a selection marker for reprogramming. The iPS cells generated in this manner were morphologically similar to ES cells but were thought to be incompletely reprogrammed because, when injected into blastocysts, the cells did not contribute to chimera formation.[102] Remarkably, when the drug selection promoter *fbx15* was substituted by *Nanog* or *Oct4* promoters, the activities of which are essential in the pluripotent state, germline-competent iPS cells were generated.[103,104] Subsequently, mouse iPS cells passed tetraploid complementation tests, the most stringent examination of pluripotency, and were thus shown to be functionally equivalent to mouse ES cells.[105–107] Using similar combinations of reprogramming factors, either *OCT4, SOX2, c-MYC*, and *KLF4*, or *OCT4, SOX2, NANOG*, and *LIN28*, pluripotent human iPS cells were established from adult or neonate human fibroblasts.[108,109] Regardless of the factor combination used, human iPS cells seem to be very similar to human ES cells. The cells are morphologically indistinguishable: human iPS cells express markers hitherto thought specific to human ES cells; the iPS cells can differentiate into derivatives of the three germ layers in vitro; and the iPS cells can form teratomas when injected into immunodeficient mice. However, subtle differences exist among the expression patterns of both coding and noncoding transcripts.[110] Hence it is important to keep in mind that embryo-derived pluripotent stem cells remain the gold standard for pluripotency assessment and that more work is required to learn how to generate "perfect" iPS cells for use in regenerative medicine.[111] One concern when using iPS cells for clinical purposes is the tumorigenic nature of such cells. In fact, in one report on construction of germline-competent mouse iPS cells, approximately 20% of chimeric mice suffered from tumors.[103] This was presumably due to reactivation of the exogenous oncogene *c-Myc* that was integrated into the host genome during reprogramming. Modified protocols allow iPS cells to be generated without inclusion of *c-Myc* in the reprogramming factor pool, although reprogramming efficiency decreases markedly.[112] The use of retrovirus to deliver reprogramming factors is another obvious problem to apply pluripotent iPS cells for cell-based transplantation. Retroviral vectors tend to integrate into the host genome, producing undesired and irreversible genomic modifications.[113] In fact, 1–30 proviral integration sites can be detected in each iPS clone.[102,114] Because generation of iPS cells appears to be independent of such integration events, modified procedures have been developed to bypass the viral infection step. These protocols include plasmid-mediated transient transfection,[115,116] direct delivery of recombinant proteins,[117] piggyBac transposon-mediated expression of reprogramming factors,[118] and use of nonintegrating episomal vectors.[119] It should be noted that reprogramming factors delivered by piggyBac transposons and episomal vectors can be removed once pluripotent cell lines become established; thus the host genome will be intact.

THE PROMISE OF PLURIPOTENT STEM CELLS FOR REGENERATIVE MEDICINE

Eleven years after the first human ES cell lines were derived,[65] the first test of such cells in a Phase I human clinical trial was approved by the United States Food and Drug Administration. The Geron Corporation and collaborators developed human ES cell-derived oligodendrocytes, with the objective of treating patients suffering from spinal cord injury. Oligodendrocytes are known to produce neurotrophic factors and to assist in the formation of myelin; both features are essential for neuronal regrowth after injury. Human ES cell-derived oligodendrocytes became successfully engrafted into lesion sites in a rat injury model. In addition to incorporated donor oligodendrocytes, myelinated rat axons bridging lesions were found. Rats so treated also showed improved locomotive behavior.[120]

Perhaps the most attractive source of pluripotent stem cells is from tissue biopsies of individual patients. Proof-of-principle studies in mice or rats have already yielded encouraging results. For example, in a humanized sickle-cell anemia mouse model, symptoms can be rescued after autologous transplantation of genetically corrected iPS cells.[121] When mouse iPS cells were differentiated into neural precursors, followed by transplantation into a rat Parkinson disease model, the donor cells were able to form dopamine-synthesizing neurons in vivo. In successfully engrafted animals, the disease phenotype significantly improved.[88] Differentiation of wild-type mouse iPS cells into endothelial/endothelial progenitor cells followed by transplantation into the livers of *hemophilia A* mutant animals, which normally would not survive a bleeding challenge, resulted in eventual recovery. iPS-derived donor cells showed long-term functional engraftment in animal livers.[122] Further technical challenges may be expected when translating these therapeutic strategies into treatments for human diseases. Nevertheless, generation of disease-specific iPS cells is currently underway.[123–127] If a specific locus is known to be responsible for a disease, such as Gaucher disease type III (*GBA*), Huntington disease (*HUNTINGTIN*), Lesch–Nyhan syndrome (*HPRT*),[126] or Fanconi anemia,[127] genetic correction (so-called gene targeting) can be used to revert the mutation to the wild-type before transplantation. Gene targeting is achieved by homologous recombination in human pluripotent ES cells.[128] However, the proportion of correct homologous recombination events remains as low as $1/10^6$ in human ES cells and no successful gene targeting was reported in human iPS cells until recently.[129] An engineered gene-targeting vector expressing a zinc finger nuclease increased the rate of homologous recombination by almost 200-fold in human ES and iPS cells.[129] This technique will likely facilitate development of future therapeutic gene targeting to correct mutations in patient-specific iPS cells. However, if the cause of disease is unclear, or if many factors are likely to be involved, for example, in Parkinson disease, juvenile diabetes mellitus, or Becker muscular dystrophy,[126] disease-specific iPS cells can nonetheless serve as a platform for small-molecule screening in the development of novel drugs. Recently, a drug candidate termed "kinetin," developed for treatment of familial dysautonomia (FD), has been shown to reverse mis-spliced transcription of the *IKBKAP* gene in FD-iPS-derived neural crest cells.[125] Overall, use of disease-specific iPS cells is likely to result in significant insights into the molecular mechanisms of human disorders, as well as providing a new platform for drug discovery.[124,125] With improvements in homologous recombination efficiency, correction of patient-specific mutation(s) will be a crucial step in regenerative medicine.

FUTURE PROSPECTIVE

Pluripotency is a unique state that can be captured under in vitro conditions, but is thought to be transient in vivo. However, we are far from having a complete understanding as to how the pluripotent state is established and maintained. Experiments seeking to delineate a precise relationship between ES cells and early embryos should yield new insights into how ES cells are produced.[34,130–132] It is known that a few transcription factors function as master regulators of the pluripotent state; these include *Oct4, Nanog,* and *Sox2.* An unexpected finding is that *Nanog,* when genetically ablated, is not required for maintenance of the pluripotent state, indicating that the establishment and maintenance of ES cells may involve at least partially distinct molecular pathways.[133,134] Moreover, detailed studies have shown that pluripotent stem cells exist in a dynamic metastable state.[36,37] An investigation of molecular networks active in the pluripotent state eventually led to the discovery of iPS cells.[102] With forced expression of a few master regulators even the most terminally differentiated cells can be reprogrammed into pluripotent stem cells. One intriguing question is whether dedifferentiation to the pluripotent state is an absolute requirement for conversion of one cell type into another. At least one recent study demonstrated that a shorter route is feasible, demonstrating in vivo reprogramming of pancreatic exocrine cells into beta cells.[135] It will be of great interest to determine whether the same principle can be applied to other somatic cells. If cell fates can be directly interchanged in vivo by manipulating the levels of tissue-specific master regulators, transplantation-free therapy may become possible.

REFERENCES

1. Stevens LC, Little CC. Spontaneous testicular teratomas in an inbred strain of mice. *Proc Natl Acad Sci USA*. 1954;40:1080–1087.

2. Stevens LC. Experimental production of testicular teratomas in mice. *Proc Natl Acad Sci USA*. 1964;52:654–661.

3. Kleinsmith LJ, Pierce GB Jr. Multipotentiality of single embryonal carcinoma cells. *Cancer Res*. 1964;24:1544–1551.

4. Kahan BW, Ephrussi B. Developmental potentialities of clonal in vitro cultures of mouse testicular teratoma. *J Natl Cancer Inst*. 1970;44:1015–1036.

5. Martin GR, Evans MJ. Differentiation of clonal lines of teratocarcinoma cells: formation of embryoid bodies in vitro. *Proc Natl Acad Sci USA*. 1975;72:1441–1445.

6. Brinster RL. The effect of cells transferred into the mouse blastocyst on subsequent development. *J Exp Med*. 1974;140:1049–1056.

7. Mintz B, Illmensee K. Normal genetically mosaic mice produced from malignant teratocarcinoma cells. *Proc Natl Acad Sci USA*. 1975;72:3585–3589.

8. Papaioannou VE, McBurney MW, Gardner RL, Evans MJ. Fate of teratocarcinoma cells injected into early mouse embryos. *Nature*. 1975;258:70–73.

9. Evans MJ, Kaufman MH. Establishment in culture of pluripotential cells from mouse embryos. *Nature*. 1981;292:154–156.

10. Martin GR. Isolation of a pluripotent cell line from early mouse embryos cultured in medium conditioned by teratocarcinoma stem cells. *Proc Natl Acad Sci USA*. 1981;78:7634–7638.

11. Bradley A, Evans M, Kaufman MH, Robertson E. Formation of germ-line chimaeras from embryo-derived teratocarcinoma cell lines. *Nature*. 1984;309:255–256.

12. Nagy A, Gocza E, Diaz EM, et al. Embryonic stem cells alone are able to support fetal development in the mouse. *Development*. 1990;110:815–821.

13. Smith AG, Heath JK, Donaldson DD, et al. Inhibition of pluripotential embryonic stem cell differentiation by purified polypeptides. *Nature*. 1988;336:688–690.

14. Williams RL, Hilton DJ, Pease S, et al. Myeloid leukaemia inhibitory factor maintains the developmental potential of embryonic stem cells. *Nature*. 1988;336:684–687.

15. Matsuda T, Nakamura T, Nakao K, et al. STAT3 activation is sufficient to maintain an undifferentiated state of mouse embryonic stem cells. *EMBO J*. 1999;18:4261–4269.

16. Ying QL, Nichols J, Chambers I, Smith A. BMP induction of Id proteins suppresses differentiation and sustains embryonic stem cell self-renewal in collaboration with STAT3. *Cell*. 2003;115:281–292.

17. Ying QL, Wray J, Nichols J, et al. The ground state of embryonic stem cell self-renewal. *Nature*. 2008;453:519–523.

18. Niwa H, Miyazaki J, Smith AG. Quantitative expression of Oct-3/4 defines differentiation, dedifferentiation or self-renewal of ES cells. *Nat Genet*. 2000;24:372–376.

19. Chambers I, Colby D, Robertson M, et al. Functional expression cloning of Nanog, a pluripotency sustaining factor in embryonic stem cells. *Cell*. 2003;113:643–655.

20. Mitsui K, Tokuzawa Y, Itoh H, et al. The homeoprotein Nanog is required for maintenance of pluripotency in mouse epiblast and ES cells. *Cell*. 2003;113:631–642.

21. Boyer LA, Plath K, Zeitlinger J, et al. Polycomb complexes repress developmental regulators in murine embryonic stem cells. *Nature*. 2006;441:349–353.

22. Chen X, Xu H, Yuan P, et al. Integration of external signaling pathways with the core transcriptional network in embryonic stem cells. *Cell*. 2008;133:1106–1117.

23. Dejosez M, Krumenacker JS, Zitur LJ, et al. Ronin is essential for embryogenesis and the pluripotency of mouse embryonic stem cells. *Cell*. 2008;133:1162–1174.

24. Lim LS, Loh YH, Zhang W, et al. Zic3 is required for maintenance of pluripotency in embryonic stem cells. *Mol Biol Cell*. 2007;18:1348–1358.

25. Zhang J, Tam WL, Tong GQ, et al. Sall4 modulates embryonic stem cell pluripotency and early embryonic development by the transcriptional regulation of Pou5f1. *Nat Cell Biol*. 2006;8:1114–1123.

26. Doetschman T, Gregg RG, Maeda N, et al. Targetted correction of a mutant HPRT gene in mouse embryonic stem cells. *Nature*. 1987;330:576–578.

27. Thomas KR, Capecchi MR. Site-directed mutagenesis by gene targeting in mouse embryo-derived stem cells. *Cell*. 1987;51:503–512.

28. Hooper M, Hardy K, Handyside A, Hunter S, Monk M. HPRT-deficient (Lesch–Nyhan) mouse embryos derived from germline colonization by cultured cells. *Nature*. 1987;326:292–295.

29. Kuehn MR, Bradley A, Robertson EJ, Evans MJ. A potential animal model for Lesch–Nyhan syndrome through introduction of HPRT mutations into mice. *Nature*. 1987;326:295–298.

30. Collins FS, Rossant J, Wurst W. A mouse for all reasons. *Cell*. 2007;128:9–13.

31. Brons IG, Smithers LE, Trotter MW, et al. Derivation of pluripotent epiblast stem cells from mammalian embryos. *Nature*. 2007;448:191–195.

32. Tesar PJ, Chenoweth JG, Brook FA, et al. New cell lines from mouse epiblast share defining features with human embryonic stem cells. *Nature*. 2007;448:196–199.

33. Bao S, Tang F, Li X, et al. Epigenetic reversion of post-implantation epiblast to pluripotent embryonic stem cells. *Nature*. 2009;461(7868):1292–1295.

34. Rossant J. Stem cells and early lineage development. *Cell*. 2008;132:527–531.

35. Hayashi K, Surani MA. Self-renewing epiblast stem cells exhibit continual delineation of germ cells with epigenetic reprogramming in vitro. *Development*. 2009;136(21):3549–3556.

36. Hanna J, Markoulaki S, Mitalipova M, et al. Metastable pluripotent states in NOD-mouse-derived ESCs. *Cell Stem Cell*. 2009;4:513–524.

37. Hayashi K, Lopes SM, Tang F, Surani MA. Dynamic equilibrium and heterogeneity of mouse pluripotent stem cells with distinct functional and epigenetic states. *Cell Stem Cell*. 2008;3:391–401.

38. Durcova-Hills G, Ainscough J, McLaren A. Pluripotential stem cells derived from migrating primordial germ cells. *Differentiation*. 2006;68:220–226.

39. Matsui Y, Zsebo K, Hogan BL. Derivation of pluripotential embryonic stem cells from murine primordial germ cells in culture. *Cell*. 1992;70:841–847.

40. Resnick JL, Bixler LS, Cheng L, Donovan PJ. Long-term proliferation of mouse primordial germ cells in culture. *Nature*. 1992;359:550–551.

41. Durcova-Hills G, Adams IR, Barton SC, Surani MA, McLaren A. The role of exogenous fibroblast growth factor-2 on the reprogramming of primordial germ cells into pluripotent stem cells. *Stem Cells*. 2006;24:1441–1449.

42. Labosky PA, Barlow DP, Hogan BL. Mouse embryonic germ (EG) cell lines: transmission through the germline and differences in the methylation imprint of insulin-like growth factor 2 receptor (Igf2r) gene compared with embryonic stem (ES) cell lines. *Development*. 1994;120:3197–3204.

43. Stewart CL, Gadi I, Bhatt H. Stem cells from primordial germ cells can reenter the germ line. *Dev Biol*. 1994;161:626–628.

44. Durcova-Hills G, Tang F, Doody G, Tooze R, Surani MA. Reprogramming primordial germ cells into pluripotent stem cells. *PLoS One*. 2008;3:e3531.

45. Kanatsu-Shinohara M, Inoue K, Lee J, et al. Generation of pluripotent stem cells from neonatal mouse testis. *Cell*. 2004;119:1001–1012.

46. Guan K, Nayernia K, Maier LS, et al. Pluripotency of spermatogonial stem cells from adult mouse testis. *Nature*. 2006;440:1199–1203.

47. Seandel M, James D, Shmelkov SV, et al. Generation of functional multipotent adult stem cells from GPR125+ germline progenitors. *Nature*. 2007;449:346–350.

48. Ko K, Tapia N, Wu G, et al. Induction of pluripotency in adult unipotent germline stem cells. *Cell Stem Cell*. 2009;5:87–96.

49. Shamblott MJ, Axelman J, Wang S, et al. Derivation of pluripotent stem cells from cultured human primordial germ cells. *Proc Natl Acad Sci USA*. 1998;95:13726–13731.

50. Conrad S, Renninger M, Hennenlotter J, et al. Generation of pluripotent stem cells from adult human testis. *Nature*. 2008;456:344–349.

51. Golestaneh N, Kokkinaki M, Pant D, et al. Pluripotent stem cells derived from adult human testes. *Stem Cells Dev*. 2009;18(80):1115–1126.

52. Kossack N, Meneses J, Shefi S, et al. Isolation and characterization of pluripotent human spermatogonial stem cell-derived cells. *Stem Cells*. 2009;27:138–149.

53. Mizrak SC, Chikhovskaya JV, Sadri-Ardekani H, et al. Embryonic stem cell-like cells derived from adult human testis. *Hum Reprod*. 2009;25(1):158–161.

54. Brook FA, Gardner RL. The origin and efficient derivation of embryonic stem cells in the mouse. *Proc Natl Acad Sci USA*. 1997;94:5709–5712.

55. Gardner RL, Brook FA. Reflections on the biology of embryonic stem (ES) cells. *Int J Dev Biol*. 1997;41:235–243.

56. Buehr M, Meek S, Blair K, et al. Capture of authentic embryonic stem cells from rat blastocysts. *Cell*. 2008;135:1287–1298.

57. Li P, Tong C, Mehrian-Shai R, et al. Germline competent embryonic stem cells derived from rat blastocysts. *Cell*. 2008;135:1299–1310.

58. Hogan B, Fellous M, Avner P, Jacob F. Isolation of a human teratoma cell line which expresses F9 antigen. *Nature*. 1977;270:515–518.

59. Andrews PW, Damjanov I, Simon D, et al. Pluripotent embryonal carcinoma clones derived from the human teratocarcinoma cell line Tera-2. Differentiation in vivo and in vitro. *Lab Invest*. 1984;50:147–162.

60. Andrews PW, Goodfellow PN, Shevinsky LH, Bronson DL, Knowles BB. Cell-surface antigens of a clonal human embryonal carcinoma cell line: morphological and antigenic differentiation in culture. *Int J Cancer*. 1982;29:523–531.

61. Solter D, Knowles BB. Monoclonal antibody defining a stage-specific mouse embryonic antigen (SSEA-1). *Proc Natl Acad Sci USA*. 1978;75:5565–5569.

62. Andrews PW, Banting G, Damjanov I, Arnaud D, Avner P. Three monoclonal antibodies defining distinct differentiation antigens associated with different high molecular weight polypeptides on the surface of human embryonal carcinoma cells. *Hybridoma*. 1984;3:347–361.

63. Thomson JA, Kalishman J, Golos TG, et al. Isolation of a primate embryonic stem cell line. *Proc Natl Acad Sci USA*. 1995;92:7844–7848.

64. Thomson JA, Kalishman J, Golos TG, Durning M, Harris CP, Hearn JP. Pluripotent cell lines derived from common marmoset (*Callithrix jacchus*) blastocysts. *Biol Reprod*. 1996;55:254–259.

65. Thomson JA, Itskovitz-Eldor J, Shapiro SS, et al. Embryonic stem cell lines derived from human blastocysts. *Science*. 1998;282:1145–1147.

66. Amit M, Carpenter MK, Inokuma MS, et al. Clonally derived human embryonic stem cell lines maintain pluripotency and proliferative potential for prolonged periods of culture. *Dev Biol*. 2000;227:271–278.

67. Brandenberger R, Wei H, Zhang S, et al. Transcriptome characterization elucidates signaling networks that control human ES cell growth and differentiation. *Nat Biotechnol*. 2004;22:707–716.

68. Klimanskaya I, Chung Y, Meisner L, Johnson J, West MD, Lanza R. Human embryonic stem cells derived without feeder cells. *Lancet*. 2005;365:1636–1641.

69. Wang L, Li L, Menendez P, Cerdan C, Bhatia M. Human embryonic stem cells maintained in the absence of mouse

embryonic fibroblasts or conditioned media are capable of hematopoietic development. *Blood*. 2005;105:4598–4603.

70. Xu C, Rosler E, Jiang J, et al. Basic fibroblast growth factor supports undifferentiated human embryonic stem cell growth without conditioned medium. *Stem Cells*. 2005;23:315–323.

71. Xu RH, Peck RM, Li DS, Feng X, Ludwig T, Thomson JA. Basic FGF and suppression of BMP signaling sustain undifferentiated proliferation of human ES cells. *Nat Methods*. 2005;2:185–190.

72. Levenstein ME, Ludwig TE, Xu RH, et al. Basic fibroblast growth factor support of human embryonic stem cell self-renewal. *Stem Cells*. 2006;24:568–574.

73. Wang G, Zhang H, Zhao Y, et al. Noggin and bFGF cooperate to maintain the pluripotency of human embryonic stem cells in the absence of feeder layers. *Biochem Biophys Res Commun*. 2005;330:934–942.

74. Xu RH, Chen X, Li DS, et al. BMP4 initiates human embryonic stem cell differentiation to trophoblast. *Nat Biotechnol*. 2002;20:1261–1264.

75. Ludwig TE, Levenstein ME, Jones JM, et al. Derivation of human embryonic stem cells in defined conditions. *Nat Biotechnol*. 2006;24:185–187.

76. Pebay A, Wong RC, Pitson SM, et al. Essential roles of sphingosine-1-phosphate and platelet-derived growth factor in the maintenance of human embryonic stem cells. *Stem Cells*. 2005;23:1541–1548.

77. Bendall SC, Stewart MH, Menendez P, et al. IGF and FGF cooperatively establish the regulatory stem cell niche of pluripotent human cells in vitro. *Nature*. 2007;448:1015–1021.

78. Watanabe K, Ueno M, Kamiya D, et al. A ROCK inhibitor permits survival of dissociated human embryonic stem cells. *Nat Biotechnol*. 2007;25:681–686.

79. Sato N, Meijer L, Skaltsounis L, Greengard P, Brivanlou AH. Maintenance of pluripotency in human and mouse embryonic stem cells through activation of Wnt signaling by a pharmacological GSK-3-specific inhibitor. *Nat Med*. 2004;10:55–63.

80. Itskovitz-Eldor J, Schuldiner M, Karsenti D, et al. Differentiation of human embryonic stem cells into embryoid bodies compromising the three embryonic germ layers. *Mol Med*. 2000;6:88–95.

81. Zhang SC, Wernig M, Duncan ID, Brustle O, Thomson JA. In Vitro differentiation of transplantable neural precursors from human embryonic stem cells. *Nat Biotechnol*. 2001;19:1129–1133.

82. Kehat I, Kenyagin-Karsenti D, Snir M, et al. Human embryonic stem cells can differentiate into myocytes with structural and functional properties of cardiomyocytes. *J Clin Invest*. 2001;108:407–414.

83. Levenberg S, Golub JS, Amit M, Itskovitz-Eldor J, Langer R. Endothelial cells derived from human embryonic stem cells. *Proc Natl Acad Sci USA*. 2002;99:4391–4396.

84. Reubinoff BE, Itsykson P, Turetsky T, et al. Neural progenitors from human embryonic stem cells. *Nat Biotechnol*. 2001;19:1134–1140.

85. Kaufman DS, Hanson ET, Lewis RL, Auerbach R, Thomson JA. Hematopoietic colony-forming cells derived from human embryonic stem cells. *Proc Natl Acad Sci USA*. 2001;98:10716–10721.

86. Vodyanik MA, Bork JA, Thomson JA, Slukvin II Human embryonic stem cell-derived CD34$^+$ cells: efficient production in the coculture with OP9 stromal cells and analysis of lymphohematopoietic potential. *Blood*. 2005;105:617–626.

87. Mummery C, Ward D, van den Brink CE, et al. Cardiomyocyte differentiation of mouse and human embryonic stem cells. *J Anat*. 2002;200:233–242.

88. Wernig M, Zhao JP, Pruszak J, et al. Neurons derived from reprogrammed fibroblasts functionally integrate into the fetal brain and improve symptoms of rats with Parkinson's disease. *Proc Natl Acad Sci USA*. 2008;105:5856–5861.

89. Fukuda H, Takahashi J, Watanabe K, et al. Fluorescence-activated cell sorting-based purification of embryonic stem cell-derived neural precursors averts tumor formation after transplantation. *Stem Cells*. 2006;24:763–771.

90. Chambers SM, Fasano CA, Papapetrou EP, Tomishima M, Sadelain M, Studer L. Highly efficient neural conversion of human ES and iPS cells by dual inhibition of SMAD signaling. *Nat Biotechnol*. 2009;27:275–280.

91. Briggs R, King TJ. Transplantation of living nuclei from blastula cells into enucleated frogs' eggs. *Proc Natl Acad Sci USA*. 1952;38:455–463.

92. Gurdon JB. The developmental capacity of nuclei taken from intestinal epithelium cells of feeding tadpoles. *J Embryol Exp Morphol*. 1962;10:622–640.

93. Gurdon JB, Laskey RA, Reeves OR. The developmental capacity of nuclei transplanted from keratinized skin cells of adult frogs. *J Embryol Exp Morphol*. 1975;34:93–112.

94. Eggan K, Baldwin K, Tackett M, et al. Mice cloned from olfactory sensory neurons. *Nature*. 2004;428:44–49.

95. Hochedlinger K, Jaenisch R. Monoclonal mice generated by nuclear transfer from mature B and T donor cells. *Nature*. 2002;415:1035–1038.

96. Inoue K, Wakao H, Ogonuki N, et al. Generation of cloned mice by direct nuclear transfer from natural killer T cells. *Curr Biol*. 2005;15:1114–1118.

97. Li J, Ishii T, Feinstein P, Mombaerts P. Odorant receptor gene choice is reset by nuclear transfer from mouse olfactory sensory neurons. *Nature*. 2004;428:393–399.

98. Wilmut I, Schnieke AE, McWhir J, Kind AJ, Campbell KH. Viable offspring derived from fetal and adult mammalian cells. *Nature*. 1997;385:810–813.

99. Wakayama T, Perry AC, Zuccotti M, Johnson KR, Yanagimachi R. Full-term development of mice from enucleated oocytes injected with cumulus cell nuclei. *Nature*. 1998;394:369–374.

100. Kawase E, Yamazaki Y, Yagi T, Yanagimachi R, Pedersen RA. Mouse embryonic stem (ES) cell lines established from neuronal cell-derived cloned blastocysts. *Genesis*. 2000;28:156–163.

101. Byrne JA, Pedersen DA, Clepper LL, et al. Producing primate embryonic stem cells by somatic cell nuclear transfer. *Nature*. 2007;450:497–502.

102. Takahashi K, Yamanaka S. Induction of pluripotent stem cells from mouse embryonic and adult fibroblast cultures by defined factors. *Cell*. 2006;126:663–676.

103. Okita K, Ichisaka T, Yamanaka S. Generation of germline-competent induced pluripotent stem cells. *Nature*. 2007;448:313–317.

104. Wernig M, Meissner A, Foreman R, et al. In Vitro reprogramming of fibroblasts into a pluripotent ES-cell-like state. *Nature*. 2007;448:318–324.

105. Boland MJ, Hazen JL, Nazor KL, et al. Adult mice generated from induced pluripotent stem cells. *Nature*. 2009;461:91–94.

106. Kang L, Wang J, Zhang Y, Kou Z, Gao S. iPS cells can support full-term development of tetraploid blastocyst-complemented embryos. *Cell Stem Cell*. 2009;5:135–138.

107. Zhao XY, Li W, Lv Z, et al. iPS cells produce viable mice through tetraploid complementation. *Nature*. 2009;461:86–90.

108. Takahashi K, Tanabe K, Ohnuki M, Induction of pluripotent stem cells from adult human fibroblasts by defined factors. *Cell*. 2007;131:861–872.

109. Yu J, Vodyanik MA, Smuga-Otto K, et al. Induced pluripotent stem cell lines derived from human somatic cells. *Science*. 2007;318:1917–1920.

110. Chin MH, Mason MJ, Xie W, et al. Induced pluripotent stem cells and embryonic stem cells are distinguished by gene expression signatures. *Cell Stem Cell*. 2009;5:111–123.

111. Hyun I, Hochedlinger K, Jaenisch R, Yamanaka S. New advances in iPS cell research do not obviate the need for human embryonic stem cells. *Cell Stem Cell*. 2007;1:367–368.

112. Nakagawa M, Koyanagi M, Tanabe K, et al. Generation of induced pluripotent stem cells without Myc from mouse and human fibroblasts. *Nat Biotechnol*. 2008;26:101–106.

113. Jaenisch R, Fan H, Croker B. Infection of preimplantation mouse embryos and of newborn mice with leukemia virus: tissue distribution of viral DNA and RNA and leukemogenesis in the adult animal. *Proc Natl Acad Sci USA*. 1975;72:4008–4012.

114. Aoi T, Yae K, Nakagawa M, et al. Generation of pluripotent stem cells from adult mouse liver and stomach cells. *Science*. 2008;321:699–702.

115. Kaji K, Norrby K, Paca A, Mileikovsky M, Mohseni P, Woltjen K. Virus-free induction of pluripotency and subsequent excision of reprogramming factors. *Nature*. 2009;458:771–775.

116. Okita K, Nakagawa M, Hyenjong H, Ichisaka T, Yamanaka S. Generation of mouse induced pluripotent stem cells without viral vectors. *Science*. 2008;322:949–953.

117. Zhou H, Wu S, Joo JY, et al. Generation of induced pluripotent stem cells using recombinant proteins. *Cell Stem Cell*. 2009;4:381–384.

118. Woltjen K, Michael IP, Mohseni P, et al. piggyBac transposition reprograms fibroblasts to induced pluripotent stem cells. *Nature*. 2009;458:766–770.

119. Yu J, Hu K, Smuga-Otto K, et al. Human induced pluripotent stem cells free of vector and transgene sequences. *Science*. 2009;324:797–801.

120. Keirstead HS, Nistor G, Bernal G, et al. Human embryonic stem cell-derived oligodendrocyte progenitor cell transplants remyelinate and restore locomotion after spinal cord injury. *J Neurosci*. 2005;25:4694–4705.

121. Hanna J, Wernig M, Markoulaki S, et al. Treatment of sickle cell anemia mouse model with iPS cells generated from autologous skin. *Science*. 2007;318:1920–1923.

122. Xu D, Alipio Z, Fink LM, et al. Phenotypic correction of murine hemophilia A using an iPS cell-based therapy. *Proc Natl Acad Sci USA*. 2009;106:808–813.

123. Dimos JT, Rodolfa KT, Niakan KK, et al. Induced pluripotent stem cells generated from patients with ALS can be differentiated into motor neurons. *Science*. 2008;321:1218–1221.

124. Ebert AD, Yu J, Rose FF Jr, et al. Induced pluripotent stem cells from a spinal muscular atrophy patient. *Nature*. 2009;457:277–280.

125. Lee G, Papapetrou EP, Kim H, et al. Modelling pathogenesis and treatment of familial dysautonomia using patient-specific iPSCs. *Nature*. 2009;461:402–406.

126. Park IH, Arora N, Huo H, et al. Disease-specific induced pluripotent stem cells. *Cell*. 2008;134:877–886.

127. Raya A, Rodriguez-Piza I, Guenechea G, et al. Disease-corrected haematopoietic progenitors from Fanconi anaemia induced pluripotent stem cells. *Nature*. 2009;460:53–59.

128. Zwaka TP, Thomson JA. Homologous recombination in human embryonic stem cells. *Nat Biotechnol*. 2003;21:319–321.

129. Zou J, Maeder ML, Mali P, et al. Gene targeting of a disease-related gene in human induced pluripotent stem and embryonic stem cells. *Cell Stem Cell*. 2009;5:97–110.

130. Evans M, Hunter S. Source and nature of embryonic stem cells. *C R Biol*. 2002;325:1003–1007.

131. Nichols J, Silva J, Roode M, Smith A. Suppression of Erk signalling promotes ground state pluripotency in the mouse embryo. *Development*. 2009;136:3215–3222.

132. Zwaka TP, Thomson JA. A germ cell origin of embryonic stem cells? *Development*. 2005;132:227–233.

133. Chambers I, Silva J, Colby D, et al. Nanog safeguards pluripotency and mediates germline development. *Nature*. 2007;450:1230–1234.

134. Silva J, Nichols J, Theunissen TW, et al. Nanog is the gateway to the pluripotent ground state. *Cell*. 2009;138:722–737.

135. Zhou Q, Brown J, Kanarek A, Rajagopal J, Melton DA. In Vivo reprogramming of adult pancreatic exocrine cells to beta-cells. *Nature*. 2008;455:627–632.

HUMAN PLURIPOTENT CELLS FOR REGENERATIVE MEDICINE: POTENTIAL APPLICATIONS FOR REGENERATIVE MEDICINE

27

Christopher C. Ford and Darrell N. Kotton

INTRODUCTION

Many fields of biomedical research have focused on the long-term goals of either maintaining an organism's homeostasis despite advancing age, or repairing tissues damaged by disease or injury. These diverse research areas have more recently been cumulatively referred to as the field of regenerative medicine, a vast research area that increasingly includes the fields of stem cell biology, tissue engineering, developmental biology, and gene therapy. The unique capacities of stem cells to indefinitely self-renew and differentiate have long suggested their potential to advance regenerative medicine. Pluripotent stem cells, in particular, offer unprecedented opportunities to model embryonic development, or to be employed in vivo to reconstitute tissues damaged by disease, inherited mutations, or advancing age.

This chapter will review potential applications of human pluripotent stem cells for regenerative therapies. It should be noted, however, that regenerative approaches to treating disease do not necessarily require stem cells. Indeed, alternative approaches focus on utilizing differentiated progenitors or more mature cells based on the observation that some differentiated, functional cell types are able to proliferate and survive when transplanted in vivo. However, the most notable and robust regenerative medicine application in clinical medicine to date employs stem cells in the procedure commonly referred to as "bone marrow transplantation" (BMT) or peripheral stem cell transplantation (PSCT). BMT or PSCT utilizes a multipotent hematopoietic stem cell within the bone marrow or circulating blood for transplantation in order to indefinitely reconstitute all blood lineages in human or animal recipients.[1-4] Hematopoietic stem cell transplantation emphasizes the advantage of employing stem cells for tissue reconstitution, since the classical stem cell properties of "self-renewal" and "differentiation" allow even a single transplanted hematopoietic stem cell to permanently reconstitute all blood lineages of an animal after transplantation into a myeloablated recipient.[5,6]

In contrast to the hematopoietic system, well-characterized multipotent, tissue committed stem cells in most other organs are either not yet known, poorly characterized, or difficult to purify for regenerative therapies. A potential solution to this limitation is the use of pluripotent stem cells that may be coaxed in culture into a potentially unlimited supply of any tissue-committed progenitor or mature

Human Stem Cell Technology and Biology, edited by Stein, Borowski, Luong, Shi, Smith, and Vazquez
Copyright © 2011 Wiley-Blackwell.

differentiated cell type for use in regenerative medicine.[7] It should be noted that, beyond regenerative therapies, pluripotent stem cells may also be harnessed as ex vivo drug screening vehicles or ex vivo model systems of development and disease pathogenesis. While the focus of this chapter is the potential in vivo application of pluripotent stem cells, these other potential ex vivo applications are addressed in other chapters of this text.

SOURCES OF PLURIPOTENT STEM CELLS

Several types of stem cells have been isolated displaying the functional differentiation capacity referred to as "pluripotency." Pluripotency is often defined as the capacity of a cell or its progeny to differentiate into multiple cell types characteristic of more than one germ layer. A more stringent definition of pluripotency requires that a cell or its progeny be capable of differentiating into *all* cell types (excluding extraembryonic tissues) in an organism. This latter definition requires the cell pass stringent functional tests of pluripotency in transplantation assays, including the capacity to (1) form a teratoma after transplantation, (2) contribute to all embryonic germ layers (chimerism) after injection into a developing blastocyst, (3) display the potential to contribute to the germline by giving rise to progeny of future generations, or (4) give rise to an entire organism after injection into a tetraploid embryo (tetraploid complementation assay). It is important to note that ethical and legal limitations prevent human candidate cell populations from undergoing the above four tests. Mouse embryonic stem (ES) cells isolated from the inner cell mass (ICM) of the developing mouse blastocyst are broadly accepted as meeting all the above criteria of pluripotency. Many other cell types have also recently been proposed as having some capacity to form multiple cell types from multiple germ layers (i.e., pluripotency). However, many of these claims are still controversial[8] and there is not yet consensus in the scientific literature as to whether cells such as bone marrow stem cells,[9] amniotic stem cells,[10] mesenchymal stem cells,[11] and neural stem cells[12] truly display pluripotency meeting these criteria.

The isolation of ES cells, the gold standard of pluripotent stem cells today, was the culmination of many years of research deriving other pluripotent cell lines from embryonic tissues.[13] For example, the study of teratocarcinomas[14,15] since the 1950s led to the isolation of pluripotent embryonal carcinoma (EC) cell lines[16–18] and subsequently to the derivation of embryonic germ (EG) cells from primordial germ cells (PGCs).[19,20] Like ES cells, these other pluripotent cell lines remain undifferentiated in culture and undergo differentiation into multiple cell types of multiple germ layers after injection into blastocysts.[21–23] Because few attempts at regenerative therapies with these alternate pluripotent lines have been published, the remainder of this chapter will focus on ES cells or cells that are reprogrammed to be highly similar to ES cells in terms of pluripotency and self-renewal. Because only teratoma assays and in vitro differentiation studies using human cells are ethically and legally permissible tests of pluripotency, it should be noted that the pluripotency of human ES cells is accepted based on these assays as well as based on extrapolations of the results obtained using ES cells of other species.

EMBRYONIC STEM CELLS

Mouse embryonic stem (ES) cells exhibiting properties of self-renewal and pluripotency were first isolated in 1981 by two independent labs.[24,25] It was not until 1998 that human ES cells were isolated by Thomson and colleagues.[26] ES cells are typically isolated by culturing preimplantation blastocyst-staged embryos to allow outgrowth of cells from the blastocyst inner cell mass. The inner cell mass of an embryo gives rise to all tissues of an animal (with the exception of extraembryonic tissues, such as the placenta), and ES cells can similarly give rise to all somatic and germline cell types. Hence although ES cells are an artificial laboratory creation, their resemblance to either inner cell mass cells or epiblast cells of the early embryo means they can be viewed, in theory, as primordial undifferentiated cells with the developmental potential to undergo directed differentiation into any desired cell type (please see other chapters of this text for a full review of ES cells). Indeed, many publications from investigators around the world demonstrate the proof-of-principle of being able to coax human or rodent ES cells into an abundance of desired differentiated, functional cell types for in vitro studies or for transplantation strategies designed to regenerate diseased tissues in animal models.[27,28] Despite the unprecedented potential of ES cells in regenerative medicine, the application of human ES cells for clinical therapies is limited by several obstacles. These include ethical concerns regarding the utilization of human embryos for scientific research, the limited supply of embryos, and the technical difficulty of deriving human ES cells from each embryo. In addition, transplantation of allogeneic ES-derived cells faces the potential of rejection by a recipient immune system, if they are brought to therapeutic use.

SOMATIC CELL NUCLEAR TRANSFER

The limitations facing ES cell-based therapeutics has encouraged the development of several reprogramming strategies in an attempt to generate primitive/embryonic pluripotent stem cells that might be functionally similar to ES cells while avoiding the ethical, technical, and allogeneic limitations of blastocyst-derived ES cells. Somatic cell nuclear transfer (SCNT) is one approach that was first attempted in the 1950s using frog eggs and was performed to test for nuclear equivalency.[29] SCNT involves the transfer of the nucleus of a somatic cell into the enucleated oocyte cytoplasm with the goal of reprogramming the transferred nucleus into an embryonic epigenetic state. The resulting cells produced have the nuclear genetic makeup of the somatic donor cell, raising the possibility of employing them for autologous transplantation without fear of immunologic rejection. Achievements in SCNT have included the cloning of several animals such as mice,[30] nonhuman primates,[31] and others including the much publicized sheep, Dolly.[32] In addition to the cloning of animals, ES cells can also be obtained from preimplantation blastocysts that result from SCNT.[33-36] Similar to ES cells derived through traditional methods, SCNT-derived ES cells are pluripotent as evidenced by their capacity to contribute to mouse diploid chimeras[36] and their differentiation potential in the more stringent assay of tetraploid embryo complementation.[33] To date, SCNT is a technically challenging procedure that is still inefficient[37] but has been used to make cloned animals as well as autologous, isogeneic animal ES cells. The use of SCNT for the production of human embryos and ES cells has not yet been successful.

REPROGRAMMING THROUGH CELLULAR FUSION

An alternative strategy for reprogramming somatic cells is engineered cellular fusion of a somatic cell with a pluripotent cell, such as an ES cell.[38] Cellular fusion is the process of two cells joining together through the merging of their membranes creating a single cell. The resulting cell has the chance to become phenotypically equal to either of the fusing cells or some combination of the two. The most recognizable occurrence is the merging of myoblasts to form myotubes containing multiple nuclei or the fusing of sperm and egg cells. Cell fusion has been shown to occur between cells of different lineages and can result in a new tetraploid cell with the epigenetic profile and phenotype of one of the lineages.[39] For the specific goal of generating pluripotent stem cells in vitro, cell fusion has been performed between ES cells and somatic cells to produce cells expressing characteristics of ES cells but containing the somatic cell's genome.[38,39] Although pluripotent cells formed from the cell fusion of somatic cells with ES cells may be differentiated into all three germ layers, the use of ES cells in cell fusion is limited by some of the technical and ethical hurdles that have limited ES cell research. In addition, the resulting fused cells are typically tetraploid, presenting additional technical challenges for investigators to maintain stable derivatives of the cells in culture.

INDUCED PLURIPOTENT STEM CELLS

The demonstration that a differentiated somatic cell nucleus could be reprogrammed back into an embryonic state after fusion or SCNT indicated that the protein products responsible for nuclear reprogramming were present in the cytoplasm of primordial embryonic cells such as oocytes or ES cells. Thus investigators began studies designed to identify the candidate sets of factors present in primordial embryonic cells or oocytes that might be responsible for nuclear reprogramming. In 2006, Takahashi and Yamanaka identified that retroviral delivery of transgenes encoding four transcription factors, Oct4, Klf4, Sox2, and c-Myc, to mouse fibroblasts resulted in extensive reprogramming of somatic cells and the induction of pluripotency.[40] The resulting cells were named induced pluripotent stem (iPS) cells. Within a year the process had been refined in mouse cells to the point where the transcriptome, epigenome, and functional capacities of the reprogrammed iPS cells were highly similar to ES cells, including the capacity to perform well in teratoma assays and the potential to give rise to chimeric mice after blastocyst transplantation.[41] The most dramatic demonstration that mouse iPS cells were functionally highly similar to ES cells in terms of their pluripotency was their capacity to pass the stringent in vivo assays of germline competence after injection into mouse blastocysts[41–43] and, most recently, their capacity to give rise to mice derived entirely from iPS cells in tetraploid complementation assays.[44]

Although it had taken 17 years for investigators to extend the techniques for mouse ES cell isolation to the isolation of human ES cells, it took only 1 short year after the discovery of mouse iPS cells to achieve the generation of human iPS cells with a combination of Yamanaka's original four factors,[45] or in a simultaneous publication by Thomson and colleagues, a combination of Oct4, Sox2, Nanog, and Lin28.[46]

The reprogramming of cells into iPS cells has since been achieved in both mouse and human cells with fewer than four factors, with c-Myc being dispensable in many studies,[47,48] and in some tissues expressing high endogenous levels of the factors (such as neural stem cells) single factor reprogramming with only Oct4 has been demonstrated.[49] In addition, small molecules have been found that can replace some of the reprogramming factors. For example, valproic acid allows for the reprogramming of primary human fibroblasts with only Oct4 and Sox2.[50–52]

In marked contrast to SCNT and cell fusion strategies, the reprogramming protocol for generating iPS cells has proved to be remarkably simple and technically easy to reproduce in many labs around the world. This is observed in the numerous different ways in which iPS cells have been created. The majority of mouse and human iPS cells to date have been generated using integrating retroviral or lentiviral vectors[40,42,43,45,53–58]; however, several new notable strategies for deriving iPS cells have now been published including the use of single polycistronic viral vectors,[59–62] cre-excisable integrating viral vectors,[63] and transposon/transposase-based vectors.[64,65] Insertional mutagenesis is not required for the derivation of iPS cells, based on notable recently published approaches for generating iPS cells using a variety of nonintegrating strategies, such as plasmid transfection,[41,66] adenoviral infection,[67] or transfection of the reprogramming proteins themselves.[68,69] Remarkably, reprogramming induced by the delivery of the specified transcription factors is clearly a self-sustaining process as was evident based on the observation that retrovirally delivered reprogramming transgenes were conveniently silenced in the iPS cells as the endogenous loci encoding the reprogramming factors were activated. The self-sustained kinetics of reprogramming has since been detailed in studies demonstrating that only approximately 10 days of doxycycline-induced lentiviral expression of reprogramming transgenes is required for progression of fibroblasts toward self-sustained reprogramming into stable iPS cell lines.[55,56,70]

iPS cells overcome many of the hurdles of ES cells. For example, autologous iPS cells can be derived potentially from any individual; thus they have the potential to be patient-specific and would be compatible with the immune system. The ethical barriers that have limited ES cell research may be avoided since oocytes and embryos are not required for the generation of iPS cells. Indeed, many of the human somatic cell types that can be reprogrammed into iPS cells, such as keratinocytes,[53,56,59] dermal fibroblasts, and blood cells,[71] may be harvested by minimally invasive techniques.

Patient-specific iPS cells offer the opportunity to create pluripotent cell lines from those individuals carrying a genetic profile that leads to disease. The study of diseased cell lines provides a chance to develop in vitro models of the pathogenesis of inherited disease and allows in vitro tests of the effectiveness of future therapeutic approaches, such as drug screens and toxicology studies. Directed differentiation of iPS cells is particularly attractive as this has the potential to provide a limitless supply of disease-specific, differentiated cells. These iPS-derived differentiated cells would solve a significant current barrier in regenerative medicine research, since access to differentiated cells was previously limited by the fact that many primary cells cannot easily be expanded in culture, and other culture-adapted cell lines have changed their epigenetic profiles, have been derived from cancerous cells, or have been otherwise genetically modified to become immortal. Furthermore, the restrictions on ES cell research and the limited supply of embryos has limited the production of ES lines with genetic variations, leaving a void that iPS cells have been quick to fill.

Notable initial experiments deriving disease-specific iPS cells were completed utilizing postnatal somatic cells from ALS patients, including fibroblasts obtained from an 82 year old, suggesting that age of the source was not a limitation. The therapeutic and ex vivo modeling potential of the resulting cells was displayed as the iPS cells were differentiated into motor neurons.[54] iPS cells were also subsequently generated from individuals with 10 different diseases including Down syndrome, Huntington disease, Parkinson disease, and type I diabetes.[72] Despite a growing number of iPS cell lines generated from an increasing number of diseases (Fanconi anemia,[73] familial dysautonomia,[74] spinal muscular atrophy,[75] and acquired blood disorders[76]), the challenge now facing this nascent field is to effectively differentiate the cells into phenotypically relevant cell types displaying an observable diseased state.[74,77]

POTENTIAL OF PLURIPOTENT CELLS IN REGENERATIVE MEDICINE

Much excitement regarding pluripotent stem cells derives from their potential use as an unlimited source of healthy functional cells for transplantation (see Fig. 27.1). The protocols for directed differentiation of ES cells into specified lineages are quickly being transferred to iPS cells. iPS cells have already been differentiated into cardiomyocytes,[78] insulin-producing pancreatic islet-like cells,[79,80] hepatocytes,[81]

Potential of iPS Cell Research

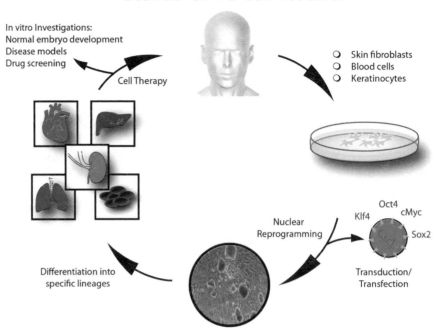

FIGURE 27.1. *Pluripotent stem cells derive from their potential use as an unlimited source of healthy functional cells for transplantation.*

blood cells,[82] endothelium,[78] neurons,[54] and retinal cells.[83] As with prior publications of ES cell differentiation, the challenge remains to prove through functional in vivo replacement assays that a defined differentiated lineage of cells has truly been derived from pluripotent stem cells.

Studies have only recently been initiated in animal models of degenerative disease or injury to test whether functional improvements might result from iPS cell-based therapies, and most of these studies adapt transplantation protocols originally tested using ES cell derivatives (see Chapter 28). For example, in rodent models of Parkinson disease, transplanted dopaminergic neurons derived from embryonic stem cells[84] or iPS cells[85] have been shown to integrate into the host brain and lead to an improvement in motor function of treated recipients. iPS cells have also been shown to improve cardiac function in a mouse model of acute myocardial infarction, although the mechanism responsible for this phenomenon remains unclear.[86] Additional studies have successfully employed iPS cells as in vivo gene delivery vehicles (reviewed in pluripotent cells for Gene Therapy).

Although translation of iPS cell-based therapies for human clinical trials is still a number of years away, the first US clinical trial using human ES cells, sponsored by Geron Corporation, is now paving the way for future use of pluripotent stem cells. In January 2009, Geron received FDA approval for a clinical trial using differentiated human ES cells.[87] Geron intends to deliver precursors of oligodendrocytes (GRNOPC1) derived from human ES cells into subjects with acute spinal cord injury. The cells must be injected within 2 weeks of injury with the aim of repairing myelin sheaths of damaged neurons as observed in preclinical studies performed in rats.[88] The initiation of this landmark clinical trial is an important first step in understanding the complexities of bringing pluripotent stem cell-based therapies to the clinic. The findings from this study will also help to define the oncogenic risk of transplanting the cells, since teratoma formation is one feared complication that is observed to occur in rodent models if residual undifferentiated stem cells remain among the cells that have been carefully differentiated in vitro prior to transplantation.[89] Since the risks of teratoma formation apply to transplantation of any pluripotent stem cell population, including all types of reprogrammed cells (e.g., iPS cells), all investigators working on pluripotent stem cell-based therapies anxiously await the safety results from Geron's Phase I clinical trial.

PLURIPOTENT CELLS FOR GENE THERAPY

Pluripotent stem cells also offer great promise as gene therapy delivery vehicles.[90] Gene therapy attempts to introduce genes in vivo in order to correct a genetic disease or enhance a cell's natural functions. Some examples of inherited diseases for which gene therapy is applicable include cystic fibrosis, hemophilia, severe combined immunodeficiency (SCID), and sickle cell anemia.[91] Although multiple methods have been developed previously to deliver genes to cells in vivo, results from gene therapy clinical trials have often been disappointing as gene expression levels in vivo have been low or short-lived.[92,93]

While integrating viral vectors allow more stable gene delivery than non-integrating methods, viral integrations in transplanted hematopoietic stem cells, for example, can lead to insertional mutagenesis as has occurred in gene therapy trials for SCID-X1.[94] When stem cells are not targeted for gene therapy, the target cells are often terminally differentiated cells with limited in vivo self-renewal capacity and finite lifespan that has also contributed to short-lived in vivo gene expression.

The culturing of pluripotent stem cells allows for the alteration of the genome in vitro before transplantation into a patient or animal. Much more efficient gene delivery methods have been developed for in vitro application, and extensive screening of the genetically manipulated cells for safety and toxicity is possible in vitro prior to transplantation. Consequently, several "proof-of-principle" experiments using ES or iPS cell-based gene therapies have already been tested in rodent models of human inherited/genetic diseases. For example, an immunodeficient mouse null for the Rag2 gene was successfully treated using tail tip fibroblast cells from this mouse that were reprogrammed through SCNT to derive genetically autologous ES cells.[95] The resulting ES cells had their deficient Rag2 gene corrected by homologous recombination, followed by the derivation of hematopoietic progenitors from the gene-corrected ES cells, which were transplanted into the irradiated recipient Rag2 null mouse to successfully treat the immunodeficiency.

In a more recent example of a similar approach, a humanized knock-in mouse model of sickle cell anemia was treated successfully by gene-corrected iPS cells. In that study, iPS cells were generated from tail tip fibroblasts of the sickle cell adult mouse and the $h\beta^S$ alleles were corrected by homologous recombination with a human β^A wild-type globin gene. The gene-corrected iPS cells were then differentiated into hematopoietic progenitor cells for transplantation into irradiated mice. Progeny of the transplanted cells were observed for up to 12 weeks in vivo and hematological and systemic parameters of sickle cell anemia improved substantially and were comparable to those in control mice.[96]

The list of genetic diseases now being treated with iPS cells in mouse modeling experiments is quickly growing. Hemophilia due to factor VIII deficiency, for example, was recently treated by reconstituting endothelial cells in a factor VIII-deficient mouse with iPS cell-derived endothelial cells isolated from a wild-type mouse.[97] In terms of treating genetic diseases with human cells, most recently, human iPS cells generated from individuals with Fanconi anemia (FA) were isolated for functional gene therapy experiments. Attempts to correct phenotypes associated with FA were performed both before and after reprogramming with lentiviral vectors encoding FANCA and FANCD2 genes. Both gene-corrected iPS cells and gene-corrected iPS cell-derived hematopoietic cells showed functional FA pathways in testing, including in vivo after transplantation into mice.[73]

FUTURE CHALLENGES FOR PLURIPOTENT STEM CELL-BASED REGENERATIVE MEDICINE

Human ES cell research has been conducted for over ten years, resulting in many discoveries regarding mechanisms that regulate self-renewal, pluripotency, and differentiation. While protocols for the controlled and reproducible directed differentiation of human ES cells are being developed, these protocols will now require precise optimization and appropriate functional testing of the phenotypes being derived, rather than simply the marker gene expression analyses performed in most studies to date. The discovery of iPS cells has expanded the pace and scope of pluripotent stem cell research and a new challenge is now assessing whether iPS cells are identical to ES cells or just highly similar. This is an important distinction, as it remains unclear whether reprogrammed cells are indeed completely identical to their ES cell counterparts.

As in vitro work continues to progress rapidly, allowing the directed differentiation of desired cells from pluripotent stem cells, the highest hurdle limiting regenerative therapies will quickly become how to safely and durably deliver the resulting cells in vivo for clinical therapies. The new ES cell-based therapeutic trial by Geron Corporation will provide valuable information on the safety and durability of one transplantation approach. In addition, a growing collaboration with biomedical engineers will be required to devise novel biocapsules, scaffolds, or other methods for successful engraftment of the pluripotent stem cell derivatives to be transplanted. Finally, as therapeutics come to fruition, investigators will need to employ bioreactors and novel methods for rapidly producing the massive quantities of cells required for each human transplantation, and techniques will need to be ready for reproducibly deriving human cells with minimal, if any, xenogeneic presence of animal products or feeder cells. Expectations for therapeutic applications of human pluripotent stem cells are very high and there is well-deserved excitement and hope among scientists that the rapid pace of research in the field will be able to meet these expectations with time.

REFERENCES

1. Becker AJ, McCulloch CE, Till JE. Cytological demonstration of the clonal nature of spleen colonies derived from transplanted mouse marrow cells. *Nature*. 1963; 197:452–454.

2. Ford CE, Hamerton JL, Barnes DW, Loutit JF. Cytological identification of radiation-chimaeras. *Nature*. 1956;177(4506):452–454.

3. Lorenz E, Uphoff D, Reid TR, Shelton E. Modification of irradiation injury in mice and guinea pigs by bone marrow injections. *J Natl Cancer Inst*. 1951;12(1):197–201.

4. Nowell PC, Cole LJ, Habermeyer JG, Roan PL. Growth and continued function of rat marrow cells in X-radiated mice. *Cancer Res*. 1956;16(3):258–261.

5. Morrison SJ, Uchida N, Weissman IL. The biology of hematopoietic stem cells. *Annu Rev Cell Dev Biol*. 1995; 11:35–71.

6. Matsuzaki Y, Kinjo K, Mulligan RC, Okano H. Unexpectedly efficient homing capacity of purified murine hematopoietic stem cells. *Immunity*. 2004;20(1):87–93.

7. Klimanskaya I, Rosenthal N, Lanza R. Derive and conquer: sourcing and differentiating stem cells for therapeutic applications. *Nat Rev Drug Discov*. 2008;7(2):131–142.

8. Wagers AJ, Weissman IL. Plasticity of adult stem cells. *Cell*. 2004;116(5):639–648.

9. Krause DS, Theise ND, Collector MI, et al. Multi-organ, multi-lineage engraftment by a single bone marrow-derived stem cell. *Cell*. 2001;105(3):369–377.

10. De Coppi P, Bartsch G Jr, Siddiqui MM, et al. Isolation of amniotic stem cell lines with potential for therapy. *Nat Biotechnol*. 2007;25(1):100–106.

11. Jiang Y, Jahagirdar BN, Reinhardt RL, et al. Pluripotency of mesenchymal stem cells derived from adult marrow. *Nature*. 2002;418(6893):41–49.

12. Clarke DL, Johansson CB, Wilbertz J, et al. Generalized potential of adult neural stem cells. *Science*. 2000;288(5471):1660–1663.

13. Hochedlinger K, Plath K. Epigenetic reprogramming and induced pluripotency. *Development*. 2009;136(4):509–523.

14. Stevens LC. Origin of testicular teratomas from primordial germ cells in mice. *J Natl Cancer Inst*. 1967;38(4):549–552.

15. Stevens LC, Little CC. Spontaneous testicular teratomas in an inbred strain of mice. *Proc Natl Acad Sci USA*. 1954;40(11):1080–1087.

16. Finch BW, Ephrussi B. Retention of multiple developmental Potentialities by cells of a mouse testicular Teratocarcinoma during prolonged culture in vitro and their extinction upon hybridization with cells of permanent lines. *Proc Natl Acad Sci USA*. 1967;57(3):615–621.

17. Kahan BW, Ephrussi B. Developmental potentialities of clonal in vitro cultures of mouse testicular teratoma. *J Natl Cancer Inst*. 1970;44(5):1015–1036.

18. Kleinsmith LJ, Pierce GB Jr. Multipotentiality of single embryonal carcinoma cells. *Cancer Res*. 1964;24:1544–1551.

19. Matsui Y, Zsebo K, Hogan BL. Derivation of pluripotential embryonic stem cells from murine primordial germ cells in culture. *Cell*. 1992;70(5):841–847.

20. Resnick JL, Bixler LS, Cheng L, Donovan PJ. Long-term proliferation of mouse primordial germ cells in culture. *Nature*. 1992;359(6395):550–551.

21. Bradley A, Evans M, Kaufman MH, Robertson E. Formation of germ-line chimaeras from embryo-derived teratocarcinoma cell lines. *Nature*. 1984;309(5965):255–256.

22. Brinster RL. The effect of cells transferred into the mouse blastocyst on subsequent development. *J Exp Med*. 1974;140(4):1049–1056.

23. Mintz B, Illmensee K. Normal genetically mosaic mice produced from malignant teratocarcinoma cells. *Proc Natl Acad Sci USA*. 1975;72(9):3585–3589.

24. Evans MJ, Kaufman MH. Establishment in culture of pluripotential cells from mouse embryos. *Nature*. 1981;292(5819):154–156.

25. Martin GR. Isolation of a pluripotent cell line from early mouse embryos cultured in medium conditioned by teratocarcinoma stem cells. *Proc Natl Acad Sci USA*. 1981;78(12):7634–7638.

26. Thomson JA, Itskovitz-Eldor J, Shapiro SS, et al. Embryonic stem cell lines derived from human blastocysts. *Science*. 1998;282(5391):1145–1147.

27. Keller GM. In vitro differentiation of embryonic stem cells. *Curr Opin Cell Biol*. 1995;7(6):862–869.

28. Murry CE, Keller G. Differentiation of embryonic stem cells to clinically relevant populations: lessons from embryonic development. *Cell*. 2008;132(4):661–680.

29. Briggs R, King T. Transplantation of living nuclei from blastula cells into enucleated frog's eggs. *PNAS* 1952;38(5):455–463.

30. Wakayama T, Perry AC, Zuccotti M, Johnson KR, Yanagimachi R. Full-term development of mice from enucleated oocytes injected with cumulus cell nuclei. *Nature*. 1998;394(6691):369–374.

31. Byrne JA, Pedersen DA, Clepper LL, et al. Producing primate embryonic stem cells by somatic cell nuclear transfer. *Nature*. 2007;450(7169):497–502.

32. Wilmut I, Schnieke AE, McWhir J, Kind AJ, Campbell KH. Viable offspring derived from fetal and adult mammalian cells. *Nature*. 1997;385(6619):810–813.

33. Hochedlinger K, Jaenisch R. Monoclonal mice generated by nuclear transfer from mature B and T donor cells. *Nature*. 2002;415(6875):1035–1038.

34. Kawase E, Yamazaki Y, Yagi T, Yanagimachi R, Pedersen RA. Mouse embryonic stem (ES) cell lines established from neuronal cell-derived cloned blastocysts. *Genesis*. 2000;28(3–4):156–163.

35. Munsie MJ, Michalska AE, O'Brien CM, Trounson AO, Pera MF, Mountford PS. Isolation of pluripotent embryonic stem cells from reprogrammed adult mouse somatic cell nuclei. *Curr Biol*. 2000;10(16):989–992.

36. Wakayama T, Tabar V, Rodriguez I, Perry AC, Studer L, Mombaerts P. Differentiation of embryonic stem cell lines generated from adult somatic cells by nuclear transfer. *Science*. 2001;292(5517):740–743.

37. Markoulaki S, Meissner A, Jaenisch R. Somatic cell nuclear transfer and derivation of embryonic stem cells in the mouse. *Methods*. 2008;45(2):101–114.

38. Cowan CA, Atienza J, Melton DA, Eggan K. Nuclear reprogramming of somatic cells after fusion with human embryonic stem cells. *Science*. 2005;309(5739):1369–1373.

39. Sullivan S, Eggan K. The potential of cell fusion for human therapy. *Stem Cell Rev*. 2006;2(4):341–349.

40. Takahashi K, Yamanaka S. Induction of pluripotent stem cells from mouse embryonic and adult fibroblast cultures by defined factors. *Cell*. 2006;126(4):663–676.

41. Okita K, Nakagawa M, Hyenjong H, Ichisaka T, Yamanaka S. Generation of mouse induced pluripotent stem cells without viral vectors. *Science*. 2008;322(5903):949–953.

42. Maherali N, Sridharan R, Xie W, et al. Directly reprogrammed fibroblasts show global epigenetic remodeling and widespread tissue contribution. *Cell Stem Cell*. 2007;1(1):55–70.

43. Wernig M, Meissner A, Foreman R, et al. In vitro reprogramming of fibroblasts into a pluripotent ES-cell-like state. *Nature*. 2007;448(7151):318–324.

44. Zhao XY, Li W, Lv Z, et al. iPS cells produce viable mice through tetraploid complementation. *Nature*. 2009;461(7260):86–90.

45. Takahashi K, Tanabe K, Ohnuki M, et al. Induction of pluripotent stem cells from adult human fibroblasts by defined factors. *Cell*. 2007;131(5):861–872.

46. Yu J, Vodyanik MA, Smuga-Otto K, et al. Induced pluripotent stem cell lines derived from human somatic cells. *Science*. 2007;318(5858):1917–1920.

47. Nakagawa M, Koyanagi M, Tanabe K, et al. Generation of induced pluripotent stem cells without Myc from mouse and human fibroblasts. *Nat Biotechnol*. 2008;26(1):101–106.

48. Wernig M. c-Myc is dispensable for direct reprogramming of mouse fibroblasts. *Cell Stem Cell*. 2008;2(1):10–12.

49. Kim JB, Greber B, Arauzo-Bravo MJ, et al. Direct reprogramming of human neural stem cells by OCT4. *Nature*. 2009;461(7264):649–643.

50. Lyssiotis CA, Foreman RK, Staerk J, et al. Reprogramming of murine fibroblasts to induced pluripotent stem cells with chemical complementation of Klf4. *Proc Natl Acad Sci USA*. 2009;106(22):8912–8917.

51. Huangfu D, Maehr R, Guo W, et al. Induction of pluripotent stem cells by defined factors is greatly improved by small-molecule compounds. *Nat Biotechnol*. 2008;26(7):795–797.

52. Huangfu D, Osafune K, Maehr R, et al. Induction of pluripotent stem cells from primary human fibroblasts with only Oct4 and Sox2. *Nat Biotechnol*. 2008;26(11):1269–1275.

53. Aasen T, Raya A, Barrero MJ, et al. Efficient and rapid generation of induced pluripotent stem cells from human keratinocytes. *Nat Biotechnol*. 2008;26(11):1276–1284.

54. Dimos JT, Rodolfa KT, Niakan KK, et al. Induced pluripotent stem cells generated from patients with ALS can be differentiated into motor neurons. *Science*. 2008;321(5893):1218–1221.

55. Hockemeyer D, Soldner F, Cook EG, Gao Q, Mitalipova M, Jaenisch R. A drug-inducible system for direct reprogramming of human somatic cells to pluripotency. *Cell Stem Cell*. 2008;3(3):346–353.

56. Maherali N, Ahfeldt T, Rigamonti A, Utikal J, Cowan C, Hochedlinger K. A high-efficiency system for the generation and study of human induced pluripotent stem cells. *Cell Stem Cell*. 2008;3(3):340–345.

57. Okita K, Ichisaka T, Yamanaka S. Generation of germline-competent induced pluripotent stem cells. *Nature*. 2007;448(7151):313–317.

58. Park I-H, Zhao R, West JA, et al. Reprogramming of human somatic cells to pluripotency with defined factors. *Nature*. 2008;451(7175):141–146.

59. Carey BW, Markoulaki S, Hanna J, et al. Reprogramming of murine and human somatic cells using a single polycistronic vector. *Proc Natl Acad Sci USA*. 2009;106(1):157–162.

60. Chang C-W, Lai Y-S, Pawlik KM, et al. Polycistronic lentiviral vector for "hit and run" reprogramming of adult skin fibroblasts to induced pluripotent stem cells. *Stem Cells (Dayton, Ohio)*. 2009;27(5):1042–1049.

61. Sommer CA, Stadtfeld M, Murphy GJ, Hochedlinger K, Kotton DN, Mostoslavsky G. Induced pluripotent stem cell generation using a single lentiviral stem cell cassette. *Stem Cells (Dayton, Ohio)*. 2009;27(3):543–549.

62. Shao L, Feng W, Sun Y, et al. Generation of iPS cells using defined factors linked via the self-cleaving 2A sequences in a single open reading frame. *Cell Res*. 2009;19(3):296–306.

63. Soldner F, Hockemeyer D, Beard C, et al. Parkinson's disease patient-derived induced pluripotent stem cells free of viral reprogramming factors. *Cell*. 2009;136(5):964–977.

64. Kaji K, Norrby K, Paca A, Mileikovsky M, Mohseni P, Woltjen K. Virus-free induction of pluripotency and subsequent excision of reprogramming factors. *Nature*. 2009;458(7239):771–775.

65. Woltjen K, Michael IP, Mohseni P, et al. piggyBac transposition reprograms fibroblasts to induced pluripotent stem cells. *Nature*. 2009;458(7239):766–770.

66. Gonzalez F, Barragan Monasterio M, Tiscornia G, et al. Generation of mouse-induced pluripotent stem cells by transient expression of a single nonviral polycistronic vector. *Proc Natl Acad Sci USA*. 2009;106(22):8918–8922.

67. Stadtfeld M, Nagaya M, Utikal J, Weir G, Hochedlinger K. Induced pluripotent stem cells generated without viral integration. *Science*. 2008;322(5903):945–949.

68. Kim D, Kim CH, Moon JI, et al. Generation of human induced pluripotent stem cells by direct delivery of reprogramming proteins. *Cell Stem Cell*. 2009;4(6):472–476.

69. Zhou H, Wu S, Joo JY, et al. Generation of induced pluripotent stem cells using recombinant proteins. *Cell Stem Cell*. 2009;4(5):381–384.

70. Brambrink T, Foreman R, Welstead GG, et al. Sequential expression of pluripotency markers during direct reprogramming of mouse somatic cells. *Cell Stem Cell*. 2008;2(2):151–159.

71. Loh YH, Agarwal S, Park IH, et al. Generation of induced pluripotent stem cells from human blood. *Blood*. 2009;113(22):5476–5479.

72. Park I-H, Arora N, Huo H, et al. Disease-specific induced pluripotent stem cells. *Cell*. 2008;134(5):877–886.

73. Raya A, Rodriguez-Piza I, Guenechea G, et al. Disease-corrected haematopoietic progenitors from Fanconi anaemia induced pluripotent stem cells. *Nature*. 2009;460(7251):53–59.

74. Lee G, Papapetrou EP, Kim H, et al. Modelling pathogenesis and treatment of familial dysautonomia using patient-specific iPSCs. *Nature*. 2009;461(7262):402–406.

75. Ebert AD, Yu J, Rose FF Jr, et al. Induced pluripotent stem cells from a spinal muscular atrophy patient. *Nature*. 2009;457(7227):277–280.

76. Ye Z, Zhan H, Mali P, et al. Human induced pluripotent stem cells from blood cells of healthy donors and patients with acquired blood disorders. *Blood*. 2009;114(27):5473–5480.

77. Lengerke C, Daley GQ. Disease models from pluripotent stem cells. *Ann NY Acad Sci*. 2009;1176:191–196.

78. Narazaki G, Uosaki H, Teranishi M, et al. Directed and systematic differentiation of cardiovascular cells from mouse induced pluripotent stem cells. *Circulation*. 2008;118(5):498–506.

79. Tateishi K, He J, Taranova O, Liang G, D'Alessio AC, Zhang Y. Generation of insulin-secreting islet-like clusters from human skin fibroblasts. *J Biol Chem*. 2008;283(46):31601–31607.

80. Maehr R, Chen S, Snitow M, et al. Generation of pluripotent stem cells from patients with type 1 diabetes. *Proc Natl Acad Sci USA*. 2009;106:15768–15773.

81. Song Z, Cai J, Liu Y, et al. Efficient generation of hepatocyte-like cells from human induced pluripotent stem cells. *Cell Res*. 2009;19:1233–1242.

82. Choi KD, Yu J, Smuga-Otto K, et al. Hematopoietic and endothelial differentiation of human induced pluripotent stem cells. *Stem Cells*. 2009;27(3):559–567.

83. Buchholz DE, Hikita ST, Rowland TJ, et al. Derivation of functional retinal pigmented epithelium from induced pluripotent stem cells. *Stem Cells*. 2009;27(10):2427–2434.

84. Kim JH, Auerbach JM, Rodriguez-Gomez JA, et al. Dopamine neurons derived from embryonic stem cells function in an animal model of Parkinson's disease. *Nature*. 2002;418(6893):50–56.

85. Wernig M, Zhao J-P, Pruszak J, et al. Neurons derived from reprogrammed fibroblasts functionally integrate into the fetal brain and improve symptoms of rats with Parkinson's disease. *Proc Natl Acad Sci USA*. 2008;105(15):5856–5861.

86. Nelson TJ, Martinez-Fernandez A, Yamada S, Perez-Terzic C, Ikeda Y, Terzic A. Repair of acute myocardial infarction by human stemness factors induced pluripotent stem cells. *Circulation*. 2009;120(5):408–416.

87. Alper J. Geron gets green light for human trial of ES cell-derived product. *Nat Biotechnol*. 2009;27(3):213–214.

88. Keirstead HS, Nistor G, Bernal G, et al. Human embryonic stem cell-derived oligodendrocyte progenitor cell transplants remyelinate and restore locomotion after spinal cord injury. *J Neurosci*. 2005;25(19):4694–4705.

89. Defrancesco L. Fits and starts for Geron. *Nat Biotechnol*. 2009;27(10):877.

90. Conrad C, Gupta R, Mohan H, et al. Genetically engineered stem cells for therapeutic gene delivery. *Curr Gene Ther*. 2007;7(4):249–260.

91. Zabner J, Couture LA, Gregory RJ, Graham SM, Smith AE, Welsh MJ. Adenovirus-mediated gene transfer transiently corrects the chloride transport defect in nasal epithelia of patients with cystic fibrosis. *Cell*. 1993;75(2):207–216.

92. Grubb BR, Pickles RJ, Ye H, et al. Inefficient gene transfer by adenovirus vector to cystic fibrosis airway epithelia of mice and humans. *Nature*. 1994;371(6500):802–806.

93. Young LS, Searle PF, Onion D, Mautner V. Viral gene therapy strategies: from basic science to clinical application. *J Pathol*. 2006;208(2):299–318.

94. Hacein-Bey-Abina S, Garrigue A, Wang GP, et al. Insertional oncogenesis in 4 patients after retrovirus-mediated gene therapy of SCID-X1. *J Clin Invest*. 2008;118(9):3132–3142.

95. Rideout WM 3rd, Hochedlinger K, Kyba M, Daley GQ, Jaenisch R. Correction of a genetic defect by nuclear transplantation and combined cell and gene therapy. *Cell*. 2002;109(1):17–27.

96. Hanna J, Wernig M, Markoulaki S, et al. Treatment of sickle cell anemia mouse model with iPS cells generated from autologous skin. *Science*. 2007;318(5858):1920–1923.

97. Xu D, Alipio Z, Fink LM, et al. Phenotypic correction of murine hemophilia A using an iPS cell-based therapy. *Proc Natl Acad Sci USA*. 2009;106(3):808–813.

THERAPEUTIC APPLICATIONS OF HUMAN EMBRYONIC STEM CELLS

28

Shi-Jiang Lu, Irina Klimanskaya, Edmund Mickunas, and Robert Lanza

INTRODUCTION

Human embryonic stem cells (hESCs) promise to provide a potentially inexhaustible source of tissue for regenerative medicine.[1] The ability of hESCs to differentiate into derivatives of all three embryonic germ layers is well established, and significant progress has been made toward controlled in vitro differentiation of hESCs into specific replacement cell types, including neurons,[2] retinal pigment epithelium (RPE),[3] insulin-producing cells,[4] hepatocytes,[5] cardiomyocytes,[6] and hematopoietic cells,[7] among others.

However, before hESC derivatives can be used in the clinic, it is important to understand the steps involved, as well as the risks and challenges associated with the production of therapeutic products. hESC derivatives have shown therapeutic potential across a range of disease conditions in animals: neural progenitors transplanted into the brains of rats with Parkinson disease generated functional dopaminergic neurons[2]; hESC-derived cardiomyocytes have shown in vivo functional integration in rats with infarcted hearts and in a large-animal (pig) model of bradycardia (slow heart rate) (see Ref. 8 for review); and retinal pigment epithelium derived from hESCs have been shown to preserve visual function in animal models of macular degeneration.[9–11] In addition, insulin-producing cells derived from hESCs were able to secrete insulin and normalize blood glucose levels upon transplantation into diabetic mice (see Ref. 8 for review). When hESC-derived hemangioblasts were injected into rats with spontaneous type II diabetes or mice with ischemia/reperfusion injury of the retina, the cells migrated to the site of injury and showed robust reparative function of the damaged vasculature.[12] The cells also showed a similar regenerative capacity in NOD/SCID β 2 $-/-$ mouse models of both myocardial infarction (50% reduction in mortality rate) and hind limb ischemia, with restoration of blood flow in the latter model to near normal levels.[12]

For hESC derivatives, the ability to create banks of hESC lines with matched histocompatibility or reduced incompatibility could potentially reduce or eliminate the need for immunosuppressive drugs and/or immunomodulatory protocols. For example, (O)Rh-negative hESC lines could be used for the generation of universal red blood cells.

Progress is also being made to solve the challenges of using hESC lines in the clinic. Although at the time of this writing clinical studies have not yet started,

Human Stem Cell Technology and Biology, edited by Stein, Borowski, Luong, Shi, Smith, and Vazquez
Copyright © 2011 Wiley-Blackwell.

Investigational New Drug Applications (IND) for Phase I clinical trials have been approved by the FDA for Geron (a California-based biotechnology company) for the use of hESC-derived oligodendrocyte cells or progenitors for treating patients with spinal cord injury. Clinical Trial will begin by late 2010. Additionally, Advanced Cell Technology (a biotechnology company located in Massachusetts) submitted an IND for the treatment of Stargardt macular degeneration using retinal pigment epithelium (RPE) cells derived from hESC.

This chapter will focus on hESC-derived RPE cells, hematopoietic cells, and hemangioblasts, which could be among the first clinical applications of hESC technology. We will also address issues that impact the development of these products for the purposes of treatment and commercialization.

RETINAL PIGMENT EPITHELIUM

Retinal pigment epithelium (RPE) is a tissue that underlies the photoreceptor layer in the eye and serves a very important function in photoreceptor "life support" as the photoreceptors have no direct blood supply. Among the RPE's most important functions are phagocytosis of shed segments of a photoreceptor and vitamin A metabolism and delivery. When the RPE cannot perform these functions, the photoreceptors degenerate, eventually leading to full or partial loss of vision. For instance, mutation in the enzyme RPE65, which is critical for the synthesis of 11-cis vitamin A[13] in the retinal visual cycle, results in Leber congenital amaurosis,[14] which has been successfully treated in genetic mouse models and in RPE65-deficient dogs with gene therapy.[15] Death of RPE leading to photoreceptor death occurs in macular degeneration and Stargardt disease,[16] both resulting in vision impairment or blindness.

Human RPE cells have been studied for their ability to prevent vision loss in animal models and in human patients for over a decade with promising results.[17] These studies have been carried out with fetal or adult cadaver RPE, but there are multiple limitations associated with both sources, so the possibility of producing a consistent reliable characterized and safe RPE product from an infinite supply such as hESCs would be a very welcome opportunity for treatment of several retinal diseases.

Differentiation and Purification

hESCs are known to commit to neural fate by default when they spontaneously differentiate in the absence of other instructive cues,[18] and while following this pathway, many of them develop into the retinal lineage and subsequently into RPE, as was previously shown in our laboratory followed by several other groups around the world.[3,11,19–21] Differentiating cultures of hESCs or embryoid body (EB) outgrowth usually consists of multiple cell layers, but the unique morphology of these cells allows easy recognition of them in differentiating cultures as "freckles" that comprise cobblestone pigmented epithelial cells. Such clusters can easily be isolated by hand picking, eliminating other cell types present in the culture and thus creating the ability to obtain high-purity RPE cultures. There were a few reports describing generation of RPE from human induced pluripotent stem cells (hiPSCs),[22,23] but although these cells seem to be very similar to hESC-derived RPE cells (hESC-RPE) or even fetal RPE, more studies need to be performed to determine the long-term behavior of such cells in culture (over multiple passages) and in transplantation models. Therefore we will focus our discussion on hESC-RPE.

RPE originates in early vertebrate development from the same neuroectodermal progenitor as the neural retina, and under certain conditions can "transdifferentiate" into neural progenitor-like cells.[24–28]

Like their in vivo counterparts, fetal or adult RPE, hESC-RPE, even after differentiation to its "terminal" phenotype, can still proliferate in culture when given space and conditions for growth: its proliferation is accompanied by depigmentation, loss of epithelial morphology and RPE markers, and acquisition of the features of neural/retinal progenitors, such as spindle-like shape and molecular markers such as tubulin β-III, nestin, or Pax6.[3] RPE cells can undergo several passages, transdifferentiating to neural progenitor-like cells in a manner similar to RPE originating in vivo, and once they become confluent, they redifferentiate into mature pigmented epithelial cells. This is, in fact, a "lucky" combination of unique RPE morphology (dark pigmented epithelial cells that are easily detectable in culture of multiple cell types produced by differentiating ES cells) and the ability to proliferate in culture undergoing multiple population doublings. Even though the initial differentiation efficiency of hESCs to RPE is very low (in most cases below 1%), even a small-scale culture that fits onto one standard CO_2 incubator shelf could provide enough cells for a Phase I clinical trial.

Characterization

Characterization of hESC-RPE shows that in the mature state, the cells express molecular markers of RPE found in vivo, such as bestrophin, a membrane localized 68-kDa product of the Best vitelliform macular dystrophy gene, CRALBP, 36-kDa cellular retinaldehyde-binding protein, RPE65, a 65-kDa cytoplasmic protein involved in retinoid metabolism, pigment epithelium derived factor, PEDF, a secreted 48-kDa protein with angiostatic properties, and MITF. We found that bestrophin and CRALBP can be immunolocalized in "mature" cultures highlighting the cell boundaries while at an earlier stage of differentiation it is only seen in a diffuse pattern. RPE65 mRNA has been detected at much higher levels in long-term (4–6 weeks) differentiated cultures, and Pax6, found in 100% cells during their proliferative stage, becomes downregulated and is found only in small cell clusters after several weeks of differentiation in culture.[3] Comparative analysis of gene expression of hESC-RPE and the in vivo counterpart, fetal human RPE, showed a high degree of similarity, Moreover, hESC-RPE had a greater transcriptional identity to fetal RPE than an ARPE-19 (588 vs. 364 genes) or D407 (849 vs. 373 genes), the latter two cell lines established from adult human RPE.[3] In vitro assessment of functionality of RPE is commonly performed by phagocytosis assay, where latex beads of rod outer segments (ROSs) are used. We have shown that hES-RPE cells are capable of phagocytosis of both latex beads and ROSs,[3] and other groups confirmed these findings with hESC-RPE that has been generated in their laboratories[11,20] and with hiPSC-RPE.[22]

Reliability and Reproducibility

Although differentiation of hESCs in our model produces rather low numbers of RPE cells (below 1%) and is rather slow (6–8 weeks before the first pigmented clusters appear), these cells retain the potential for multiple divisions. On average, one six-well plate of differentiating hESCs can yield 10–12 million RPE cells at p1, passageable at a 1 : 3–1 : 6 ratio up to p5-p6 before multiple cells begin to show signs of senescence and lose RPE morphology and molecular markers. This differentiation has been very reproducible in our laboratory and to date we have produced RPE from more than 20 different hESC lines in over 100 independent experiments.

Functionality and Safety in Animal Models

The ultimate functionality test of an hESC-RPE is its behavior in animal models. There is a rat model of macular degeneration known as the Royal College of

Surgeons (RCS) rat. These animals have a mutation of the MERTK enzyme, which is involved in RPE-specific phagocytosis, and their RPE is unable to phagocytose the shed rod segments. This results in photoreceptor degeneration and subsequent vision deterioration that becomes detectable at about 3 months of age and progresses to blindness by 6 months.[29]

The ELOVL4 mouse is a model for Stargardt disease, a form of macular dystrophy, associated with a mutation in the elongation of the very long chain fatty acids (ELOVL4) gene. In this model RPE cells accumulate undigested phagosomes and lipofuscin, and over time, the photoreceptor degeneration occurs in the central retina.[16,30]

hESC-RPE derived in our laboratory is capable of extensive photoreceptor rescue in the RCS rat model. Electroretinograms (ERGs) performed at 60 days of age showed that hESC-RPE transplanted animals had significantly better a-wave, b-wave, and cone mediated responses over untreated and sham-injected animals. Histological examination demonstrated extensive photoreceptor rescue 5–7 cells deep in the outer nuclear layer, while untreated animals had this layer reduced to one cell at 100 days of age. There were no signs of uncontrolled cell growth or tumor formation.[9] Similar results were reported for hES-RPE generated in a defined system using nicotinamide, factors from the TGF-β superfamily.[11] Long-term survival and safety were assessed in RCS rats and in the ELOVL4 mouse model using hESC-RPE generated under GMP conditions.[10] After 220 days there were no signs of tumors or uncontrolled cell growth, and the functional measurements in RCS rats at 60 days were near normal.[10] These and other results provide support for moving hESC-derived RPE cells into human clinical trials.

HEMATOPOIETIC CELLS

Differentiation of hESCs into hematopoietic cells has been extensively investigated in vitro for the past decade. The directed hematopoietic differentiation of hESCs has been successfully achieved in vitro by means of two different types of culture systems. One of these employs co-cultures of hESCs with stromal feeder cells, in serum-containing medium.[31,32] The second type of procedure employs suspension culture conditions in ultralow cell binding plates, in the presence of cytokines with/without serum[33–35]; its endpoint is the formation of EBs. Hematopoietic precursors as well as mature, functional progenies representing erythroid, myeloid, macrophage, megakaryocytic, and lymphoid lineages have been identified in both of the above differentiating hESC culture systems.[31,32,34,36–40] Although hESC-derived hematopoietic stem cells (HSCs) for bone marrow (BM) reconstitution have been a subject of great interest for both basic and clinical researchers, hematopoietic cells derived from hESCs, however, exhibit limited repopulating activity after transplantation into immunodeficient mice. BM repopulation is invariably low, only rarely exceeding 1–2%, in all reported studies.[37,40–42]. These studies suggest that the imminent clinical use of hematopoietic cells derived from hESCs will likely be red blood cells (RBCs) and platelets in the area of transfusion medicine. Both mature RBCs and platelets are terminally differentiated and carry no genetic materials, which decreases the safety concerns for hESC-derived cell products. For RBCs, establishment/identification of a hESC line with a genotype of (O)Rh-negative will enable the generation of universal types of RBCs that can be transfused in all patients, eliminating the immunogenicity concerns.

Red Blood Cells

Limitations in the supply of RBCs can have potentially life-threatening consequences for patients with massive blood loss due to trauma or surgery, or for those who suffer from diseases that cause severe anemia. Unfortunately, the supply of transfusable RBCs, especially "universal" donor type (O)Rh-negative, is often insufficient, particularly in the battlefield environment due to the lack of blood type information and the limited time required for life-saving transfusion. Although alternative sources of progenitors for the generation of large-scale transfusable RBCs have been investigated, including cord blood, BM, and peripheral blood,[43–47] it is clear that even after expansion and differentiation, these progenitors represent donor-limited sources of RBCs. Moreover, the low prevalence of O(−) blood type in the general population (<8% in Western countries and <0.3% in Asia) further intensifies the consequences of blood shortages for emergency situations where blood typing may not be possible. Derivation of erythroid cells from hESCs may translate into a new and safer source of RBCs for transfusions.

The in vitro differentiation and expansion of pure erythrocytes using mouse ES cells as starting material has been achieved. With no additional exogenous cytokines, co-culture of murine ES cells with OP9, a mouse stromal cell line deficient in macrophage colony stimulating factor (M-CSF), is sufficient to induce erythroid cell production.[48] The presence of M-CSF has inhibitory effects on the differentiation of murine ES cells to blood cells other than macrophages. Erythropoiesis of murine ES cells cultured with OP9 comes in two waves: the first wave includes large nucleated cells, deemed to be primitive erythrocytes, and the second wave includes either small nucleated erythroblasts or enucleated mature blood cells morphologically identical to definitive erythrocytes, containing only adult globins.[49]

An EB-based culture system can result in as much as 5 millionfold expansion of murine ES cell-generated erythroblasts within 10 weeks.[50] EBs are cultured in the presence of serum, dissociated and purified in serum-free liquid culture, and terminally differentiated using erythroid cytokines. Erythrocytes generated using this method express adult hemoglobin and are enucleated when exposed to a high concentration of Epo, transferrin, and insulin. When these erythroblasts are transplanted into mice, they show faster maturation and superior morphology of enucleated cells compared to those cultured with stromal cells.

This generation of erythrocytes has also been achieved from hESCs either by EB formation[34] or co-culturing with stromal cells followed by isolation of CD34+ cells and further expansion/differentiation.[51–53] Chang et al.[34] generated erythroid cells from hESCs by isolating and expanding nonadherent cells of day-14 EBs in a span of 15–56 days of culture time. The definitive-like, but nucleated erythroid cells obtained from the above approach, however, coexpressed high levels of embryonic ε- and fetal γ-globins with little or no adult β-globin. Olivier et al.[51] and Qiu et al.[52] have developed a method for a relatively large-scale ((0.5–$5) \times 10^7$ cells) production of erythroid cells from hESCs. In their method, hESCs were co-cultured for 14–35 days with human fetal liver stromal cells (FHB-hTERT) to produce CD34+ cells that were seeded in a four-step culture system. In steps 1 and 2, cocktails of cytokines were used to promote the proliferation and maturation of erythroid precursors. In steps 3 and 4, erythroid cells were transferred onto mouse BM stromal cells (MS5) to facilitate terminal maturation. Similar to the results observed by Chang et al.,[34] the mostly erythroid cells expressed mainly the embryonic ε- and fetal γ-globins; only a trace amount of the adult β-globin gene was detected by real-time PCR (mRNA level), but not by HPLC (protein level) analysis. Recently, Ma et al.[53] showed that, by using immunostaining with globin chain specific monoclonal antibodies, almost 100% of hESC-derived erythrocytes expressed the adult β-globin chain after co-culture with murine fetal liver derived stromal cells (mFLCs) in vitro, suggesting erythrocytes derived from hESCs are capable of switching on the expression of definitive adult β-globin chain. However, the majority of the erythrocytes obtained by co-culturing with mFLCs were nucleated.

A critical scientific and clinical issue is whether hESC-derived erythroid cells can be produced at a sufficiently large scale and can be matured in vitro to generate enucleated erythrocytes. We recently developed a strategy that efficiently and reproducibly generates hemangioblasts and functional erythroid cells from hESCs.[12,54] This differentiation system uses a defined serum-free medium and eliminates the use of feeder cells. Using hemangioblasts as an intermediate, we have generated functional RBCs (blood types A, B, O, and both RhD+ and RhD−) on a large scale from multiple hESC lines under serum-free conditions suitable for scale-up and clinical translation.[54] Three steps are critical for the efficient scale-up of RBCs: (1) generation of hemangioblasts with high efficiency and high density, without disruption of their colony forming environment; (2) expansion of hemangioblasts to erythroblasts in a high cell density; and (3) culture of erythroblasts in semisolid medium containing methylcellulose to provide optimal conditions for maximum expansion and erythroid purity. We have used this approach to generate 10^{10}–10^{11} pure erythroid cells from one six-well plate of hESCs ($\approx 1 \times 10^7$ cells), which is over a thousandfold more efficient than previously reported.[51] We have also demonstrated that the oxygen equilibrium curves of the hESC-derived erythroid cells are comparable to normal transfusible RBCs and respond to changes in pH and 2,3-diphosphoglycerate. Importantly, the cells undergo multiple maturation events in vitro, including a progressive decrease in size and increase in glycophorin A expression, and chromatin and nuclear condensation. This process resulted in the extrusion of the pycnotic nucleus in up to 60% of the cells. The enucleated erythrocytes appeared morphologically identical to normal RBCs with a diameter of approximately 6–8 μm. These cells also possess the capacity to express the adult definitive β-globin chain upon further maturation in vitro. Globin-chain-specific PCR and immunofluorescent analysis showed that the cells' expression of β-globin increased from 0% to 15% after in vitro culture. The results show that it is feasible to differentiate and mature hESCs into functional oxygen-carrying erythrocytes on a clinically applicable scale. The identification of a hESC line with an O(−) (Rh−) genotype would permit the production of ABO and RhD compatible (and pathogen-free) "universal donor" RBCs. While considerable effort is still needed to bring hESC-derived RBCs to clinical trials, these efforts certainly provide a promising direction.

Megakaryocytes and Platelets

Like RBCs, limitations in the supply of platelets can also have potentially life-threatening consequences for transfusion-dependent patients with unusual/rare blood types, particularly those who are alloimmunized, and patients with cancer or leukemia who, as often happens, develop platelet alloimmunity. Frequent transfusion of platelets is clinically necessary because the half-life of transfused human platelets is 4–5 days. Moreover, platelets from volunteer donor programs are at the constant risk of supply limitations and contamination with various pathogens. Platelets cannot be stored frozen; thus the ability to generate platelets in vitro would provide significant advances for platelet replacement therapy in clinical settings. For more than a decade, human HSCs (CD34+) from bone marrow, cord blood, or peripheral blood have been studied for megakaryocyte (MK) and platelet generation. With the combinations of cytokines, growth factors, and/or stromal feeder cells, functional platelets are produced from HSCs with significant success.[47,55] However, HSCs are still collected from donors and have limited expansion capacity under current culture conditions, which will likely prevent the large-scale production and future clinical applications. Platelets derived from hESCs have the potential to improve transfusion medicine, and importantly, like RBCs, platelets do not have a nucleus and only minimal genetic material, and can be irradiated before transfusion to effectively eliminate any contaminating cell, such as an undifferentiated hESC, so that safety concerns may be significantly decreased.

In recent years, ESCs have been used as alternative sources for in vitro generation of megakaryocytes and platelets. Eto et al.[56] and Fujimoto et al.[57] have developed OP9 stromal-based methods to produce megakaryocytes and platelets in vitro from mouse ESCs. Recently, Nishikii et al.[58] demonstrated that c-Kit + integrin αIIb+ cells isolated from murine ESCs differentiate with high efficiency into megakaryocytes and platelets in the presence of TPO and stromal cells, and inhibition of metalloproteinases in cultures of c-Kit + integrin αIIb+ primitive cells substantially improve the production of functional platelets. Similar OP9 stromal systems have been adapted to generate megakaryocytes from hESCs with characteristic polyploid nuclei, cytoskeletal proteins, and the ability to signal through integrin αIIbβ3.[59,60] After longer co-culture with stromal cells in the presence of TPO these megakaro- cytes were able to generate platelets with morphology and function similar to those isolated from fresh plasma. However, the above reported methods for generating megakaryocytes and platelets from hESCs are problematic for potential clinical applications, because (1) the yields of megakaryocytes/platelets from hESCs are very low, (2) they require the presence of undefined animal stromal cells (e.g., OP9), and (3) using these methods makes it difficult to achieve large-scale cell production.

As described above, we have successfully generated erythroid cells on a large scale from hESC-derived hemangioblasts, which prompted us to investigate whether hemangioblasts can also be used as an intermediate to generate megakaryocytes and platelets, since erythroid cells and megakaryocytes share a common progenitor during hematopoietic development. Therefore hemangioblasts derived from hESCs were cultured in serum-free medium supplemented with defined growth factors including high doses of TPO; greater than 90% of cells expressed CD41a, a marker of megakaryocytes, and a majority of the cells also stained positive for CD42b. These polyploid cells reach up to 70–90 μm in diameter and express vWF in their cytoplasmic granules, indicating they are mature megakaryocytes. Approximately 1×10^8 CD41$^+$ megakaryocytes were generated from 1×10^6 hESCs, which is a magnitude more efficient than the most recently reported method using stromal cells and serum. Importantly, these in vitro derived megakaryocytes underwent terminal differentiation by forming functional platelets that are no different from those from human peripheral blood. Although tremendous effort will be needed to achieve clinical scale generation of functional platelets from hESCs, this serum- and stromal-free but more efficient system is suitable for scale-up production.

HEMANGIOBLASTS

Hemangioblasts are progenitors with the capacity to differentiate into hematopoietic and vascular cells[61]; thus they are excellent candidates for cell therapy in both blood and vascular diseases. Myocardial infarction, stroke, coronary artery disease, ischemic limbs caused by diabetes, and diabetic retinopathy are devastating and life-threatening ischemic diseases that are primarily caused by vascular dysfunction. The ability to repair vascular damage could have a profound impact on these major diseases afflicting humans. Adoptive transfer of endothelial precursor cells has previously been shown to restore blood flow and increase capillary density, decreasing limb loss and facilitating recovery from myocardial injury. Therefore identification and isolation of progenitors from an inexhaustible source, such as hESCs or iPSCs, will be desirable for cell replacement therapy.

The formation and regeneration of functional vasculature requires both endothelial cells (ECs) and vascular smooth muscle cells (SMCs). Blood vessels are typically composed of two major cell types: the inner endothelium, a thin layer of ECs that separate the blood from tissues, and an outer layer of mural cells (pericytes and vascular SMCs) that protect the fragile channels from rupture and help control blood flow.[62] Although ECs play an essential role in vasculogenesis and angiogenesis, they alone cannot complete the process of vessel growth and development. The formation of a mature and functional vascular network requires communication between ECs and SMCs.[62] Vascular SMCs play a critical role in structural and functional support of the vascular network by stabilizing nascent endothelial vessels during vascular development and blood vessel growth. Although functional ECs and SMCs have been identified and isolated from hESC differentiation cultures by several groups using different systems,[63–70] a progenitor cell population that can differentiate into both ECs and SMCs, and that can be propagated and expanded indefinitely, would be important in the treatment of human diseases caused by deficient vessel growth.

Hemangioblasts have been identified in differentiation cultures of mouse, nonhuman primate, and human ESCs.[12,71–76] We recently developed a simple strategy to efficiently and reproducibly generate hemangioblasts from multiple hESC lines on a large scale.[12] These cells express gene signatures characteristic of hemangioblasts and can differentiate into ECs and hematopoietic cells as well as SMCs.[12,77,78] The ECs derived from hESC-hemangioblasts expressed CD31, VE-cad, and vWF, uptake of Dil-Ac-LDL, and formed a vascular network on Matrigel™. Similarly, SMCs derived from hESC-hemangioblasts expressed markers of SMCs, contracted in response to carbachol stimulation, and formed vasculature-like networks in alignment with endothelial cells on Matrigel. When hESC-hemangioblasts were injected into animals with spontaneous type II diabetes or ischemia/reperfusion (I/R) injury of the retina, they homed to the site of injury and showed robust reparative function of the damaged vasculature. The cells also showed a similar regenerative capacity in NOD/SCID $\beta 2 - /-$ mouse models of both myocardial infarction (50% reduction in mortality rate) and hind limb ischemia, with restoration of blood flow in the latter model to near normal levels.[12] Fluorescence immunocytochemistry showed that the vascular lumens were surrounded by human ECs and SMCs in both diabetic and I/R damaged retinas. Similarly, confocal microscopy confirmed the incorporation of human ECs and SMCs into the lumens of microvessels in the infarcted heart and ischemic limb tissues, and the presence of vascular organization in intramuscular areas with human cells.[12,78] These results demonstrate that hemangioblasts generated from hESCs differentiate into functional ECs and SMCs both in vitro and in vivo, and can provide a potentially inexhaustible source of cells for the treatment of human vascular diseases such as diabetic retinopathy, stroke, myocardial infarction, and in ischemic limbs.

ISSUES ASSOCIATED WITH hESC-DERIVED CELL PRODUCTS

Cell products derived from hESCs face many hurdles, both in the laboratory and manufacturing suite, prior to reaching approval for use in the clinic. Major issues are presented below.

Cell Sources

Most of the existing hESC lines have not been derived under Good Manufacturing Practices (GMP) conditions and utilized undefined animal products and feeder cells, which could potentially contaminate them with animal pathogens and viruses. It has also been reported that hESCs maintained on mouse embryonic fibroblast (MEF) feeders contain the nonhuman sialic acid N-glycolylneuraminic acid (Neu5Gc),[79,80] and that animal sources of Neu5Gc can cause a potential immunogenic reaction with human complement. The culturing of hESCs on MEF feeders prevents complete elimination of animal Neu5Gc and raises concerns for the potential clinical applications of cells generated from hESC lines maintained under these conditions.

Although the US Food and Drug Administration (FDA) allows the use of such cells, the agency classifies them as xenotransplantation products, which require more extensive testing and in the clinical setting require more specific monitoring. In many other countries, cells exposed to animal products are banned from clinical use altogether. Therefore in order to comply with government regulations, new hESC lines will need to be produced under GMP conditions using defined materials that can be traced back to the origin or using xeno-free products in manufacturing cell products for future clinical application.

Safety

One of the most important characteristics of undifferentiated hESCs is teratoma formation, which has been used as a standard test for their pluripotency.[81] This characteristic is a serious concern: if the derivative of hESCs is not completely free of pluripotent cells, the possibility exists that tumor formation may occur if contaminated undifferentiated or underdifferentiated hESCs remain in the final product.[82,83] Any cell products derived from hESCs must be free of undifferentiated cells and should be rigorously tested for tumor formation potential in appropriately designed animal studies.

Although tumorigenicity is the main safety concern, biodistribution, toxicity, and stability for hESC-derived cells must also be evaluated. Toxicologic tests should evaluate acute and chronic toxicity in major organs, at injection sites, and in the blood. The issue of stability for cells derived from hESCs is whether these cells will become "unstable" or transformed/dedifferentiated over time, leading to the formation of tumors.

Product Consistency

Successful clinical application of cell products derived from hESCs will require efficient and controlled differentiation of hESCs toward a specific cell type and the generation of homogeneous cell populations that are transplantable. Differentiation of hESCs toward a specific cell type has been extensively investigated in vitro, and precursors as well as differentiated progenies representing various tissues, such as heart, pancreas, liver, lung, eye, digestive tract, and neuronal and hematopoietic systems have been identified in differentiating hESC cultures. However, the efficient and controlled differentiation of hESCs into homogeneous cell populations has not been previously achieved, except for RPE and erythroid cells.[3,77]

Other concerns in manufacturing a hESC-derived product are the consistent generation of sufficient cell numbers (billion cells for most tissues and >100 billion cells for blood cells) under GMP conditions using standardized protocols, and the development of validated tests to monitor consistency not only throughout the entire production cycle but also to ensure batch-to-batch consistency. Ensuring consistency of hESC-derived cell products will be a challenging task due to the nature of biological systems, especially for hESCs, which in theory can be propagated and expanded in vitro indefinitely, but numerous studies have demonstrated that late passages of hESCs acquire genetic abnormalities and change their differentiation potentials.[84–86] Furthermore, cell line to cell line variations impose another challenge for the consistency of hESC-derived cell products.[12,87]

Immunogenicity

Although reports indicate that hESCs and their derivatives may be less immunogenic, they can still induce immune rejection in immunodeficient animals.[88] Therefore the immune status of hESC-derived cell products will need to be tested and, if possible, to be matched with the recipients; otherwise immunosuppression will be required. This would require the availability of many hESC lines. However, if the site of injection is an immune privileged one, such as the eye, no or less-aggressive immunosuppression protocols may be satisfactory.

Potency

Potency is defined as the specific ability or capacity of a therapeutic product to elicit an expected result, as indicated by appropriate laboratory tests or by adequately controlled clinical data obtained through the administration of the product in the manner intended[21] (CFR 600.3(s)).

Since multipotent stem cells are expected to undergo additional differentiation following patient administration, prior to mediating a measurable clinical effect, the problem of deciding what to evaluate as an index of potency, and how to correlate particular measurements with a stem cell product's final outcome, is a significant challenge. Measurement of the potency of a stem cell-based product usually is very difficult because of a lack of appropriate in vitro or in vivo assays, and alternative approaches that assess surrogate biomarkers or activities providing an estimate of potency may be considered adequate for initiation of early phase clinical trials. However, to comply with regulations for FDA approval for commercialization, development of a valid potency assay that is used for release of the final formulated stem cell-based product is required. It is expected that a suitable potency assay be in place prior to commencing a pivotal Phase III clinical trial to confirm safety and determine efficacy of the investigational stem cell-based product.

REGULATORY CONSIDERATIONS

Several concerns have been enumerated in the previous section. Once these issues have been satisfactorily addressed, appropriate submissions must be made to regulatory bodies, which have responsibility for approval for use in clinical trials and ultimately for approval of the product for commercialization. In the United States, an IND should be submitted; outside the United States appropriate submissions will also be required.

There exist relevant guidance documents available from the various regulatory bodies. Review of these guidance documents is essential for a sponsor to understand current issues that need to be addressed prior to submission.

An IND is a request for permission to be granted by the FDA to proceed with patient administration of an investigational stem cell-based product. Essential items included in the IND submission document are the following:

- Detailed information that describes the manufacture and characterization of the investigational stem cell-based therapy
- Compilation of results obtained from preclinical studies conducted in animals
- Well-designed, prospectively written, adequately controlled clinical protocol

Since the development of stem cell-based products is an extremely complex process, it is important that sponsors request a premeeting with CBER/FDA before submitting a completed IND application for evaluation. During the pre-IND meeting, sponsors are able to receive input from CBER/FDA staff that represents current thinking with respect to regulatory expectations that will be addressed by the information provided in an IND application. It is very important that the sponsor clarify all comments made by the Agency; often sponsors may interpret statements and recommendations made incorrectly and this may lead to costly delays in getting to the clinic.

PRECLINICAL ASSESSMENTS

Proof of Concept

Proof-of-concept demonstration for a stem cell-based product in an animal model of human disease is encouraged for the following reasons:

- To establish a rationale that supports initiation of clinical trials
- To permit concurrent measurement of bioactivity and safety endpoints
- To explore the dose–response relationship between a stem cell-based product and an activity/safety outcome
- To facilitate optimization of the route of administration

Demonstrating that a restorative effect occurs following implantation of an investigational stem cell-based product into an animal model of human disease provides circumstantial evidence for anatomical/functional integration within the host physiology. If corroborated by histopathological data, this proof-of-concept information serves to provide support for hypotheses underlying the mechanism of action of a stem cell-based product. It is important that the clinically relevant route of administration and intended clinical delivery system are used in the animal models to the extent feasible. Immune tolerance to the implanted investigational human stem cell product and comparability between cell phenotypes are important considerations when determining whether the use of an analogous stem cell-based product derived from animals is necessary and/or beneficial.

Toxicological Assessment

Establishment of the safety of an implanted stem cell-based product is crucial to the development program, and regulatory bodies require presentation of appropriate safety data for evaluation prior to allowing clinical trial commencement.

Choice of animal models, whether normal animals or those that represent the disease model, is critical in providing evidence of safety. Endpoints are established prior to evaluating the cell product and under ideal circumstances are conducted according to Good Laboratory Practices (GLP).

Examples of toxicity endpoints are selected that focus on the potential for both local and systemic effects to occur. It is important that appropriate regulatory input be provided; otherwise the data from preclinical studies may not be sufficient for a regulatory body to determine safety, thus hindering the product from moving into the clinic. At a minimum, the following areas should be evaluated in the studies:

- Implant site reaction
- Any inflammatory response observed in target or nontarget tissues
- Differentiation/phenotype expression of the stem cell-based product postimplant
- Capacity for the cells to survive, migrate, and integrate into the host physiology postimplant
- Observed behavioral changes
- Morphological alterations in either target or nontarget tissues
- Evidence of tumorigenicity or ectopic tissue formation

Depending on the specifics of the cell product or clinical programs, additional evaluations may be necessary.

REFERENCES

1. Lanza R, et al. (eds). *Essentials of Stem Cell Biology*, 2nd ed. New York: Academic Press;2009.

2. Roy NS, Cleren C, Singh SK, Yang L, Beal MF, Goldman SA. Functional engraftment of human ES cell-derived dopaminergic neurons enriched by coculture with telomerase-immortalized midbrain astrocytes. *Nat Med*. 2006;12:1259–1268.

3. Klimanskaya I, Hipp J, Rezai KA, West M, Atala A, Lanza R. Derivation and comparative assessment of retinal pigment epithelium from human embryonic stem cells using transcriptomics. *Cloning Stem Cells*. 2004;6:217–245.

4. Damour KA, Bang AG, Eliazer S, et al. Production of pancreatic hormone-expressing endocrine cells from human embryonic stem cells. *Nat Biotechnol*. 2006;24:1392–1401.

5. Agarwal S, Holton KL, Lanza R. Efficient differentiation of functional hepatocytes from human embryonic stem cells. *Stem Cells*. 2008;26:1117–1127.

6. Laflamme MA, Chen KY, Naumova AV, et al. Cardiomyocytes derived from human embryonic stem cells in pro-survival factors enhance function of infarcted rat hearts. *Nat Biotechnol*. 2007;25:1015–1024.

7. Tian X, Kaufman DS. Differentiation of embryonic stem cells towards hematopoietic cells: progress and pitfalls. *Curr Opin Hematol*. 2008;15:312–318.

8. Klimanskaya I, Rosenthal N, Lanza R. Derive and conquer: sourcing and differentiating stem cells for therapeutic applications. *Nat Rev Drug Discov*. 2008;7:131–142.

9. Lund RD, Wang S, Klimanskaya I, et al. Human embryonic stem cell-derived cells rescue visual function in dystrophic RCS rats. *Cloning Stem Cells*. 2006;8:189–199.

10. Lu B, Malcuit C, Wang S, et al. Long-term safety and function of RPE from human embryonic stem cells in preclinical models of macular degeneration. *Stem Cells*. 2009;27:2126–2135.

11. Idelson M, Alper R, Obolensky A, et al. Directed differentiation of human embryonic stem cells into functional retinal pigment epithelium cells. *Cell Stem Cell*. 2009;5:396–408.

12. Lu SJ, Feng Q, Caballero S, et al. Generation of functional hemangioblasts from human embryonic stem cells. *Nat Methods*. 2007;4:501–509.

13. Redmond TM, Yu S, Lee E, et al. Rpe65 is necessary for production of 11-cis-vitamin A in the retinal visual cycle. *Nat Genet*. 1998;20:344–351.

14. Gu SM, Thompson DA, Srikumari CR, et al. Mutations in RPE65 cause autosomal recessive childhood-onset severe retinal dystrophy. *Nat Genet*. 1997;17:194–197.

15. Koenekoop RK. Successful RPE65 gene replacement and improved visual function in humans. *Ophthalmic Genet*. 2008;29:89–91.

16. Karan G, Lillo C, Yang Z, et al. Lipofuscin accumulation, abnormal electrophysiology, and photoreceptor degeneration in mutant ELOVL4 transgenic mice: a model for macular degeneration. *Proc Natl Acad Sci USA*. 2005;102:4164–4169.

17. Baker PS, Brown GC. Stem-cell therapy in retinal disease. *Curr Opin Ophthalmol*. 2009;20:175–181.

18. Munoz-Sanjuan I, Brivanlou AH. Neural induction, the default model and embryonic stem cells. *Nat Rev Neurosci*. 2002;3:271–280.

19. Gong J, Sagiv O, Cai H, Tsang SH, Del Priore LV. Effects of extracellular matrix and neighboring cells on induction of human embryonic stem cells into retinal or retinal pigment epithelial progenitors. *Exp Eye Res*. 2008;86:957–965.

20. Carr AJ, Vugler A, Lawrence J, et al. Molecular characterization and functional analysis of phagocytosis by human embryonic stem cell-derived RPE cells using a novel human retinal assay. *Mol Vis*. 2009;15:283–295.

21. Osakada F, Ikeda H, Mandai M, et al. Toward the generation of rod and cone photoreceptors from mouse, monkey and human embryonic stem cells. *Nat Biotechnol*. 2008;26:215–224.

22. Buchholz DE, Hikita ST, Rowland TJ, et al. Derivation of functional retinal pigmented epithelium from induced pluripotent stem cells. *Stem Cells*. 2009;27:2427–2434.

23. Hirami Y, Osakada F, Takahashi K, et al. Generation of retinal cells from mouse and human induced pluripotent stem cells. *Neurosci Lett*. 2009;458:126–131.

24. Sakami S, Etter P, Reh TA. Activin signaling limits the competence for retinal regeneration from the pigmented epithelium. *Mech Dev*. 2008;125:106–116.

25. Pacheco-Dominguez RL, Palma-Nicolas JP, Lopez E, Lopez-Colome AM. The activation of MEK-ERK1/2 by glutamate receptor-stimulation is involved in the regulation of RPE proliferation and morphologic transformation. *Exp Eye Res*. 2008;86:207–219.

26. Yoshii C, Ueda Y, Okamoto M, Araki M. Neural retinal regeneration in the anuran amphibian *Xenopus laevis* post-metamorphosis: transdifferentiation of retinal pigmented epithelium regenerates the neural retina. *Dev Biol*. 2007;303:45–56.

27. Liang L, Yan RT, Ma W, Zhang H, Wang SZ. Exploring RPE as a source of photoreceptors: differentiation and integration of transdifferentiating cells grafted into embryonic chick eyes. *Invest Ophthalmol Vis Sci*. 2006;47:5066–5074.

28. Arresta E, Bernardini S, Bernardini E, Filoni S, Cannata SM. Pigmented epithelium to retinal transdifferentiation and Pax6 expression in larval *Xenopus laevis*. *J Exp Zool A Comp Exp Biol*. 2005;303:958–967.

29. Girman SV, Wang S, Lund RD. Time course of deterioration of rod and cone function in RCS rat and the effects of subretinal cell grafting: a light- and dark-adaptation study. *Vision Res*. 2005;45:343–354.

30. Vasireddy V, Jablonski MM, Khan NW, et al. Elovl4 5-bp deletion knock-in mouse model for Stargardt-like macular degeneration demonstrates accumulation of ELOVL4 and lipofuscin. *Exp Eye Res*. 2009;89:905–912.

31. Kaufman DS, Hanson ET, Lewis RL, Auerbach R, Thomson JA. Hematopoietic colony-forming cells derived from

human embryonic stem cells. *Proc Natl Acad Sci USA.* 2001;98:10716–10721.

32. Lu S-J, Li F, Vida L, Honig GR. CD34$^+$CD38$^-$ hematopoietic precursors derived from human embryonic stem cells exhibit an embryonic gene expression pattern. *Blood.* 2004;103:4134–4141.

33. Chadwick K, Wang L, Li L, et al. Cytokines and BMP-4 promote hematopoietic differentiation of human embryonic stem cells. *Blood.* 2003;102:906–915.

34. Chang KH, Nelson AM, Cao H, et al. Definitive-like erythroid cells derived from human embryonic stem cells coexpress high levels of embryonic and fetal globins with little or no adult globin. *Blood.* 2006;108:1515–1523.

35. Tian X, Morris JK, Linehan JL, Kaufman DS. Cytokine requirements differ for stroma and embryoid body-mediated hematopoiesis from human embryonic stem cells. *Exp Hematol.* 2004;32:1000–1009.

36. Vodyanik MA, Bork JA, Thomson JA, Slukvin II. Human embryonic stem cell-derived CD34$^+$ cells: efficient production in the coculture with OP9 stromal cells and analysis of lymphohematopoietic potential. *Blood.* 2005;105:617–626.

37. Wang L, Menendez P, Shojaei F, et al. Generation of hematopoietic repopulating cells from human embryonic stem cells independent of ectopic HOXB4 expression. *J Exp Med.* 2005;201:1603–1614.

38. Qiu C, Hanson E, Olivier E, et al. Differentiation of human embryonic stem cells into hematopoietic cells by coculture with human fetal liver cells recapitulates the globin switch that occurs early in development. *Exp Hematol.* 2005;33:1450–1458.

39. Zhan X, Dravid G, Ye Z, et al. Functional antigen-presenting leucocytes derived from human embryonic stem cells in vitro. *Lancet.* 2004;364:163–171.

40. Ledran MH, Krassowska A, Armstrong L, et al. Efficient hematopoietic differentiation of human embryonic stem cells on stromal cells derived from hematopoietic niches. *Cell Stem Cell.* 2008;3:85–98.

41. Tian X, Woll PS, Morris JK, Linehan JL, Kaufman DS. Hematopoietic engraftment of human embryonic stem cell-derived cells is regulated by recipient innate immunity. *Stem Cells.* 2006;24:1370–1380.

42. Narayan AD, Chase JL, Lewis RL, et al. Human embryonic stem cell-derived hematopoietic cells are capable of engrafting primary as well as secondary fetal sheep recipients. *Blood.* 2006;107:2180–2183.

43. Leberbauer C, Boulme F, Unfried G, Huber J, Beug H, Mullner EW. Different steroids co-regulate long-term expansion versus terminal differentiation in primary human erythroid progenitors. *Blood.* 2005;105:85–94.

44. Giarratana MC, Kobari L, Lapillonne H, et al. Ex vivo generation of fully mature human red blood cells from hematopoietic stem cells. *Nat Biotechnol.* 2005;23:69–74.

45. Miharada K, Hiroyama T, Sudo K, Nagasawa T, Nakamura Y. Efficient enucleation of erythroblasts differentiated in vitro from hematopoietic stem and progenitor cells. *Nat Biotechnol.* 2006;24:1255–1256.

46. Sullenbarger B, Bahng JH, Gruner R, Kotov N, Lasky LC. Prolonged continuous in vitro human platelet production using three-dimensional scaffolds. *Exp Hematol.* 2009;37:101–110.

47. Matsunaga T, Tanaka I, Kobune M, et al. Ex vivo large-scale generation of human platelets from cord blood CD34$^+$ cells. *Stem Cells.* 2006;24:2877–2887.

48. Nakano T, Kodama H, Honjo T. Generation of lymphohematopoietic cells from embryonic stem cells in culture. *Science.* 1994;265:1098–1101.

49. Leder A, Kuo A, Shen MM, Leder P. In situ hybridization reveals co-expression of embryonic and adult alpha globin genes in the earliest murine erythrocyte progenitors. *Development.* 1992;116:1041–1049.

50. Carotta S, Pilat S, Mairhofer A, et al. Directed differentiation and mass cultivation of pure erythroid progenitors from mouse embryonic stem cells. *Blood.* 2004;104:1873–1880.

51. Olivier EN, Qiu C, Velho M, Hirsch RE, Bouhassira EE. Large-scale production of embryonic red blood cells from human embryonic stem cells. *Exp Hematol.* 2006;34:1635–1642.

52. Qiu C, Olivier EN, Velho M, Bouhassira EE. Globin switches in yolk-sac-like primitive and fetal-like definitive red blood cells produced from human embryonic stem cells. *Blood.* 2008;111:2400–2408.

53. Ma F, Ebihara Y, Umeda K, et al. Generation of functional erythrocytes from human embryonic stem cell-derived definitive hematopoiesis. *Proc Natl Acad Sci USA.* 2008;105:13087–13092.

54. Lu SJ, Feng Q, Park JS, et al. Biologic properties and enucleation of red blood cells from human embryonic stem cells. *Blood.* 2008;112:4475–4484.

55. Guerriero R, Mattia G, Testa U, et al. Stromal cell-derived factor 1alpha increases polyploidization of megakaryocytes generated by human hematopoietic progenitor cells. *Blood.* 2001;97:2587–2595.

56. Eto K, Murphy R, Kerrigan SW, et al. Megakaryocytes derived from embryonic stem cells implicate CalDAG-GEFI in integrin signaling. *Proc Natl Acad Sci USA.* 2002;99:12819–12824.

57. Fujimoto TT, Kohata S, Suzuki H, Miyazaki H, Fujimura K. Production of functional platelets by differentiated embryonic stem (ES) cells in vitro. *Blood.* 2003;102:4044–4051.

58. Nishikii H, Eto K, Tamura N, et al. Metalloproteinase regulation improves in vitro generation of efficacious platelets from mouse embryonic stem cells. *J Exp Med.* 2008;205:1917–1927.

59. Gaur M, Kamata T, Wang S, Moran B, Shattil SJ, Leavitt AD. Megakaryocytes derived from human embryonic stem cells: a genetically tractable system to study megakaryocytopoiesis and integrin function. *J Thromb Haemost.* 2006;4:436–442.

60. Takayama N, Nishikii H, Usui J, et al. Generation of functional platelets from human embryonic stem cells in vitro via ES-sacs, VEGF-promoted structures that concentrate hematopoietic progenitors. *Blood.* 2008;111:5298–5306.

61. Choi K. The hemangioblast: a common progenitor of hematopoietic and endothelial cells. *J Hematother Stem Cell Res*. 2002;11:91–101.

62. Conway EM, Collen D, Carmeliet P. Molecular mechanisms of blood vessel growth. *Cardiovasc Res*. 2001;49:507–521.

63. Levenberg S, Golub JS, Amit M, Itskovitz-Eldor J, Langer R. Endothelial cells derived from human embryonic stem cells. *Proc Natl Acad Sci USA*. 2002;99:4391–4396.

64. Chen T, Bai H, Shao Y, et al. SDF-1/CXCR4 signaling modifies the capillary-like organization of human embryonic stem cell-derived endothelium in vitro. *Stem Cells*. 2007;25:392–401.

65. Cho SW, Moon SH, Lee SH, et al. Improvement of postnatal neovascularization by human embryonic stem cell derived endothelial-like cell transplantation in a mouse model of hindlimb ischemia. *Circulation*. 2007;116:2409–2419.

66. Oyamada N, Itoh H, Sone M, et al. Transplantation of vascular cells derived from human embryonic stem cells contributes to vascular regeneration after stroke in mice. *J Transl Med*. 2008;6:54.

67. Wang ZZ, Au P, Chen T, et al. Endothelial cells derived from human embryonic stem cells form durable blood vessels in vivo. *Nat Biotechnol*. 2007;25:317–318.

68. Huang H, Zhao X, Chen L, et al. Differentiation of human embryonic stem cells into smooth muscle cells in adherent monolayer culture. *Biochem Biophys Res Commun*. 2006;351:321–327.

69. Sone M, Itoh H, Yamahara K, et al. Pathway for differentiation of human embryonic stem cells to vascular cell components and their potential for vascular regeneration. *Arterioscler Thromb Vasc Biol*. 2007;27:2127–2134.

70. Ferreira LS, Gerecht S, Shieh HF, et al. Vascular progenitor cells isolated from human embryonic stem cells give rise to endothelial and smooth muscle like cells and form vascular networks in vivo. *Circ Res*. 2007;101:286–294.

71. Choi K, Kennedy M, Kazarov A, Papadimitriou JC, Keller G. A common precursor for hematopoietic and endothelial cells. *Development*. 1998;125:725–732.

72. Kennedy M, Firpo M, Choi K, et al. A common precursor for primitive erythropoiesis and definitive haematopoiesis. *Nature*. 1997;386:488–493.

73. Wang L, Li L, Shojaei F, et al. Endothelial and hematopoietic cell fate of human embryonic stem cells originates from primitive endothelium with hemangioblastic properties. *Immunity*. 2004;21:31–41.

74. Zambidis ET, Peault B, Park TS, Bunz F, Civin CI. Hematopoietic differentiation of human embryonic stem cells progresses through sequential hematoendothelial, primitive, and definitive stages resembling human yolk sac development. *Blood*. 2005;106:860–870.

75. Umeda K, Heike T, Yoshimoto M, et al. Identification and characterization of hemoangiogenic progenitors during cynomolgus monkey embryonic stem cell differentiation. *Stem Cells*. 2006;24:1348–1358.

76. Kennedy M, D'Souza SL, Lynch-Kattman M, Schwantz S, Keller G. Development of the hemangioblast defines the onset of hematopoiesis in human ES cell differentiation cultures. *Blood*. 2007;109:2679–2687.

77. Lu SJ, Luo C, Holton K, Feng Q, Ivanova Y, Lanza R. Robust generation of hemangioblastic progenitors from human embryonic stem cells. *Regen Med*. 2008;3:693–704.

78. Lu SJ, Ivanova Y, Feng Q, Luo C, Lanza R. Hemangioblasts from human embryonic stem cells generate multilayered blood vessels with functional smooth muscle cells. *Regen Med*. 2009;4:37–47.

79. Martin MJ, Muotri A, Gage F, Varki A. Human embryonic stem cells express an immunogenic nonhuman sialic acid. *Nat Med*. 2005;11:228–232.

80. Ludwig TE, Bergendahl V, Levenstein ME, Yu J, Probasco MD, Thomson JA. Feeder-independent culture of human embryonic stem cells. *Nat Methods*. 2006;3:637–646.

81. Hentze H, Soong PL, Wang ST, Phillips BW, Putti TC, Dunn NR. Teratoma formation by human embryonic stem cells: evaluation of essential parameters for future safety studies. *Stem Cell Res*. 2009;2:198–210.

82. Brederlau A, Correia AS, Anisimov SV, et al. Transplantation of human embryonic stem cell-derived cells to a rat model of Parkinson's disease: effect of in vitro differentiation on graft survival and teratoma formation. *Stem Cells*. 2006;24:1433–1440.

83. Kroon E, Martinson LA, Kadoya K, et al. Pancreatic endoderm derived from human embryonic stem cells generates glucose-responsive insulin-secreting cells in vivo. *Nat Biotechnol*. 2008;26:443–452.

84. Mitalipova MM, Rao RR, Hoyer DM, et al. Preserving the genetic integrity of human embryonic stem cells. *Nat Biotechnol*. 2005;23:19–20.

85. Draper JS, Smith K, Gokhale P, et al. Recurrent gain of chromosomes 17q and 12 in cultured human embryonic stem cells. *Nat Biotechnol*. 2004;22:53–54.

86. Spits C, Mateizel I, Geens M, et al. Recurrent chromosomal abnormalities in human embryonic stem cells. *Nat Biotechnol*. 2008;26:1361–1363.

87. Osafune K, Caron L, Borowiak M, et al. Marked differences in differentiation propensity among human embryonic stem cell lines. *Nat Biotechnol*. 2008;26:313–315.

88. Swijnenburg RJ, Schrepfer S, Govaert JA, et al. Immunosuppressive therapy mitigates immunological rejection of human embryonic stem cell xenografts. *Proc Natl Acad Sci USA*. 2008;105:12991–12996.

HUMAN STEM CELLS FOR DRUG SCREENING, TOXICITY, SPECIFICITY, AND OFF-TARGET EFFECTS

29

Arnaud Lacoste and Mark Burcin

HUMAN EMBRYONIC STEM CELLS IN THE CONTEXT OF MODERN DRUG DISCOVERY

In order to be successful, modern drug development strategies must take into account a wide range of factors. For example, duration of research and development (R&D) cycles, size of clinical trials, rates of attrition as drugs progress through the development pipeline, and time required for approval have a major impact on the qualitative and quantitative outcome of the drug discovery process. A successful strategy aims at reducing costs associated with drug discovery and development while ensuring that therapeutics are as efficient and safe as possible. Since clinical trial size and approval processes are optimized to ensure the patients' benefit, they can rarely be modified. However, drug discovery approaches can be optimized to reduce costs and improve drug quality.

Currently, most if not all pharmaceutical companies use a target-based approach to drug discovery (Fig. 29.1). The process starts with the identification of biological targets that play a key role in the onset or progression of a given disease. This target identification phase is followed by a hit identification phase, where therapeutic agents that modulate the activity of the target are discovered, typically using a high-throughput screening approach. Hits then enter the drug development pipeline, where they are modified to improve efficacy and maximize safety. Until recently, this drug discovery and development process has remained relatively similar across all pharmaceutical companies. However, as the industry is now attacking diseases of increasing complexity, a trend toward a deeper understanding of pathway biology and the molecular basis of human diseases is beginning to emerge. Perhaps pioneered by Novartis and the Novartis Institute for Biomedical Research, this novel approach adds a dimension to traditional drug discovery. Instead of looking for isolated drug targets, researchers now map entire networks of biological pathways and identify which one(s) could be safely and efficiently targeted to treat diseases. In a context of extreme commercial pressure, this strategy may appear risky because of the important upfront financial investment needed to support exploratory phases of this mapping effort. However, recent results suggest that this approach to drug discovery will be sustainable in the long run because one signaling pathway is often associated with many diseases. For example, Novartis compound RAD001, which was originally FDA approved as an immunosuppressant to prevent the rejection of

Human Stem Cell Technology and Biology, edited by Stein, Borowski, Luong, Shi, Smith, and Vazquez
Copyright © 2011 Wiley-Blackwell.

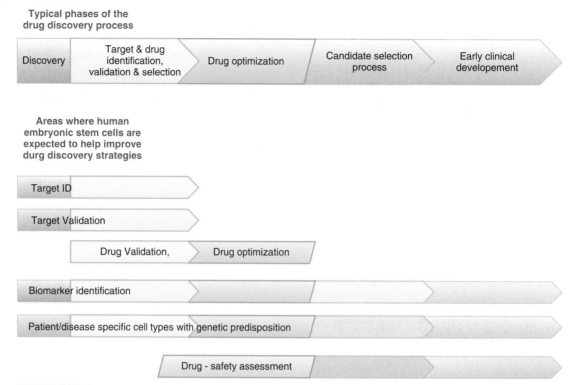

FIGURE 29.1. *Utilization of hESC-derived cell types during drug development.*

organ transplants, now shows clinical value for the treatment of multiple cancers, tuberous sclerosis, and glaucoma. This example illustrates how, once a drug has been developed to control a particular signaling node and treat a given disease, the signaling map indicates what other diseases could be treated by controlling the same node. In other words, in-depth understanding of pathway biology makes it possible to develop drugs that treat several diseases, including rare illnesses for which traditional drug discovery approaches may be financially unsustainable.

It has now become clear that whether a pharmaceutical company favors the traditional single target identification approach or the novel pathway mapping strategy, the increasing complexity of the drug discovery process requires technologies and biological models that, more than ever, improve productivity and exhibit relevance to patients' illnesses. Therefore in the past two decades, pharmaceutical companies have added robotized high-throughput systems, quantitative biology platforms, pharmacogenetics, and automated imaging to the drug discovery toolbox. More recently, the pharmaceutical industry has started exploring whether human embryonic stem cells could help develop in vitro models that are directly relevant to human diseases. Such models would help map disease-relevant signaling networks, improve the R&D process, and reduce drug attrition rates. Ultimately, these improvements should help design novel drug discovery strategies that benefit patients and healthcare systems, while sustaining a productive pharmaceutical drug discovery pipeline.

BASIC REQUIREMENTS FOR HUMAN EMBRYONIC STEM CELL-BASED MODELS IN PHARMACEUTICAL SETTINGS

Human Embryonic Stem Cell-Derived Models as New Drug Discovery Tools

The design of functional in vitro assays that recapitulate in vivo physiological processes is central to the drug discovery and development process. In most cases, functional assays use cell culture models derived from naturally occurring tumors or artificially immortalized cell types. These cellular models have been widely used in the pharmaceutical industry because immortalization makes them expandable and amenable to large-scale screening campaigns and toxicology studies. However, these cellular systems are often derived from animals, not human tissues, and as a cause or consequence of their immortalization, they tend to carry genetic mutations or karyotypic abnormalities. As a result, their phenotype is usually different from that of endogenous cells in a normal or diseased human organ. For example, they often fail to express appropriate signaling pathways or molecular networks downstream of a receptor of interest.

Primary cell cultures offer an alternative to immortalized cell systems, but in many cases, they are difficult to generate in large quantities. In addition, they tend to lose their original phenotype during the isolation process or as a result of maintenance in vitro. For example, primary satellite cell cultures isolated from skeletal muscle offer an alternative to the use of immortalized C2C12 cells in muscle regeneration studies. However, as a result of isolation procedures and in vitro culture, satellite cells rapidly become activated, enter the cell cycle, and differentiate toward committed myoblasts that differ from endogenous satellite cells.

Human embryonic stem cells are currently attracting a lot of attention from the pharmaceutical industry because they exhibit many of the advantages but only a few of the challenges associated with immortalized or primary cell systems. For example, their pluripotent nature suggests that they could be used to derive many, if not all, differentiated cell types of the human body. In addition, their ability to self-renew implies that large quantities of human embryonic stem cells could be generated in vitro to derive differentiated cell types at a scale that is compatible with the drug discovery and development process. However, to successfully implement human embryonic stem cell-based technologies into the pharmaceutical drug development process, a number of challenges must be addressed. First, we must establish reliable methods for large-scale expansion of undifferentiated, human embryonic stem cells. Second, we need to identify conditions that direct the differentiation of human embryonic stem cells toward specific cell types of interest. Third, we need to design functional assays that can be used for target identification, high-throughput screens, and/or toxicology studies.

Culture of Human Embryonic Stem Cells for Drug Discovery Platforms

Since the first derivation of human embryonic stem cells,[1] a number of studies have shown that conditions used for the maintenance of mouse embryonic stem cells are not appropriate for human embryonic stem cell culture. For example, activation of LIF/Stat3 signaling, which is essential for the maintenance of mouse embryonic stem cells, does not support self-renewal of human embryonic stem cells.[2,3] Instead, FGF and activin/nodal signaling are required for human embryonic stem cell self-renewal.[4] Furthermore, BMP4, which blocks mouse embryonic stem cell differentiation along the neuroectoderm default pathway,[5,6] induces trophectoderm differentiation in human embryonic stem cells.[7] So far, only Wnt signaling has been reported to help maintain the pluripotent state in both human and mouse embryonic stem cells,[3] although Wnt activity alone is not sufficient to maintain pluripotency in human cells.[8] These studies highlight the need for culture techniques that are specific for the maintenance of human embryonic stem cells. Currently, these cells are maintained on irradiated or mitomycin-C treated embryonic fibroblasts, and cell passaging is performed manually in order to remove cells undergoing spontaneous differentiation. This process is labor intensive and difficult to expand for large-scale pharmaceutical screening platforms. Recently, feeder-free human embryonic stem cell cultures have been successfully developed on

Matrigel;™ however, this culture technique still requires manual passaging for the removal of spontaneously differentiating cell clusters. Ideally, cultures in suspension would enable large-scale expansion of human embryonic stem cells; but these cells tend to form differentiating embryoid bodies when grown in suspension and they exhibit poor viability when kept in a dissociated state. Recent research has shown that culture of human embryonic stem cells in the presence of Rho-associated kinase (ROCK) inhibitor, Y-27632, markedly diminishes dissociation-induced apoptosis.[9] This constitutes an important step toward the development of feeder-free suspension cultures, which can be used for large-scale expansion of undifferentiated human embryonic stem cells.

Differentiation of Human Embryonic Stem Cells for Drug Discovery

Our current knowledge of early embryonic development is useful for differentiating human embryonic stem cells into many cell types of interest, including neurons, ventricular and atrial cardiomyocytes, many cells of the hematopoietic lineages, and pancreatic insulin-producing cells. However, before we can reproducibly generate defined cell types at scales and levels of purity that are compatible with pharmaceutical settings, we need to increase our understanding of stem cell biology. In addition, most human embryonic stem cell-derived cell types exhibit embryonic or fetal characteristics, and, in many cases, knowledge is lacking on how mature cells could be obtained. This constitutes a significant barrier toward the application of human embryonic stem cells to drug discovery.

However, progress is being made to improve differentiation protocols and facilitate the generation of homogeneous cell preparations from human embryonic stem cells. For example, recent research has shown that the synergistic action of two inhibitors of SMAD signaling, Noggin and SB431542, is sufficient to induce rapid and complete neural conversion of 80% of human embryonic stem cells under adherent and feeder-free culture conditions.[10] Interestingly, this finding is a direct result of early frog studies that identified bone morphogenic protein (BMP) signaling inhibitors as neural-inducing factors in the Spemann organizer.[11,12] This example illustrates how basic research on pathway biology will eventually help design models that improve drug discovery strategies. However, such success stories in the field of human embryonic stem cell-derived models remain rare. In most cases, disease-relevant cell types can only be obtained in relatively small quantities and in highly heterogeneous cultures. Currently, most differentiation protocols start with the generation of human embryonic stem cell aggregates called "embryoid bodies." "Embryoid" is probably an inadequate term to describe these aggregates because, unlike real embryos, embryoid bodies are devoid of organizers and their differentiation proceeds in a largely chaotic fashion. This leads to the generation of highly heterogeneous cultures where cell types that would never come in contact during normal development can influence each other's differentiation and possibly lead to the formation of cellular structures that are not relevant to normal human biology. However, during the differentiation process, many cell types of interest emerge and one of the main goals of current human embryonic stem cell research is to find ways to sort them and generate pure preparations of disease-relevant cell types. In certain cases, selection of a cell type of interest is relatively straightforward. For example, following embryoid body-mediated differentiation, human embryonic stem cell-derived heart cells form easily identifiable beating clusters that can be manually selected and subcultured to obtain 60% pure cardiomyocyte cultures.[13] In most cases, however, the cell type of interest cannot be distinguished from other cells present in embryoid body-derived cultures and the use of transgenic selectable markers or reporter systems would be preferable. Unfortunately, transgenesis in human embryonic stem cells is not as advanced as in mouse and, despite recent improvements in gene delivery technologies,[14–16] selectable markers and reporter systems are not routinely used. Undoubtedly, future technological improvements in this domain will help generate large quantities of human embryonic stem cell derivatives that can be used by the pharmaceutical industry in target identification, high-throughput screens, and toxicology studies.

HUMAN EMBRYONIC STEM CELL-DERIVED MODELS FOR TARGET ID

Despite the substantial investment into human embryonic stem cell-based cell models, their use for new drug target identification and validation remains relatively unexplored. One important advantage for using human embryonic stem cell-derived cell lineages during preclinical drug development is the potential to induce specific disease pathologies within the relevant human cell type. For example, human embryonic stem cell-derived cell lines could be infected with cDNA or siRNA libraries. Afterwards, a specific pathology could be induced and screens designed to detect a protective readout in pools of modified cells (Fig. 29.2). Using disease-relevant cell types opens up the opportunity for identifying targets that selectively regulate distinct pathways. So, rather than focusing on selected targets such as kinases or GPCRs, one could identify novel targets in an unbiased way and validate the interaction of known targets within disease-relevant pathways. During recent years, transgenic mouse models have been extensively used for target validation. Genetic knockout or overexpression of specific genes in mice has been especially useful for studying the mechanism of disease onset and progression. Human embryonic stem cell-derived cell types could confirm that results observed in knockout or transgenic rodents translate to humans and that the observed target is cell autonomous. For inherited disorders, the ability to generate induced pluripotent stem cells (iPS cells) further extends the opportunity to develop patient-specific cell models carrying genetic mutations responsible for a defined disease. By using such "diseased" cell types, the effect of a specific mutation could be better understood in a pathways-centric view. For example, the generation of dopamineric neurons from human embryonic stem cells opened the opportunity to identify and validate targets for Parkinson disease. The overexpression of alpha-synuclein, a contributor of the pathology of Parkinson disease, led to cytotoxicity in human embryonic stem cell-derived dopaminergic neurons.[17] Such a disease-specific in vitro model reflects in vivo pathogenesis and is an ideal tool for target validation. Recent success generating human embryonic stem cell-derived cardiomyocytes provides further opportunities to yield beneficial information for cardiac-related disease.[13,18,19] In the future, human cardiomyocytes could be used to induce pathologies such as hypertrophy or oxidative stress and screened for protective readouts. Another promising, recent development is the generation of large numbers of endocrine, pancreatic beta cells.[20] The use of human embryonic stem cell-derived beta cells could provide models to identify novel targets and

FIGURE 29.2. *Novel targets by gain or loss of function screens.*

possibly enable new approaches to the treatment of diabetes. For example, such new therapies could make use of (1) cell replacement to restore function to the damaged pancreatic islets and/or (2) drug treatment to stimulate the body's own regenerative capabilities in the pancreas by promoting survival, migration/homing, proliferation, and differentiation of endogenous pancreatic beta or progenitor cells. However, such approaches will require identification of renewable cell sources of functional beta cells, an improved ability to manipulate their proliferation and differentiation, and a better understanding of the signaling pathways that control pancreatic endocrine-cell fate and beta-cell development. Cell-based phenotypic and pathway-specific screens of human embryonic stem cell-derived beta cells using cDNAs, siRNAs, synthetic small molecules, and natural products can be used to identify targets that selectively regulate beta-cell protection and expansion in diabetic patients. These targets may also provide new insight into the molecular mechanisms of endocrine pancreas development and could ultimately facilitate therapeutic applications of stem cells in diabetes and the development of small molecule therapeutics to stimulate in vivo regeneration of the pancreatic islet.

HUMAN EMBRYONIC STEM CELL-DERIVED MODELS FOR HIGH-THROUGHPUT SCREENS

Together with combinatorial chemistry, high-throughput screens have considerably expanded our ability to evaluate the biological activity and therapeutic potential of small molecular entities. In fact, high-throughput screening has now become the most widely used approach for hit identification in medium and large pharmaceutical companies. However, the proportion of high-throughput screen-based hits that progress to lead optimization remains low (less than 1%). In addition, the development of high-throughput screens has so far failed to reverse the downward trend in the number of new chemical entities that reach the market. This screening approach has been employed for only two decades, and since 10–15 years are usually required for the discovery and development of new drugs, it is probably too early to fully evaluate the return on investment associated with high-throughput screens. Nevertheless, in an effort to increase productivity, pharmaceutical companies continuously strive to improve their screening strategies. In recent years, automated systems, novel assay technologies, and extensive miniaturization have been integrated within screening platforms to increase discovery productivity and maximize screening fidelity. It is in this context of intense commercial pressure and continuous technological customization that human embryonic stem cell-based approaches are entering the field of high-throughput screening.

As discussed above, large-scale culture of human embryonic stem cells and/or reliable protocols for their differentiation into disease-relevant cell types is required for the application of human embryonic stem cell-based approaches to any aspect of drug discovery, such as target identification, hit identification, lead development, and toxicology studies. However, incorporating the use of human embryonic stem cells into high-throughput screening platforms presents additional key challenges. First, protocols must be established to make human embryonic stem cells and their derivatives amenable to high-density assays. High-throughput screens are typically performed in 96-, 384-, or 1536-well microtiter plate formats. A recent survey of the drug discovery community indicated that more than 60% of high-throughput screens are now performed in the 384-well format and the 1536-well plate is currently the fastest growing format.[21] As liquid handling technologies improve and plate reader development continues, it is likely that, in the near future, most, if not all, high-throughput screens will be performed using the 1536-well plate format and higher density formats are already emerging. For example, the use of 3456-well plates and microfluidic chip arrays has already been reported. This trend toward higher density and increased miniaturization presents logistical advantages and reduces limitations associated with reagent cost. Therefore robust protocols will need to be developed for human embryonic stem cell-based models to become compatible with automated platforms and miniaturized formats. This will present unique challenges. For example, one can anticipate that the requirement for feeder layers for the maintenance of human embryonic stem cells and some of their derivatives, such as dopaminergic and motor neurons, may present a barrier toward assay automation and miniaturization. In addition, human embryonic stem cells and certain human embryonic stem cell-derived models change phenotype when cultured at high densities. When grown at lower densities, they form colonies and only cover part of the culture substrate. This is likely to cause well-to-well heterogeneity that may also be limiting in automated, image analysis-based screens. Until now, screening campaigns using human embryonic stem cells have been performed manually or in semiautomated fashion at a small scale.[22] Additional efforts are required for human embryonic stem cell-based models to become compatible with the high-density screening formats commonly used by the drug discovery industry.

The use of human embryonic stem cell-based approaches in high-throughput screening campaigns will also require reliable assays to probe disease-relevant cellular and molecular processes. Recent advances in gene targeting and gene delivery technologies[14–16] greatly facilitate the generation of transgenic human embryonic stem cell lines carrying cellular biosensors such as fluorescent or luminescent proteins. Since human embryonic stem cells can self-renew indefinitely, such lines should allow the drug discovery community to generate large quantities of disease-relevant cell types that are homogeneously labeled with a marker of interest. This should eliminate logistical constraints associated with large-scale transgene delivery in primary cell models. Transgenic human embryonic stem cell lines would also facilitate the large-scale production of cellular models that are amenable to high-content screening platforms, a complex screening format where reporter transgenes, high-resolution fluorescence microscopy, and automated image analysis are combined to monitor biological processes at a subcellular scale.

In conclusion, although major challenges must be overcome before human embryonic stem cell-based models can be used reliably in high-throughput screening campaigns, ongoing scientific and technological advances provide hope that, in the future, such models will play a key role in pharmaceutical discovery efforts.

DERIVING HUMAN CELLULAR MODELS FOR TOXICOLOGY STUDIES

Because species-specific differences limit the predictability of animal models, toxicity is often observed in late phases of drug development. For this reason developing more predictive human cell models is mandatory. For example, primary hepatocytes isolated from human liver are used as a key source in drug discovery. Because of their high-level expression of drug-metabolizing enzymes, they constitute an important tool for the investigation of drug metabolism toxicology. However, due to the difficulties in obtaining sufficient quantities of human hepatocytes, the industry has to rely on animal models instead.[23] As an alternative source, studies have shown that such cells could be isolated from fetal liver or differentiated human embryonic stem cells.[24,25] However, despite the fact that each cell derivative shows some selected hepatic function, differentiation into fully functional and predictive hepatocytes has yet to be accomplished. In addition, the human liver does not consist exclusively of hepatocytes and, for this reason, many other cell types could contribute to hepatotoxicity. To address this, the generation of simple three-dimensional liver tissue organoids from human embryonic stem cells could improve the accuracy in predicting human toxicity.

Embryo toxicity can be the negative consequence of various drugs on the reproductive cycle, which then translates into adverse effects on the embryo. To avoid this outcome, animal studies are used to predict toxic effects of drugs during embryo development. Embryonic stem cells with their potential to differentiate toward all cell lineages provide a potent cell model for identifying potential embryo toxic compounds during drug development.[26,27] The embryonic stem cell test (EST) by the European Center for Validation of Alternative Methods was the first assay established to predict teratogens. The underlying principle of the EST lies in the disruption of cardiomyocyte differentiation during mouse ESC differentiation and has been shown to have a very good correlation with in vivo studies. Using human embryonic stem cells for developing additional embryotoxicity assays seems to be the next logical step. By generating gene signatures of several lineage-specific differentiation processes, adverse effects on embryos could be detected by the disturbance of the specific gene expression profile after drug incubation. In addition, the modifications of proteins as well as miRNA signatures could be used for predicting embryotoxicity. To evaluate a range of potential toxicants that impair normal human embryonic development, the identification of drugs that have a negative effect on fertility or trigger predisposition to disease would also be beneficial. Here human embryonic stem cell technology could provide a predictive cell model system by reducing the use of animal numbers for toxicology testing and eliminating species-specific misinterpretations.

Human embryonic stem cell technology, for example, is well suited to provide a powerful enhancement to well-established cardiotoxicity assays. Throughout the drug discovery process, cardiotoxicity is one of the most important caveats during the development of new chemical entities. As drug-induced cardiotoxicity can result in sudden death, the US Food and Drug Administration (FDA) put in place assay guidelines for testing drug candidates in vitro and in vivo for their potential to evoke ventricular repolarization disturbances. Drugs blocking the hERG channel can cause ventricular arrhythmias in the clinic and are all shown to be blockers of IKr, the rapid component of delayed rectifier potassium current. In recent years the identification of such cardiac complications has led to the withdrawal of several drugs from the market because, under certain circumstances, they cause potentially fatal cardiac arrhythmia called torsade de pointes (TdP). The underlying cellular mechanisms of such cardiac complications can be quite diverse but involve, in most cases, the modulation of cardiac ion channels/receptors. For this reasons, in vitro assays for assessing the hERG channel activity have been widely used in the pharmaceutical industry during preclinical, cardiac safety profiling.[28] Even as the number of assays has increased over the past years, the predictive value of these assays is limited by false positives and negatives, and there is no consensus on how to define the safety margins based on in vitro data, which would accurately translate to the human.[29] In general, the biggest challenge for the pharmaceutical industry is that none of the assays alone is predictive for the clinical situation and in most cases the first time a new molecule faces a human cardiomyocyte is in the clinic. For this reason, developing assays that are more predictive for humans would be a big step forward to develop safe drugs faster and get them into the clinic quicker. One of the biggest opportunities in this area will be the generation of genetically distinct stem cell lines in order to validate drug response and toxicity within specific patient populations carrying various genetic predispositions.

Human primary cardiomyocytes are not suitable for drug discovery and development applications due to the limitations of donor material and difficulties associated with isolation of cardiomyocytes from the heart tissue. As a result, there is a clear need for novel in vitro models representing changes in cardiac electrophysiology to aid the discovery of new effective therapies.[30,31] Human embryonic stem cell-derived cardiomyocytes could provide an unlimited source of cell material that could be used

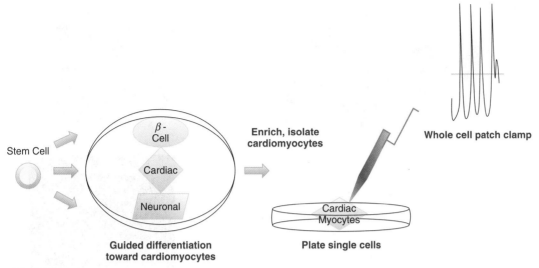

FIGURE 29.3. *hESC-derived cardiomyocytes for drug safety assessment.*

for preclinical, cardio-safety assessment (Fig. 29.3). The development of the mammalian heart involves complex networks of signaling pathways that temporally and spatially regulate cell growth, proliferation, specification, and differentiation toward cardiac cell lineages. Conventional methods for differentiating human embryonic stem cells into cardiomyocytes result in low and inconsistent yields, making it difficult to establish cellular models. Developing improved differentiation protocols by inducing defined, developmental transitions would offset such limitations. To systematically and consistently improve human embryonic stem cell differentiation toward the cardiac lineage,[13] we have recently described a protocol that substantially improves the differentiation of human embryonic stem cell into cardiomyocytes by implementing the following critical steps: (1) the enzymatic and physical disruption of undifferentiated human embryonic stem cells to generate EBs; (2) the early and transient treatment of human embryonic stem cells with Wnt3a; and (3) the gradual reduction of serum and insulin levels. During the first step of this protocol undifferentiated human embryonic stem cells undergo enzymatic and physical disruption into small cell aggregates. These small cell aggregates then develop into suitably sized embryoid bodies, which prove to be crucial for an improved cardiomyogenesis of human embryonic stem cells. Optimizing the exposure window of Wnt3a during the second step also provided substantial enhancement to the differentiation procedure. We found that early and brief exposure with Wnt3a during the first 48 hours of differentiation was critical for preferentially reprogramming human embryonic stem cell toward mesoderm and endoderm, excluding ectoderm. In addition to mesendoderm induction, Wnt3a significantly improved the formation of embryoid bodies, either by increasing cell viability or cell proliferation. During the third step of the differentiation process, we augmented cardiomyogenesis from human embryonic stem cells even further by gradually reducing serum and insulin levels. Under these conditions, we were able to generate significant quantities of beating clusters that were substantially enriched in cardiomyocytes ($\sim$50% enrichment) and capable of cultivation under serum-free conditions for up to 60 days. Then, in order to utilize these cells for cardio-safety assessment, cardiomyocytes were further purified by subjecting the cell population at differentiation day 30 to a Percoll™ gradient (Fig. 29.4). As an alternative approach, beating cardiomyocyte containing clusters can be manually microdissected, a labor-intensive but highly efficient approach. After isolation and separation into single cells, the human embryonic stem cell-derived cardiomyocytes were subjected to whole-cell patch-clamp assays

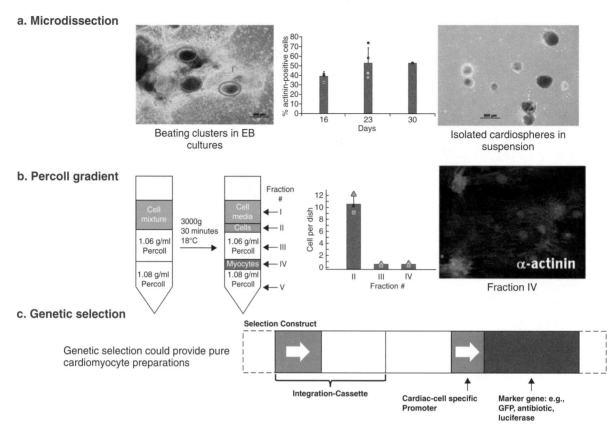

a. Microdissection

Beating clusters in EB cultures

Isolated cardiospheres in suspension

b. Percoll gradient

Fraction IV

c. Genetic selection

Genetic selection could provide pure cardiomyocyte preparations

d. FACS sorting

FACS based methods to isolate cardiac precursor cell populations

FIGURE 29.4. *Enrichment strategies for hESC-derived cell types.*

(current-clamp/voltage-clamp). From these measurements the action potentials (APs) in spontaneously beating human embryonic stem cell-derived cardiomyocytes were distinct (atrial-, nodal, ventricular-like) and comparable with those reported in the literature. In addition, APs could be triggered by a depolarizing step, in which Ca^{2+} channel blockade depressed spontaneous activity and hERG channel blockade led to action potential prolongation and early after-depolarization. These studies demonstrate that human embryonic stem cell-derived cardiomyocytes can be used to advantage in cardio-safety assessments. In the future, the utilization of genetic selection methods to generate pure ventricular human cardiomyocytes will provide an opportunity to use these cells in combination with automated electrophysiology platforms and will allow development of cost-effective and highly predictive assays for investigating QT interval prolongation and proarrhythmia liability of novel compounds.[32,33] Future key activities will need to confirm whether these phenotypic, relatively immature cardiomyocytes derived from human embryonic stem cells are functionally equivalent to native human cardiomyocytes.

INDUCED PLURIPOTENT STEM CELLS: MODELING HUMAN DISEASE FOR DRUG DISCOVERY

The pharmaceutical industry has shown interest in human induced pluripotent stem (iPS) cells before their initial description in scientific journals. In fact, whether the first human iPS cell patent was filed by Shinya Yamanaka from Kyoto University or Kazuhiro Sakurada from the drug company Bayer Yakuhin became a matter of debate[34] soon after peer-reviewed articles reported that Oct4, Sox2, Klf4, and c-myc could reprogram human fibroblasts into pluripotent stem cells.[35,36] In the past two years, several pharmaceutical companies have integrated reprogramming approaches in their R&D strategies to explore the potential of iPS cells for drug development. The general expectation is that patient-specific iPS cell lines will help generate large quantities of disease-relevant cell types for target identification, high-throughput screens, and toxicology studies. If iPS cell-based approaches can meet this expectation, they could trigger a revolution in drug discovery. However, before iPS cells can reach their full potential, a number of challenges must be overcome. First, as with human embryonic stem cells, large quantities of homogeneous iPS cells and their derivatives must be generated in a reproducible manner. To date, there has not been any public report of iPS cells or their derivatives being generated at a billion-cell scale. In addition, we are still lacking established standards and quality control criteria ensuring that iPS cell-derived products are homogeneous and exhibit a consistent phenotype at screening scale. Such standards have been established for human embryonic stem cell research,[37] but even if many of the same standards can be used for iPS cells, specific criteria may need to be defined for reprogrammed cells. For example, qualitative and quantitative assessment of reprogramming efficiency varies among different research groups.[38] This implies that the differentiation potential of iPS cell lines is likely to vary across different laboratories. This will have an impact on iPS cell-based drug discovery efforts. Therefore the issues of quality control and consistency will need to be addressed thoroughly in the near future.

In the meantime, however, small-scale experiments performed by academic research groups provide hope that iPS cells will soon play an important role during the drug discovery process. For example, iPS cell lines have been derived from patients affected by genetically inherited neurodegenerative disorders such as amyotrophic lateral sclerosis (ALS), caused by mutations in *SOD-1*, and spinal muscular atrophy (SMA), caused by a defect in SMN1 function. These reprogrammed pluripotent cells have been successfully differentiated into motor neurons, the specific cell type that degenerates during disease progression.[39,40] In the SMA study, the authors examined whether compounds such as valproic acid and tobramycin, which are known to increase SMN1 levels, were able to protect motor neurons from typical defects observed in SMA patients. They found that two days of drug treatment led to a significant improvement in SMN1-deficient cell functions.[40]

More recently, academic researchers used a similar approach to test drug candidates for the treatment of familial dysautonomia, a fatal peripheral neuropathy caused by a mutation in the gene encoding IκB kinase complex-associated protein (*IKBKAP*). They found that the plant hormone kinetin was able to rescue the IKBKAP defect in iPS cell-derived peripheral neurons and neural crest cells. Such proof-of-concept studies suggest that iPS cell derivatives will be useful for drug discovery, especially in the context of compound screens and drug validation studies.

FUTURE DIRECTIONS

Human embryonic stem cell- and iPS cell-based technologies are still new and, to a large extent, their potential for drug discovery remains to be fully explored. This field of research will need to overcome formidable challenges before it can significantly contribute to drug discovery and improve the practice of medicine. Importantly, in order to meet current expectations from the pharmaceutical industry, human embryonic stem cell models will need to reproducibly mirror human biology. To achieve this goal, efficient differentiation protocols for generating pure and physiologically relevant cell culture models are needed. In addition, human embryonic stem cell-based models will need to be scalable and amenable to high-throughput screens. Despite these numerous challenges, stem cell research is one of the fastest growing fields in contemporary biology. Technologies, such as the reprogramming of somatic cells using defined factors, which seemed realistic to only a few scientists a couple of years ago, have now become almost routine tools in stem cell research. In such a rapidly evolving context, it is difficult to precisely predict what directions this field will follow and how it will contribute to drug discovery. It seems likely that our current knowledge of human embryonic stem cell and iPS cell biology will open novel avenues in key areas of drug development such as target identification, hit validation and toxicology studies. This will probably happen in the relatively near future (2–5 years). But predicting how and to what extent human embryonic stem cells and iPS cells will ultimately alter the practice of medicine is less straightforward. Certainly, if scientific progress in the field of stem cell biology continues at its current pace, we can expect that a revolution in drug discovery is underway. Already, researchers are able to generate organ-like structures in vitro[41] in formats that could potentially be adapted to high-throughput screening platforms. Such advances suggest that, in the next few years, the combination of stem cell research together with quantitative biology, molecular genetics, epigenetics, and theoretical modeling will enable us to generate models of increased complexity and unprecedented relevance to human diseases. This should considerably improve the pharmaceutical industry's ability to create cost-effective, safe, and more efficient therapeutics.

REFERENCES

1. Thomson JA, Itskovitz-Eldor J, Shapiro SS, et al. Embryonic stem cell lines derived from human blastocysts. *Science*. 1998;282:1145–1147.

2. Humphrey RK, Beattie GM, Lopez AD, et al. Maintenance of pluripotency in human embryonic stem cells is STAT3 independent. *Stem Cells*. 2004;22:522–530.

3. Sato N, Sanjuan IM, Heke M, et al. Molecular signature of human embryonic stem cells and its comparison with the mouse. *Dev Biol*. 2003;260:404–413.

4. Vallier L, Alexander M, Pedersen RA. Activin/nodal and FGF pathways cooperate to maintain pluripotency of human embryonic stem cells. *J Cell Sci*. 2005;118:4495–4509.

5. Munoz-Sanjuan I, Brivanlou AH. Neural induction, the default model and embryonic stem cells. *Nat Rev Neurosci*. 2002;3:271–280.

6. Ying QL, Nichols J, Chambers I, et al. BMP induction of Id proteins suppresses differentiation and sustains embryonic stem cell self-renewal in collaboration with STAT3. *Cell*. 2003;115:281–292.

7. Xu RH, Chen X, Li DS, et al. BMP4 initiates human embryonic stem cell differentiation to trophoblast. *Nat Biotechnol*. 2002;20:1261–1264.

8. James D, Levine AJ, Besser D, et al. TGFbeta/activin/nodal signaling is necessary for the maintenance of pluripotency in human embryonic stem cells. *Development*. 2005;132:1273–1282.

9. Watanabe K, Ueno M, Kamiya D, et al. A ROCK inhibitor permits survival of dissociated human embryonic stem cells. *Nat Biotechnol*. 2007;25:681–686.

10. Chambers SM, Fasano CA, Papapetrou EP, et al. Highly efficient neural conversion of human ES and iPS cells by dual inhibition of SMAD signaling. *Nat Biotechnol*. 2009;27:275–280.

11. Sasai Y, Lu B, Steinbeisser H, et al. *Xenopus chordin: a novel dorsalizing factor activated by organizer-specific homeobox genes. Cell*. 1994;79:779–790.

12. Hemmati-Brivanlou A, Kelly OG, Melton DA. Follistatin, an antagonist of activin, is expressed in the Spemann organizer and displays direct neuralizing activity. *Cell*. 1994;77:283–295.

13. Tran TH, Wang X, Browne C, et al. Wnt3a-induced mesoderm formation and cardiomyogenesis in human embryonic stem cells. *Stem Cells*. 2009;27:1869–1878.

14. Bu L, Jiang X, Martin-Puig S, et al. Human ISL1 heart progenitors generate diverse multipotent cardiovascular cell lineages. *Nature*. 2009;460:113–117.

15. Hockemeyer D, Soldner F, Beard C, et al. Efficient targeting of expressed and silent genes in human ESCs and iPSCs using zinc-finger nucleases. *Nat Biotechnol*. 2009;27:851–857.

16. Lacoste A, Berenshteyn F, Brivanlou AH. An efficient and reversible transposable system for gene delivery and lineage-specific differentiation in human embryonic stem cells. *Cell Stem Cell*. 2009;5:332–342.

17. Gasser T. Molecular pathogenesis of Parkinson disease: insights from genetic studies. *Expert Rev Mol Med*. 2009;11: e22.

18. Burridge PW, Anderson D, Priddle H, et al. Improved human embryonic stem cell embryoid body homogeneity and cardiomyocyte differentiation from a novel V-96 plate aggregation system highlights interline variability. *Stem Cells*. 2007;25:929–938.

19. Freund C, Ward-van Oostwaard D, Monshouwer-Kloots J, et al. Insulin redirects differentiation from cardiogenic mesoderm and endoderm to neuroectoderm in differentiating human embryonic stem cells. *Stem Cells*. 2008;26:724–733.

20. Baetge EE. Production of beta-cells from human embryonic stem cells. *Diabetes Obes Metab*. 2008;10(Suppl 4): 186–194.

21. High-throughput screening 2007: HighTech Business Decision, Reports new strategies, success rates, and the use of enabling technologies. HighTech Business Decisions; 2007.

22. Desbordes SC, Placantonakis DG, Ciro A, et al. High-throughput screening assay for the identification of compounds regulating self-renewal and differentiation in human embryonic stem cells. *Cell Stem Cell*. 2008;2:602–612.

23. Cross DM, Bayliss MK. A commentary on the use of hepatocytes in drug metabolism studies during drug discovery and development. *Drug Metab Rev*. 2000;32:219–240.

24. Hay DC, Fletcher J, Payne C, et al. Highly efficient differentiation of hESCs to functional hepatic endoderm requires ActivinA and Wnt3a signaling. *Proc Natl Acad Sci USA*. 2008;105:12301–12306.

25. Jensen J, Hyllner J, Bjorquist P. Human embryonic stem cell technologies and drug discovery. *J Cell Physiol*. 2009;219:513–519.

26. Rohwedel J, Guan K, Hegert C, et al. Embryonic stem cells as an in vitro model for mutagenicity, cytotoxicity and embryotoxicity studies: present state and future prospects. *Toxicol in vitro*. 2001;15:741–753.

27. Krtolica A, Ilic D, Genbacev O, et al. Human embryonic stem cells as a model for embryotoxicity screening. *Regen Med*. 2009;4:449–459.

28. Pollard CE, Valentin JP, Hammond TG. Strategies to reduce the risk of drug-induced QT interval prolongation: a pharmaceutical company perspective. *Br J Pharmacol*. 2008;154:1538–1543.

29. Redfern WS, Carlsson L, Davis AS, et al. Relationships between preclinical cardiac electrophysiology, clinical QT interval prolongation and torsade de pointes for a broad range of drugs: evidence for a provisional safety margin in drug development. *Cardiovasc Res*. 2003;58:32–45.

30. Nass RD, Aiba T, Tomaselli GF, et al. Mechanisms of disease: ion channel remodeling in the failing ventricle. *Nat Clin Pract Cardiovasc Med*. 2008;5:196–207.

31. Marrus SB, Nerbonne JM. Mechanisms linking short- and long-term electrical remodeling in the heart . . . is it a stretch? *Channels (Austin)*. 2008;2.

32. He, J-Q, Ma. J, Anson B. Use of pluripotent stem cell-derived cardiomyocytes to understand mechanisms of cardiotoxic compounds. *Cell Notes*. 2009;23:10–12.

33. Cezar GG. Can human embryonic stem cells contribute to the discovery of safer and more effective drugs ? *Curr Opin Chem Biol*. 2007;11:405–409.

34. Cyranoski D. Japan ramps up patent effort to keep iPS lead. *Nature*. 2008;453:962–963.

35. Takahashi K, Tanabe K, Ohnuki M, et al. Induction of pluripotent stem cells from adult human fibroblasts by defined factors. *Cell*. 2007;131:861–872.

36. Masaki H, Ishikawa T, Takahashi S, et al. Heterogeneity of pluripotent marker gene expression in colonies generated in human iPS cell induction culture. *Stem Cell Res*. 2007;1:105–115.

37. Brivanlou AH, Gage FH, Jaenisch R, et al. Stem cells. Setting standards for human embryonic stem cells. *Science*. 2003;300:913–916.

38. Maherali N, Hochedlinger K. Guidelines and techniques for the generation of induced pluripotent stem cells. *Cell Stem Cell*. 2008;3:595–605.

39. Dimos JT, Rodolfa KT, Niakan KK, et al. Induced pluripotent stem cells generated from patients with ALS can be differentiated into motor neurons. *Science*. 2008;321:1218–1221.

40. Ebert AD, Yu J, Rose FF Jr, et al. Induced pluripotent stem cells from a spinal muscular atrophy patient. *Nature*. 2009;457:277–280.

41. Sato T, Vries RG, Snippert HJ, et al. Single Lgr5 stem cells build crypt-villus structures in vitro without a mesenchymal niche. *Nature*. 2009;459:262–265.

INTERFACING STEM CELLS WITH GENE THERAPY

30

Christian Mueller and Terence R. Flotte

INTRODUCTION

As the field of gene therapy has progressed through the last three decades, the idea of disease correction at the genetic level has inevitably merged with the field of stem cell therapy. The obvious advantage of a combined approach is that it could allow one to regenerate tissues after they are already damaged by a disease, while genetically protecting them from recurrence of the disease, and still avoiding immunologic rejection of the regenerated tissues. Alternatively, gene modification could provide an additional control mechanism for the use of cells whose regenerative capacity might well lead to undesired consequences, such as tumor formation. Overall, the convergence of these fields has expanded the possibilities of disease targets and has led to promising proof-of-concept studies and early phase clinical trials. In this chapter, we will summarize some of the past accomplishments and methods of the stem cell and gene therapy interface, as well as describe some of the technologies and research currently underway.

Stem and progenitor cells of a number of different types may be employed in a combined stem cell/gene therapy (Table 30.1). Bone marrow can be a source of adult stem cells, mesenchymal stem cell (MSCs), and hematopoietic stem cells (HSCs). Gene therapy was originally pioneered in the latter hematopoietic stem cell population as gene therapy for single gene disorders affecting HSCs and gene-marking during HSC transplantation were attempted by numerous investigators. HSCs are characterized by their ability to regenerate all blood lineages over extended periods of time, essentially providing a self-renewing pool of cells. Since these cells give rise to the cell populations that comprise the immune system, genetic defects that are inherited in this stem cell population are known as primary immunodeficiencies.[1] The most severe of these disorders is aptly named severe combined immunodeficiency (SCID), a disorder in which T lymphocyte development and, in certain circumstances, B lymphocyte and natural killer cell development are compromised.[2,3] While HSC transplantation has proved to be highly successful at treating SCID, it is invariably limited in most cases by the lack of genetically matched donor cells.[4] Due to the lower survival rate in mismatched individuals and the associated toxicities with the chemotherapeutic agents used to increase HSC engraftment in these individuals, this became a natural first target for stem cell gene therapy. Particular attention was paid to SCID disorders in which correction of a few HSCs imparted a survival and proliferative advantage to the corrected stem cells. Due to the self-renewing and exponential replicating capacity

Human Stem Cell Technology and Biology, edited by Stein, Borowski, Luong, Shi, Smith, and Vazquez
Copyright © 2011 Wiley-Blackwell.

TABLE 30.1 VARIOUS TYPES OF PROGENITOR AND STEM CELLS THAT COULD BE USED IN COMBINED STEM CELL/GENE THERAPY

Progenitor/Stem Cell Type	Examples of Vector Modifications	Examples of Clinical Applications
Hematopoietic stem cells (HSCs)	Moloney murine leukemia virus (MLV); lentivirus; rAAV7, 8, 9	MLV-IL-2RG for XSCID MLV-ADA for ADA-SCID
Mesenchymal stem cells (MSCs)	rAAV2 lentivirus	Bone regeneration
Human embryonic stem cells (hESCs)	MLV-HSV-TK	Cardiac regeneration
Induced pluirpotent stem cells (iPS cells)	Lentivirus	Potential for hemoglobinopathy

of stem cells, the approach required that gene transfer to these cells accomplish integration into the host genome to confer gene correction stability from one daughter cell generation to the next. To date, most of the clinical trials using gene transfer to HSCs have been conducted using either murine gamma retroviruses or HIV-1 based lentiviral vectors which are capable of integration.

STEM CELL GENE TRANSFER

Gene transfer to HSCs has produced some dramatic clinical successes, followed by eventual serious complications. Replication-deficient viral vectors, in which the viral gene has been replaced by the therapeutic gene of interest, are often used as gene therapy agents due to their efficiency at introducing DNA into the host cell. The Moloney murine leukemia virus (MLV) is a gamma retrovirus that has been used extensively as a vector for gene therapy with HSCs.[5] This vector has the ability to deliver inserts up to 8 kb in length, can generate high titer stocks, and can easily be pseudotyped by replacing its wild-type envelope proteins with those of other viruses, to alter its tropism. One of the drawbacks of MLV is its inability to penetrate the nuclear envelope, which limits the usefulness of the vector in nonreplicating cells. This limitation has been addressed in HSC gene therapy by inducing mitosis in the HSC cells ex vivo, thereby allowing the MLV vectors to access the chromosomes and to integrate its proviral genome.[6] Other integrating viral vectors have been developed to bypass this limitation including HIV- or SIV-based lentiviral vectors and human foamy virus vectors.[7] These vectors have a more stable preintegration complex that is able to cross the nuclear membrane, thus allowing them to transduce both dividing and nondividing cells.[8] Regardless, most of the gene transfer for HSC-related disorders is performed ex vivo in controlled laboratory settings. While bone marrow harvest from the iliac crests has been the historic source for HSC for ex vivo gene transfer, this method is being replaced by the induction of HSC mobilization into the peripheral blood via the use of cytokine therapy such as granulocyte-macrophage colony stimulating factor (GM-CSF) and G-CSF. In this manner one can obtain a large number of primitive HSCs for ex vivo gene transfer with minimal discomfort to the patient. Peripheral blood harvest is followed by enrichment for CD34+ cells, which are then cultured on a fibronectin-coated surface and stimulated with FMS-like tyrosine kinase 3 (FLT3) ligand, stem cell (SCF), interleukin-6 (IL-6), and either thrombopoietin or interleukin-3 (IL-3) in the presence of the MLV vectors for 72–96 hours or 24–48 hours for lentiviral vectors.[9–11] Cells are then autologously transplanted into the patient, where they migrate to the bone marrow to initiate hematopoiesis. If incoming cells do not have a competitive advantage in repopulating the blood lineages as is present in some SCID disorders, patients may be subjected to a radiation and/or chemotherapy regimen to create "space" for the incoming stem cells and increase the success of engraftment.

RETROVIRUS-MEDIATED HEMATOPOIETIC STEM CELL GENE THERAPY

X-Linked SCID

Approximately 50% of the SCID cases are related to a mutation in the gene encoding for the IL-2 receptor gamma chain (IL-2RG). This form of SCID, termed X-linked SCID (X-SCID) due to the chromosomal location of the gene, was one of the first candidates for HSC gene therapy. The downstream effects of a mutation in the IL-2RG gene are substantial, because in addition to its necessity for IL-2 signaling, this receptor also interacts with other receptor subunits for the cytokines IL-4, IL-7, IL-9, IL-15, and IL-21.[12–15] As a consequence, the impaired signaling of all these cytokines leads to an absence of T and NK lymphocyte cells and dysfunctional B lymphocyte cells. In the late 1990s a team from the Necker Hospital in France led by Fischer and Cavazzano-Calvo performed the first HSC gene therapy clinical trial with unambiguous clinical benefits.[16] Eleven boys with X-SCID received autologous CD34$^+$ bone marrow cells, transduced by an MLV vector carrying the gamma-chain of the IL-2 receptor under the transcriptional control of the MLV long terminal repeat promoter. Due to the inherent competitive advantage that the transduced cells had, they were reinfused into the children without the necessity for prior chemo or radiotherapy. Eventually, in all but one of the patients, the vast majority of circulating T and NK lymphocyte cells contained an MLV vector sequence within 8 weeks of the transplant. These cells behaved normally in response to mitogens and T cell receptor antigen stimulation, suggesting that HSC correction had occurred.[16,17] Most patients were eventually able to discontinue immunoglobulin therapy. Another recently described trial included two patients with X-linked adrenoleukodystrophy. In this case autologous CD34$^+$ cells were removed from the patients, genetically corrected ex vivo with a lentiviral vector encoding wild-type ABCD1, and then reinfused into the patients after they had received myeloablative treatment.[18]

Adenosine Deaminase Deficiency SCID

Other concurrent HSC gene therapy clinical trials in the 1990s involved the treatment of patients with SCID resulting from adenosine deaminase (ADA) enzyme deficiency. ADA catalyzes the deamination of deoxyadenosine and adenosine to deoxyinosine and inosine; thus the deficiency results in the toxic accumulation of purine metabolites in early lymphoid progenitors, leading to apoptosis and the eventual SCID phenotype. The patients' phenotype may vary but is generally characterized by a low number of T, B, and NK lymphocytes. As in the case with X-SCID, treatment with allogeneic HSC transplants is highly successful for ADA-SCID, but mismatched transplants have poor survival outcomes. Exogenous enzyme replacement with polyethlyne glycol-conjugated bovine ADA (PEG-ADA) is employed with regular intramuscular injections, leading to temporary detoxification of body tissues. Despite these beneficial clinical interventions for ADA-SCID, the long-term prognosis for these children remains unclear and thus permanent correction by gene therapy is a sensible approach. Ironically, it was the success of the

PEG-ADA that led to the largely unsuccessful MLV gene therapy trials for ADA-SCID.[19–22] The concomitant use of the PEG-ADA therapy in gene therapy trials was believed to diminish the survival or competitive advantage of gene-corrected cells. Regardless of this, important milestones were reached with these trials; careful follow-up on patients receiving autologous transfer of ADA-corrected peripheral blood lymphocytes demonstrated that corrected lymphocytes were able to survive in circulation for up to 10 years.[23] One trial reported that withdrawal of PEG-ADA from a patient who received various infusions of corrected autologous peripheral blood lymphocytes caused a jump in detectable transduced T lymphocytes cells from 10% to nearly 100%, lending credence to the inefficiency of engraftment in the presence of enzyme replacement.[23] Based on this concept, trials that followed used bone marrow conditioning regimens to create the "space" for engraftment and expansion of transplanted cells. The mild myelosuppressant busulfan was administered on two consecutive days prior to transplantation of CD34$^+$ MLV corrected HSCs, resulting in excellent immune reconstitution in the 13 patients treated.[17]

SIDE EFFECTS OF INITIAL STEM CELL GENE THERAPY TRIALS

The initial landmark successes of these clinical gene therapy trials were eventually followed by serious complications that were mainly oncogenic in nature. To better understand the origins of such serious adverse events, it is imperative to appreciate that the stability of the MLV vectors through chromosomal integration in and of itself brings the risk of insertional mutagenesis. In fact, further study of gamma retroviruses integrated on CD34$^+$ HSCs revealed a predilection for integration in or near proto-oncogenes.[24] In the X-SCID MLV trials, within the first three years of clinical follow-up, three of the patients had developed acute T-cell leukemias. Further molecular analysis of the cancer cells revealed clonal vector integration near the LMO2 proto-oncogene.[25,26] MLV vector integration in or near this gene resulted in increased expression of LMO2, a gene that is known to participate in human leukemogenesis by chromosomal translocation.[27] Eventually five gene therapy treated patients out of a total of 30 (both X-SCID and ADA-SCID trials combined) developed acute T-cell leukemias; one died whereas four went into complete remission following chemotherapy. The development of leukemias in the SCID trials may have been missed due to the fact that preliminary safety data for the transplant of CD34$^+$ MLV transduced HSCs was conducted on advanced cancer patients with short follow-up times. Although it is not entirely clear why only some patients developed these complications, it is no longer possible to assume that MLV proviral integration into the host's genome is random. In fact, large-scale sequencing for retroviral insertions and analysis of MLV-infected cell lines have revealed clustering at the 5′ end of genes.[28,29] Thus in the design of future vectors, it is important to consider the incorporation of DNA insulator sequences and self-inactivating vectors to diminish the enhancer properties of viral LTR or, more ideally, the targeting of safe regions in the genome.

ADENO ASSOCIATED VIRUS (AAV), SITE-SPECIFIC INTEGRATION, HOMOLOGOUS RECOMBINATION, AND OTHER MODES OF PERSISTENCE

There are a number of potential advantages and disadvantages as to which particular viral vector is chosen to transduce a stem or progenitor cell for gene therapy (Table 30.2). To date, the adeno associated virus (AAV) is the only mammalian virus observed to integrate into a defined site in the human chromosome, the

	TABLE 30.2	POTENTIAL ADVANTAGES AND DISADVANTAGES OF GENE THERAPY VECTORS FOR STEM CELL MODIFICATION

Vector Class	Potential Advantages	Potential Disadvantages
Gammaretrovirus	Integration leads to permanent transduction	Insertional mutagenesis leads to leukemia syndrome
Lentivirus	Integration in G_0 stem cells	Insertional mutagenesis
Adeno associated virus	Safe, multiple serotypes to choose from	Small packaging capacity although strategies to overcome this are available

AAVS1 site (19q13.4).[30–33] These findings, while quite remarkable in the initial studies with wild-type AAV2 in immortalized cell lines, were not reproduced with the recombinant vectors.[34–36] One component of this discordance is the role of the AAV Rep protein in binding to the AAVS1 site and to sequences within the AAV2 genome. Indeed, approaches in which the Rep protein was supplied *in trans* along with rAAV vectors did result in some site-specific integration. The efficiency of this process has not been very high in vivo, however. Additional factors appear to favor the formation of high molecular weight concatemeric episomes over vector integration.[37] Among these is a natural mechanism involving the double-strand break repair pathway, which either inhibits integration or competes with integration. The net effect is that site-specific integration based on the AAV-Rep-mediated process has yet to be effectively exploited as a therapeutic strategy.

AAV vectors may also facilitate homologous recombination under circumstances where genomic sequences are used. This approach has been used to accomplish true gene repair by substituting normal sequences in place of mutant sequences within the natural gene locus.[38,39] The great promise of this approach is that, if successful, the natural mechanisms for control of gene expression would be retained. While this has been accomplished with a number of cell lines, it has not yet been proved that this can be accomplished with stem cells at sufficient efficiency to be useful for therapy. Nonetheless, this promising approach is likely to be pursued vigorously in the future.

Interestingly, improvements of more conventional rAAV vectors have greatly increased the efficiency of transduction and, unexpectedly, resulted in long-term persistence of rAAV genomes through a combination of episomal persistence and random-site integration. These improvements have included the use of alternative serotype capsids, mutant capsids, and self-complementary (sc) vector genomes.[40–44] Most remarkable have been the findings with AAV serotype 1, 2, 7, 8, and 9 capsids and their ability to transduce murine HSCs. In these studies, robust expression was present and could be serially transferred from mouse to mouse by syngeneic HSC transplant.[45] If such an approach could be scaled up to humans, the potential for genetic correction of primary immune deficiencies and hemoglobinopathies could be substantial.

MESENCHYMAL STEM CELL GENE THERAPY

Mesenchymal stem cells (MSCs) have also generated widespread interest as potential therapeutic agents because of their ability to generate numerous types of mesodermal-derived tissue types (cardiac and skeletal muscle, bone, cartilage, and fat) and for their poorly understood ability to suppress immunologic responses. Numerous proof-of-concept studies have demonstrated the ability of genetically enhanced MSCs to exert potentially therapeutic effects. One such example is the use of MSCs transduced with a rAAV2-bone morphogenic peptide-2 (BMP2) vector in a mouse model of osteopenia. The BMP2 exerts an enhanced osteogenic effect. Furthermore, transient expression of the murine alpha-4 integrin in these cells was shown to enhance homing of the transduced MSCs to the bone.[46] Similar studies have indicated the ability of MSCs to regenerate cardiac muscle, articular cartilage, and other tissues, with or without gene modification, depending on the specific clinical context.

PROSPECTS

Therapeutic Modification of Autologous Stem Cells

Embryonic stem (ES) cells from preimplantation embryos can be propagated in vitro and give rise to all three embryonic germ layers. These cells could serve as a potentially limitless source of transplantable tissue if not for two major hurdles: one is socially enforced while the other is naturally occurring. The socioethical moral dilemmas surrounding the use, harvest, and development of stem cells are self-imposed and may eventually be solved. However, the necessity for a genetically or immune matched transplantable tissue arising from ES cells is a natural and biological constraint on the eventual success of this technology. Although tremendous effort has been put into deriving such immune compatible cells, success has been limited until very recently when induced pluripotent stem (iPS) cells were created by transducing embryonic and adult fibroblasts with a set of defined transcription factors.[47] Conveniently, this alternative has also bypassed the ethical dilemmas associated with ES cell technology since iPS cells can be created from somatic adult cells. This novel method is striking in that it can convert somatic cells directly into pluripotent cells, in a manner that is completely independent of the ES cell realm. This phenomenon is possible through the reversal of epigenetic changes in the DNA methylation state and chromatin organization that maintain the differentiated state of a somatic cell. This is brought about by the exogenous delivery of various combinations of developmental transcription factors. Yamanaka first reported the reprogramming of adult mouse tail tip fibroblasts by introducing the four transcription factors Oct4, Sox2, Klf4, and c-Myc via retroviral delivery, which led to the generation of iPS colonies 2 weeks post-transduction. This epigenetic reprogramming is the basis for iPS cell technology as it implies the possibility of converting any given somatic cell lineage to other lineages through iPS cell induction. This type of reprogramming was not possible in human cells before; thus iPS cells create an exciting research avenue with multiple clinical applications.

The potential of utilizing gene augmentation in the context of iPS cell-mediated regeneration was identified in its earliest iterations. The use of self-derived iPS cells versus allogeneic ES cells allows one to avoid immune rejection even without immunosuppressive drugs. In many instances, this advantage can be further exploited if the iPS cell-derived tissues can be genetically modified so as to avoid recurrence of the underlying disease. This is most obviously the case in genetic diseases but could be advantageous in a number of acquired conditions as well. An initial proof of this concept was accomplished in the context of gene therapy for hemoglobinopathies.[48] Investigators used a humanized sickle cell anemia mouse model. Cells were reprogrammed to become stem cells and then differentiated into HSCs after gene correction of the beta globin locus. Subsequent autologous transplantation of those cells resulted in a long-term cure of the hemoglobinopathy in this model. An analogous approach could be used with any number of genetic diseases. The example of how it might be used in alpha-1 antitrypsin deficiency is provided (Fig. 30.1).

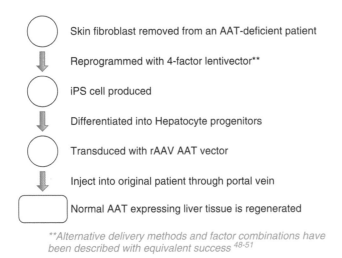

Skin fibroblast removed from an AAT-deficient patient

↓ Reprogrammed with 4-factor lentivector**

iPS cell produced

↓ Differentiated into Hepatocyte progenitors

Transduced with rAAV AAT vector

↓ Inject into original patient through portal vein

Normal AAT expressing liver tissue is regenerated

**Alternative delivery methods and factor combinations have been described with equivalent success [48-51]

FIGURE 30.1. *Anticipated clinical design of combined iPS cell-based gene therapy approach.*

Genetic control of stem cell therapies

One advantage of genetic alteration of stem cells that would apply to any clinical use, is the ability to control or limit the ultimate fate of such cells or their progeny. The potential for stem cells to generate tumors is well recognized. Indeed, the generation of teratomas is one of the accepted experimental indicators that a cell population truly is totipotent. More malignant tumors may also be generated, such as teratocarcinomas and malignant tumors of the desired therapeutic lineage that one is ultimately trying to therapeutically regenerate. The most straightforward way to approach the possibility of tumorigenesis from stem cell transplant is to use gene transfer vectors to modify the transplanted cells in a manner that would render them uniquely sensitive to elimination at some later point at which the risk for generating a cancer has increased.[52] One typical approach has been to transduce cells with herpes simplex virus–thymidine kinase (HSV-TK), which renders the cells sensitive to the prodrug ganciclovir. In these uses, HSV-TK has either been expressed constitutively or from the Oct4 promoter, which will specifically be expressed within teratomas. A variety of different vectors have been used for this purpose, and given the ability to perform this manipulation ex vivo, there would appear to be numerous suitable options. As an alternative to the prodrug approach, one might use directly toxic genes or genes that make cells optimal targets for immune-mediated elimination.

CONCLUSION

Human ES cells have recently entered the clinical trial stage for spinal cord injury, while ongoing work is proceeding rapidly with adult-derived progenitor cells, such as HSCs and MSCs. Clinical trials of iPS cells are likely to start in the near future. Unless there are compelling and unexpected safety concerns from preclinical studies, it seems probable that some of these stem cell trials will include a gene modification component at some point, whether it is intended to prevent the recurrence of a disease in an autologous iPS cell, to provide a control mechanism for an ES cell, or perhaps even to provide some form of enhancement to a cell-based therapy. Multiplying the potential of such therapies by combining them seems logical at this point in their development. Only time and experience will reveal how narrow or broad the therapeutic niche will be for advanced therapeutics in the future.

REFERENCES

1. Fischer A. Human primary immunodeficiency diseases: a perspective. *Nat Immunol*. 2004;5(1):23–30.

2. Fischer A, Le Deist F, Hacein-Bey-Abina S, et al. Severe combined immunodeficiency. A model disease for molecular immunology and therapy. *Immunol Rev*. 2005;203:98–109.

3. Buckley RH. Advances in the understanding and treatment of human severe combined immunodeficiency. *Immunol Res*. 2000;22(2–3):237–251.

4. Antoine C, Muller S, Cant A, et al. Long-term survival and transplantation of haemopoietic stem cells for immunodeficiencies: report of the European experience 1968-99. *Lancet*. 2003;361(9357):553–560.

5. Pages JC, Bru T. Toolbox for retrovectorologists. *J Gene Med*. 2004;6(Suppl 1):S67–S82.

6. Miller DG, Adam MA, Miller AD. Gene transfer by retrovirus vectors occurs only in cells that are actively replicating at the time of infection. *Mol Cell Biol*. 1990;10(8):4239–4242.

7. Vassilopoulos G, Josephson NC, Trobridge G. Development of foamy virus vectors. *Methods Mol Med*. 2003;76:545–564.

8. Lewis PF, Emerman M. Passage through mitosis is required for oncoretroviruses but not for the human immunodeficiency virus. *J Virol*. 1994;68(1):510–516.

9. Dao MA, Hannum CH, Kohn DB, Nolta JA. FLT3 ligand preserves the ability of human CD34$^+$ progenitors to sustain long-term hematopoiesis in immune-deficient mice after ex vivo retroviral-mediated transduction. *Blood*. 1997;89(2):446–456.

10. Kiem HP, Andrews RG, Morris J, et al. Improved gene transfer into baboon marrow repopulating cells using recombinant human fibronectin fragment CH-296 in combination with interleukin-6, stem cell factor, FLT-3 ligand, and megakaryocyte growth and development factor. *Blood*. 1998;92(6):1878–1886.

11. Wu T, Kim HJ, Sellers SE, et al. Prolonged high-level detection of retrovirally marked hematopoietic cells in nonhuman primates after transduction of CD34$^+$ progenitors using clinically feasible methods. *Mol Ther*. 2000;1(3):285–293.

12. Takeshita T, Asao H, Ohtani K, et al. Cloning of the gamma chain of the human IL-2 receptor. *Science*. 1992;257(5068):379–382.

13. Kumaki S, Kondo M, Takeshita T, Asao H, Nakamura M, Sugamura K. Cloning of the mouse interleukin 2 receptor gamma chain: demonstration of functional differences between the mouse and human receptors. *Biochem Biophys Res Commun*. 1993;193(1):356–363.

14. Noguchi M, Yi H, Rosenblatt HM, et al. Interleukin-2 receptor gamma chain mutation results in X-linked severe combined immunodeficiency in humans. *Cell*. 1993;73:147–157. *J Immunol*. 2008; 181(9): 5817–5827.

15. Noguchi M, Yi H, Rosenblatt HM, et al. Interleukin-2 receptor gamma chain mutation results in X-linked severe combined immunodeficiency in humans. *Cell*. 1993;73(1):147–157.

16. Cavazzana-Calvo M, Hacein-Bey S, de Saint Basile G, et al. Gene therapy of human severe combined immunodeficiency (SCID)-X1 disease. *Science*. 2000;288(5466):669–672.

17. Aiuti A, Slavin S, Aker M, et al. Correction of ADA-SCID by stem cell gene therapy combined with nonmyeloablative conditioning. *Science*. 2002;296(5577):2410–2413.

18. Cartier N, Hacein-Bey-Abina S, Bartholomae CC, et al. Hematopoietic stem cell gene therapy with a lentiviral vector in X-linked adrenoleukodystrophy. *Science*. 2009;326(5954):818–823.

19. Blaese RM, Culver KW, Miller AD, et al. T lymphocyte-directed gene therapy for ADA- SCID: initial trial results after 4 years. *Science*. 1995;270(5235):475–480.

20. Kohn DB, Hershfield MS, Carbonaro D, et al. T lymphocytes with a normal ADA gene accumulate after transplantation of transduced autologous umbilical cord blood CD34$^+$ cells in ADA-deficient SCID neonates. *Nat Med*. 1998;4(7):775–780.

21. Hoogerbrugge PM, van Beusechem VW, Fischer A, et al. Bone marrow gene transfer in three patients with adenosine deaminase deficiency. *Gene Ther*. 1996;3(2):179–183.

22. Bordignon C, Notarangelo LD, Nobili N, et al. Gene therapy in peripheral blood lymphocytes and bone marrow for ADA- immunodeficient patients. *Science*. 1995;270(5235):470–475.

23. Muul LM, Tuschong LM, Soenen SL, et al. Persistence and expression of the adenosine deaminase gene for 12 years and immune reaction to gene transfer components: long-term results of the first clinical gene therapy trial. *Blood*. 2003;101(7):2563–2569.

24. Cattoglio C, Facchini G, Sartori D, et al. Hot spots of retroviral integration in human CD34$^+$ hematopoietic cells. *Blood*. 2007;110(6):1770–1778.

25. Hacein-Bey-Abina S, Von Kalle C, Schmidt M, et al. LMO2-associated clonal T cell proliferation in two patients after gene therapy for SCID-X1. *Science*. 2003;302(5644):415–419.

26. Hacein-Bey-Abina S, von Kalle C, Schmidt M, et al. A serious adverse event after successful gene therapy for X-linked severe combined immunodeficiency. *N Engl J Med*. 2003;348(3):255–256.

27. Dik WA, Nadel B, Przybylski GK, et al. Different chromosomal breakpoints impact the level of LMO2 expression in T-ALL. *Blood*. 2007;110(1):388–392.

28. Ciuffi A, Mitchell RS, Hoffmann C, et al. Integration site selection by HIV-based vectors in dividing and growth-arrested IMR-90 lung fibroblasts. *Mol Ther*. 2006;13(2):366–373.

29. Mitchell RS, Beitzel BF, Schroder AR, et al. Retroviral DNA integration: ASLV, HIV, and MLV show distinct target site preferences. *PLoS Biol*. 2004;2(8):E234.

30. Ghosh A, Yue Y, Shin JH, Duan D. Systemic trans-splicing adeno-associated viral delivery efficiently transduces the heart of adult mdx mouse, a model for Duchenne muscular dystrophy. *Hum Gene Ther*. 2009;20(11):1319–1328.

31. Kotin RM, Linden RM, Berns KI. Characterization of a preferred site on human chromosome 19q for integration of adeno-associated virus DNA by non-homologous recombination. *EMBO J*. 1992;11(13):5071–5078.

32. Kotin RM, Menninger JC, Ward DC, Berns KI. Mapping and direct visualization of a region-specific viral DNA integration site on chromosome 19q13-qter. *Genomics*. 1991;10(3):831–834.

33. Kotin RM, Siniscalco M, Samulski RJ, et al. Site-specific integration by adeno-associated virus. *Proc Natl Acad Sci USA*. 1990;87(6):2211–2215.

34. Flotte TR, Afione SA, Zeitlin PL. Adeno-associated virus vector gene expression occurs in nondividing cells in the absence of vector DNA integration. *Am J Respir Cell Mol Biol*. 1994;11(5):517–521.

35. Afione SA, Conrad CK, Kearns WG, et al. In vivo model of adeno-associated virus vector persistence and rescue. *J Virol*. 1996;70(5):3235–3241.

36. Kearns WG, Afione SA, Fulmer SB, et al. Recombinant adeno-associated virus (AAV-CFTR) vectors do not integrate in a site-specific fashion in an immortalized epithelial cell line. *Gene Ther*. 1996;3(9):748–755.

37. Hester ME, Song S, Miranda CJ, Eagle A, Schwartz PH, Kaspar BK. Two factor reprogramming of human neural stem cells into pluripotency. *PLoS One*. 2009;4(9):e7044.

38. Russell DW, Hirata RK. Human gene targeting by viral vectors. *Nat Genet*. 1998;18(4):325–330.

39. Russell DW, Hirata RK, Inoue N. Validation of AAV-mediated gene targeting. *Nat Biotechnol*. 2002;20(7):658.

40. Gao G, Alvira MR, Somanathan S, et al. Adeno-associated viruses undergo substantial evolution in primates during natural infections. *Proc Natl Acad Sci USA*. 2003;100(10):6081–6086.

41. Gao GP, Alvira MR, Wang L, Calcedo R, Johnston J, Wilson JM. Novel adeno-associated viruses from rhesus monkeys as vectors for human gene therapy. *Proc Natl Acad Sci USA*. 2002;99(18):11854–11859.

42. McCarty DM, Monahan PE, Samulski RJ. Self-complementary recombinant adeno-associated virus (scAAV) vectors promote efficient transduction independently of DNA synthesis. *Gene Ther*. 2001;8(16):1248–1254.

43. Fu H, Muenzer J, Samulski RJ, et al. Self-complementary adeno-associated virus serotype 2 vector: global distribution and broad dispersion of AAV-mediated transgene expression in mouse brain. *Mol Ther*. 2003;8(6):911–917.

44. Gao G, Vandenberghe LH, Wilson JM. New recombinant serotypes of AAV vectors. *Curr Gene Ther*. 2005;5(3):285–297.

45. Maina N, Han Z, Li X, et al. Recombinant self-complementary adeno-associated virus serotype vector-mediated hematopoietic stem cell transduction and lineage-restricted, long-term transgene expression in a murine serial bone marrow transplantation model. *Hum Gene Ther*. 2008;19(4):376–383.

46. Kumar S, Nagy TR, Ponnazhagan S. Therapeutic potential of genetically modified adult stem cells for osteopenia. *Gene Ther*. 2010;17(1):105–116.

47. Takahashi K, Yamanaka S. Induction of pluripotent stem cells from mouse embryonic and adult fibroblast cultures by defined factors. *Cell*. 2006;126(4):663–676.

48. Lyssiotis CA, Foreman RK, Staerk J, et al. Reprogramming of murine fibroblasts to induced pluripotent stem cells with chemical complementation of Klf4. *Proc Natl Acad Sci USA*. 2009;106(22):8912–8917.

49. Huangfu D, Osafune K, Maehr R, et al. Induction of pluripotent stem cells from primary human fibroblasts with only Oct4 and Sox2. *Nat Biotechnol*. 2008;26(11):1269–1275.

50. Okita K, Nakagawa M, Hyenjong H, Ichisaka T, Yamanaka S. Generation of mouse induced pluripotent stem cells without viral vectors. *Science*. 2008;322(5903):949–953.

51. Kaji K, Norrby K, Paca A, Mileikovsky M, Mohseni P, Woltjen K. Virus-free induction of pluripotency and subsequent excision of reprogramming factors. *Nature*. 2009;458(7239):771–775.

52. Kiuru M, Boyer JL, O'Connor TP, Crystal RG. Genetic control of wayward pluripotent stem cells and their progeny after transplantation. *Cell Stem Cell*. 2009;4(4):289–300.

TISSUE ENGINEERING FOR STEM CELL MEDIATED REGENERATIVE MEDICINE

31

Janet Zoldan, Thomas P. Kraehenbuehl, Abigail K. R. Lytton-Jean, Robert S. Langer, and Daniel G. Anderson

INTRODUCTION

Human embryonic stem cells (hESCs), derived from the inner cell mass of preimplantation embryos, have been shown to give rise to stable pluripotent cell lines that appear to proliferate indefinitely under specific culture conditions. Upon injection into immune deficient mice, undifferentiated hESCs produce teratomas comprised of multiple cell types, demonstrating their pluripotent potential. In vitro aggregation of hESCs into clusters of cells, termed embryoid bodies (EBs), allows spontaneous and random differentiation of hESCs into multiple tissue lineages. These cell lineages represent all three of the germ layers: endoderm, ectoderm, and mesoderm.[1,2] Since the derivation of the first hESC line by Thomson and co-workers,[1] various lineages have been derived, including neurons,[3,4] cardiomyocytes,[5,6] smooth muscle cells, hematopoietic cells,[7] osteogenic cells,[8] hepatocytes,[9] insulin-producing cells,[10] keratinocytes,[11] and endothelial cells.[12]

Recent studies in cell and tissue engineering have focused on the use of adult stem cells isolated from the blood,[13,14] bone marrow,[15–20] and adult or fetal tissue.[21,22] These cells represent a promising source for human cells because they can be obtained from human sources without a full organ donation. However, their clinical applications are hampered by limited availability, scarcity of certain cell populations, and poor isolation process. Thus the potential of having a limitless supply of human cells provided by hESCs is a significant advantage for cell and tissue engineering application. There are, however, many challenges and obstacles to overcome before this vision can be achieved. The most urgent challenge is developing efficient methods to further differentiate hESCs into specific cell types. This chapter will outline recent advances in cell and tissue engineering applications using hESC-derived cells by individual analysis of the three germ layers. The concentric organization of the germinal layers is a hallmark in embryonic development since all adult tissue is derived from these three germ layers.

ECTODERMAL TISSUE

The ectoderm, the outermost germ layer, gives rise to the skin covering, and to the nervous system.

Human Stem Cell Technology and Biology, edited by Stein, Borowski, Luong, Shi, Smith, and Vazquez
Copyright © 2011 Wiley-Blackwell.

Neurons

In the absence of appropriate intercellular signals specifying alternate cellular fates, hESCs are directed toward neuronal differentiation.[23,24] Recent years have seen studies describing the effects of a plethora of growth factors and feeder cells on maximizing the differentiation of hESCs into neuronal cells including functional neurons, glial cells, and oligodendrocytes. Most commonly, formation of EBs[4,25–27] and co-culturing with mouse stromal lines (PA6[26,28] and MS5[29–32]) have been used. Increased commitment to a neuronal fate during EB differentiation required supplementing the culture medium with certain mitogenic factors such as epidermal growth factor (EGF), basic fibroblast growth factor (bFGF), and Noggin, a bone morphogenic protein (BMP) antagonist known to suppress neuronal differentiation.[30,33–36] In co-culture systems with mouse stromal lines, in vitro neural differentiation occurs in response to signals from stromal cells, which is yet to be fully understood. Formation of neural rosettes is a morphologic marker of hESC differentiation to neural cells. Here, neuroectodermal columnar cells arrange radially to form a rosette-like structure

Dopaminergic (DA) neurons that express tyrosine hydroxylase (TH) are of special interest because of their degeneration in Parkinson disease. Different growth factors are known to be involved in the development and maintenance of the midbrain DA system and have been used to induce the DA differentiation in vitro. The most commonly used growth factors are FGF8 and sonic hedgehog (HSS) (reviewed by Correia et al.[37]). In addition, co-cultures with amniotic membrane matrix[38] and human midbrain astrocytes[39] have also been shown to promote dopamine-like differentiation. Although studies have various degrees of efficiency (highest of which are 40%[9,10] to 60%[29,39,40]), most of the hESC-derived DA neurons release dopamine and exhibit some electrophysiological properties.[29,39–42]

Implantation of in vitro produced human DA neurons have not yet provided therapeutic effect in 6-hydroxydopamine (6-OHDA)-lesioned rodents,[27,28,30,32–34,42,43] mostly due to the low number of TH+ cells found in implanted grafts.[26,28,33,43] The most efficient protocols led to the survival of only 2.7% (10732 cells) TH+ grafted cells and therefore little behavioral recovery. If fully differentiated cells are implanted, the survival rate is too low.[40] However, if less differentiated cells are used, teratoma formation, overgrowth, and the presence of primitive neuroepithelia are commonly observed even after a relatively short time period (8–13 weeks).[27,30,39,43] In addition, current differentiation approaches produce a mixed population of DA neurons with forebrain and midbrain phenotypes. Other important factors that can compromise cell survival include hypoxia, oxidative stress, and trophic factor withdrawal. This raises the question of whether hESC-derived DA neural cells can functionally engraft in a safe manner and how this can be accomplished.

Motor neurons represent another specialized class of neurons essential for the control of body movement. Motor neuron loss can occur due to spinal trauma or a wide range of neurological disorders including amyotrophic lateral sclerosis and spinal muscular atrophy. Recent studies have demonstrated that the in vitro derivation of motor neurons from hESCs is restricted to the earliest stages of neural induction.[44,45] Current protocols, based on EBs[44,45] or co-culture with stromal cells,[46] are followed by exposure to defined morphogens such a SHH or retinoic acid (RA), which act as ventralizing and caudalizing factors, respectively.[44–46] These cells were electrophysiologically active, formed action potentials in response to depolarizing currents, secreted acetylcholine, and displayed clustered acetylcholine receptors in myotubes. Human ESC-derived motor neurons that were implanted into chick spinal cords exhibited long-distance axon growth outside the central nerve system (CNS), reaching peripheral targets in the trunk musculature. In an adult rat spinal cord model these cells survived (6 weeks postimplantation, 14% of cells were positive for choline acetyl transferase) and maintained phenotype.[46]

Oligodendrocytes are myelin-forming cells found in the central nervous system. They wrap around the axons of neurons and facilitate conduction of electrical signals between neurons. Oligodendrocyte injury and/or dysfunction are responsible for demyelinating diseases such as multiple sclerosis, Alzheimer disease, multisystem degeneration, and oligodendrogliomas. Various growth factors and signaling molecules have been tested for their ability to mediate hESC-derived oligodendrocyte differentiation.[47–50] Nistor et al.[47] demonstrated for the first time that hESCs can be induced to form high yielding oliogdendroglial lineage cultures (85%). Upon implantation into the shiverer mouse model of dymyelination, these cells proliferated and expressed myelin basic protein.[47] In rat spinal cord injuries an inflammatory response was necessary for hESC-derived oligodendrocyte progenitor cells to survive, proliferate, and differentiate into mature oligodendrocyte. Rats receiving hESC-derived oligodendrocyte progenitors 7 days after injury showed substantial improvements in motor function compared to those receiving undifferentiated hESCs or oligodendrocytes 10 months after injury[51] or in rats with minor injuries.[52] These results highlight the potential use of hESCs for neuronal engineering.

The interplay between cellular behavior, material properties, and architecture emerges as a new tool for directing hESC differentiation in general, and specifically into neuronal fate. We have shown that RA and neurotrophins induced neural differentiation and formation of neural tube-like rosettes when hESCs are cultured on three-dimensional porous polymer scaffolds (examined by nestin and β-tubilibIII) (Fig. 31.1). Neurotrophine supplementation enhanced neuronal differentiation and maturation of rosette-like structures as well as promoted formations of vascular structures throughout the tissue constructs. A vascularized neuronal tissue will be advantageous for in vivo applications.[53,54] Recently, nanofiber 3D scaffolds, mimicking the structure of extracellular matrixes, were reported to direct hESC differentiation toward a neuronal fate.[55] Nanofiber scaffolds composed of matrix peptides were able to reconnect axons in a severed optic tract hamster model.[56] Control over surface patterning and development of novel biomaterials may improve neuronal tissue engineering.

Cornea

In the adult retina, Müller glia cells act as endogenous progenitors in response to injury but are too few in number to restore function after damage has occurred. Thus transplantation of donor cells to replace damaged or lost cells is a promising approach for regeneration therapy. Partial restoration of visual function was reported in humans after transplantation of autologous retinal pigment epithelium (RPE) layers.[57] The RPE plays a crucial role in the maintenance and function of the retina and photoreceptors. However, the scarcity of donors limits the application of this therapy. Human ESCs may serve as an unlimited source of RPE cells. To date, hESC studies have focused on the derivation of subsets of retinal cell populations with emphasis on the production of retinal progenitors,[58,59] mature RPE cells,[60,61] and photoreceptors.[62]

The spontaneous differentiation of RPE cells from hESCs occurs at a relatively low efficiency (<1%).[60] The yield of RPE-like cells was recently improved by using Wnt and Nodal antagonists, giving rise to 35% RPE-like cells after 8 weeks culture.[62] Similar differentiation efficiencies were reported upon augmentation with nicotinamide and activin.[63] Osakada et al.[64] have recently developed a defined differentiation procedure using only chemicals (casein kinase I inhibitor CKI-7, ALK4 inhibitor SB-431542, and Rho-associated kinase inhibitor Y-27632) in serum-free and feeder-free floating aggregate culture to induce retinal progenitors

NGF

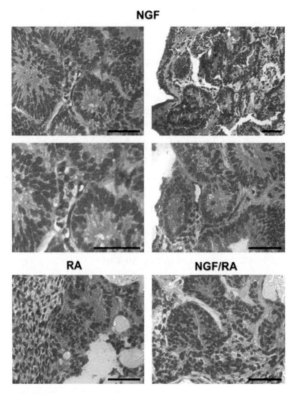

FIGURE 31.1. *Human ESC differentiation into three-dimensional (3D) neural rosette-like structures. Differentiating hESCs from 4-day-old EBs were seeded onto scaffolds and grown for 2 weeks in control medium or medium supplemented with nerve growth factor (NGF), RA, or a combination of NGF and RA (NGF/RA). Endothelial vessel-like network (immunostained with anti-CD31) formed in constructs treated with NGF. Scale bars: 50 μm. (Reproduced with permission from Levenberg et al.[54] Copyright © 2005, Mary Ann Leibert, Inc.)*

(18%). These cells expressed RPE markers, formed positive tight junctions, and exhibited phagocytic functions. Subsequent treatment with RA and taurine induced photoreceptors (6.5%) that express recoverin, rhodopsin, and genes involved in phototransduction. Additionally, co-culture of hESCs with postnatal day 1 rat retinal cells directed differentiation into bipolar and photoreceptor progenitor cells.[65]

Lund et al.[66] were the first to demonstrate the potential of hESC-derived RPE cells to attenuate retinal degeneration and to preserve visual function after subretinal implantation. Although grafting efficiencies were not stated, hESC-derived RPE-like cells were detected 80 days postimplantation, in the subretinal space without migration into the retina. More recently, Idelson et al.[63] showed that implantation of hESC-derived RPE cells into an animal model with retinal degeneration rescued retinal function. The engrafted cells were capable of absorbing the visual pigment rhodopsin, suggesting that these cells were capable of phagocytic function in vivo, one of the most important functions of RPE cells. After 19 weeks, 89% of the implanted cells expressed RPE65, suggesting that the majority of cells within the grafts were RPE-like cells. Less than half of the cells within the grafts were pigmented, suggesting the presence of immature RPE-like cells. Lamba et al.[67] implanted hESC-derived cells with an 80% retinal progenitor population into wild-type and Crx-deficient mice (model for Leber's congential amaurosis). When implanted into adult or newborn wild-type mice, cells integrated into all of the retinal layers and expressed markers appropriate for the lamina in which they had settled. When implanted into the subretinal space of adult Crx-deficient mice, the cells integrated into the retina and restored light response to otherwise unresponsive animals. After 3 weeks, 36% of the cells expressed recoverin. In both reports, cell populations contained mixed mature and immature cells. Improved purification of the pigmented cells prior to implantation may improve the homogeneity of pigmented cells within the grafts. Although teratoma formation was not observed, additional long-term studies are required to confirm the safety of hESC-derived cells.

Skin

Skin is the first line of defense that protects the body from dehydration, injury, and infection. The outer thinner layer, known as the epidermis, is a source of embryonic ectoderm and is composed primarily of epithelial cells. The outermost cells are keratinocytes. As the cells mature and migrate to the surface, they form keratin, which becomes an effective barrier to environmental hazards such as infection and excess water evaporation. The dermis, lying just underneath the epidermis, is composed of extracellular matrix (ECM) and countless blood vessels that provide vital nutrients to the outer skin.[68,69]

The functions of the skin are compromised when it is damaged. Although effective, traditional skin grafting (autologous or allogeneic) procedures have disadvantages. Proliferation and maintenance are limited as skin grafts are composed of either acellular components or adult skin cells and lack prevascularization that is essential for graft survival. In addition, these substitutes are unable to fully reproduce the structures and functions of uninjured skin.

Human ESCs and their progeny have been investigated as a source of cells for engineered skin. This has motivated the generation of epithelial lineages from hESCs through various methods.[70–72] Spontaneous differentiation of hESCs into early keratinocyte progenitors (marked by K14 and p63 expression) has been shown.[11] Human ESCs and EBs, grown on gelatin in the presence of keratinocyte media, differentiated into the mature epidermal keratinocyte lineage (expressing involucrin and filaggrin markers), albeit at low efficacies (2%).[73] Growth factor supplementation (such as BMP-4[70] and EGF[74]) induced differentiation into ectoderm and skin-related keratinocyte lineages. Co-culture of keratinocyte progenitors with 3T3 cells supported terminal differentiation to involucrin-expressing keratinocyte.[11] However, stratification of these cells has not yet been reported. Recently, treatment of undifferentiated hESCs with RA, in the presence of endogenous BMP signaling, has been demonstrated as an effective means of producing high-purity populations of 97% epithelial cells.[72]

To assess cell function and tissue morphogenesis, progenitor cells have been cultivated in complex microenvironments with basement membrane analogs or at the air–liquid interface (ALI). Air contact results in proliferating keratinocytes oriented toward the medium with differentiated cornfield keratinocytes localizing at the air–medium interface. Recently, Hewitt et al.[75] derived cell populations with properties of ectodermal and mesoderm cells. Sequential cultivation of both cell populations on collagen gels at the ALI generated a multilayered epithelial-like tissue that expressed keratins (K12) and basement membrane proteins. However, no stratification was observed in this system, indicating that these precursors had not yet obtained a basal epithelial phenotype. On the other hand, successful stratification and mature keratinocytes (expressing keratin 10, involucrin, and filaggrin) were reported upon co-culturing keratinocyte progenitors with human dermal fibroblasts on collagen gels at the ALI.[76]

Only one study examined in vivo formation of keratinocytes. Human ESCs injected into severe combined immunodeficient (SCID) mice formed nodules containing early stage keratinocytes (identified by expression of K14 and p63 markers).[11] This demonstrates that wound sites can contain signals to guide partially differentiated hESCs into mature keratinocytes. However, further studies are required to assess whether in vivo signals will be sufficient to guide stem cell development to differentiated lineages involved in skin generation or toward teratoma formation.

ENDODERMAL TISSUE

The innermost layer is the endoderm, lining the alimentary canal and giving rise to the organs associated with the digestion system.

Liver

The liver plays a major role in the body's metabolism and has a number of vital functions including glycogen storage, decomposition of red blood cells, plasma protein synthesis, hormone production, and detoxification. Hence patients with acute hepatic failure or end-stage liver disease must receive liver transplants to survive. Liver and hepatocyte transplantation has become an accepted treatment for liver failure. Although clinically used, sources for hepatocytes and organ donors are limited. Thus pluripotent hESCs have been investigated as a possible alternative source for hepatocytes.

Several studies have examined the differentiation of mouse embryonic stem cells (mESCs) and hESCs in an attempt to induce hepatic cell formation (reviewed in Refs. 77 and 78). EB-induced differentiation of hESCs can result in hepatic differentiation and this has been further enhanced by a variety of growth factors.[53,79–82] Based on cDNA microarray characterization, Albumin (Alb) was identified as a hepatocyte-specific enriched gene in EBs and was used to isolate the hepatic-like cells. Following introduction of an Alb promoter regulated by GFP reporter gene into hESCs, GFP-positive clusters of hepatic-like cells were detected in 20-day-old EBs (6%). Results of immunostaining showed that most cells expressed Alb while some cells still expressed the earlier protein, β-fetoprotein (AFP).[80] Three-dimensional culture systems of both biodegradable[53] and nondegradable[83,84] polymeric scaffolds were utilized to promote EBs to differentiate into fetal liver-like cells[53] or more mature functional hepatocytes (secreting Alb and loading ammonia and lidocaine).[83,84]

An alternative means of directing hepatic differentiation is to sequentially induce differentiation, first to definitive endoderm (DE) followed by directed differentiation into hepatic cells.[85–88] These strategies vary from study to study, specifically in the starting cell population. Some studies used EBs that were later plated onto diverse matrices.[9,80–82] Others used hESCs that differentiated on mouse embryonic fibroblast (MEF) feeder layers[85] or mouse mesodermal-derived cells (M15),[87] while other hESCs cultures were differentiated in the absence of a feeder layer.[89] Several nonspecific (sodium butyrate[9] and dimethyl sulfoxide (DMSO)[89] and specific inducing factors (aFGF, hepatocyte growth factor, oncostatin M, and dexamethasone)[81,85,89] have been investigated. These inducing factors have been used in the culture of primary hepatocytes and were linked to hepatic development. In several reports, BMP has been used in conjunction with FGF because the latter signaling is enhanced by signaling from mesenchymal cells of the neighboring septum transversum.[85,87] Collagen has been investigated as a matrix for growing cells, so as to mimic the collagen containing connective tissue that the liver bud migrates toward in the septum transversum mesenchyme.

The characterization of the hESC-derived hepatic cells was carried out by analyzing endodermal gene expression using metabolic and morphologic assays. Still, the most efficient way to examine the functionality of hESC-derived hepatocytes is by examining their ability to restore liver function in vivo. Cai et al.[85] were the first to show that day-18 differentiated hESC-derived hepatocytes were capable of engrafting (0.05%) in the spleen and migrating to the liver of immune-comprised CCl4-treated mice. Similar results were reported by Agrawal et al.,[86] who examined a more immature population of hESC-derived DE cells (70%). Human ESC-derived definitive endoderm (DE) was injected into mice and integrated into the adult liver followed by differentiation into liver cells in vivo. Basma et al.[88] showed that hESC-derived hepatocytes were capable of homing to and engrafting normally in the liver, expanding in response to a physiological proliferation signal, and secreting functional liver-specific proteins [i.e., human Alb and human alpha 1 anti-trypsin (AAT)] after engraftment. Twelve days postimplantation, engrafted cells exhibited mixed populations of mature Alb expressing hepatocytes (55%) and fetal AFP positive cells (12%). However, in the long term, teratomas developed in the liver, spleen, and peritoneal cavity.

Attempting to reduce undifferentiated cell populations, hESC-derived hepatocytes were sorted for surface asialoglycoprotein-receptor expression (ASGR), a definitive feature of hepatocytes (approximately 18–26%). Implantation of sorted hepatocytes into the spleen of one immune suppressed Nagase analbuminemic rat, which underwent a 70% partial hepatectomy at the time of implantation, resulted in increased secretion of Alb and AAT. Although the implanted cell population was enriched for hepatocytes, evidence for tumor development (histologically consistent with well-differentiated adenocarcinoma) was observed in the peritoneum 55 days after surgery.

These studies establish the feasibility of hepatic differentiation of hESCs and suggest induction strategies. Interestingly, in all differentiation protocols, although the hESC-derived hepatocytes exhibit characteristics of mature hepatocytes (gene expression/protein profile and hepatic functions), they also appeared to retain some immature characteristics such as relatively low levels of expression of cytochrome P450 transcripts and continued expression of AFP, a marker of fetal rather than adult hepatocytes. Thus current methods have several challenges including inefficient differentiation resulting in heterogeneous cultures that contain many cell types and low proportions of hepatocytes. It has been suggested that these limitations are in part due to inadequate characterization of the hESC-derived hepatic cells.[77] Most of the genes and metabolic enzymes used to identify hepatocytes are not solely expressed by the liver, except AFP, which is expressed only in the fetal liver. Hepatocytes that stop expressing AFP can be considered adult hepatocytes. The synthesis of urea can also be used to test for hepatocytes as this function is performed only by mature hepatocytes. Recently, CYP7A1 has been shown as a marker for mature hepatocytes in both mouse-derived[78] and human-derived[87] hepatocytes. In the few studies that used an in vivo assay to characterize cell differentiation, incorporation of implanted cells into the liver of the mice and the production of Alb by those cells was tested. It is yet to be discovered how well hESC-derived hepatocytes will replace adult hepatocytes when implanted into the liver. Further research is required to develop robust differentiation processes and standardize characterization of mature hepatocytes.

Pancreas

The spontaneous differentiation of hESCs into insulin-expressing cells has been examined.[90–92] Various studies have investigated the ability to induce hESCs to differentiate into nestin-producing neural precursors by exposure to conditions that result in clusters of insulin-containing cells.[10,93] These reports have not affirmed the in vitro production of fully functional β-cells that can secrete physiologically sufficient amounts of insulin in response to glucose. The recent reports from D'Amour et al.[94] and Kroon et al.[95] describe differentiation of a pancreatic lineage in vitro from hESCs. Cells were first differentiated into definitive endoderm (DE) and subsequently into pancreatic cell types. DE induction was achieved by treating differentiating hESCs with high concentrations of activin A in combination with Wnt3a.[94–96] Initial differentiation steps toward DE required low serum levels since serum components contain activators of the phosphatidylinositol 3-kinase pathway that inhibit the induction of DE.[97] DE cell population was then further purified by sorting for CXCR4 expressing cell, a cell marker for human pancreatic cells.[98,99] Mimicking normal pancreagenesis development, the activity of SHH, a member of the Hedgehog signaling pathway, was suppressed via treatment with cyclopamine, resulting in formation of *PDX1* expressing cells, the earliest marker of pancreatic epithelium. Retinoic acid was added to promote the commitment of the PDX1-positive cells toward the endocrine lineage over pancreatic exocrine cells.[94,95] The percentages of insulin-positive cells were relatively low (7%) and the production of C-peptide in response to glucose was low. In addition, expression of *MAFA*, a transcription factor normally upregulated during the final stages of β-cell formation, was not observed. All of these data indicate that insulin-producing cells differentiated in vitro have not progressed to fully matured, functional β-cells.[94]

Implantation of hESC-derived pancreatic precursors resulted in their maturation into insulin-positive cells capable of rescuing hyperglycemic mice.[95] Removal of the implanted cells caused recurrence of the diabetic phenotype. Although the proportion of insulin-producing cells was not analyzed in the implanted precursor population, 87 days postexplantation, grafts contained more than 50% endocrine cells. Since the implanted cell population was not purified, it is not surprising that teratomas were formed. It will be interesting to examine the in vivo long-term behavior of these cells where current adult islet transplantation has failed.

Jiang et al.[99] used feeder-free cultures of hESCs and a serum-free protocol for multistep differentiation of insulin-like clusters (ILCs) from a DE population. Within these ILCs, cells testing positive for markers of the endocrine pancreas plus ductal epithelium were found. In addition to insulin, ILCs also contain human C-peptide and glucagon-positive cells, and release C-peptide in response to elevated concentrations of glucose. This population of 2–8% of insulin-producing cells produced significantly higher amounts of C-peptide in response to glucose stimulation compared to that described by D'Amour et al.[94] However, implantation of hESC-derived ILCs into diabetic mice did not significantly reduce blood glucose levels.[100] Besides the low levels of β-like cells in ILCs, it is possible that the implantation procedure used by Kroon et al.[95] (seeding cells on Matrigel-coated gel-foam scaffolds) provided a more conducive environment for cell differentiation. Teratomas were not reported, however, in the long term (>11–12 weeks) animals exhibited swollen abdomens.

To increase efficiency of pancreatic differentiation from hESCs, a more direct genetic engineering approach has been examined by overexpressing critical transcription factors in hESCs such as *Pdx1* through plasmid insertion.[101] Despite enhanced differentiation toward pancreatic lineages, expression of insulin genes could only be demonstrated when the cells differentiated in vivo within teratomas. Recently, Bernardo et al.[102] used a temporally biphasic system to regulate the expression of *Pdx1*. With the induction of the second wave of *Pdx1* expression, an increase in β-cell numbers was achieved. However, unlike β-cells in mature islets, these insulin-secreting cells express *MafB* instead of *MafA*. Furthermore, the response of these human-derived cells to glucose stimulation was not reported.

Alternatively, small molecules have been shown to facilitate cell differentiation in the absence of permanent genetic changes.[103,104] High-throughput screening of libraries containing diverse chemical compounds identified small molecules that guide specific steps during hESC differentiation into β-cells.[103,104] Combinations of small molecules (indolactam V) and growth factors (FGF10) led to derivation of more than 45% PDX1 expressing cells from a DE-rich hESC culture. Following maturation, a mixture of endocrine/exocrine and duct endocrine cells was detected by immunostaining, of which 0.5% were insulin positive. C-peptide release was not very responsive to glucose, indicating that the glucose-sensing mechanism was not fully active.

A number of groups have reported the generation of insulin-secreting cells from hESCs in vitro. However, these cells likely correspond to immature β-cells because they present low insulin content, express multiple hormones in the same cell, or show little (if any) response to glucose stimulation. A major difference between β-cells in the pancreas and β-cells generated in vitro lies in the different environments in which they reside. Pancreatic islets are complex structures that consist of multiple cell types, including different hormone secreting cells, neuronal cells, and vascular endothelial cells. It is possible that a more complex environment, mimicking the 3D interactions among β-cells as well as between mesenchymal and epithelial cells, may improve efficiency of β-cell specification and maturation.

MESODERMAL TISSUE

The middle germ layer is the mesoderm, producing the muscles, excretory organs, circulatory organs, sex organs (gonads), and internal skeleton.

Circulatory System

Hematopoietic Progenitor Cells

The stem cells that form blood and immune cells are known as hematopoietic stem cells (HSCs). HSCs differentiate into lymphoid and myeloid precursors, the two classes of precursors for the two major lineages of blood cells. Lymphoid precursors differentiate into T cells, B cells, and natural killer cells. Myeloid precursors differentiate into monocytes and macrophages, neutrophils, eosinophils, basophils, megakaryocytes, and erythrocytes.[105] HSCs occur in the bone marrow, blood, liver, and spleen but are extremely rare in any of these tissues.[106] HSC transplants are now routinely used to treat patients with cancers and other disorders of the blood and immune system. Despite the significant research on HSCs, in vitro proliferation remains limited. Thus hESCs may play a role in developing an abundant supply of HSCs grown in the lab.

The in vitro hematopoietic differentiation of hESCs has been investigated extensively and hematopoietic precursors, as well as differentiated progeny, representing erythroid, myeloid, megakaryocytic, and lymphoid lineages, have been observed in differentiating cultures of hESCs. Hematopoietic differentiation of hESCs has been achieved using two experimental approaches: formation of EBs[107–116] and co-culture with stromal cells.[7,117–123] Most of the stromal cell lines that are used to induce hESC differentiation into hematopoietic cells (MS5, S17, and OP9) are of murine origin, posing the risk of mouse-related diseases. Although at lower efficiency, hematopoietic cells could be obtained by culturing EBs on human primary bone marrow stromal cells with low-dose cytokines or culturing hESCs on human fetal liver-derived cells without cytokines.[119,124] Supplementing the culture medium with proteins[125,126] or genetically modifying hESCs, in order to overexpress genes associated with HSC self-renewal,[127,128] has been shown to increase the efficiency of hematopoietic differentiation.

The in vivo functionality of hESC-derived hematopoietic cells has been examined.[120,129–131] In general, these reports have used different differentiation procedures and cell types. Wang et al.[129] were the first to demonstrate that intrabone marrow injection of hESC-derived cells (that are PECAM$^+$ FLK1$^+$ VECAD$^+$ CD45$^-$) into SCID mice can result in a low level of human hematopoietic engraftment, 8 weeks postimplantation. Although low ($\sim$1%), long-term engraftment of hESC-derived HSCs was achieved in fetal sheep[131] and in SCID mice.[130] Recently, the origin of the feeder layer microenvironment was examined on hematopoietic differentiation and in vivo engraftment.[120] Human ESCs were co-cultured with three different stromal cell lines as well as with primary cells derived from the aorta-gonad-mesonephros (AGM) region and fetal liver. All were capable of long-term engraftment in SCID mice, yet hESCs that differentiated on mouse stromal cell lines from the AGM region exhibited the highest published engraftment efficiency of 2.06–16.26%.[120]

Further differentiation of hESC-derived HSCs into erythrocytes (red blood cells (RBCs)) is envisioned to not only relieve limitations on blood supply but also to produce these cells as a donorless universal source of RBCs for transfusions. To this end, two reports have recently shown efficient differentiation of hESCs into nucleated RBCs.[126,132] Olivier et al.[132] reported the production of early-stage (nucleated) RBCs from hESCs that have matured into enucleated (6.7%) adult globin expressing (less than 2%) RBCs. However, the system required the use of serum, and no studies were carried out to determine if the cells were functional. Lu et al.[126] were able to generate 40-fold increases in the amount of hESC-derived RBCs compared to Olivier's method by using an EB differentiation protocol and addition of HoxB4 protein (a prominent HSC mitogen). Thirty percent to 65% of hESC-derived RBCs underwent multiple differentiation events, including a progressive decrease in size, increase in expression of mature RBC markers, and chromatin/nuclear condensation. This resulted in enucleated functional erythrocytes with a diameter similar to normal RBCs. Human ESC-derived RBCs possessed oxygen equilibrium curves comparable with normal transfusable RBCs and also responded to pH changes (Bohr effect) and depletion of 2,3-diphosphoglycerate. However, further study will be needed to investigate their in vivo function.

The differentiation of hESCs into dendritic cells (DCs) has also been examined. Human ESCs can serve as a theoretically unlimited source for generating DCs. Genetically engineering hESC-derived DCs to present specific antigens offers prospects for reprogramming the immune system to stimulate either T-cell immunity against tumor-associated or viral antigens or to dampen the immune response in an autoimmune-transplant setting. Highly efficient differentiation of hESCs to DCs has been achieved (95%). A culture system without feeder cells was developed to differentiate hESCs into large numbers of mature DCs. This was achieved through both co-culturing of hESCs on OP9[123] cells and feeder-free sequential differentiation of EBs through hematopoietic and myeloid precursor stages.[116] In the former, a lin⁻ CD34⁺ CD43⁺ CD45⁺ hematopoietic cell population gave rise to an essentially homogeneous mature population of each of the following myelonic lineages: neutrophils, eosinophils, macrophages, osteoclasts, and Langerhans cells.[123] Recently, the prospects of engineering the immune system were boosted by the capability of genetically engineered hESC-dervied DCs to stimulate potent antiviral T-cell[116,122] and antitumor responses while avoiding activation of regulatory T cells.[116]

Vasculature

Engineering vascular networks holds promise for many therapeutic applications including cell implantation for repair of ischemic tissues (formation of blood vessels and heart valves), engineering artificial vessels, repair of damaged vessels, and inducing formation of blood vessel networks in engineered tissues.[133,134] A functional vascular system is essential for the formation and maintenance of most tissues in the body, and the lack of vascularization results in ischemic tissues with limited intrinsic regeneration capacity. Therefore vascularization of engineered tissues in vitro, before implantation, may enable cell viability during tissue growth, induce structural organization, and promote integration upon implantation. Endothelial and smooth muscle cells are the building blocks of blood vessels. Endothelial cells line the inner surface of blood vessels as an interface between the circulating blood and the adjacent smooth muscle layers. By contraction and relaxation, the smooth muscle cells drive the blood through the circulatory system of the body.

Human ESCs represent a potential source of vascular cells for the treatment of ischemic tissues such as after myocardial infarction. Their ability to self-renew enables the large-scale expansion in vitro, which is required for repair of large organs like the (ischemic) heart, whereas their pluripotent differentiation capacity may allow for the generation of vascular cell types (endothelial cells, smooth muscle cells) for replacing dead cells in ischemic tissues.

Early endothelial progenitor cells, isolated from differentiating mESCs and hESCs, have been shown to give rise to the cell types involved in blood vessel, endothelial, and smooth muscle cells.[135,136] Moreover, it was recently shown that embryonic endothelial cells are critical for the earliest stages of liver,[137] pancreas,[138] and neuronal[139] organogenesis. Therefore, in addition to their potential clinical applications to support angiogenesis and vasculogenesis, purified hESCs could be important for studying early human development as well as for studying direct differentiation of hESCs into various tissues.

The differentiation and isolation of endothelial and vascular cells from hESCs was recently described in several reviews.[12,140] The first hESC-derived endothelial progenitor cells were described by Levenberg et al.[141] The authors analyzed the expression pattern of endothelial markers (e.g., platelet/endothelial cell adhesion molecule-1 (PECAM-1), CD34, kinase insert domain receptor (KDR) of differentiating hESCs and demonstrated that they can differentiate into endothelial-like cells forming vascular-like structures within the EBs. The hESC-derived endothelial-like cells were isolated using PECAM-1 antibodies and fluorescence activated cell sorting (FACS). In addition, the isolated cells were able to form into tube-like structures when cultured on Matrigel. The subcutaneous implantation of hESC-derived endothelial-like cells into SCID mice resulted in formation of hESC-derived microvessels containing mouse blood cells, indicating functional integration with host vasculature to a certain degree.

Ferreira et al.[142] reported the derivation of endothelial- and smooth muscle-like cells from hESC-derived vascular progenitor cells (CD34$^+$). The addition of vascular endothelial growth factor-165 (VEGF-165) (50 ng/mL) resulted in endothelial-like cells, whereas a medium supplemented with platelet-derived growth factor-BB (PDGF-BB) (50 ng/mL) induced formation of smooth muscle-like cells. Subcutanous implantation into SCID mice revealed that both cell types contributed to the formation of human microvasculatures, with a few microvessels containing mouse blood cells. Similar results were reported by Wang et al.[143] using a 2D culture system that bypassed EB formation. These CD34$^+$ progenitor cells were able to form some vascular networks in vivo only when co-implanted with mouse mesenchymal stem cells. Stable perfused blood vessels were detected over a period of 159 days.

Yang et al.[136] recently demonstrated that hESC-derived KDR(low)/C-KIT(CD117)(neg) cells can give rise to endothelial-like cells, smooth muscle-like cells, and cardiomyocytes. The combinations of activin A, BMP4, bFGF, VEGF, and dickkopf homolog 1 (DKK1) in serum-free media induced the generation of these KDR(low)/C-KIT(CD117)(neg) cell populations. They report a short-term improvement (after 2 weeks) of cardiac structure and function after injection into infarcted hearts of immunosuppressed mice.

Despite these promising results, when injecting hESC-derived vascular-like cells directly into preclinical animal models, a prolonged therapeutic effect is limited by the low efficiency of incorporation within the recipient's vasculature (less than 3% of injected cells may engraft, mainly due to cell death).[144,145] Subsequently, biomaterials were engineered with the goal of preventing *anoikis* and improving functional engraftment. Biodegradable materials were employed as cell carriers and as cell growth matrices[144–146] or, alternatively, as protective environments for the controlled release of bioactive cytokines.[147–149]

Ferreira et al.[148] reported the efficient vascular differentiation of hESCs using a dextran-based bioactive hydrogel. These gels were engineered to display insoluble RGD adhesion signals or soluble factors (VEGF) loaded onto PLGA microspheres. The RGD adhesion ligands, derived from fibronectin, are known to stimulate integrins $\alpha_v\beta_3$ and $\alpha_5\beta_1$, which are relevant in early cardiovascular development and maintenance, whereas VEGF was shown to induce differentiation of hESCs into endothelial and hematopoietic cells. All examined gels increased the fraction of cells expressing VEGF receptor KDR/Flk-1, a vascular marker, compared to spontaneously differentiated EBs.

Levenberg et al.[150] have engineered vascularized skeletal muscle tissue using highly porous, biodegradable PLLA/PLGA scaffolds consisting of myoblasts, embryonic fibroblasts, and hESC-derived endothelial progenitor cells. In vitro, they demonstrated that embryonic fibroblasts promoted formation and stabilization of the endothelial vessels. Three different models were used to assess the therapeutic potential: (1) subcutaneous implantation in the back of immune suppressed mice, (2) intramuscular implantation into the quadricep muscle of nude rats, and (3) replacement of the anterior abdominal muscle segment of nude mice with the construct. They report that the prevascularization in vitro can improve the vascularization and survival of the 3D muscle-like construct after implantation into SCID mice. This approach may thus have potential applications in tissue engineering, but may also provide an in vitro model system to study developmental vascularization processes.

Heart

The prevalence of cardiovascular disease and its related mortality necessitate the development of novel approaches for repairing damaged hearts. Despite successful pharmacologic approaches, to prevent or limit cardiovascular disease, the restoration of function to the damaged heart remains challenging. Because of limited intrinsic regeneration capabilities and a shortage of donors for heart transplantation, cell-based therapies have been investigated as a potential therapy for damaged cardiomyocytes. A number of autologous cell types (skeletal myoblasts, bone marrow mononuclear cells, mesenchymal stem cells, and endothelial progenitor cells) are currently being tested in clinical trials.[151,152] Skeletal myoblasts form well-defined myotubes; however, these were not shown to form electrical junctions with host myocardium and thus are unlikely to participate in functional synctium.[153] Clinical trials with adult stem cells, derived from the bone marrow, were inconclusive.[154] Functional and structural cardiac improvement may be related to paracrine effects rather than true replacement of the muscle.[155,156] Fetal and neonatal cardiomyocyte donor cells integrate structurally and functionally into the host myocardium,[157,158] but a main bottleneck of their clinical application is the allocation of a sufficient quantity of cells.

On average, an infarct in humans destroys $\sim$1 billion cardiomyocytes.[159] A possible solution to the above-mentioned cell-sourcing shortage may be the use of hESCs. The generation of cardiomyocytes from differentiating EBs was originally described by Kehat et al.[5] Contracting cardiac foci were first observed at 27–30 days of differentiation in 10% of EBs. Likewise, spontaneous differentiation of EBs into cardiomyocytes was also observed by others.[160,161] An alternative approach for the derivation of cardiomyocytes was co-culture of hESCs with a mouse visceral endoderm-like cell line (END-2).[162] This resulted in beating areas in approximately 35% of the cells after 12 days in co-culture. In terms of more defined factors that can direct differentiation to cardiomyocytes, four families of peptide growth factors and their combination with or absence of serum have been widely implicated in promoting cardiogenesis, namely, BMPs (members of the TGF-b superfamily), Wnts, and FGFs.[163,164] Xu et al.[6] demonstrated enhancement of cardiac differentiation (threefold increase compared to controls) in hESCs by using 5-aza-2′-deoxycytidene. Up to 70% of cardiomyocytes were further enriched by utilizing Percoll gradient purification. Recently, treating hESCs with BMP4 and retinoic acid, followed by Percoll purification, led to a highly rich cardiomyocytes population (71–95%).[165] Transgenic selection may offer an alternative route to enrichment.[166–168]

Three-dimensional cultures of cardiomyocytes on synthetic scaffolds[169] or in suspension (suspended cells agglomerated into a disk under rotation)[170] promoted cardiac cell proliferation, maturation, and myofibrillar organization. These constructs beat spontaneously and synchronously in culture and were shown to transmit intracellular calcium transients.

Several reports examined the in vivo engraftment and functionality of hESC-derived cardiomyocytes. Dissected contracting regions of adherent EBs were able to restore pacing activity in immune-suppressed pigs with experimental heart block.[171] Xue and colleges[172] implanted GFP-expressing human EBs into the left ventricle of immune-suppressed guinea pig hearts and observed propagating action potential waves that originated from the site of cell engraftment. Both of these studies suggested that electromechanical coupling occurred between donor and host cells; however, engraftment efficiencies were not quantified. Engraftment dropped significantly when beating EBs were transplanted in ischemic nude rat hearts[165,173,174] compared to uninjured ones[175] (18% and 90%, respectively). To increase engraftment and cardiomyocyte survival, the cardiomyocte population was enriched (growth factor supplementation followed by Percoll purification) and then embedded in Matrigel containing a cocktail of proteins and chemicals. After 4 weeks, 80% of the infarct zone was remuscularized and cardiac function was improved, as evidenced by greater thickening (2.5-fold) of infarcted walls in hearts receiving the hESC-derived cardiomyocytes compared to controls. However, engraftment was probably poor as only rare contacts between host and transplanted grafts were demonstrated. Similar results were also reported when implanted populations contained only 20–25% cardiomyocytes.[176] Improvement in cardiac function was only observed 4 weeks postimplantation but not at longer periods (12 weeks). Thus the observed short-term effects of hESC-derived cardiomyocyte implantation on cardiac function might be due to paracrine effects and not to actual engraftment.

The heart is composed of 30% cardiac myocytes (representing 70% of heart mass) and 70% nonmyocytes, including endothelial cells, smooth muscle cells, and fibroblasts. Cardiac regeneration is apparently not only a matter of cardiac myocyte addition and therefore a co-culture will be more realistic in reconstruction of the myocardium. A pivotal issue in cardiac tissue engineering is the generation of thick myocardial structures in vitro and eventual maintenance in vivo. Vascularization is likely to be a prerequisite to achieve these goals. Vascularized tissue engineered cardiac muscle has been developed via triculture with endothelial progenitor and fibroblasts that either support angiogenesis or vasculogenesis, leading to increased cardiomyocte survival and proliferation both in vitro[26] and in vivo[169,177] (Fig. 31.2).

The cardiomyocyte populations derived so far are immature and exhibit robust proliferative capacity both in vitro and after implantation, implying that delivery of an initially subtherapeutic cell dose can further differentiate in vivo.[173,174] It is yet to be seen whether these populations will generate teratomas. It will be important to evaluate engraftment efficiencies from paracrine effects as well as to examine the long-term effects of transplanted cardiomyocytes.

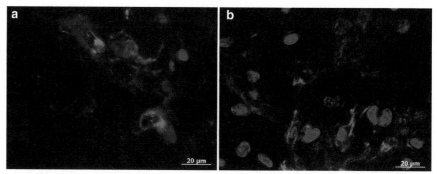

FIGURE 31.2. *In vitro engineered vascularized cardiac muscle (using a triculture of hESC-derived cardiomyocytes, human umbilical vein endothelial cells, and mouse embryonic fibroblasts). Vascular structures are stained in green for the expression of VWF. hESC-derived cardiomyocytes are stained in red for Troponin I expression. DAPI was used for nuclear staining in blue. (a) Vessels are organized in close proximity to cardiomyocytes. (b) Mature cardiomyocytes exhibit sarcomeric patterning. (Generously provided by Shulamit Levenberg and Ayelet Lesman.)*

Bone and Cartilage

Bone

The reconstruction of large bone defects caused by trauma and disease (e.g., tumor extraction) is a major issue in orthopedic surgery. Although bone tissue has a comparatively high regenerative capacity, this self-repairing process can fail when the bone defects are too large or the natural healing capabilities are not sufficient. Most of the current strategies used to repair large bone defects use bone grafting materials such as autografts, allografts, or prosthetic implanting materials. However, challenges with these approaches such as their limited availability, potential disease transmission by bone grafts, poor biocompatibility, and failure of prosthetic implants have motivated further investigation into alternative strategies such as tissue engineering.

Recently, mesenchymal stem cells (MSCs) or osteoprogenitor cells, derived from human bone marrow or adipose tissues, have been studied as potential cell sources for autologous bone tissue engineering.[178,179] MSCs have been reported to facilitate bone repair in various bone defects (reviewed in Refs. 180 and181). Yet the use of MSCs for tissue engineering and bone repair has some limitations. For example, the self-renewal and proliferative abilities of MSCs are limited and decrease as donor ages increase. In addition, the relative frequency of MSC occurrence in the marrow stroma is low (1 in 100,000 cells) and indicates that only a limited proportion of cells can differentiate into the osteogenic lineage. Therefore the development of alternative cell sources is still needed to completely satisfy the various demands of skeletal tissue engineering.

Several studies have attempted to establish osteogenic progenitor cells from differentiated EBs using osteogenic supplements (OS) containing ascorbic acid (an extracellular synthesis promoter), β-glycerophosphate (induces mineralization), and dexamethasone.[8,182,183] Co-culturing of hESCs with human primary bone-derived cells (hPBDs) without the use of any additional exogenous factors was sufficient for the differentiation of hESCs into an osteogenic cell population.[184] Karp et al.[185] and others[186] have shown that osteogenic differentiation is sevenfold enhanced when EB formation was omitted. Functional outcomes such as the capacity to form bone-like formation was ascribed to mineralized nodules and was mostly assessed by calcium and phosphorus staining with Alizarin Red and von Kossa, respectively. To avoid identification of dystrophic mineralization, electron microscopy and Fourier transform infrared analysis should be used to verify normal mineralization. Using

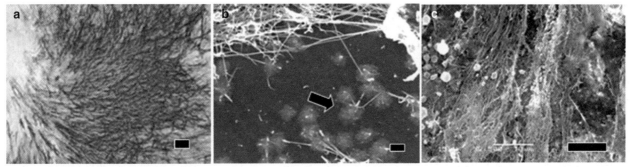

FIGURE 31.3. *Human ESC-derived bone through EB formation and osteogenic supplementation. (a) Alkaline phosphatase staining of bone nodules following 35 days in culture (scale bars: 100 μm). Scanning electron micrograph of (b) cement line globules (black arrows) observed at the matrix/culture dish interface (scale bars: 1 μm) and (c) mineralized matrix (scale bars: 10 μm). (Reproduced with permission from Karp et al.[185] Copyright © 2006, Stem Cells, AlphaMed Press).*

these techniques, Karp et al.[185] demonstrated that the bone matrix formed by hESC-derived osteogenic progenitors included cement line matrix and mineralized collagen, which displayed similar structure and composition to hydroxyapatite and human bone (Fig. 31.3).

Since most protocols yielded low osteogenic differentiation efficiencies, only a few researchers have examined in vivo behavior. Osteogenic progenitors (without sorting) were seeded on synthetic 3D scaffolds and implanted subcutaneously into immune-deficient mice. Osteogenic progenitors, seeded on poly-D,L-lactide (PDLLA) scaffolds, were able to form mineralized tissue identified by von Kossa staining and human osteocalcin.[182] Kim et al.[187] used a porous PLGA scaffold coated by hydroxyapatite (HA) and reported a significant increase in all osteogenic genes and mineralization (analyzed by hematoxylene and eosin staining and X-ray). These were higher when scaffolds were loaded with BMP-2.[187] Tissues formed in these studies were reported to be all bone, presumably due to the fact that the starting cell population had a higher fraction of ostegenic progenitors (20% versus less than 1%). Interestingly, neither of these reports discusses the formation of teratomas, although seeded cell populations contained a mix of immature and mature cells of various origins, implying that in vivo differentiation is apparently more efficient than in vitro. Although these results are promising, it will be interesting to examine these cells in models where functionality of the derived bone tissue can be assessed.

Cartilage

Cartilage is a specialized tissue consisting of chondrocytes and an organic matrix comprised of water, proteoglycans, and type II collagen. Long-term repair and regeneration of cartilage defects remains an elusive goal. Chondrogenic differentiation of hESCs was achieved through differentiation of EBs and then direct co-culturing with primary chondrocytes,[188,189] fibrochondrocytes,[189] or high-density seeding and BMP2 supplementation.[190] In all reports, differentiated cells formed chondrogenic modules that secreted extracellular matrix containing glycosaminoglycans rich in collagen II costained with SOX-9. In vivo implantation of hESC-derived chondrocytes resulted in tissue that macroscopically and phenotypically resembled cartilage.[188]

Several studies have suggested the possibility of obtaining mature mesenchymal cell types from hESCs and their further differentiation into adipose, cartilage, and bone tissue.[191–193] However, the cellular and molecular mechanisms underlying the formation of mesenchymal precursors and the onset of further differentiation, specifically to chondrogenesis, remain obscure. In light of this notion, Tremoleda et al.[194] compared the osteogenic potential of hESCs to bone marrow-derived stem cells in an in vivo model (diffusion chamber). Both cell lines were subjected to the same differentiation procedure, primarily exposure to OS media. The in vivo directed differentiation of hESCs to form bone using a diffusion chamber model was reported to be equivalent to bone material formed by hMSCs. The pretreatment of hESCs with osteogenic factors induced and maintained their differentiation along the osteogenic pathway. Osteoprogenitors generated from hESCs maintained their phenotypic stability during in vivo incubation without developing any tumors or teratoma within the 79-day period of implantation.

Identifying biomaterials that can mimic the in vivo environment, or present the appropriate cues (stiffness, growth factors, proteins, etc.) for promoting oestogenic and chondrogenic differentiation, may lead to advances in hESC-derived bone and cartilage tissue engineering.

SUMMARY AND FUTURE CHALLENGES

This chapter discusses hESCs as a source for human cells and their therapeutic potential in cell and tissue engineering. In the last two decades, hESC research has focused on examining the ability of hESCs to differentiate into many cell types. However, for many differentiated cell populations, current protocols have suboptimal efficiencies and can produce a mixture of mature and immature cell populations. The two exceptions are the neuronal and hematopoietic lineages. Current research is focused on devising efficient differentiation protocols to produce large numbers of specialized cell populations, thereby bringing us closer to realizing the potential of hESCs as an unlimited source of human cells.

In vivo experiments have demonstrated that mixtures of cell populations can further differentiate into a mature subtype and, in some cases, even improve functionality. Yet in some cases, these can lead to the formation of teratomas. Because the undifferentiated cells induce teratomas, tumor formation might be avoided by devising methods for removing any undifferentiated hESCs prior to implantation. For this to be possible, further studies are required to identify specific surface markers and the kinetics of their expression during the different stages of differentiation. Ultimately, it will be necessary to both identify the optimal stage(s) of differentiation for implantation and demonstrate that the implanted hESC-derived cells can survive, integrate, and function in the recipient.

During embryogenesis, cells and organs are developed in a unique 3D environment and are affected by paracrine effects on neighboring cells/tissue. It is possible that attempting to recapture the differentiation process of hESCs by 2D culture and supplementation with cytokines and chemicals is an oversimplification of what in reality is a complex process. Considering that tissues are composed of various cell types, a multiculture system will advance the field of hESC-engineered tissue. Human ESCs provide an opportunity to derive both multilineage progenitor cells from a common pluripotent source, and to combine these into distinct yet interactive tissue compartments. This approach has great potential to better understand how intercellular interactions stimulate tissue assembly and organization to generate functional tissues, streamlining progenitor cell acquisition and tissue fabrication.

The novelty of 3D culture systems is manifested in the ability of hESCs to differentiate within scaffolds into a variety of cell types and generate organ-like structures. In implantation studies, cell survival improved upon encapsulation in hydrogels or scaffolds. Recent reports on the interplay between matrix rigidity, structure and architecture, and stem cell differentiation further highlight the active role that 3D scaffolds can take in mimicking the microenvironment favorable for specific tissue differentiation. In this light, properly harnessing the potential of hESCs to develop into functional tissues, in an engineered 3D tissue environment, would provide an important means of generating complex tissues for implantation.

The ideal therapeutic hESC-derived cells should not trigger immune rejection (or require immune suppression). Human ESCs are weakly immunogenic, expressing only moderate amounts of major histocompatibility complex (MHC) class I and no MHC class II proteins.[195] The potential immunological rejection of hESC-derived cells might be avoided by the generation of a library of hESCs possessing the necessary immunological diversity.[196] A number of exciting studies have demonstrated that a cocktail of only four different transcription factors is sufficient to reprogram somatic cells into induced pluirpotent stem (iPS) cells.[195,197] The iPS cells, similar to hESCs, can differentiate into cells of all three germ layers, suggesting that patient-specific stem cell populations may become a reality.[198,199] In support of this notion, iPS cells have recently been generated from cohorts of human patients suffering from a variety of diseases,[200] including type 1 diabetes[201] and ALS.[202] Yet deciphering the signals communicated by hESC-derived progenitors in distinct tissue compartments may provide new tissue engineering approaches to fabricate complex tissues as well as their application to newly discovered pluripotent stem cell sources such as induced iPS cells to generate patient-specific tissues.

REFERENCES

1. Thomson JA, Itskovitz-Eldor J, Shapiro SS, et al. Embryonic stem cell lines derived from human blastocysts. *Science*. 1998;282:1145–1147.

2. Reubinoff BE, Pera MF, Fong CY, et al. Embryonic stem cell lines from human blastocysts: somatic differentiation in vitro. *Nat Biotechnol*. 2000;18:399–404.

3. Reubinoff BE, Itsykson P, Turetsky T, et al. Neural progenitors from human embryonic stem cells. *Nat Biotechnol*. 2001;19:1134–1140.

4. Zhang SC, Wernig M, Duncan ID, et al. In vitro differentiation of transplantable neural precursors from human embryonic stem cells. *Nat Biotechnol*. 2001;19:1129–1133.

5. Kehat I, Kenyagin-Karsenti D, Snir M, et al. Human embryonic stem cells can differentiate into myocytes with structural and functional properties of cardiomyocytes. *J Clin Invest*. 2001;108:407–414.

6. Xu C, Police S, Rao N, et al. Characterization and enrichment of cardiomyocytes derived from human embryonic stem cells. *Circ Res*. 2002;91:501–508.

7. Kaufman DS, Hanson ET, Lewis RL, et al. Hematopoietic colony-forming cells derived from human embryonic stem cells. *Proc Natl Acad Sci USA*. 2001;98:10716–10721.

8. Sottile V, Thomson A, and McWhir J. In vitro osteogenic differentiation of human ES cells. *Cloning Stem Cells*. 2003;5:149–155.

9. Rambhatla L, Chiu CP, Kundu P, et al. Generation of hepatocyte-like cells from human embryonic stem cells. *Cell Transplant*. 2003;12:1–11.

10. Segev H, Fishman B, Ziskind A, et al. Differentiation of human embryonic stem cells into insulin-producing clusters. *Stem Cells*. 2004;22:265–274.

11. Green H, Easley K, Iuchi S. Marker succession during the development of keratinocytes from cultured human embryonic stem cells. *Proc Natl Acad Sci USA*. 2003;100:15625–15630.

12. Levenberg S, Zoldan J, Basevitch Y, et al. Endothelial potential of human embryonic stem cells. *Blood*. 2007;110:806–814.

13. Kaushal S, Amiel GE, Guleserian KJ, et al. Functional small-diameter neovessels created using endothelial progenitor cells expanded ex vivo. *Nat Med*. 2001;7:1035–1040.

14. Shirota T, He H, Yasui H, et al. Human endothelial progenitor cell-seeded hybrid graft: proliferative and antithrombogenic potentials in vitro and fabrication processing. *Tissue Eng*. 2003;9:127–136.

15. Chiu RC. Bone-marrow stem cells as a source for cell therapy. *Heart Fail Rev*. 2003;8:247–251.

16. Martin I, Shastri VP, Padera RF, et al. Selective differentiation of mammalian bone marrow stromal cells cultured on three-dimensional polymer foams. *J Biomed Mater Res*. 2001;55:229–235.

17. Pelled G, Ge T, Aslan H, et al. Mesenchymal stem cells for bone gene therapy and tissue engineering. *Curr Pharm Des*. 2002;8:1917–1928.

18. Mezey E, Chandross KJ, Harta G, et al. Turning blood into brain: cells bearing neuronal antigens generated in vivo from bone marrow. *Science*. 2000;290:1779–1782.

19. Noel D, Djouad F, Jorgense C. Regenerative medicine through mesenchymal stem cells for bone and cartilage repair. *Curr Opin Invest Drugs*. 2002;3:1000–1004.

20. Perin EC, Geng YJ, Willerson JT. Adult stem cell therapy in perspective. *Circulation*. 2003;107:935–938.

21. Park KI, Teng YD, Snyder EY. The injured brain interacts reciprocally with neural stem cells supported by scaffolds to reconstitute lost tissue. *Nat Biotechnol*. 2002;20:1111–1117.

22. Powell C, Shansky J, Del Tatto M, et al. Tissue-engineered human bioartificial muscles expressing a foreign recombinant protein for gene therapy. *Hum Gene Ther*. 1999;10:565–577.

23. Vugler A, Lawrence J, Walsh J, et al. Embryonic stem cells and retinal repair. *Mech Dev*. 2007;124:807–829.

24. Dhara SK, Stice SL. Neural differentiation of human embryonic stem cells. *J Cell Biochem*. 2008;105:633–640.

25. Carpenter MK, Inokuma MS, Denham J, et al. Enrichment of neurons and neural precursors from human embryonic stem cells. *Exp Neurol*. 2001;172:383–397.

26. Park S, Lee KS, Lee YJ, et al. Generation of dopaminergic neurons in vitro from human embryonic stem cells treated with neurotrophic factors. *Neurosci Lett*. 2004;359:99–103.

27. Schulz TC, Palmarini GM, Noggle SA, et al. Directed neuronal differentiation of human embryonic stem cells. *BMC Neurosci*. 2003;4:27.

28. Zeng X, Cai J, Chen J, et al. Dopaminergic differentiation of human embryonic stem cells. *Stem Cells*. 2004;22:925–940.

29. Perrier AL, Tabar V, Barberi T, et al. Derivation of midbrain dopamine neurons from human embryonic stem cells. *Proc Natl Acad Sci USA*. 2004;101:12543–12548.

30. Sonntag KC, Pruszak J, Yoshizaki T, et al. Enhanced yield of neuroepithelial precursors and midbrain-like dopaminergic neurons from human embryonic stem cells using the bone morphogenic protein antagonist noggin. *Stem Cells*. 2007;25:411–418.

31. Hong S, Kang UJ, Isacson O, et al. Neural precursors derived from human embryonic stem cells maintain long-term proliferation without losing the potential to differentiate into all three neural lineages, including dopaminergic neurons. *J Neurochem*. 2008;104:316–324.

32. Ko JY, Park CH, Koh HC, et al. Human embryonic stem cell-derived neural precursors as a continuous, stable, and on-demand source for human dopamine neurons. *J Neurochem*. 2007;103:1417–1429.

33. Ben-Hur T, Idelson M, Khaner H, et al. Transplantation of human embryonic stem cell-derived neural progenitors improves behavioral deficit in Parkinsonian rats. *Stem Cells*. 2004;22:1246–1255.

34. Iacovitti L, Donaldson AE, Marshall CE, et al. A protocol for the differentiation of human embryonic stem cells into dopaminergic neurons using only chemically defined human additives: studies in vitro and in vivo. *Brain Res*. 2007;1127:19–25.

35. Pera MF, Andrade J, Houssami S, et al. Regulation of human embryonic stem cell differentiation by BMP-2 and its antagonist noggin. *J Cell Sci*. 2004;117:1269–1280.

36. Chiba S, Lee YM, Zhou W, et al. Noggin enhances dopamine neuron production from human embryonic stem cells and improves behavioral outcome after transplantation into Parkinsonian rats. *Stem Cells*. 2008;26:2810–2820.

37. Correia AS, Anisimov SV, Li JY, et al. Growth factors and feeder cells promote differentiation of human embryonic stem cells into dopaminergic neurons: a novel role for fibroblast growth factor-20. *Front Neurosci*. 2008;2:26–34.

38. Ueno M, Matsumura M, Watanabe K, et al. Neural conversion of ES cells by an inductive activity on human amniotic membrane matrix. *Proc Natl Acad Sci USA*. 2006;103:9554–9559.

39. Roy NS, Cleren C, Singh SK, et al. Functional engraftment of human ES cell-derived dopaminergic neurons enriched by coculture with telomerase-immortalized midbrain astrocytes. *Nat Med*. 2006;12:1259–1268.

40. Cho MS, Lee YE, Kim JY, et al. Highly efficient and large-scale generation of functional dopamine neurons from human embryonic stem cells. *Proc Natl Acad Sci USA*. 2008;105:3392–3397.

41. Yan Y, Yang D, Zarnowska ED, et al. Directed differentiation of dopaminergic neuronal subtypes from human embryonic stem cells. *Stem Cells*. 2005;23:781–790.

42. Park CH, Minn YK, Lee JY, et al. In vitro and in vivo analyses of human embryonic stem cell-derived dopamine neurons. *J Neurochem*. 2005;92:1265–1276.

43. Brederlau A, Correia AS, Anisimov SV, et al. Transplantation of human embryonic stem cell-derived cells to a rat model of Parkinson's disease: effect of in vitro differentiation on graft survival and teratoma formation. *Stem Cells*. 2006;24:1433–1440.

44. Li XJ, Du ZW, Zarnowska ED, et al. Specification of motoneurons from human embryonic stem cells. *Nat Biotechnol*. 2005;23:215–221.

45. Singh Roy N, Nakano T, Xuing L, et al. Enhancer-specified GFP-based FACS purification of human spinal motor neurons from embryonic stem cells. *Exp Neurol*. 2005;196:224–234.

46. Lee H, Shamy GA, Elkabetz Y, et al. Directed differentiation and transplantation of human embryonic stem cell-derived motoneurons. *Stem Cells*. 2007;25:1931–1939.

47. Nistor GI, Totoiu MO, Haque N, et al. Human embryonic stem cells differentiate into oligodendrocytes in high purity and myelinate after spinal cord transplantation. *Glia*. 2005;49:385–396.

48. Kang SM, Cho MS, Seo H, et al. Efficient induction of oligodendrocytes from human embryonic stem cells. *Stem Cells*. 2007;25:419–424.

49. Hu Z, Li T, Zhang X, et al. Hepatocyte growth factor enhances the generation of high-purity oligodendrocytes from human embryonic stem cells. *Differentiation*. 2009;78:177–184.

50. Gil JE, Woo DH, Shim JH, et al. Vitronectin promotes oligodendrocyte differentiation during neurogenesis of human embryonic stem cells. *FEBS Lett*. 2009;583:561–567.

51. Keirstead HS, Nistor G, Bernal G, et al. Human embryonic stem cell-derived oligodendrocyte progenitor cell transplants remyelinate and restore locomotion after spinal cord injury. *J Neurosci*. 2005;25:4694–4705.

52. Cloutier F, Siegenthaler MM, Nistor G, et al. Transplantation of human embryonic stem cell-derived oligodendrocyte progenitors into rat spinal cord injuries does not cause harm. *Regen Med*. 2006;1:469–479.

53. Levenberg S, Huang NF, Lavik E, et al. Differentiation of human embryonic stem cells on three-dimensional polymer scaffolds. *Proc Natl Acad Sci USA*. 2003;100:12741–12746.

54. Levenberg S, Burdick JA, Kraehenbuehl T, et al. Neurotrophin-induced differentiation of human embryonic

stem cells on three-dimensional polymeric scaffolds. *Tissue Eng*. 2005;11:506–512.

55. Chao TI, Xiang S, Chen CS, et al. Carbon nanotubes promote neuron differentiation from human embryonic stem cells. *Biochem Biophys Res Commun*. 2009;384:426–430.

56. Ellis-Behnke RG, Liang YX, You SW, et al. Nano neuro knitting: peptide nanofiber scaffold for brain repair and axon regeneration with functional return of vision. *Proc Natl Acad Sci USA*. 2006;103:5054–5059.

57. da Cruz L, Chen FK, Ahmado A, et al. RPE transplantation and its role in retinal disease. *Prog Retin Eye Res*. 2007;26:598–635.

58. Banin E, Obolensky A, Idelson M, et al. Retinal incorporation and differentiation of neural precursors derived from human embryonic stem cells. *Stem Cells*. 2006;24:246–257.

59. Lamba DA, Karl MO, Ware CB, et al. Efficient generation of retinal progenitor cells from human embryonic stem cells. *Proc Natl Acad Sci USA*. 2006;103:12769–12774.

60. Klimanskaya I, Hipp J, Rezai KA, et al. Derivation and comparative assessment of retinal pigment epithelium from human embryonic stem cells using transcriptomics. *Cloning Stem Cells*. 2004;6:217–245.

61. Vugler A, Carr AJ, Lawrence J, et al. Elucidating the phenomenon of HESC-derived RPE: anatomy of cell genesis, expansion and retinal transplantation. *Exp Neurol*. 2008;214:347–361.

62. Osakada F, Ikeda H, Mandai M, et al. Toward the generation of rod and cone photoreceptors from mouse, monkey and human embryonic stem cells. *Nat Biotechnol*. 2008;26:215–224.

63. Idelson M, Alper R, Obolensky A, et al. Directed differentiation of human embryonic stem cells into functional retinal pigment epithelium cells. *Cell Stem Cell*. 2009;5:396–408.

64. Osakada F, Jin ZB, Hirami Y, et al. In vitro differentiation of retinal cells from human pluripotent stem cells by small-molecule induction. *J Cell Sci*. 2009;122:3169–3179.

65. Zhao X, Liu J, Ahmad I. Differentiation of embryonic stem cells into retinal neurons. *Biochem Biophys Res Commun*. 2002;297:177–184.

66. Lund RD, Wang S, Klimanskaya I, et al. Human embryonic stem cell-derived cells rescue visual function in dystrophic RCS rats. *Cloning Stem Cells*. 2006;8:189–199.

67. Lamba DA, Gust J, Reh TA. Transplantation of human embryonic stem cell-derived photoreceptors restores some visual function in Crx-deficient mice. *Cell Stem Cell*. 2009;4:73–79.

68. Alonso L, Fuchs E. Stem cells of the skin epithelium. *Proc Natl Acad Sci USA*. 2003;100(Suppl 1):11830–11835.

69. Aberdam D. Derivation of keratinocyte progenitor cells and skin formation from embryonic stem cells. *Int J Dev Biol*. 2004;48:203–206.

70. Aberdam E, Barak E, Rouleau M, et al. A pure population of ectodermal cells derived from human embryonic stem cells. *Stem Cells*. 2008;26:440–444.

71. Iuchi S, Dabelsteen S, Easley K, et al. Immortalized keratinocyte lines derived from human embryonic stem cells. *Proc Natl Acad Sci USA*. 2006;103:1792–1797.

72. Metallo CM, Ji L, de Pablo JJ, et al. Retinoic acid and bone morphogenetic protein signaling synergize to efficiently direct epithelial differentiation of human embryonic stem cells. *Stem Cells*. 2008;26:372–380.

73. Ji L, Allen-Hoffmann BL, de Pablo JJ, et al. Generation and differentiation of human embryonic stem cell-derived keratinocyte precursors. *Tissue Eng*. 2006;12:665–679.

74. Li X, Chen Y, Scheele S, et al. Fibroblast growth factor signaling and basement membrane assembly are connected during epithelial morphogenesis of the embryoid body. *J Cell Biol*. 2001;153:811–822.

75. Hewitt KJ, Shamis Y, Carlson MW, et al. Three-dimensional epithelial tissues generated from human embryonic stem cells. *Tissue Eng Part A*. 2009;15:3417–3426.

76. Metallo CM, Azarin SM, Moses LE, et al. Human embryonic stem cell-derived keratinocytes exhibit an epidermal transcription program and undergo epithelial morphogenesis in engineered tissue constructs. *Tissue Eng Part A*. 2010;16:213–223.

77. Lavon N, Benvenisty N. Study of hepatocyte differentiation using embryonic stem cells. *J Cell Biochem*. 2005;96:1193–1202.

78. Asahina K, Teramoto K, Teraoka H. Embryonic stem cells: hepatic differentiation and regenerative medicine for the treatment of liver disease. *Curr Stem Cell Res Ther*. 2006;1:139–156.

79. Schuldiner M, Yanuka O, Itskovitz-Eldor J, et al. Effects of eight growth factors on the differentiation of cells derived from human embryonic stem cells. *Proc Natl Acad Sci USA*. 2000;97:11307–11312.

80. Lavon N, Yanuka O, Benvenisty N. Differentiation and isolation of hepatic-like cells from human embryonic stem cells. *Differentiation*. 2004;72:230–238.

81. Schwartz RE, Linehan JL, Painschab MS, et al. Defined conditions for development of functional hepatic cells from human embryonic stem cells. *Stem Cells Dev*. 2005;14:643–655.

82. Shirahashi H, Wu J, Yamamoto N, et al. Differentiation of human and mouse embryonic stem cells along a hepatocyte lineage. *Cell Transplant*. 2004;13:197–211.

83. Soto-Gutierrez A, Navarro-Alvarez N, Rivas-Carrillo JD, et al. Construction and transplantation of an engineered hepatic tissue using a polyaminourethane-coated nonwoven polytetrafluoroethylene fabric. *Transplantation*. 2007;83:129–137.

84. Soto-Gutierrez A, Navarro-Alvarez N, Rivas-Carrillo JD, et al. Differentiation of human embryonic stem cells to hepatocytes using deleted variant of HGF and poly-amino-urethane-coated nonwoven polytetrafluoroethylene fabric. *Cell Transplant*. 2006;15:335–341.

85. Cai J, Zhao Y, Liu Y, et al. Directed differentiation of human embryonic stem cells into functional hepatic cells. *Hepatology*. 2007;45:1229–1239.

86. Agarwal S, Holton KL, Lanza R. Efficient differentiation of functional hepatocytes from human embryonic stem cells. *Stem Cells*. 2008;26:1117–1127.

87. Shiraki N, Umeda K, Sakashita N, et al. Differentiation of mouse and human embryonic stem cells into hepatic lineages. *Genes Cells*. 2008;13:731–746.

88. Basma H, Soto-Gutierrez A, Yannam GR, et al. Differentiation and transplantation of human embryonic stem cell-derived hepatocytes. *Gastroenterology*. 2009;136:990–999.

89. Hay DC, Zhao D, Ross A, et al. Direct differentiation of human embryonic stem cells to hepatocyte-like cells exhibiting functional activities. *Cloning Stem Cells*. 2007;9:51–62.

90. Assady S, Maor G, Amit M, et al. Insulin production by human embryonic stem cells. *Diabetes*. 2001;50:1691–1697.

91. Brolen GK, Heins N, Edsbagge J, et al. Signals from the embryonic mouse pancreas induce differentiation of human embryonic stem cells into insulin-producing beta-cell-like cells. *Diabetes*. 2005;54:2867–2874.

92. Xu X, Kahan B, Forgianni A, et al. Endoderm and pancreatic islet lineage differentiation from human embryonic stem cells. *Cloning Stem Cells*. 2006;8:96–107.

93. Baharvand H, Jafary H, Massumi M, et al. Generation of insulin-secreting cells from human embryonic stem cells. *Dev Growth Differ*. 2006;48:323–332.

94. D'Amour KA, Bang AG, Eliazer S, et al. Production of pancreatic hormone-expressing endocrine cells from human embryonic stem cells. *Nat Biotechnol*. 2006;24:1392–1401.

95. Kroon E, Martinson LA, Kadoya K, et al. Pancreatic endoderm derived from human embryonic stem cells generates glucose-responsive insulin-secreting cells in vivo. *Nat Biotechnol*. 2008;26:443–452.

96. Yao S, Chen S, Clark J, et al. Long-term self-renewal and directed differentiation of human embryonic stem cells in chemically defined conditions. *Proc Natl Acad Sci USA*. 2006;103:6907–6912.

97. McLean AB, D'Amour KA, Jones KL, et al. Activin a efficiently specifies definitive endoderm from human embryonic stem cells only when phosphatidylinositol 3-kinase signaling is suppressed. *Stem Cells*. 2007;25:29–38.

98. D'Amour KA, Agulnick AD, Eliazer S, et al. Efficient differentiation of human embryonic stem cells to definitive endoderm. *Nat Biotechnol*. 2005;23:1534–1541.

99. Jiang J, Au M, Lu K, et al. Generation of insulin-producing islet-like clusters from human embryonic stem cells. *Stem Cells*. 2007;25:1940–1953.

100. Eshpeter A, Jiang J, Au M, et al. In vivo characterization of transplanted human embryonic stem cell-derived pancreatic endocrine islet cells. *Cell Prolif*. 2008;41:843–858.

101. Lavon N, Yanuka O, Benvenisty N. The effect of over-expression of Pdx1 and Foxa2 on the differentiation of human embryonic stem cells into pancreatic cells. *Stem Cells*. 2006;24:1923–1930.

102. Bernardo AS, Cho CH, Mason S, et al. Biphasic induction of Pdx1 in mouse and human embryonic stem cells can mimic development of pancreatic beta-cells. *Stem Cells*. 2009;27:341–351.

103. Borowiak M, Maehr R, Chen S, et al. Small molecules efficiently direct endodermal differentiation of mouse and human embryonic stem cells. *Cell Stem Cell*. 2009;4:348–358.

104. Chen S, Borowiak M, Fox JL, et al. A small molecule that directs differentiation of human ESCs into the pancreatic lineage. *Nat Chem Biol*. 2009;5:258–265.

105. Akashi K, Traver D, Miyamoto T, et al. A clonogenic common myeloid progenitor that gives rise to all myeloid lineages. *Nature*. 2000;404:193–197.

106. Weissman IL. Stem cells: units of development, units of regeneration, and units in evolution. *Cell*. 2000;100:157–168.

107. Chadwick K, Wang L, Li L, et al. Cytokines and BMP-4 promote hematopoietic differentiation of human embryonic stem cells. *Blood*. 2003;102:906–915.

108. Cerdan C, Rouleau A, Bhatia M. VEGF-A165 augments erythropoietic development from human embryonic stem cells. *Blood*. 2004;103:2504–2512.

109. Zambidis ET, Peault B, Park TS, et al. Hematopoietic differentiation of human embryonic stem cells progresses through sequential hematoendothelial, primitive, and definitive stages resembling human yolk sac development. *Blood*. 2005;106:860–870.

110. Tian X, Morris JK, Linehan JL, et al. Cytokine requirements differ for stroma and embryoid body-mediated hematopoiesis from human embryonic stem cells. *Exp Hematol*. 2004;32:1000–1009.

111. Daley GQ. From embryos to embryoid bodies: generating blood from embryonic stem cells. *Ann NY Acad Sci*. 2003;996:122–131.

112. Ng ES, Davis RP, Azzola L, et al. Forced aggregation of defined numbers of human embryonic stem cells into embryoid bodies fosters robust, reproducible hematopoietic differentiation. *Blood*. 2005;106:1601–1603.

113. Chang KH, Nelson AM, Cao H, et al. Definitive-like erythroid cells derived from human embryonic stem cells coexpress high levels of embryonic and fetal globins with little or no adult globin. *Blood*. 2006;108:1515–1523.

114. Karlsson KR, Cowley S, Martinez FO, et al. Homogeneous monocytes and macrophages from human embryonic stem cells following coculture-free differentiation in M-CSF and IL-3. *Exp Hematol*. 2008;36:1167–1175.

115. Saeki K, Nakahara M, Matsuyama S, et al. A feeder-free and efficient production of functional neutrophils from human embryonic stem cells. *Stem Cells*. 2009;27:59–67.

116. Su Z, Frye C, Bae KM, et al. Differentiation of human embryonic stem cells into immunostimulatory dendritic cells under feeder-free culture conditions. *Clin Cancer Res*. 2008;14:6207–6217.

117. Milhem M, Mahmud N, Lavelle D, et al. Modification of hematopoietic stem cell fate by 5aza 2'deoxycytidine and trichostatin A. *Blood*. 2004;103:4102–4110.

118. Vodyanik MA, Bork JA, Thomson JA, et al. Human embryonic stem cell-derived CD34$^+$ cells: efficient production

in the coculture with OP9 stromal cells and analysis of lymphohematopoietic potential. *Blood*. 2005;105:617–626.

119. Wang J, Zhao HP, Lin G, et al. In vitro hematopoietic differentiation of human embryonic stem cells induced by co-culture with human bone marrow stromal cells and low dose cytokines. *Cell Biol Int*. 2005;29:654–661.

120. Ledran MH, Krassowska A, Armstrong L, et al. Efficient hematopoietic differentiation of human embryonic stem cells on stromal cells derived from hematopoietic niches. *Cell Stem Cell*. 2008;3:85–98.

121. Slukvin II, Vodyanik MA, Thomson JA, et al. Directed differentiation of human embryonic stem cells into functional dendritic cells through the myeloid pathway. *J Immunol*. 2006;176:2924–2932.

122. Senju S, Suemori H, Zembutsu H, et al. Genetically manipulated human embryonic stem cell-derived dendritic cells with immune regulatory function. *Stem Cells*. 2007;25:2720–2729.

123. Choi KD, Vodyanik MA, Slukvin II. Generation of mature human myelomonocytic cells through expansion and differentiation of pluripotent stem cell-derived lin-CD34$^+$CD43$^+$CD45$^+$ progenitors. *J Clin Invest*. 2009;119:2818–2829.

124. Qiu C, Olivier EN, Velho M, et al. Globin switches in yolk sac-like primitive and fetal-like definitive red blood cells produced from human embryonic stem cells. *Blood*. 2008;111:2400–2408.

125. Liu YX, Ji L, Yue W, et al. Cells extract from fetal liver promotes the hematopoietic differentiation of human embryonic stem cells. *Cloning Stem Cells*. 2009;11:51–60.

126. Lu SJ, Feng Q, Park JS, et al. Biologic properties and enucleation of red blood cells from human embryonic stem cells. *Blood*. 2008;112:4475–4484.

127. Lee GS, Kim BS, Sheih JH, et al. Forced expression of HoxB4 enhances hematopoietic differentiation by human embryonic stem cells. *Mol Cells*. 2008;25:487–493.

128. Bowles KM, Vallier L, Smith JR, et al. HOXB4 overexpression promotes hematopoietic development by human embryonic stem cells. *Stem Cells*. 2006;24:1359–1369.

129. Wang L, Menendez P, Shojaei F, et al. Generation of hematopoietic repopulating cells from human embryonic stem cells independent of ectopic HOXB4 expression. *J Exp Med*. 2005;201:1603–1614.

130. Tian X, Woll PS, Morris JK, et al. Hematopoietic engraftment of human embryonic stem cell-derived cells is regulated by recipient innate immunity. *Stem Cells*. 2006;24:1370–1380.

131. Narayan AD, Chase JL, Lewis RL, et al. Human embryonic stem cell-derived hematopoietic cells are capable of engrafting primary as well as secondary fetal sheep recipients. *Blood*. 2006;107:2180–2183.

132. Olivier EN, Qiu C, Velho M, et al. Large-scale production of embryonic red blood cells from human embryonic stem cells. *Exp Hematol*. 2006;34:1635–1642.

133. Liew A, Barry F, O'Brien T. Endothelial progenitor cells: diagnostic and therapeutic considerations. *Bioessays*. 2006;28:261–270.

134. Niklason LE. Techview: medical technology. Replacement arteries made to order. *Science*. 1999;286:1493–1494.

135. Yamashita J, Itoh H, Hirashima M, et al. Flk1-positive cells derived from embryonic stem cells serve as vascular progenitors. *Nature*. 2000;408:92–96.

136. Yang D, Zhang ZJ, Oldenburg M, et al. Human embryonic stem cell-derived dopaminergic neurons reverse functional deficit in parkinsonian rats. *Stem Cells*. 2008;26:55–63.

137. Matsumoto K, Yoshitomi H, Rossant J, et al. Liver organogenesis promoted by endothelial cells prior to vascular function. *Science*. 2001;294:559–563.

138. Lammert E, Cleaver O, Melton D. Induction of pancreatic differentiation by signals from blood vessels. *Science*. 2001;294:564–567.

139. Shen Q, Goderie SK, Jin L, et al. Endothelial cells stimulate self-renewal and expand neurogenesis of neural stem cells. *Science*. 2004;304:1338–1340.

140. Bai H, Wang ZZ. Directing human embryonic stem cells to generate vascular progenitor cells. *Gene Ther*. 2008;15:89–95.

141. Levenberg S, Golub JS, Amit M, et al. Endothelial cells derived from human embryonic stem cells. *Proc Natl Acad Sci USA*. 2002;99:4391–4396.

142. Ferreira LS, Gerecht S, Shieh HF, et al. Vascular progenitor cells isolated from human embryonic stem cells give rise to endothelial and smooth muscle like cells and form vascular networks in vivo. *Circ Res*. 2007;101:286–294.

143. Wang ZZ, Au P, Chen T, et al. Endothelial cells derived from human embryonic stem cells form durable blood vessels in vivo. *Nat Biotechnol*. 2007;25:317–318.

144. Niklason LE, Gao J, Abbott WM, et al. Functional arteries grown in vitro. *Science*. 1999;284:489–493.

145. Zisch AH, Lutolf MP, Ehrbar M, et al. Cell-demanded release of VEGF from synthetic, biointeractive cell ingrowth matrices for vascularized tissue growth. *FASEB J*. 2003;17:2260–2262.

146. Ehrbar M, Metters A, Zammaretti P, et al. Endothelial cell proliferation and progenitor maturation by fibrin-bound VEGF variants with differential susceptibilities to local cellular activity. *J Control Release*. 2005;101:93–109.

147. Langer R. New methods of drug delivery. *Science*. 1990;249:1527–1533.

148. Ferreira LS, Gerecht S, Fuller J, et al. Bioactive hydrogel scaffolds for controllable vascular differentiation of human embryonic stem cells. *Biomaterials*. 2007;28:2706–2717.

149. Lee KY, Peters MC, Anderson KW, et al. Controlled growth factor release from synthetic extracellular matrices. *Nature*. 2000;408:998–1000.

150. Levenberg S, Rouwkema J, Macdonald M, et al. Engineering vascularized skeletal muscle tissue. *Nat Biotechnol*. 2005;23:879–884.

151. Murry CE, Field LJ, Menasche P. Cell-based cardiac repair: reflections at the 10-year point. *Circulation*. 2005;112:3174–3183.

152. Laflamme MA, Zbinden S, Epstein SE, et al. Cell-based therapy for myocardial ischemia and infarction: pathophysiological mechanisms. *Annu Rev Pathol*. 2007;2:307–339.

153. Leobon B, Garcin I, Menasche P, et al. Myoblasts transplanted into rat infarcted myocardium are functionally isolated from their host. *Proc Natl Acad Sci USA*. 2003;100:7808–7811.

154. Rosenzweig A. Cardiac cell therapy—mixed results from mixed cells. *N Engl J Med*. 2006;355:1274–1277.

155. Dai W, Hale SL, Martin BJ, et al. Allogeneic mesenchymal stem cell transplantation in postinfarcted rat myocardium: short- and long-term effects. *Circulation*. 2005;112:214–223.

156. Gnecchi M, He H, Noiseux N, et al. Evidence supporting paracrine hypothesis for Akt-modified mesenchymal stem cell-mediated cardiac protection and functional improvement. *FASEB J*. 2006;20:661–669.

157. Yao M, Dieterle T, Hale SL, et al. Long-term outcome of fetal cell transplantation on postinfarction ventricular remodeling and function. *J Mol Cell Cardiol*. 2003;35:661–670.

158. Zimmermann WH, Melnychenko I, Wasmeier G, et al. Engineered heart tissue grafts improve systolic and diastolic function in infarcted rat hearts. *Nat Med*. 2006;12:452–458.

159. Gepstein L. Derivation and potential applications of human embryonic stem cells. *Circ Res*. 2002;91:866–876.

160. Xu C, He JQ, Kamp TJ, et al. Human embryonic stem cell-derived cardiomyocytes can be maintained in defined medium without serum. *Stem Cells Dev*. 2006;15:931–941.

161. He JQ, Ma Y, Lee Y, et al. Human embryonic stem cells develop into multiple types of cardiac myocytes: action potential characterization. *Circ Res*. 2003;93:32–39.

162. Mummery C, Ward-van Oostwaard D, Doevendans P, et al. Differentiation of human embryonic stem cells to cardiomyocytes: role of coculture with visceral endoderm-like cells. *Circulation*. 2003;107:2733–2740.

163. Pal R. Embryonic stem (ES) cell-derived cardiomyocytes: a good candidate for cell therapy applications. *Cell Biol Int*. 2009;33:325–336.

164. Capi O, Gepstein L. Myocardial regeneration strategies using human embryonic stem cell-derived cardiomyocytes. *J Control Release*. 2006;116:211–218.

165. Laflamme MA, Chen KY, Naumova AV, et al. Cardiomyocytes derived from human embryonic stem cells in prosurvival factors enhance function of infarcted rat hearts. *Nat Biotechnol*. 2007;25:1015–1024.

166. Anderson D, Self T, Mellor IR, et al. Transgenic enrichment of cardiomyocytes from human embryonic stem cells. *Mol Ther*. 2007;15:2027–2036.

167. Gallo P, Grimaldi S, Latronico MV, et al. A lentiviral vector with a short troponin-I promoter for tracking cardiomyocyte differentiation of human embryonic stem cells. *Gene Ther*. 2008;15:161–170.

168. Huber I, Itzhaki I, Caspi O, et al. Identification and selection of cardiomyocytes during human embryonic stem cell differentiation. *FASEB J*. 2007;21:2551–2563.

169. Caspi O, Lesman A, Basevitch Y, et al. Tissue engineering of vascularized cardiac muscle from human embryonic stem cells. *Circ Res*. 2007;100:263–272.

170. Stevens KR, Pabon L, Muskheli V, et al. Scaffold-free human cardiac tissue patch created from embryonic stem cells. *Tissue Eng Part A*. 2009;15:1211–1222.

171. Kehat I, Khimovich L, Caspi O, et al. Electromechanical integration of cardiomyocytes derived from human embryonic stem cells. *Nat Biotechnol*. 2004;22:1282–1289.

172. Xue T, Cho HC, Akar FG, et al. Functional integration of electrically active cardiac derivatives from genetically engineered human embryonic stem cells with quiescent recipient ventricular cardiomyocytes: insights into the development of cell-based pacemakers. *Circulation*. 2005;111:11–20.

173. Tomescot A, Leschik J, Bellamy V, et al. Differentiation in vivo of cardiac committed human embryonic stem cells in postmyocardial infarcted rats. *Stem Cells*. 2007;25:2200–2205.

174. Dai W, Field LJ, Rubart M, et al. Survival and maturation of human embryonic stem cell-derived cardiomyocytes in rat hearts. *J Mol Cell Cardiol*. 2007;43:504–516.

175. Laflamme MA, Gold J, Xu C, et al. Formation of human myocardium in the rat heart from human embryonic stem cells. *Am J Pathol*. 2005;167:663–671.

176. van Laake LW, Passier R, Monshouwer-Kloots J, et al. Human embryonic stem cell-derived cardiomyocytes survive and mature in the mouse heart and transiently improve function after myocardial infarction. *Stem Cell Res*. 2007;1:9–24.

177. Stevens KR, Kreutziger KL, Dupras SK, et al. Physiological function and transplantation of scaffold-free and vascularized human cardiac muscle tissue. *Proc Natl Acad Sci USA*. 2009;106:16568–16573.

178. Jaiswal N, Haynesworth SE, Caplan AI, et al. Osteogenic differentiation of purified, culture-expanded human mesenchymal stem cells in vitro. *J Cell Biochem*. 1997;64:295–312.

179. Salgado AJ, Coutinho OP, Reis RL. Bone tissue engineering: state of the art and future trends. *Macromol Biosci*. 2004;4:743–765.

180. Caplan AI. Review: mesenchymal stem cells: cell-based reconstructive therapy in orthopedics. *Tissue Eng*. 2005;11:1198–1211.

181. Arinzeh TL. Mesenchymal stem cells for bone repair: preclinical studies and potential orthopedic applications. *Foot Ankle Clin*. 2005;10:651–665, viii.

182. Bielby RC, Boccaccini AR, Polak JM, et al. In vitro differentiation and in vivo mineralization of osteogenic cells derived from human embryonic stem cells. *Tissue Eng*. 2004;10:1518–1525.

183. Cao T, Heng BC, Ye CP, et al. Osteogenic differentiation within intact human embryoid bodies result in a marked increase in osteocalcin secretion after 12 days of in vitro culture, and formation of morphologically distinct nodule-like structures. *Tissue Cell*. 2005;37:325–334.

184. Ahn SE, Kim S, Park KH, et al. Primary bone-derived cells induce osteogenic differentiation without exogenous factors in human embryonic stem cells. *Biochem Biophys Res Commun*. 2006;340:403–408.

185. Karp JM, Ferreira LS, Khademhosseini A, et al. Cultivation of human embryonic stem cells without the embryoid body step enhances osteogenesis in vitro. *Stem Cells*. 2006;24:835–843.

186. Karner E, Unger C, Sloan AJ, et al. Bone matrix formation in osteogenic cultures derived from human embryonic stem cells in vitro. *Stem Cells Dev*. 2007;16:39–52.

187. Kim S, Kim S, Lee S, et al. In vivo bone formation from human embryonic stem cell-derived osteogenic cells in poly(D,L-lactic-*co*-glycolic acid)/hydroxyapatite composite scaffolds. *Biomaterials*. 2008;29:1043–1053.

188. Vats A, Bielby RC, Tolley N, et al. Chondrogenic differentiation of human embryonic stem cells: the effect of the micro-environment. *Tissue Eng*. 2006;12:1687–1697.

189. Hoben GM, Willard VP, Athanasiou KA. Fibrochondrogenesis of hESCs: growth factor combinations and cocultures. *Stem Cells Dev*. 2009;18:283–292.

190. Toh WS, Yang Z, Liu H, et al. Effects of culture conditions and bone morphogenetic protein 2 on extent of chondrogenesis from human embryonic stem cells. *Stem Cells*. 2007;25:950–960.

191. Barberi T, Willis LM, Socci ND, et al. Derivation of multipotent mesenchymal precursors from human embryonic stem cells. *PLoS Med*. 2005;2:e161.

192. Lian Q, Lye E, Suan Yeo K, et al. Derivation of clinically compliant MSCs from CD105+, CD24− differentiated human ESCs. *Stem Cells*. 2007;25:425–436.

193. Hwang NS, Varghese S, Zhang Z, et al. Chondrogenic differentiation of human embryonic stem cell-derived cells in arginine-glycine-aspartate-modified hydrogels. *Tissue Eng*. 2006;12:2695–2706.

194. Tremoleda JL, Forsyth NR, Khan NS, et al. Bone tissue formation from human embryonic stem cells in vivo. *Cloning Stem Cells*. 2008;10:119–132.

195. Drukker M, Katchman H, Katz G, et al. Human embryonic stem cells and their differentiated derivatives are less susceptible to immune rejection than adult cells. *Stem Cells*. 2006;24:221–229.

196. Taylor CJ, Bolton EM, Pocock S, et al. Banking on human embryonic stem cells: estimating the number of donor cell lines needed for HLA matching. *Lancet*. 2005;366:2019–2025.

197. Stadtfeld M, Nagaya M, Utikal J, et al. Induced pluripotent stem cells generated without viral integration. *Science*. 2008;322:945–949.

198. Takahashi K, Yamanaka S. Induction of pluripotent stem cells from mouse embryonic and adult fibroblast cultures by defined factors. *Cell*. 2006;126:663–676.

199. Wernig M, Meissner A, Foreman R, et al. In vitro reprogramming of fibroblasts into a pluripotent ES-cell-like state. *Nature*. 2007;448:318–324.

200. Park IH, Arora N, Huo H, et al. Disease-specific induced pluripotent stem cells. *Cell*. 2008;134:877–886.

201. Zhang D, Jiang W, Liu M, et al. Highly efficient differentiation of human ES cells and iPS cells into mature pancreatic insulin-producing cells. *Cell Res*. 2009;19:429–438.

202. Dimos JT, Rodolfa KT, Niakan KK, et al. Induced pluripotent stem cells generated from patients with ALS can be differentiated into motor neurons. *Science*. 2008;321:1218–1221.

GLOSSARY

Acetylation   The process of introducing acetyl groups to molecules such as proteins. Acetylation is typically associated with gene activation.

Activin A   A protein complex that is involved in cell survival, differentiation, and growth.

Adeno Associated Virus   Referring to one of several members of the dependovirus group of parvoviruses. These are small (20–27-nm diameter) nonenveloped DNA viruses, very common in humans and nonhuman primates, which are both persistent and nonpathogenic, making them useful for human gene therapy applications.

Algorithm   A precise step-by-step computational procedure for performing a task.

Aliquots   Portions of a solution.

Allogeneic   Members of the same species that are genetically different.

Allografts   Transplantation of cells, tissues, or organs, sourced from a genetically nonidentical member of the same species as the recipient.

Alloimmunity   A condition in which the body gains immunity, from another individual of the same species, against its own cells. Alloimmunity can occur in the recipient after transfusions of fluids such as blood or plasma or in the recipient after allografts.

Aneuploidy   Having a chromosome number that is not a multiple of the haploid number characteristic of the species, usually caused by loss or duplication of chromosomes.

Anoikis   Apoptosis (a form of programmed death) induced by loss of or inappropriate cell adhesion to surrounding extracellular matrix (ECM).

Antimycotic   An antifungal agent; inhibits the growth of fungi.

Arginine   An alpha amino acid. It is conditionally a nonessential amino acid, meaning that the human body can manufacture it.

Array   A collection of different genomic or transcriptional elements placed onto a physical substrate, which can be used analytically to assess different states, such as gene expression or binding of factors to DNA.

Aspirate   To remove a liquid by suction.

Autografts   Transfer of tissue or cells from one part of the body to another part of the body on the same patient.

Autologous Transplantation   A procedure in which the donor's tissue is transplanted to that same donor.

Autosome   The chromosomes not involved with sex determination. There are 22 pairs of autosomes in each human cell.

Avertin   2-2-2-Tribromoethanol, an injected anesthetic used for deeper sedation of laboratory animals.

Basic Fibroblast Growth Factor   Also known as bFGF, FGF2, or FGF-ß. Supports prolonged hESC growth in an undifferentiated state in both feeder and feeder-free conditions. Currently, the exact mechanism by which it inhibits differentiation is unknown.

Biocidal   A chemical substance with the potential to kill different forms of living organisms such as fungi, insects, algae, and germs.

Biosafety Level 2 (BSL-2)   The Centers for Disease Control and Prevention of the United States established four levels of biocontainment. The biosafety level 2 classified research includes work with agents of moderate potential hazard to both personnel and the environment. This could include work with bacteria and viruses that cause mild disease in humans or that are difficult to contract via aerosol.

Blastocyst   A structure formed on day 4–5 after fertilization in humans. A blastocyst consists of two cell types: the inner cell mass (ICM) and an outer layer known as the trophoblast. The inner cell mass will form the embryo and the trophoblast will later form the placenta.

Blastomere   An embryonic cell from the two-cell stage to the blastocyst stage.

Human Stem Cell Technology and Biology, edited by Stein, Borowski, Luong, Shi, Smith, and Vazquez
Copyright © 2011 Wiley-Blackwell.

Bone Morphogenetic Proteins (BMPs) A group of growth factors and cytokines that are known to play a role in the development of bone and cartilage.

Bone Morphogenetic Protein 2 (BMP2) A secreted protein that has been implicated in the development of the heart, bone, and cartilage.

Bone Morphogenetic Protein 4 (BMP4) A secreted protein that has an important role in the development of bone, cartilage, and muscle. In embryonic development it has a role in differentiation of the ectoderm, the germ layer that becomes the nervous system and the outer layers of the skin.

Cardiomyocytes Heart muscle cells.

Caudalizing Factors Signaling molecules that promote caudal specification of the central nervous system.

Cavitation A fluid-filled cyst.

cDNA Complementary DNA synthesized from a messenger RNA template in a reaction catalyzed by the enzyme reverse transcriptase.

Cellular Fusion The joining of two individual cells into one cell, typically by merging of their cell membranes. The resulting cell can display properties of one or both original cells.

Centromeric Describes a centromere (a region of DNA on the chromosome where two sister chromatids join) generally found near the middle of a chromosome. During metaphase of mitosis, centromeres can be identified as a constriction at the point of the mitotic spindle.

CGH (Comparative Genomic Hybridization) A molecular-cytogenetic assay that analyzes the copy number changes, by determining gains or losses of DNA content.

Chimera An organism composed of cells originating from more than one embryo.

Chimerism A measurement of how many cells in a tissue or organism originate from cells of more than one distinct genetic backgrounds. For example, a chimeric animal consists of cells that derive from two or more populations of genetically distinct cells originating in different zygotes. A chimeric tissue (e.g., blood chimerism) originates from two or more genetically distinct cells, such as the blood stystem of an individual who has received a bone marrow transplantation that does not completely replace all the cells of that tissue. In Greek mythology the chimera was a monstrous animal composed of the parts of multiple animals—part lion, part snake, and part goat.

Chromatin Complexes of DNA and proteins that are densely packaged into chromosomes.

Chromatin Immunoprecipitation (ChIP) A method used to identify DNA sequences that are associated with specific proteins or protein modifications using antibodies and PCR analysis.

Chromatin Remodeling Dynamic structural changes to chromatin that influence gene expression. These changes range from specific sequences to the whole genome.

Chromatography The technique of separating a mixture based on the subtle differences of partition coefficient of the components within the mixture.

Clustering A data analysis tool in which similar obsevations are grouped into sets, or clusters, which aids in recognizing patterns in datasets.

c-myc An oncogene that encodes a transcription factor reported to bind sequences found in promoters of numerous genes and to recruit histone acetyltransferases. C-myc is thought to regulate 15% of all genes.

Co-culture The growth of distinct cell types in a combined culture. For example, the culture of human embryonic stem cells in the presence of mouse embryonic fibroblasts is a co-culture.

Collagen The primary structural protein of connective tissue in animals, and also the most abundant protein in mammals. Gelatin is derived from the partial hydrolysis of collagen.

Collagenase A group of proteolytic enzymes that target connective tissue and decompose collagen and gelatin.

Colloid Characteristic proteinaceous hormone deposition within the thyroid containing thyroglobulin and other substances.

Concatemeric Episomes DNA forms consisting of long repeating chains of nucleotides existing in the nucleus of cells separate and distinct from the cell's chromosomes.

Conditioned Medium Cell culture medium exposed to certain cells, for example, fibroblasts, and subsequently harvesting and used to maintain or differentiate other cells, such as stem cells. Typically, conditioned medium is harvested after being in contact with cells for several days and contains growth factors and other proteins secreted by the cells.

Confluent/Confluence In tissue culture, the percentage to which adherent cells cover the growth surface (e.g., six-well plate).

Confocal Microscopy A type of microscopy using lasers of different wavelengths to optically section specimens, which are fluorescently labeled.

Constitutive Promoter Unregulated segment of DNA that allows continuous transcription of its gene.

Construct A sequence of DNA designed for a specific purpose.

CpG Islands Regions of DNA in which the percentage of cytosine and guanine nucleotides is greater than 50% and the ratio of observed to expected CG dinucleotides is greater than 60%. In most DNA, the content of C and G nucleotides is much lower than the percentages seen in CpG islands. CpG islands are often found in regions that control expression of specific genes.

Cross-hybridization The annealing of a nucleotide sequence to another nucleotide sequence to which it is only partially complementary.

Crosslinking The use of agents such as formaldehyde to bind together proteins that are close to one another in order to study molecular interactions in the cell.

Crx-Deficient Mice Genetically engineered mice lacking in expression of Crx gene. Crx, an Otx-like homeobox gene, is expressed specifically in the photoreceptors of the retina and the pinealocytes of the pineal gland. Crx has been proposed to have a role in the regulation of photoreceptor-specific genes in the eye and of pineal-specific genes in the pineal gland. Mutations in human CRX are associated with the retinal diseases.

Cryoprotectant A substance used to protect biological material when freezing.

Cryovial Used for the cryogenic storage of cells in a liquid nitrogen freezer.

Curate/Curating To act as a curator or a content specialist responsible for an institution's collections.

Cuvette A small transparent rectangular vessel made of fused quartz, plastic, or glass used to hold samples for spectroscopic analysis.

Cytogenetics The study of cellular structure and function with a focus on the molecular bases of heredity, particularly on the chromosomes of the cell.

Cytoplasm The part of a cell contained within the cell's plasma membrane external to its nucleus, and where most metabolic activity occurs. Cytoplasmic elements include cytosol (translucent fluid), membrane-bound organelles (such as mitochondria and lysosomes), and chemical substances (lipids, proteins, etc.).

Cytosines A pyrimidine base ($C_4H_5N_3O$) that is an essential constituent of the polynucleotide chain of RNA and DNA and base pairs with guanine (a purine base, $C_5H_5N_5O$) to code genetic information.

Cytoskeletal Relating to the network of fibrous protein elements, composed of microtubules and actin microfilaments, found in the cytoplasm of cells. This fibrous network provides the cell with structural and dynamic support and permits the movement of chromosomes, organelles, and the cell itself.

Cytotoxicity Relating to a cell killing property of a substance.

Deaminase Enzymes catalyzing simple hydrolysis of $C-NH_2$ bonds of purines, pyrimidines, and pterins, thus producing ammonia.

Degeneration Deterioration where functional activity is diminished or structure is impaired.

Denaturation The process of altering the molecular structure or properties of a substance by extreme conditions such as heat, acid, alkali, or ultraviolet radiation, thereby destroying or decreasing its activity.

Differential Expression Analysis The analysis of variation of protein expression levels from different samples.

Dimerization The chemical union of two molecules.

Dimers Molecules consisting of two structurally similar (homodimer) or complementary (heterodimer) molecules bound together.

Directed Differentiation Differentiation of stem cells to a particular cell type using certain culture conditions, including specific proteins, chemicals, and materials; also, induced differentiation toward a specific lineage.

Dispase A neutral protease derived from *Bacillus polymyxa*, related to trypsin but much gentler to cells. It cleaves fibronectin and collagen IV and I, but does not cleave collagen V or lamanin. It is recommended for cells cultured on BD Matrigel basement membrane matrix.

DNA Imprinting (Genomic Imprinting) A process in which specific genes from one parent are modified by methylation so that they are inactive, resulting in monoallelic expression of the same genes inherited from the other parent.

Dysplasia Abnormal growth of cells or tissues.

Ectoderm/Ectodermal Outermost of three primary embryonic germ layers that give rise to epidermis tissues, the nervous system, and external sense organs.

Electrophorese To subject macromolecules to electrophoresis for separation.

Electrophoresis A technique used for separating electrically charged macomolecules based on their migration in a gel or other medium subjected to a strong electric field.

Electrophysiological Electrical characterization of biological cells and tissues. It involves measurements of voltage change or electric current.

Electroretinogram (ERG) A recording of the electric retinal response to photostimulation, used to assess retinal status.

Eluate Also referred to as the mobile phase, it is the material emerging from a chromatographic column and consists of solutes dissolved in a liquid solvent (eluant).

Embryoid Bodies (EBs) Aggregates of cells derived from embryonic stem (ES) cells. ES cells prevented from adhering to a surface will aggregate and, upon aggregation, will begin to differentiate to recapitulate embryonic development. EBs are composed of pluripotent cells.

Embryotoxicity   The state of toxicity to an embryo, causing death or abnormal development.

ENCODE   A public consortium for the Encyclopedia of DNA Elements Project.

Endoderm/Endodermal   Innermost of the three primary embryonic germ layers; it gives rise to the gut and most of the respiratory tract.

Endotoxin   A heat-stabile toxic lipopolysaccharide found in the outer membrane of some gram-negative bacteria, which is released when bacteria are lysed, causing disease.

Enzymatic Passage   The passing or splitting of cells using enzymes such as trypsin or collagenase to dissociate cells for expansion and growing under cultured conditions.

Epiblast   The portion of the preimplantation embryo derived from the inner cell mass that undergoes gastrulation to give rise to the embryo proper and eventual fetus.

Epigenetics   The study of changes in gene function by chemical modifications of specific genes that do not involve changes in gene sequence.

Epigenome   Refers to the epigenetic state of a cell; epigenetic information including histone code and DNA methylation.

Episomal   Existing separate and distinct from the host cell chromosomes within the nucleus of a cell.

Epitope   Antigenic determinant; the region of an antigen molecule that is capable of eliciting an immune response and of binding to a specific antibody produced by this immune response.

Erythroid   Pertaining to erythrocytes (red blood cells) or their precursors.

Erythropoiesis   (**erythro = red**; **poiesis = to make**) The process by which red blood cells (erythrocytes) are produced. Erythropoiesis, which occurs almost exclusively in the bone marrow for humans, involves the maturation of a nucleated precursor into a hemoglobin-filled, nucleus-free erythrocyte that is regulated by erythropoietin, a hormone produced by the kidney.

EST Sequencing   Determination of the Expressed Sequence Tags present in a sample.

Ethylenediaminetetraacetic Acid (EDTA)   A polyamino carboxylic acid, $(CH_2N(CH_2CO_2H)_2)_2$, used for its ability to chelate metal ions such as Ca^{2+} and Fe^{3+}. Metal ions bound to EDTA remain in solution but are less reactive.

Euchromatin   Regions of chromosomes that are in a more open conformation, stain lightly, and often contain active genes.

Exogenous   External or outside.

Expression Libraries   A collection of bacterial colonies, each expressing a different protein.

FANTOM   An international scientific consortium for the Functional Annotation of the Mammalian Genome.

Fibroblast Growth Factor 4 (FGF4)   A member of the fibroblast growth factor family and a secreted protein that can act on a wide variety of cells. It is involved in cell growth and differentiation and tissue repair.

Fibronectin   A protein that connects cells to the extracellular matrix.

Filaggrins   Filament-associated proteins that bind to keratin fibers in epithelial cells.

FISH (Fluorescence In Situ Hybridization)   Method for detecting changes in chromosomes at the molecular level using fluorescently labeled probes.

Fluorochromes   Various fluorescent elements used in biological staining to generate fluorescence in a specimen.

Fluorogenic   Refers to a process in which fluorescence is produced.

Forward Scatter Detector   Flow cytometry counts, measures, and classifies cells using fluorescence surface markers by passing the cells in a stream through a laser. As the cells pass through the laser, light is scattered in all angles. The magnitude of forward, low-angle scatter is proportional to the size of the cell. The forward scatter detector converts the scattered light reading to a voltage pulse reading, and data can be quantified and plotted on a histogram: smaller cells to the left and larger cells to the right.

Functional Synctium   A functional syncytium can exist in the cardiac muscles; individual cells respond as a unit and contract together.

Gametes   Female and male germ cells.

Gametogenesis   The process of production of male or female germ cells, also known as gametes (sperm or egg).

gDNA (Genomic Deoxyribonucleic Acid)   A genome's DNA that has all coding and noncoding sequences.

Genome   Genetic make-up of an organism.

Genome Tiling   A process of sampling the genome by short tiles (~50 bp) that are fabricated into an array which can be used to assess binding of regulatory molecules or chemically modified regions of DNA using immunoprecipitation techniques.

Genomic Instability   An increased tendency of the genome to go through mutations when different processes involved in maintaining and duplicating the genome are defective.

Genomic Integration   Incorporation of foreign DNA into nuclear DNA.

Genomics/Genomic Analysis   The process of determining changes of state in genomic elements, such as control of genes by pre- and post-transcriptional control, in a global manner, rather than examining one element (e.g., a gene) at a time.

Genotyping   The unique genetic make-up of an individual that can be used to determine identity.

Germline   Inherited genetic material that comes from the sperm or egg and is transmitted to offspring.

Germline Competency   Cells capable of contributing to the germline (eggs and sperm) in chimeric animals.

Giemsa   A dye used to identify DNA.

GTG-Banding (G-Bands by Trypsin and Giemsa)   The most commonly used chromosomal banding technique. GTG-banding involves the use of trypsin (an enzyme) and Giemsa, which give a characteristic banding pattern to individual chromosomes in a cell.

Habitus   The general posture, growth, or appearance of an animal.

Hash Table   A computer data structure that indexes information efficiently by mapping from unique "key" values to their associated content.

Hemacytometer/Hemocytometer   A thick glass slide that has grids etched into it; used to count cells.

Hematopoiesis   The process of creating blood or blood cells.

HEPA Filtration   High-efficiency particulate air filtration that removes tiny particles from the air.

Hepatocyte   A liver cell.

Hepatocyte Growth Factor (HGF)   A growth factor with intense mitogenic activity on hepatocytes and primary epithelial cells through its interaction with its receptor (c-met).

hESCs (Human Embryonic Stem Cells)   Cells found only in the inner cell mass of the blastocyst stage embryo, which are the precursors to every cell type in the human body.

Heterochromatin   Regions of chromosomes that are tightly packed, stain darkly, and contain DNA that has few active genes.

HGF (Hepatocyte Growth Factor)   A secreted protein that can induce growth and motility in a variety of cells. It is also important for organ development and tissue regeneration.

High-Throughput Screening Campaigns   High-throughput screening is an automated way to conduct hundreds of thousands or millions of biological tests in multiwell formats. High-throughput screening campaigns use this approach in a systematic way to identify bioactive compounds, antibodies, or genes that modulate a particular molecular pathway. Hits obtained form high-throughput screening campaigns provide starting points for drug design. They can also help researchers to understand the role of a particular biochemical process in biology.

hiPSCs (Human Induced Pluripotent Stem Cells)   Differentiated cells that are genetically altered to an embryonic stem cell-like state. This is achieved through forced expression of genes and factors necessary for undifferentiated stem cells.

Histocompatibility   The property of having similar alleles of a set of genes reflecting the similarity in the tissues of the donor and recipient.

Histone   A group of proteins that bind to DNA to form nucleosomes (the basic building block of chromatin). Histones are proteins involved in packaging of eukaryotic DNA into chromosomes.

HMSCs (Human Mesenchymal Stem Cells)   Stem cells derived from precursor cells of mesenchymal lineage.

hPSCs (Human Pluripotent Stem Cells)   Self-renewing cells that can develop into different cell types, offering the prospect of their usage in therapeutic strategies such as tissue support, cell replacement, and regeneration.

Hydroxyapatite   A naturally occurring mineral form of calcium apatite; can be found in teeth and bones within the human body.

Hypervariable Satellite DNA   Regions of DNA that contain sequences that are repeated in tandem multiple times. The number of repeats is variable and can differ between individuals.

Hypomorphic   Causing reduced levels of gene activity.

Immortalization   The process where a cell line has the competency to permit repetitive cell division indefinitely.

Immunoblotting   A method for examining or identifying proteins through antigen–antibody specific reactions.

Immunocytochemical Staining   Use of antibody to stain and identify specific cell types.

Immunodeficient   An inherited, obtained, or induced inability to create a typical immune reaction.

Immunofluoresence Microscopy   A type of microscopy used to identify proteins by an antigen–antibody reaction, where the antibodies are conjugated to fluorochromes and viewed in a microscope with an ultraviolet (UV) light source.

Immunohistochemical   Relating to identification of proteins within cells or tissues using antibodies.

Immunomodulatory Agents   Substances (e.g., drugs) that have the ability to alter or regulate one or more immune functions; it may be an immunosuppressant or an immunostimulator based on its effect on the immune system.

Immunoprecipitation   The technique of precipitating an antigen in a sample using an antibody specific to that antigen.

Immunosuppression   Lowering of the immune response by radiation or chemical agents.

In Silico   Performed in computer software, as opposed to a living system (in vivo) or an experimental setting (in vitro).

In Vitro   A procedure that is done outside of a living organism in a controlled environment that closely matches conditions inside the organism.

In Vivo   A procedure that is done within a whole living organism.

Inoculum   A substance or bolus of cells injected into a recipient or host.

Inosine   A nucleoside formed by the deamination of adenosine.

Interactome   The complete set of interactions (physical or genetic) between biological entities in living cells.

Interphase   A phase of cell cycle where the cell is preparing for its next division by gathering nutrients for growth and reading its DNA while carrying out all of its "normal" functions.

Involucrin   A component of the keratinocyte crosslinked envelope.

iPS Cell   A type of pluripotent stem cell that has been genetically reprogrammed from an adult somatic cell to express properties of embryonic stem cells.

Irradiated   The exposure of living cells to a specific amount of radiation (e.g., gamma rays) for the purpose of damaging their DNA so that cells are mitotically inactivated.

Isoflurane   (2-Chloro-2-(difluoromethoxy)-1,1,1-trifluoro-ethane) An inhaled, halogenated form of ether anesthetic, often used at a concentration of 2–4% in oxygen to induce temporary sedation in laboratory animals.

Isoform   A slightly different amino acid sequence of the same protein.

Isogeneic   Of the same genotype or genetic background.

Isogenic   The term to describe things originating from a common source, possessing the same genetic composition.

Isotypes   Classes of antibodies that are species specific and can change as cells develop; each type is functionally specific.

Karyogram   A diagrammatic or graphical representation of chromosomes arranged in a karyotype.

Karyotyping   The primary classical cytogenetic technique used to assess the chromosomal complement of cells, including the number, morphology, and size of the mitotic chromosomes.

Kinase   A type of enzyme that adds a phosphate to amino acids on a target protein in a process called phosphorylation. The added phosphate may activate, inactivate, or allow other proteins to interact with the phoshorylated protein.

Lamin A   A gene that encodes one of the proteins of the Lamin A/C family. These intermediate filament-type proteins are the major components of nuclear membranes.

Laminar Flow   A nonturbulent layered system of air movement in a biological safety hood to maintain a sterile working environment.

Laminin   A glycoprotein that is a component of the extracellular matrix and is involved in cellular adhesion.

Lentiviral   Referring to derivatives (such as gene transfer vectors) of lentiviruses, which are a subtype of retroviruses (RNA viruses that are converted to DNA for a portion of their life cycle). Unlike some other retroviruses, lentiviruses can gain entry to the nucleus of nondividing cells.

Ligand   An ion, molecule, or group of molecules that combine with another chemical entity to form a molecular complex.

LIMS (Laboratory Information Management System)   Software used to manage laboratory samples, instruments, users, workflows, and so on.

Lyophilized   To preserve a substance without damaging its physical structure by the process of freeze-drying.

Lyse   To have undergone the process of lysis.

Lysis   The death of a cell by disrupting the cell membrane and causing its contents to be destroyed.

Manual Passage   A method of splitting hESCs into new culture vessels. This is done by manually selecting the undifferentiated areas of the culture and transferring selected cells or colonies to a new vessel.

MAPKs (Mitogen-Activated Protein Kinases)   A family of kinases that are involved in signaling pathways. They can be involved in regulating numerous functions including cell growth, differentiation, and survival. The outcome of MAPK activity depends on the individual signaling pathway initiated.

Mass Spectrometry   The method to determine the composition of a sample by generating a mass spectrum representing the mass-to-charge ratio of sample components.

Massively Parallel Signature Sequencing (MPSS)   A method for analyzing gene expression patterns by amplifying cDNAs via PCR, attaching the products to microbeads, and reading 16–20-bp sequence signatures from the beads.

Matrigel   A gelatinous protein mixture secreted from mouse sarcoma cells used as a substrate to mimic an extracellular environment for cell culture.

Mesenchymal   Of or relating to mesenchyme. Mesenchyme is composed of undifferentiated cells derived from the mesoderm.

Mesodermal   Relating to the mesoderm. One of the three primary germ layers that gives rise to many connective tissues such as bone, cartilage, muscle, and the middle layer of the dermis.

Metabolomics   The study, identification, and quantitation of the metabolic products of cells.

Metadata   Structured information used to identify and describe a resource.

Metaphase   A stage of mitosis at which the chromosomes align at the medial plane of the cell in preparation for division of genetic material.

Methylation   Addition of methyl groups to molecules such as DNA and proteins. Methylation is typically associated with gene repression.

5-Methylcytidine   Naturally occuring component of DNA that represents an epigenetic modification to DNA.

MicroRNAs (miRNAs)   Post-transcriptional regulators that are short (~22 nucleotides) RNA sequences that can bind to complementary 3/ regions of mRNA target, resulting in silencing or inactivation.

Microarray   A multiplex technology used in molecular biology that is a collection of features, usually oligonucleotide probes spatially arranged in a two-dimensional grid.

Microenvironment   The specialized environmental conditions of a relatively small and localized area.

Mitogen   Any substance that induces cell division.

Mitosis   A process involving division of a single cell into two daughter cells by duplication and segregation of an identical number of chromosomes into each daughter cell nuclei.

Mitotically   Describing the process of mitosis.

Monoallelic   Involving a single allele; alternative forms of a gene that can occupy the same locus on a particular chromosome and control the same trait are known as alleles.

Monoallelic Gene Expression   A state in which one of the two copies (or alleles) of a gene is inactive, so that only one gene is expressed.

Morphogene   Any gene involved in controlling growth and morphogenesis.

Morphogenesis   Generation of tissue organization and shape during development.

Morula   An early stage embryo that is typically formed 3–4 days after fertilization in mammals and is developed by rapid division of the zygote (fertilized egg), consisting of a spherical solid mass of blastomere.

Mosaicism   A condition that denotes the presence of two or more populations of cells with different genetic constitution (at molecular or gross chromosome level) within the same individual.

Motif   A unique but widespread molecular sequence of short nucleotides or amino acids. Motif can also describe structural elements of protein or RNA pertaining to their secondary structures.

Murine   Of or pertaining to the genus *Mus* or is a part of subfamily Murinae in the family Muridae of the animal kingdom, which consists of rats and mice.

Mutagenesis   The process involving either natural (heritable traits) or induced (exposure to radiation or chemical carcinogen) events that lead to mutation by change in the genetic constitution of an organism.

Mycoplasma   A genus of bacteria of the family Mycoplasmataceae. These bacteria lack cell walls and can be parasitic in mammals and, due to the lack of cell walls, these bacteria are mainly unaffected by common antibiotics such as penicillin.

Myeloablated   Having had the marrow destroyed or removed.

Nanog   A transcription factor (a protein that regulates expression and/or function of other genes or a network of genes) involved in regulation of self-renewal of embryonic stem cells. This protein is one of the many proteins that are mainly expressed in undifferentiated embryonic stem cells.

Neuroectoderm   A type of neural epithelium that develops from embryonic ectoderm (the outermost germ layer of the three embryonic germ layers) and forms organs of the nervous system.

Niche   A term used to explain the microenvironment in which cells, particularly stem cells, are found. It can also refer to the interdependence of the in vivo and in vitro microenvironment of stem cells, which can regulate stem cell fate.

Noncoding   Noncoding DNA is a part of DNA that does not contain instructions to code for protein. A large percentage of intergenic sequences in eukaryotes contain noncoding regions. Although some of these noncoding DNA segments are known to play a role in the regulation of coding regions, their function is largely unknown.

Oligonucleotide   A relatively short sequence of nucleic acid polymer (RNA or DNA), typically 20–30 nucleotides in length. They are usually chemically synthesized and are often used as probes to identify complementary DNA in many applications such as Southern blot or DNA microarray.

Oncogenic   Related to formation or occurrence of tumors.

Oocyte   A female germ cell.

Oocyte   An immature ovum or unfertilized egg cell formed during development of ovary, where primordial germ cells undergo a series of mitotic and meiotic cell divisions to form a female gametocyte.

Outbred The breeding of nonrelated individuals.

Pasteur Pipette A common tool used in laboratories to draw small volumes of liquid. They are typically made of long glass tubes that are tapered to a narrow point. They were named after Dr. Louis Pasteur, a French chemist and biologist.

Pathogen Any biological agent that causes a disease state. Pathogens are usually infectious agents such as bacteria or viruses, which can infect a host organism.

Pathophysiological The underlying, mechanistic basis for a disease.

Peptide Mass Fingerprinting The generation of peptides of a protein by protease digestion or other cleavage agents. After mass spectrometric analysis, the peptide mass list thus generated is analogous to a fingerprint that could be in silico compared to a database containing known protein sequences.

Perfuse Delivery of arterial blood to a capillary bed in the biological tissue.

Peripheral Stem Cell Transplantation (PSCT) Clinical approach to replacing the blood and bone marrow system in a diseased individual by transplanting blood stem cells that can be found in the circulating peripheral blood, typically after the individual received drugs designed to mobilize the stem cells into the circulation. The modern technique of PSCT often is used in place of bone marrow transplantation, where bone marrow was aspirated for transplantation.

Peritoneal Cavity The fluid-filled space between the membranes that separate the abdominal cavity from the abdominal wall.

Pharmacogenetics Pharmacogenetics aims at establishing links between interindividual variability in drug response and inherited genetic factors.

Phase Contrast A type of microscope or illumination technique used to observe unstained or transparent specimens by converting small phase shifts in the transmitted light passing through or reflected by the specimen into differences in contrasts or the intensity of the image.

Phenotype The observable traits or physical appearance of an organism that are the result of the genetic constitution and the environment of the organism.

Phosphoproteomics The branch of proteomics that specifically focuses on the study of phosphorylated proteins.

Photobleaching The photochemical decay of a fluorochrome. In microscopic examination of specimens, photobleaching of the fluorescent molecule by exposure to light can be problematic especially in time-lapse microscopy. However, this property of photobleaching can be exploited to study movement and/or diffusion of molecules using techniques such as fluorescence recovery after photobleaching (FRAP).

Photolithography A microfabrication process that uses an optical mask and photochemistry to construct microstructures such as integrated circuits and gene arrays.

Phred Score A quantitative measure assigned to each base call in automated sequencing determination to represent the quality of the base call. Phred is a base-calling program originally developed for the Human Genome Project for automated sequencing. Both the Sanger and Solexa fastq formats incorporate phred-like scores with the sequences.

Physicochemical Relating to physical and chemical properties.

Plasmid A circular DNA molecule used for introduction of foreign DNA into cells.

Ploidy The degree of recurrence of a basic set of chromosomes in an organism.

Pluripotency/Pluripotent The capacity of cells to develop or differentiate into many different cell types, which are, in turn, capable of forming different organs and tissues. In stem cell biology, it also refers to the capability of stem cells to produce cells of all three lineages of the germ layer (ectoderm, endoderm, and mesoderm) upon differentiation.

Polycistronic A nucleic acid construct containing coding sequences for several proteins.

Polyploidy The existence of more than the usual number of chromosomes or chromosome sets in cells or organisms. Most organisms normally have two sets of paired homologous chromosomes.

Post-translational Modifications (PTMs) Alteration of protein structure by one or more modifications of specific amino acid residues (e.g., phosphorylation, methylation, acetylation, sumoylation).

Prodrug The pharmacologic action of a class of drugs which results from conversion by metabolic processes within the body (biotransformation).

Promoter A region of DNA that controls the expression of an adjacent gene.

Propagation Act of multiplying or increasing in numbers.

Prostaglandin E A hormone that can trigger various signaling pathways through a wide variety of receptors. Prostaglandins, in general, are produced from fatty acids by a number of enzymes and have a short lifespan so that they will usually only act on the cell that released them or nearby cells.

Protease An enzyme that breaks down proteins.

Protein Profiling The method to achieve the identification of all the protein species within a sample.

Proteomics The large-scale study of protein components of a cell type, tissue, or whole organism.

Pseudozygote A "false" zygote incapable of completing gestation; in this case, referring to the tetraploid (4n) embryo created prior to the tetraploid complementation assay.

Puromycin An antibiotic that is derived from a fungus of the genus *Streptomyces*.

Pyrosequencing A form of sequencing by synthesis that differs from the Sanger sequencing by measuring the release of pyrophosphate during the incoporation of nucleotides in the sequencing reaction.

qPCR (Quantitative Real-Time Polymerase Chain Reaction) This is a modified PCR-based assay used to measure messenger RNA expression.

Quantile In statistics, a quantile refers to points taken at regular intervals from a bell or Gaussian curve of a random variable.

Relational Database A database consisting of separate tables with explicitly defined relationships, and whose elements may be selectively combined as the results of queries.

Replication Fork The point at which the two strands of the DNA helix are divided during DNA replication so that the strands can be copied.

Reporter A protein emitting color or fluorescence.

Retinal Pigment Epithelium (RPE) A highly specialized tissue of the retina that is adjacent to the neurosensory layer. It plays a key role in maintaining the function of the photoreceptor.

Retrovirus An RNA virus that has an enzyme called reverse transcriptase, which enables it to transcribe its RNA into DNA. The retroviral DNA can then integrate into the chromosomal DNA of the host cell to be expressed.

Rho-Associated Kinase (ROCK) Inhibitor Y-27632 Y-27632 is a selective inhibitor of p160-Rho-associated coiled-coil kinase (ROCK). Y-27632 treatment significantly improves the survival of dissociated human embryonic stem cells.

RNase Ribonuclease, a class of enzymes that break down (catalyze the hydrolysis) RNA.

rt-PCR (Reverse Transcriptase Polymerase Chain Reaction) A technique to detect and quantify mRNA (messenger RNA) by synthesizing cDNA (complementary DNA) from RNA by reverse transcription (RT) and amplification of a specific cDNA by polymerase chain reaction (PCR).

Sarcomas Connective tissue tumors formed by proliferation of mesodermal cells; these are usually highly malignant.

SCID Mice Severe combined immunodeficiency mice that are homozygous for the mutant autosomal recessive gene *scid*, which causes a lack of mature functional lymphocytes. These mice are very useful as they are receptive to human immune system transplantations.

Secretomics The study of proteins that are secreted by a cell, a tissue, or an organism.

Septum Transversum A thick mass of cranial mesenchyme that gives rise to parts of the thoracic diaphragm and the ventral mesentery of the foregut in the developed human being.

Sequencing by Ligation Refers to a DNA sequencing process that uses DNA ligase rather than DNA polymerase to lengthen and determine the sequence of the target DNA molecule.

Serial Analysis of Gene Expression (SAGE) A method for analyzing gene expression patterns by linking and cloning short (10–14 bp) sequence tags, each representing a transcript, and then determining their observed frequency by sequencing.

Severe Combined Immunodeficiency (SCID) A disease or a syndrome defined by primary immune deficiency characterized by defects in the number or function of multiple (or combined) white blood cell lineages, such as both T cells and B cells.

Signal Transduction The process by which a cell converts one kind of signal into another, resulting in a signaling cascade and the corresponding biological response.

Signaling Pathways A cascade of proteins and enzymes that provide instructions for cell function. These are often initiated through a compound such as a protein or hormone, binding to its receptor at the cell membrane and resulting in the expression of certain genes. Signaling pathways are usually complicated and can have variable effects depending on the type of cell. For example, a signaling pathway that induces growth in a particular cell type may inhibit growth in another unrelated cell type.

Signature Mapping A method for interpreting observed sequence tags by mapping them to a reference genome.

Signature Prediction A method for interpreting observed sequence tags by predicting, for each previously known transcript, which short sequence will arise from it when using a given experimental procedure.

Single Nucleotide Polymorphism (SNP) A variation of a single nucleotide of DNA, which can be useful in understanding and identifying a higher risk of disease.

Small Interfering RNA (siRNA) Also known as short interfering RNA or silencing RNA. siRNAs are a class of 20–25 nucleotide-long RNA molecules involved in the RNA interference (RNAi) pathway. Typically, siRNAs interfere with the expression of a specific gene.

SMAD Signaling SMAD proteins are homologs of both the drosophila protein, mothers against decapentaplegic (MAD), and the *Caenorhabditis elegans* protein SMA. The name is a combination of the two. SMAD signaling uses SMAD proteins as transcription factors.

Spontaneous Differentiation Differentiation of stem cells to other cell types without any external factors guiding the process.

Stargardt Macular Degeneration/Stargardt Disease An autosomal-recessive form of macular degeneration characterized by a progressive loss of central vision in one or both eyes.

Stemness The specific properties of stem cells as attributed by expression of a unique set of genes.

Subcutanous Beneath the skin.

Substrate The surface upon which an organism grows or is attached.

Supernatant The liquid overlaying sediments after pull down by centrifugation or gravity.

Taqman Applied Biosystems designed probes used for relative quantification of gene expression in real-time PCR in a fluorogenic 5′ nuclease assay.

Telomerase A ribonucleoprotein enzyme (composed of both RNA and proteins) that adds specific DNA sequence repeats (TTAGGG) at the end of a linear chromosome. These repeats help protect against chromosome degradation.

Teratocarcinoma A germ cell tumor arising from germ cells usually affecting ovary or testis.

Teratoma A tumor containing derivatives of all three embryonic germ layers: ectoderm, mesoderm, and endoderm. Teratomas may be naturally occurring or experimentally produced from pluripotent stem cells in order to assay their potency.

Tetraploid An organism with four sets of chromosomes. Most organisms are diploid: two sets of chromosomes, one from each parent.

Tetraploid Complementation A process wherein the two-celled cleaving embryo is electrofused to produce a single-celled, tetraploid (4n) embryo that is incapable of gastrulating but capable of forming the trophoblast. Pluripotent stem cells may then be introduced through the zona pellucida in order to complement the 4n embryo's inability to gastrulate, thereby forming an entire, gestationally competent embryo. Only the most pristine pluripotent cells are able to function in tetraploid complementation.

Tetraploid Embryo By electrofusion, a single cell can be generated through the fusion of two blastomeres from separate two-cell stage embryos. The resulting two-cell embryo will contain double the diploid content of DNA in each cell.

Totipotent/Totipotency The ability of a single cell to differentiate into all cells or tissues of the body including the placenta.

Transcriptome A term used to represent the set of transcribed genes (RNA) present in a given cell type or the whole organism. In a specific cell, the transcriptome represents the set of actively transcribed genes.

Transcriptomics The study of the set of all transcripts produced by a cell or a population of cells.

Transdifferentiation The ability of partially differentiated or precursor cells to differentiate into cells of other lineages, which may be outside the differentiation path of the precursor cell.

Transduce To transfer genetic material into a cell by a viral vector (a plasmid or virus that contains or carries modified genetic material).

Transfection Incorporation of exogenous DNA into a cell.

Transgene A gene or genetic material that has been introduced from one genome to another, either naturally or by genetic engineering.

Trophectoderm Cells of a preimplantation embryo that will develop into placenta.

Trophectodermal Referring to the trophectodermal tissues derived from the outer, trophoblastic layer of the preimplantation embryo.

Trophoblast A cell that forms the outer layer of a blastocyst, provides nutrients to the embryo, and facilitates implantation and formation of placenta.

Trypan Blue A vital stain used to differentiate living cells from dead cells. Living cells possess intact cell membranes that exclude certain dyes like Trypan blue but fixed or dead cells will stain blue.

Trypsin An enzyme that cuts protein molecules at specific sites in the amino acid chain. One of its main uses is to release cells growing in adherent cultures, preparing for passage.

Trypsinization The enzymatic method of dissociating cells from a substrate to pass to a new culture vessel. If trypsin is allowed to stay too long on the cells, cell viability is reduced.

Tumorigenesis The formation of tumors in the body caused by uncontrolled cell division.

Turbidity Cloudiness in liquids due to suspended particles.

Two-Dimensional Polyacrylamide Gel Electrophoresis (2D-PAGE) The gel-based technique to resolve protein species in a two-dimensional context. In the first dimension, the proteins are separated according to their individual isoelectric point (charge), and then in the second dimension, they are further separated according to their molecular weight (size).

Valproic Acid   A chemical compound, also used as an antiseizure drug administered to patients.

Vascularization   Formation of blood vessels.

Vasculogenesis   The process of blood vessel formation occurring by a de novo production of endothelial cells.

Vector   Carrier used to insert DNA sequence into a host cell.

VEGF (Vascular Endothelial Growth Factor)   A family of secreted proteins that are important for development of the circulatory system. They have also been found to act on a variety of cell types including neurons, macrophages, and kidney cells.

WNT   A family of signaling proteins that have a variety of functions. WNT signaling in stem cell biology can play roles in both stem cell differentiation and maintenance depending on the stem cell type.

Wnt Signaling   The name Wnt is a combination of Wg (the fruitfly gene wingless) and Int (the vertebrate homolog of wingless). Wnt signaling refers to signal transduction pathways that are activated by the binding of a Wnt ligand to a cell surface receptor of the Frizzled family.

X-Chromosome Inactivation   A process occurring in mammalian females in which one of the two X chromosomes is randomly inactivated so that most of its genes are not expressed. X inactivation occurs early in development. Once an X chromosome is inactivated, it remains inactive throughout the lifespan of that cell and all of its progeny.

Xenobiotics   A chemical compound that is foreign to a living organism.

Xeno-free   A term used to describe reagents or methodologies that do not contain or use components of nonhuman origin. This term is often used interchangeably with "animal-free."

Xenotransplantation   Transplantation of living cells or tissues from one species to another.

Index

Human Stem Cell Technology and Biology, edited by Stein, Borowski, Luong, Shi, Smith, and Vazquez
Copyright © 2011 Wiley-Blackwell.

CUSTOMER NOTE: IF THIS BOOK IS ACCOMPANIED BY SOFTWARE, PLEASE READ THE FOLLOWING BEFORE OPENING THE PACKAGE.

This software contains files to help you utilize the models described in the accompanying book. By opening the package, you are agreeing to be bound by the following agreement:

This software product is protected by copyright and all rights are reserved by the author, John Wiley & Sons, Inc., or their licensors. You are licensed to use this software on a single computer. Copying the software to another medium or format for use on a single computer does not violate the U.S. Copyright Law. Copying the software for any other purpose is a violation of the U.S. Copyright Law.

This software product is sold as is without warranty of any kind, either express or implied, including but not limited to the implied warranty of merchantability and fitness for a particular purpose. Neither Wiley nor its dealers or distributors assumes any liability for any alleged or actual damages arising from the use of or the inability to use this software. (Some states do not allow the exclusion of implied warranties, so the exclusion may not apply to you.)